DATE DUE			

DEMCO 38-296

DIESEL ENGINE SERVICE

Kenneth R. Babb

DIESEL ENGINE SERVICE

RESTON PUBLISHING COMPANY, INC.
A Prentice-Hall Company
Reston, Virginia

Library of Congress Cataloging in Publication Data

Babb, Kenneth.
    Diesel engine service.

    1. Diesel motor–Maintenance and repair. I. Title.
TJ799.B33 1984   621.43'68   83-24432
ISBN 0-8359-1291-4

© 1984 by Reston Publishing Company, Inc.
A Prentice-Hall Company
Reston, Virginia 22090

All rights reserved. No part of this book may be
reproduced in any way or by any means without
permission in writing from the publisher.

10 9 8 7 6 5 4 3 2 1

PRINTED IN THE UNITED STATES OF AMERICA

CONTENTS

Preface, ix

1. **Diesel Engine Development, 2**
   Diesel Engine Variety, 3
   Diesel Development, 3
   Stages of Progress, 4

2. **Basic Diesel Principles, 5**
   Engine Types, 7

3. **Diesel Engine Fuel Systems, 9**
   Introduction, 11
   Bosch Fuel System, 14
   Bosch Distributor-Type System, 17
   American Bosch Fuel System, 27
   Trouble-Shooting Fuel Systems, 38
   Stanadyne-Roosa Master Fuel System, 46
   Caterpillar Fuel System, 50
   General Motors-Detroit Diesel Fuel System, 68
   Cummins Fuel System, 84

4. **Air Induction Systems, 95**
   Functional Description, 97
   Turbochargers, 97

5. **The Cooling System, 101**
   Function Description, 103
   Oil Cooler, 114
   Cooling System Service, 118

6. **The Lubricating System, 125**
   Function Description, 127
   Uses of Oil Pressure, 130
   Scavenging Oil Pumps, 134
   Turbocharger Oil Supply and Drain, 135
   Oil Specifications, 135

7. **The Engine Block, 141**
   Functional Description, 143
   Block Features, 143
   Engine Block Service, 145
   General Repair Principles, 149

8. **Cylinder Heads, 151**
   Functional Description, 153
   Causes for Head Removal, 154

9. **Crankshaft Service, 163**
   Functional Description, 165
   Bearing Description, 165
   Crankshaft and Bearing Service, 166

10. **Cylinder, Piston, and Connection Rod Service, 177**
    Cylinder Line Replacement Caps, 179
    Cylinder Liner Installation, 183
    Piston Inspection and Cleaning Caps, 184

11. **Two-Stroke-Cycle Engines, 197**
    Liner Inspection, 199
    Crankshaft Installation, 202
    Cylinder Liner Installation—DDA Engine, 203

12. **Rocker Levers: Tappets, Followers, and Push Rods, 207**
    Rocker Lever Inspection, 209
    Valve Crossheads, 212
    Assembly Suggestions, 214
    Injection Timing on Cummins Engines, 216
    Mechanical Variable Timing, 221

13. **Gear Train and Timing Marks, 225**
    General Description, 227
    Gear Train Service, 229

14. **Vibration Dampers, 237**
    Functional Description, 239
    Vibration Damper Service, 241

15. **Cooling System Service, 247**
    Water Pump Assembly, 249
    Install Water Pump, 250
    Water Manifold Repair, 250

16. **Exhaust Manifold and Turbocharger, 257**
    Removal of Parts, 260
    Cleaning and Inspection of Exhaust Manifold, 260
    Installation of the Manifold, 260
    Installation of the Turbocharger, 262

17. **After Cooler Service, 267**
    Removal, Cleaning, and Inspection, 269
    Crossbolt Aftercooler, 272
    Detroit Diesel Aftercooler, 274

18. **Gear Cover Assembly, 279**
    General Description, 281
    Bushings, Bearings, and Seals, 281

19. **Rear Crankshaft Seal, Flywheel Housing, and Flywheel, 289**
    Flywheel, 289
    Replacement of Rear Seal, 291
    Installation of New Rear Seal, 293
    Clutch Description, 294
    Installation of Transmission, 301
    Alignment of Flywheel Housing, 303

20. **Engine Tuneup and Adjustments, 306**
    Meaning of Tuneup, 309
    Caterpillar Tuneup, 309
    Cummins Injector and Valve Adjustment, 311
    Cleaning of Air Cleaner Elements, 319

21. **Engine Testing, 321**
    Test Principles, 323
    Dynamometer Types, 323
    The Taylor Dynamometer, 327
    Electric Dynamometer Resistence Bank, 328
    Engine Cooling, 329

    Preparing for Chassis Test, 331
    Dynamometer Instrumentation, 332
    Test Procedure, 335
    Use of Chassis Dynamometer, 335
    Final Adjustment, Wash-down, and Painting, 339

22. **Auxiliary Vehicle Braking System, 341**
    Engine Retarders, 343
    Jacobs Brake, 343
    Jacobs Brakes on Detroit Diesel Engines, 352
    Jacobs Brakes on Mack and Caterpillar Engines, 356
    Other Engine Retarding Systems, 363
    Retarders on the Drive Line, 363
    Electric Retarders, 364

23. **Electrical Trouble-Shooting, 365**
    Explanation of Ohm's Law, 367
    Battery Description and Service, 368
    Electrical System Service, 370
    Cranking Motor Service, 376
    Use of Meters, 377
    General Analysis of Electrical Problems, 379
    Air Cranking Systems, 380
    Battery Charging on Vehicles, 383
    Voltage Regulator, 385
    Service on the Charging System, 387

24. **Do-It-Yourself Tools, 391**
    Introduction, 393
    Use of Torque Wrenches, 393
    Use of Power Wrenches, 394
    Use of Precision Tools, 395
    Common Tools, 397
    Parts Inspection for Wear, 401
    Tools to Make, 403
    Installation of Buttress-Type Oil Pans, 406
    Trouble-Shooting Hints, 410
    Other Tools, 413
    Salvaging Used Engine Oil, 419
    Uses of Tubing Wrenches, 421
    A Keyway, 421

Glossary, 423

Appendix, 427

ACKNOWLEDGMENTS

1. Robert Bosch Sales Corporation
2. American Bosch Division of United Technologies
3. Caterpillar Tractor Company
4. Stanodyne Division - United Technologies
5. General Motors Corporation
6. Cummins Engine Company, Inc.
7. Clayton Manufacturing Co.
8. Ingersoll-Rand Corp.
9. Go-Power Corporation
10. Taylor Dynamometer & Machine Co.
11. Ford Motor Co.
12. Chrysler Corporation
13. Lipe Rollway Corporation
14. Jacobs Manufacturing Co.
15. Neil's Detroit Diesel Co.
16. Cummins Arizona Diesel, Inc.
17. Empire Machinery Co.
18. ABC Technical & Trade School
19. Mr. P. M. Uhl

And the many good mechanics I have taught and worked with, as well as many friends whose ideas have helped.

PREFACE

Although there are many good manuals on diesel engine service, most of them are written to suit the needs of manufacturers' sales or are written in language that the line and field mechanic finds hard to follow. Manufacturers' manuals often prescribe the use of special tools and equipment sold by the factory, which may not be available on the job site. It is my purpose to describe ways to do the job that are within reach of the mechanic, without resorting to a laboratory or buying expensive tools for limited use.

Many years of experience have shown practical ways to achieve the goal of well-fitted and adjusted machinery. This book is written to help the mechanic save time and aggravation and to make the trade more attractive.

The material was gathered from my own work experience and from the contributions of hundreds of good mechanics with whom I have worked. I feel that a need exists for such a manual. The man who actually does the work needs this information.

GENERAL PRINCIPLES OF MECHANICAL WORK

The modern mechanic, assisted by many sophisticated tools, is depended on to perform all kinds of tasks, some of which are easy and some very difficult. There are some personal qualifications; the book tries to enumerate them. Good mechanics will recognize these items.

1. He must like this kind of work. It is not enough simply to make a living.
2. He must be ingenious. There are always ways to do the job.
3. He must take pride in doing a good job. It is never good enough until it is right.
4. He must remember that machines do not think. The mechanic must do the thinking.
5. He must know something about physical laws on which all design is based.
6. He must know the machine on which he is working. There is nothing wrong with using a manual, or asking others.
7. He must understand what is required. Questions should clarify any obscure points.
8. He must work in ways that protect the parts, rather than destroy them.
9. Trouble-shooting is a logical process. Analyzing the symptoms often points to the solution.
10. A good mechanic works as though he owned the machine.

These, and other principles, are the characteristics of successful mechanics. The work may be demanding; there is no place for a poor attitude. Good mechanics see the tough job as a challenge, not as an impossible assignment. Other men built the machine, and the mechanic can fix it. The mechanic is as smart as the builders.

The service trade of mechanic is a useful occupation. As time goes on, more people will find that the honest, skilled worker is a valuable person to know.

This book is to help those who are trying to make a living by repairing diesel engines used in commercial trucks.

DIESEL ENGINE SERVICE

DIESEL ENGINE DEVELOPMENT

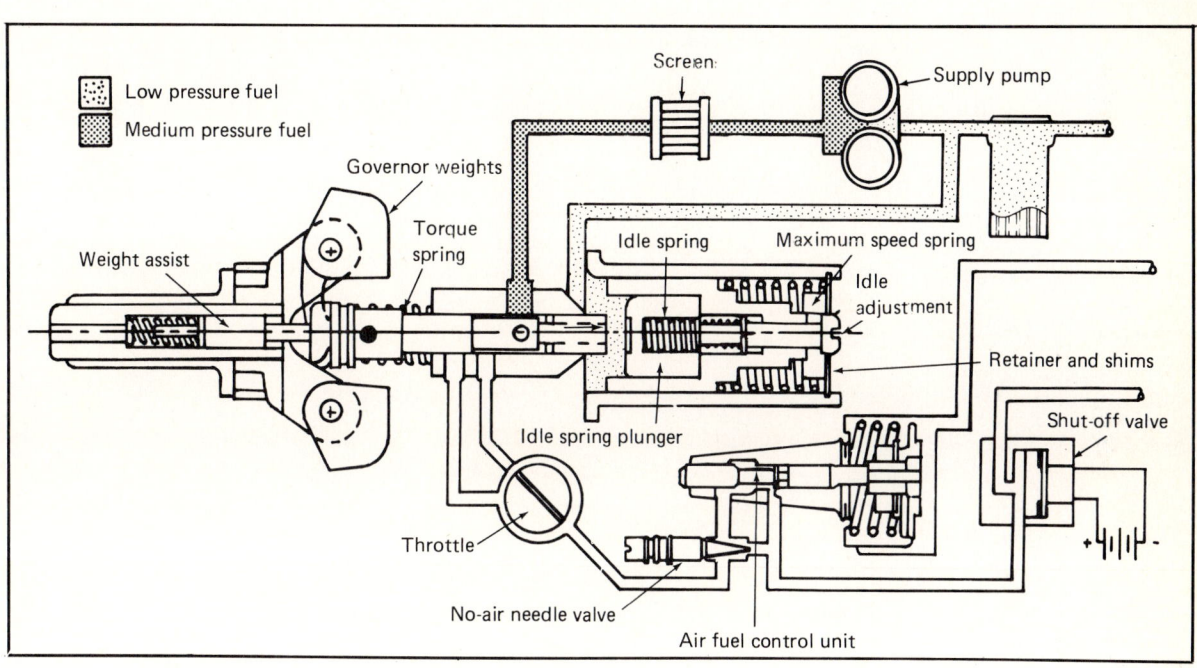

DIESEL ENGINE VARIETY

The diesel engine is made by many factories. Engine size ranges from 10 to 40,000 horsepower; applications are made to almost all kinds of mobile, marine, and stationary equipment. All diesel engines are alike in basic principles; the need for service is common to all. Differences are due to the viewpoint of designers, tradition, the wide variety of uses, and sales propaganda.

Model designations such as industrial, automotive, marine, and power generation are applied to the same basic engine. Most such differences are made by changing external components, without changing the basic engine. Variations in governing, fuel system design, and many other externals make the engine applicable to hundreds of different uses. Yet, like people, diesel engines are more alike than different. With skill and knowledge of the various systems, the work of service becomes less mysterious.

The book describes the various servicing methods needed by the components, without regard to engine make or model. It details the function, removal and installation, and adjustment of various subassemblies.

DIESEL DEVELOPMENT

The diesel engine is the result of the work and research of Dr. Rudolph Diesel, whose attempts to improve the existing internal combustion engines was attractive to such firms as Krupp and Machinen Fabric Augsberg-Nurnberg (M.A.N.). After several years of experiment, frustration, and rejection by many contemporary engineers, the first salable diesel engine was produced by M.A.N. in 1902.

Alexander Busch of the Anheuser-Busch Brewing Company obtained a license to build engines in America in 1897. The first engine built by Busch was installed in the company's brewery in St. Louis in 1898. It was not successful and was destroyed. From 1898 to 1911, engines were built and sold by the American and British Manufacturing Company of Providence, Rhode Island, under agreement with Busch. In 1911 Busch formed a company with the Sulzer Brothers of Switzerland.

Sulzer-designed engines were built in a new plant in St. Louis from 1912 to 1920. These were all large, stationary engines using petroleum fuel, which was blown into the cylinders by high-pressure air. This process is called *air injection*. Solid injection was the invention of W. T. Price of the De La Vergne Machine Company. This firm then produced engines that were successfully applied in stationary power plants for many years.

Until 1929, most diesel engines were heavy, slow-speed industrial units. The real pioneer of high-speed engines suitable for use in trucks was Clessie Cummins of Columbus, Indiana. He interested Will Erwin, also of Columbus, in his ideas; a company was organized in 1918. After several years of experimentation, the successful use of an early model in a passenger car led to the development of the popular truck engines that are accepted worldwide. The company is the largest builder of high-speed diesel engines for truck use in America and has plants in several countries.

Although Cummins has built four-stroke-cycle engines exclusively, General Motors Corporation has developed the two-stroke-cycle design to high standards and enjoys the majority of diesel-powered locomotives for railroad use.

Many experiences and much development work have produced the versatile engine of today. The continuing efforts to improve economy will result in many changes; we should soon see the use of sources that do not depend on petroleum and do not pollute the atmosphere.

STAGES OF PROGRESS

The air injection of fuel has been universally replaced with modern solid injection systems. The use of diesel-electric locomotives has replaced the steam-driven locomotives. Smaller engines have replaced gasoline as the power for highway trucks. Still-smaller engines are becoming more popular in passenger cars. Diesel engines drive power generators, pumps, air compressors, and many other uses.

For all the variety, there is one universal fact: skilled and knowledgeable mechanics will be in demand for a long time. No engine has yet been built that does not require some service.

BASIC DIESEL PRINCIPLES

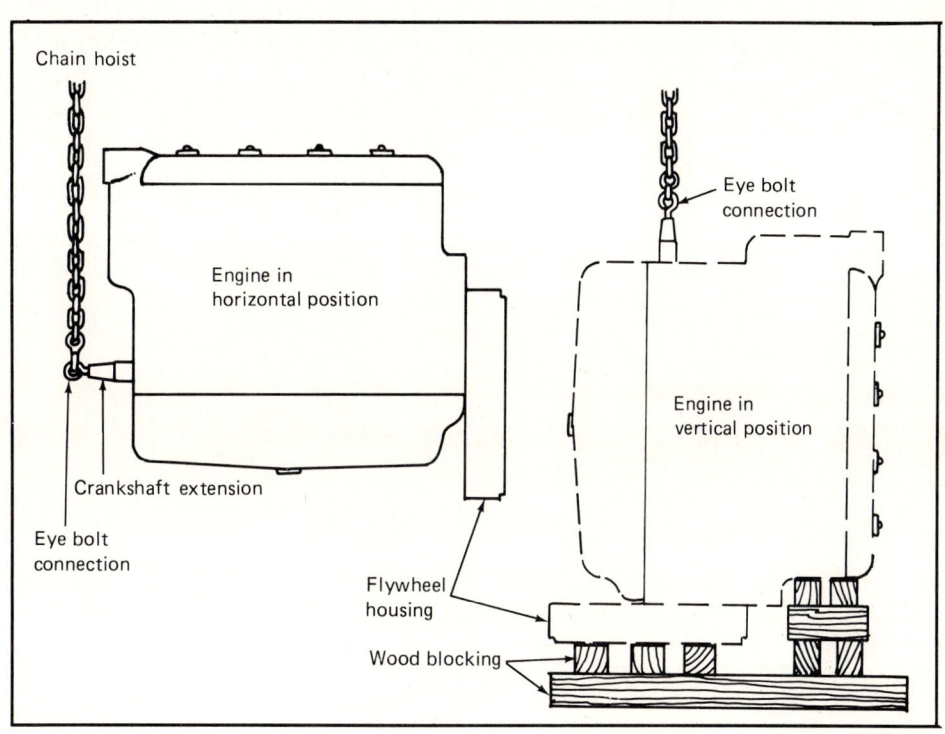

ENGINE TYPES

This manual deals strictly with the medium-sized engines used in trucks, switching locomotives, excavating machinery, and other uses. These engines range in horsepower from 100 to 600 and run at speeds up to 2,500 rpm.

Four-stroke-cycle engines take air directly into the cylinders through the air manifold and valve passages. The pistons act as pumps to move the air through the cylinder.

Two-stroke-cycle engines may use a turbocharger ahead of the blower, thus increasing air pressure on the blower intake. Because the blower is a positive-displacement pump, the air pressure is transferred to the air box surrounding the cylinders. More air can then be blown into the cylinder ports.

All diesel engines burn fuel by using the heat of compressed air. Compression ratios are high enough to heat the air to 1,200hr. -1,500 degrees, well above the burning point of the fuel. The more air available for compression, the more fuel can be burned. Thus power can be increased. The gain is limited by the maximum pressure that the engine will stand. At the same time, the extra air allows the fuel to be burned more completely, and less wasteful smoke is produced. Both the extra power and better fuel economy make the use of turbochargers attractive.

An engine without some form of air pump cannot draw in enough air to fill the cylinder completely. A pressurized air supply is advantageous. Engine efficiency is a way of expressing the amount of power produced compared to that used to turn the engine. Engines run because they get more thrust from the expanding combustion gasses than they use to compress the air during the compression stroke.

There are several designs of air cleaner, all requiring service. Air restriction due to dirty filter elements reduce power and create exhaust smoke. No matter what air cleaner is used, it must be serviced often enough to keep its restriction as low as possible.

DIESEL ENGINE FUEL SYSTEMS

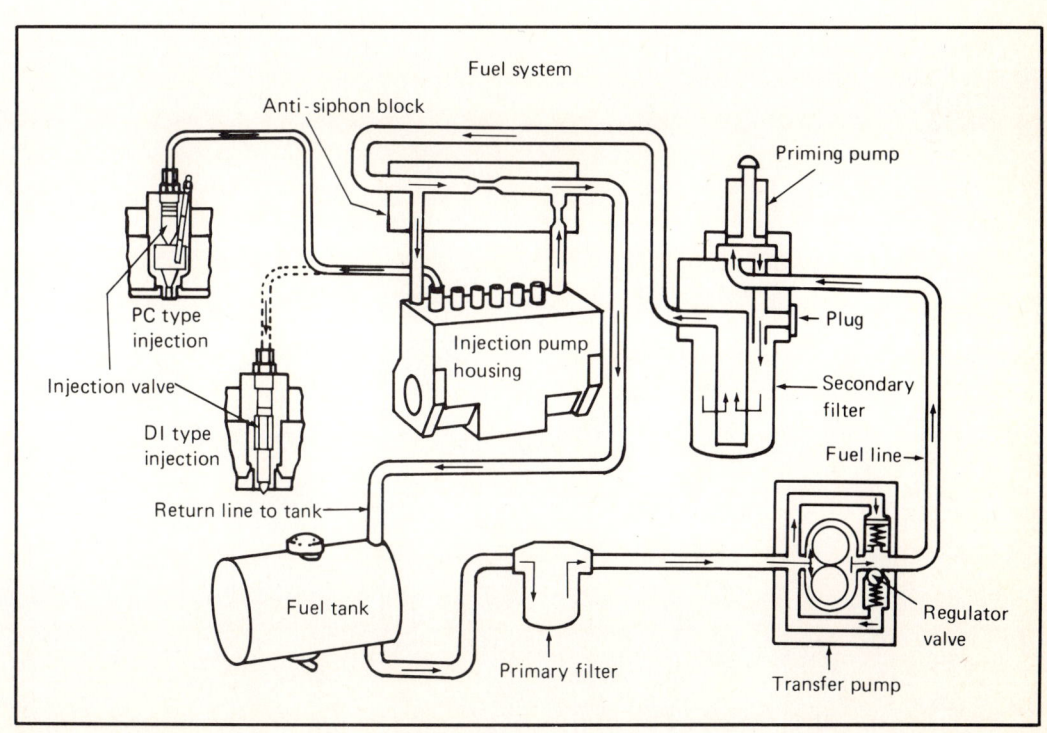

INTRODUCTION

There are many makes of diesel fuel injection equipment. They all do the same thing, in that they furnish fuel to the cylinder in the proper order, in the correct condition to burn, and in the correct amount.

Fuel quantity is usually controlled by various devices to control volume. At least one system controls the amount of fuel injected by controlling its pressure.

Speed governors vary from simple limiting governors, which control engine speed at idle and at maximum speed, to sophisticated governors, which control engine speed by sensing load.

This chapter describes the design function, keeping the fact in mind that the line and field mechanic should never be required to disassemble a fuel pump or injector in the field but should be able to determine the need for an exchange unit with accuracy. The function, installation, timing, and troubleshooting of the popular fuel systems are described. Rebuilding methods and tooling should follow the makers' manuals. [**Figs. 3-1, 3-2, 3-3, 3-4, 3-5, 3-6**] Other systems are similar.

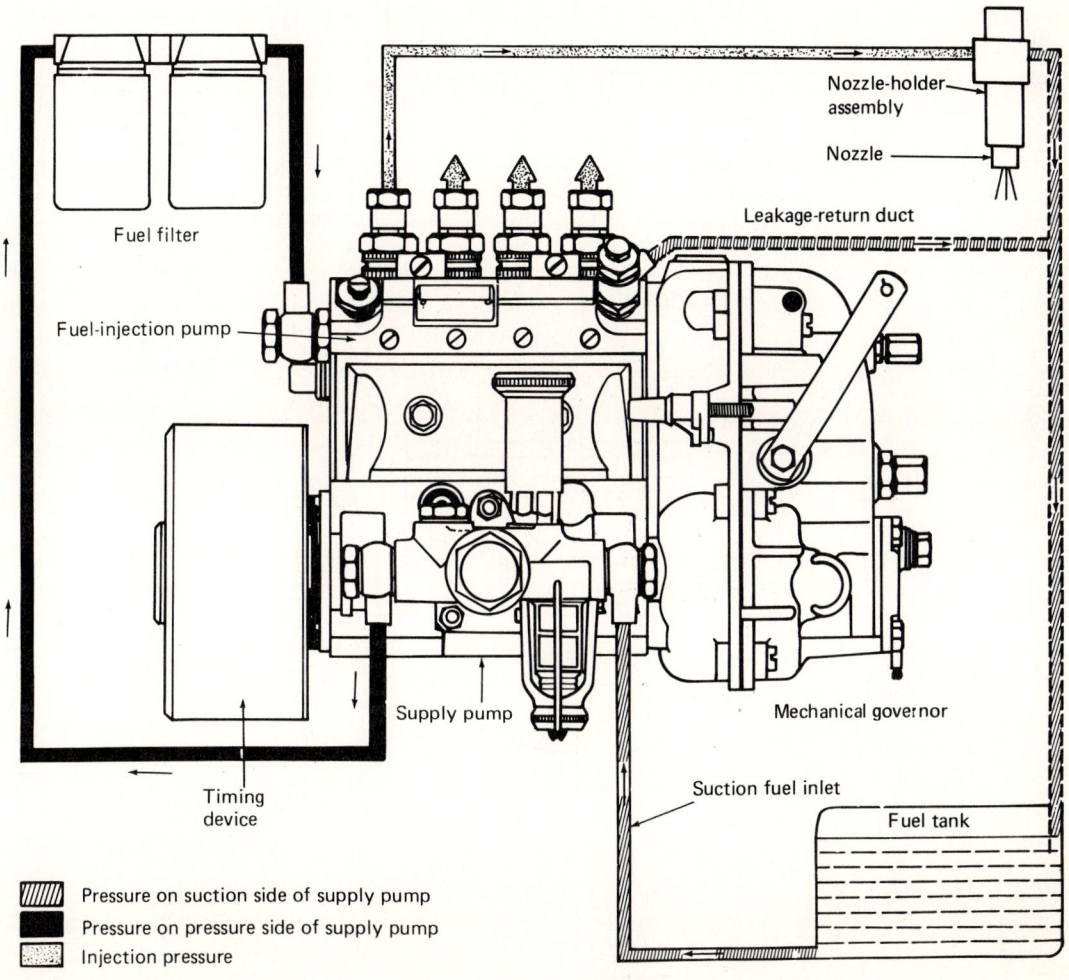

FIGURE 3-1. Robert Bosch Fuel System. *(Courtesy of Robert Bosch Sales Corp.)*

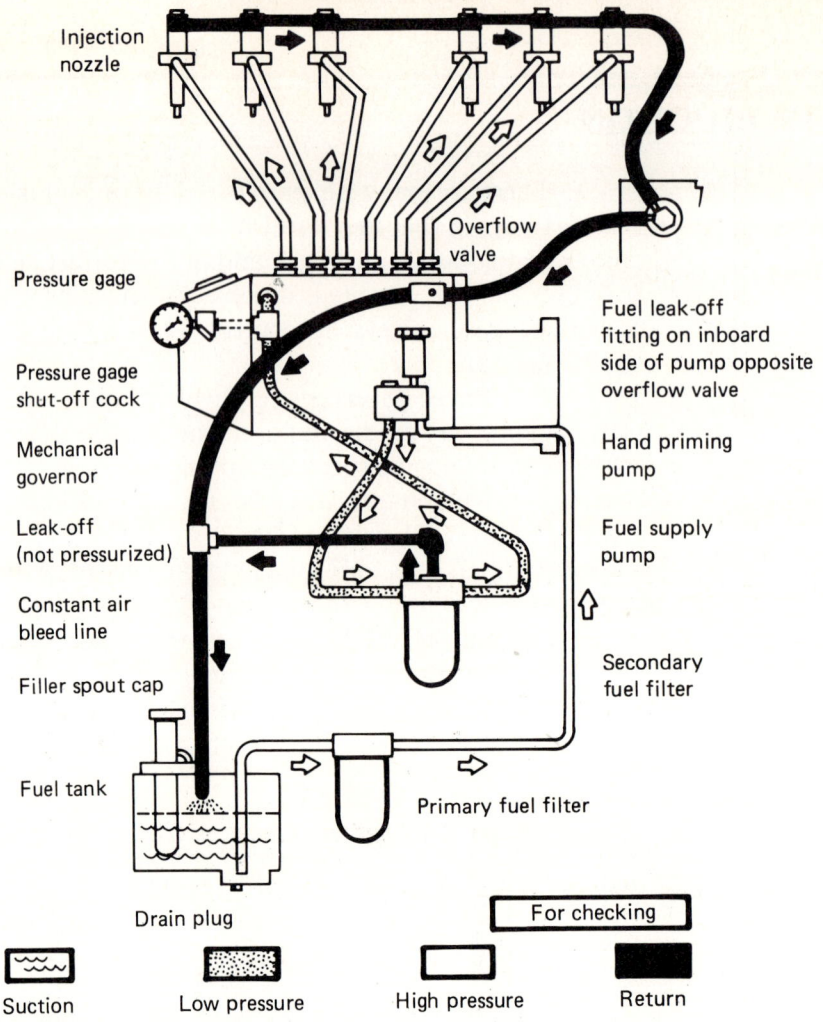

FIGURE 3-2. American Bosch Fuel System. *(Courtesy of American Bosch, Division of United Technologies)*

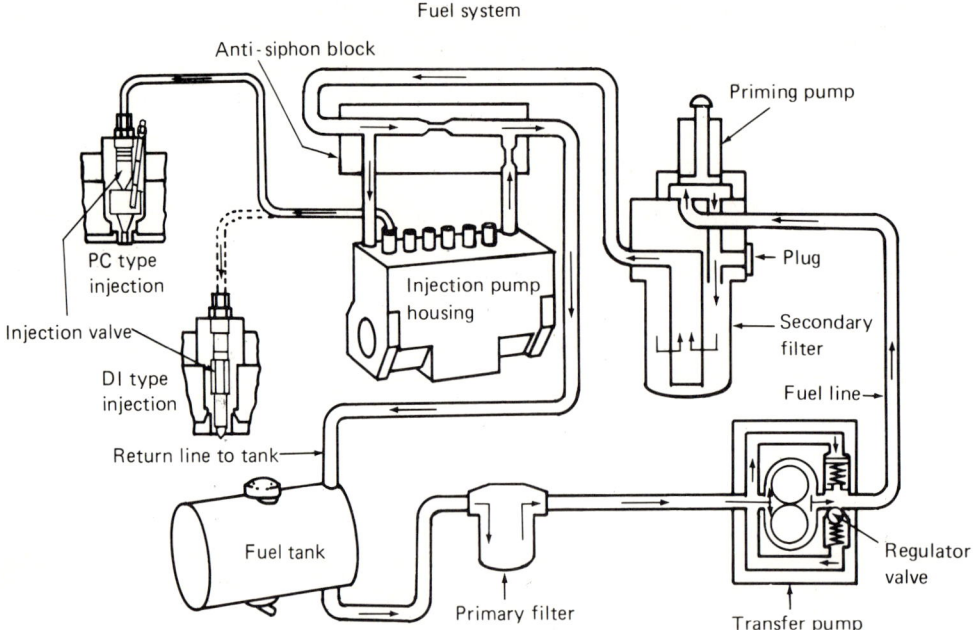

FIGURE 3-3. Caterpillar Fuel System. *(Courtesy of the Caterpillar Tractor Company)*

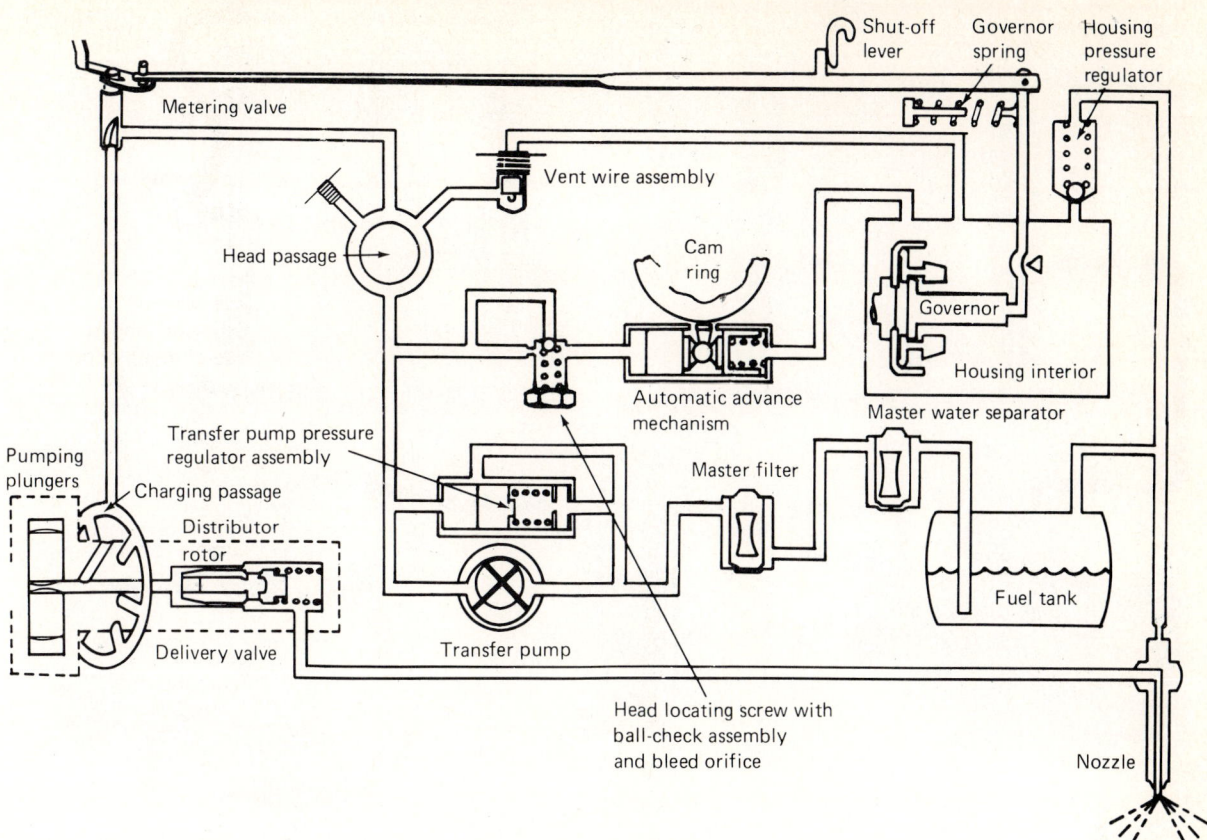

FIGURE 3-4. Roosa Master Fuel System. *(Courtesy of Stanadyne, Inc.)*

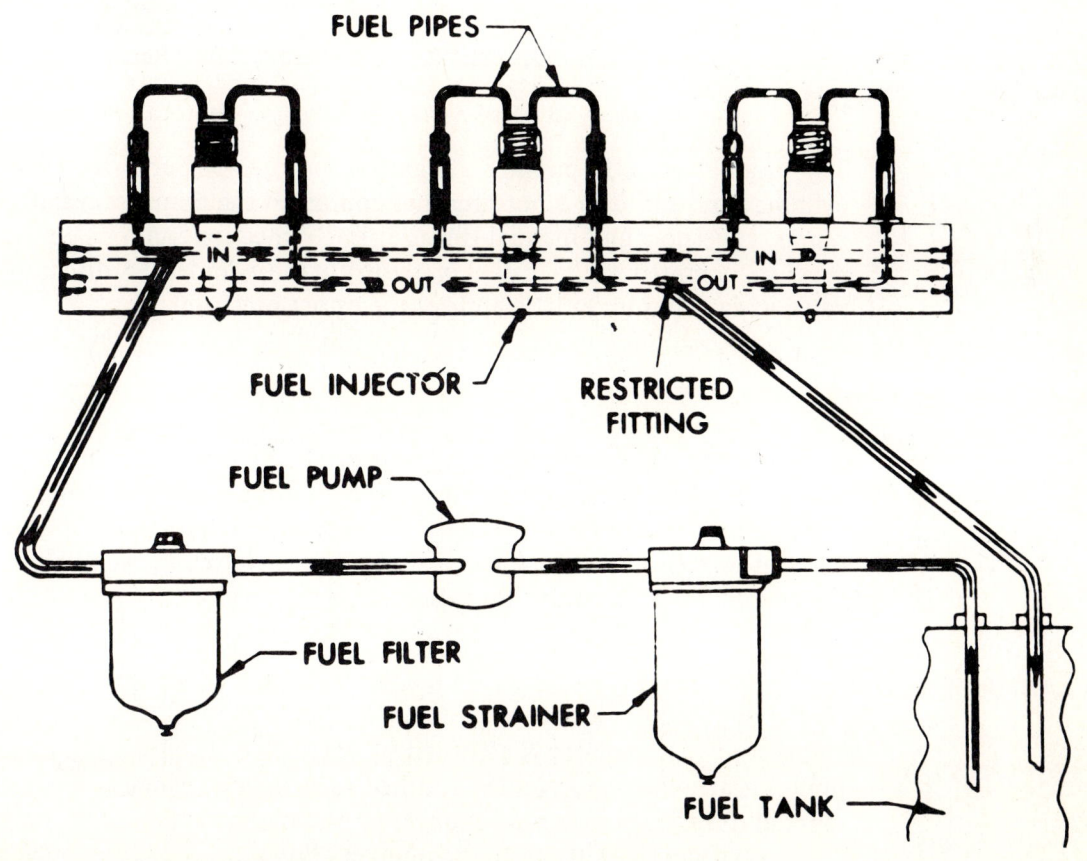

FIGURE 3-5. General Motors Fuel System. *(Courtesy of General Motors Corp.)*

14 / Diesel Engine Service

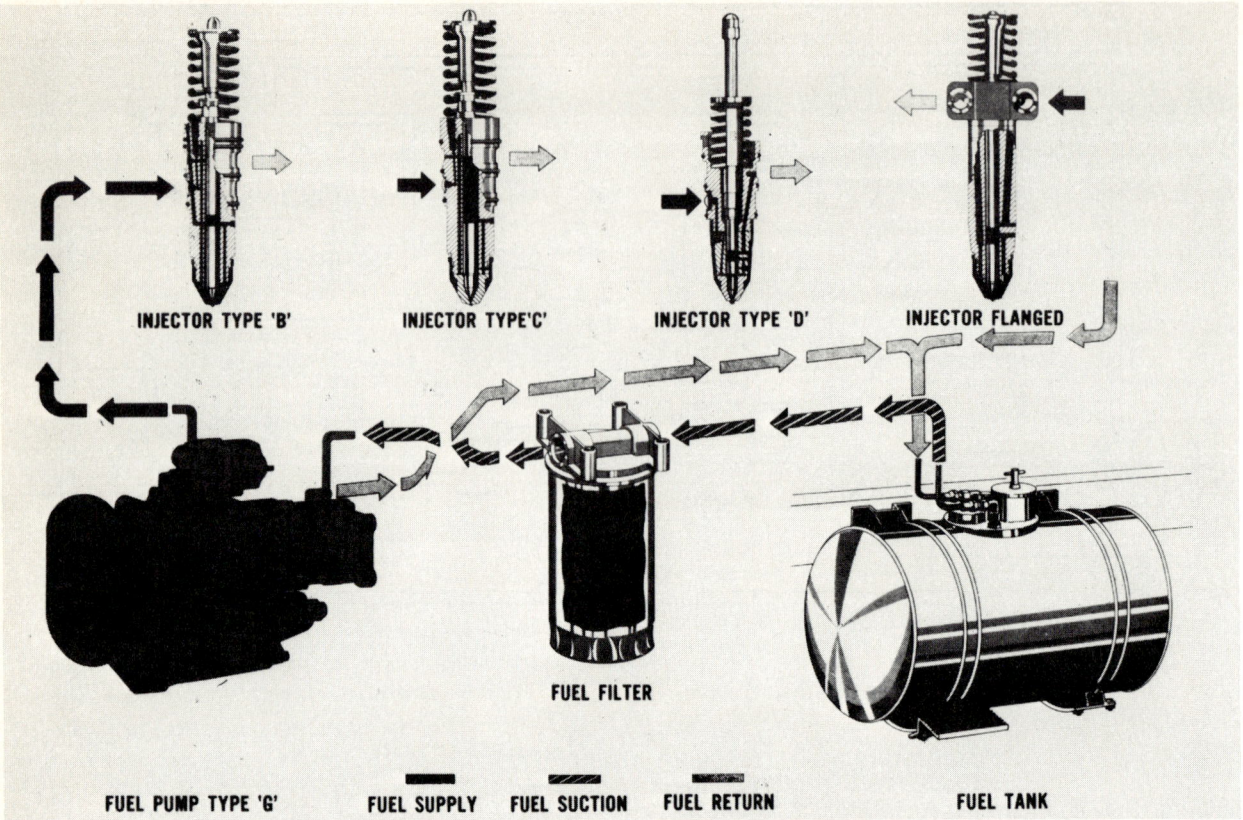

FIGURE 3-6. **Cummins Fuel System.** *(Courtesy of Cummins Engine Company, Inc.)*

BOSCH FUEL SYSTEM

The original and still popular design has an injector pump for each engine cylinder, assembled in a housing that contained the pump camshaft, supply pump, governor, and manual throttle. Each injection pump is connected to pressure-operated valves, called *injectors,* in the cylinders. A discharge check valve at each pump outlet stabilizes the outlet pressure.

Fuel Flow

Fuel from the supply pump enters a passage that opens to each injection pump inlet; a return line connects to the other end of the passage in order to return unused fuel to the tank. There is usually a suction filter on the supply pump inlet, and sometimes a secondary pressure filter between the supply pump and the fuel pump connection. Both filters require service.

Fuel Control

Control of fuel quantity is afforded by rotation of the pump plunger by its pinion gear, which engages the toothed rack or rod, which is moved by the manual throttle or governor. Plunger stroke is constant.

A scroll or helical cut on the plunger allows it to close and open ports in the barrel at different amounts of stroke, thus varying the volume of fuel dis-

Diesel Engine Fuel Systems / 15

charged. The amount of stroke when both ports are closed is called the *effective stroke*.

Fuel Supply

The supply pump, or the transfer pump, is mounted on the side of the fuel pump body and consists of a spring-loaded plunger driven by a cam on the pump camshaft. It supplies fuel under low pressure to the main feed passage, which surrounds each injection pump. The supply pump can be serviced without removing the fuel pump assembly from the engine.

In-Line Pump Governor

As with all centrifugal governors, a pair of weights is rotated by the driving shaft. The centrifugal force generated causes the weights to turn about their supporting pivot pins.[Fig. 3-7]

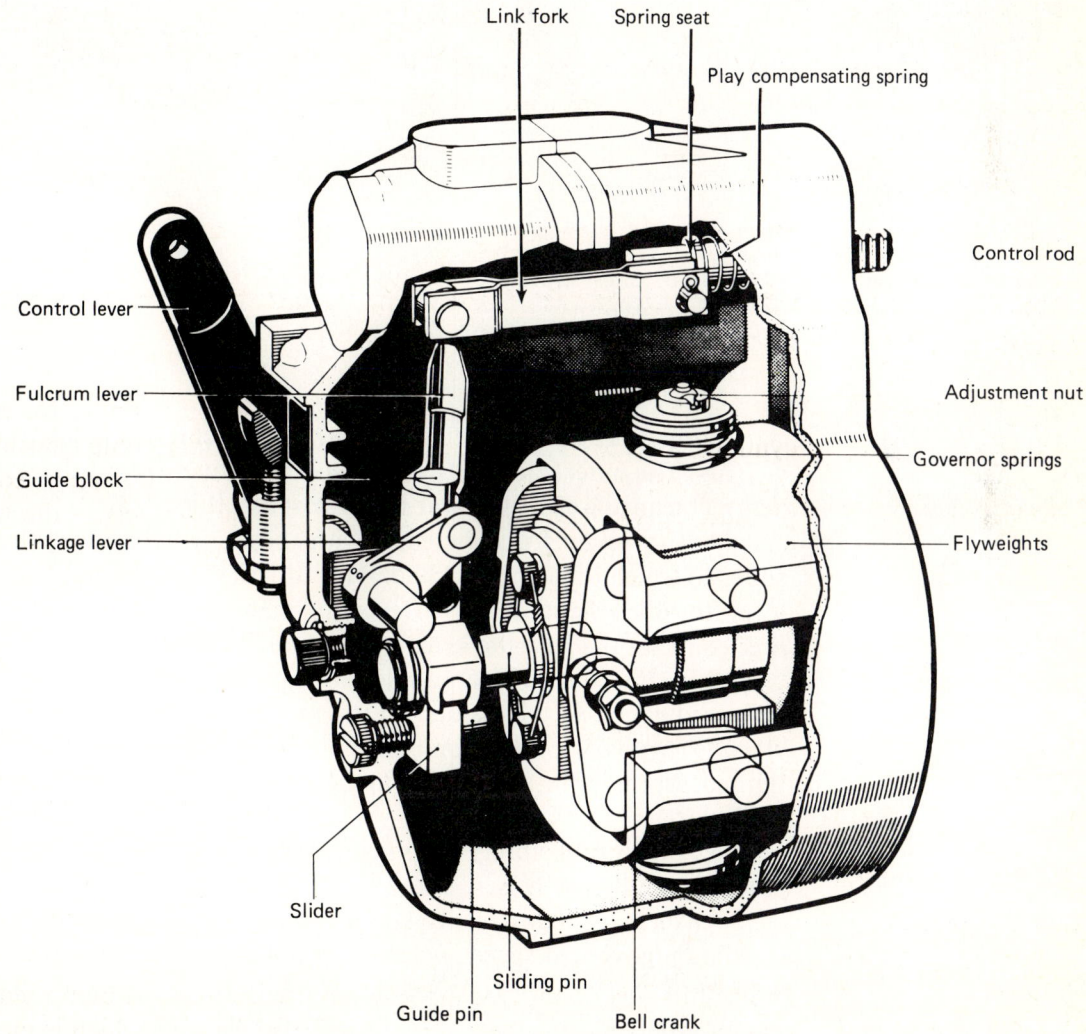

FIGURE 3-7. Robert Bosch Governor. *(Courtesy of Robert Bosch Sales Corp.)*

16 / Diesel Engine Service

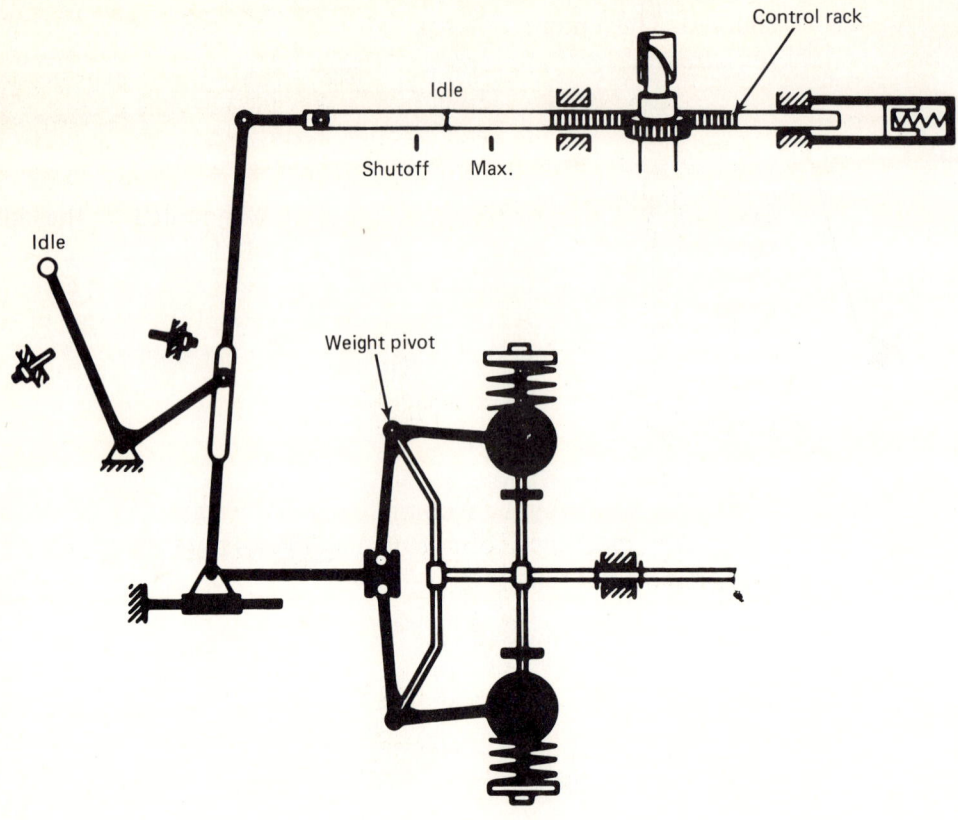

FIGURE 3-8. Governor at Idle Position. *(Courtesy of Robert Bosch Sales Corp.)*

Springs are built into the weights, which act to resist the rotation of the weights. The outer idle spring is seated on the inner side of the weight toe and acts to balance weight force at low idle speed.[**Fig. 3-8**]

As the weight force increases with speed increase, the springs move the slider to which they are attached. This moves the pivot point of the fulcrum lever, to which the rack control lever and manual control lever are attached. A link connects the fulcrum lever with the rack control.

With the control lever in idle position, the control rack is positioned to deliver idle fuel. As the manual lever is moved, the weight force moves the fulcrum lever pivot point to limit the control rack travel. Thus engine speed is limited. At this point, all springs in the weights act to resist weight force. [**Fig. 3-9**]

Engine torque ability is controlled by a small spring in the lower spring seat in the weights. It is adjusted by shims during calibration.

Any load change causes a speed change. Thus, if the engine is running at maximum governed speed, no load, and we increase load, the engine speed will drop, and the governor weights will lose force. The spring will push the sliding sleeve and yoke to move the control rack toward more fuel. When the control rack is against its stop, the engine is getting maximum fuel and is running at rated speed, full load. [**Fig. 3-9**]

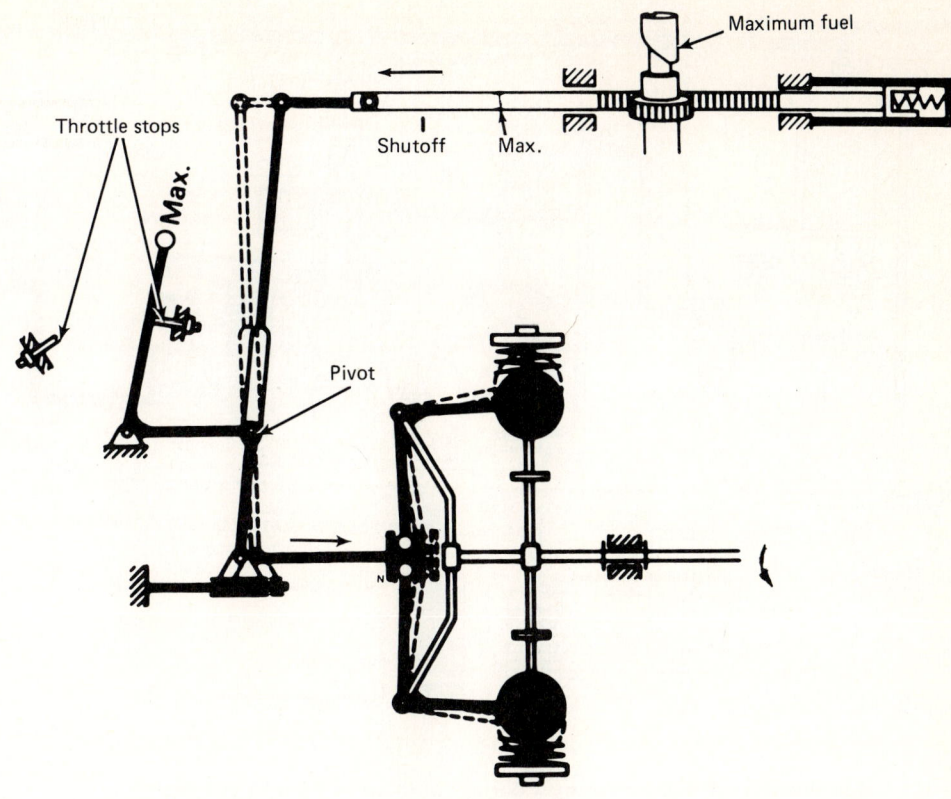

FIGURE 3-9. Governor at Full-Speed Position. *(Courtesy of Robert Bosch Sales Corp.)*

BOSCH DISTRIBUTOR-TYPE SYSTEM

Description

These compact fuel pumps are used on many automotive diesels and on smaller industrial engines. The exact arrangement varies to suit the engine builder. The basic design remains constant. [**Fig. 3-10**]

The distributor system features a single plunger, which is driven back and forth by the cams on its inner end. Its stroke is constant. The plunger is drilled with a center hole, with a cross hole at the inner end of the center hole.

A slotted hole in the plunger is indexed with the delivery valve as the plunger turns.

The control sleeve is located to cover or uncover the inner cross drilling, thus varying the effective stroke by allowing pressurized fuel to escape as the sleeve is moved along the plunger by manual throttle or governor.

The cam ring has as many cams as the number of cylinders, the pump being driven at half engine speed.

The combined rotary and back-and-forth movement of the plunger delivers injection pressure to each discharge valve in firing order.

18 / Diesel Engine Service

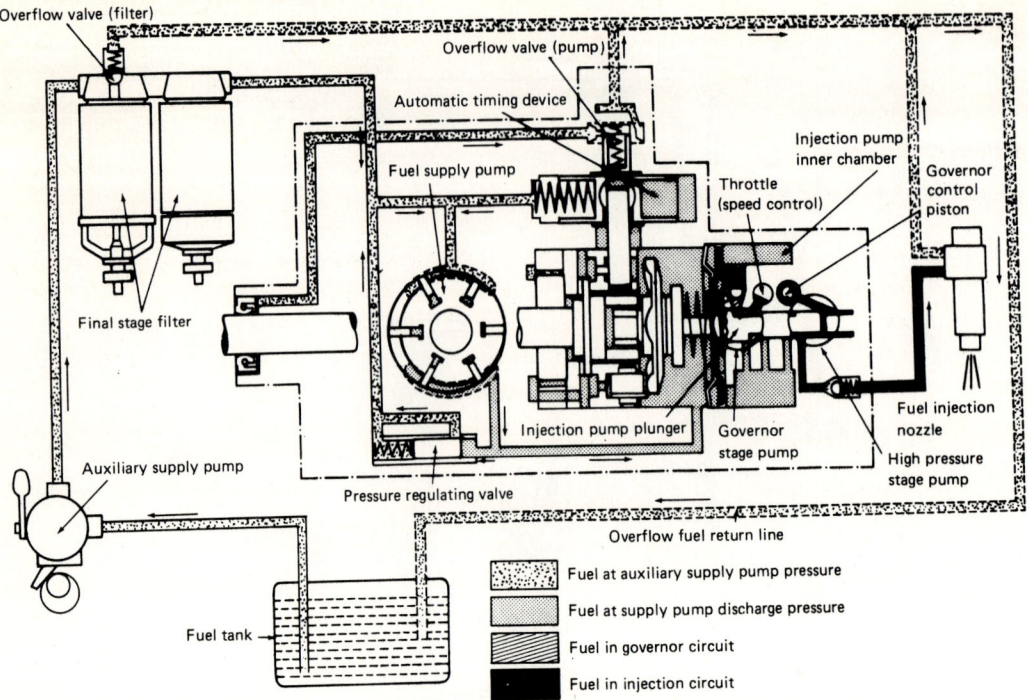

FIGURE 3-10. Robert Bosch Distributor System. *(Courtesy of Robert Bosch Sales Corp.)*

Governor Action

The centrifugal governor uses speed-sensing weight force to move the sliding sleeve against the tension of the governor spring. The manual throttle controls this spring tension, so that the governor acts at all engine speeds like a variable-speed governor. [**Fig. 3-11**]

A lever that connects to the plunger control sleeve by a ball joint is moved by the governor rod. When the throttle-adjusted governor spring is overcome by weight force as a result of engine speed increase, the governor rod pushes the lever to move the control sleeves toward an earlier opening of the cross drilling, or spill port, thus reducing effective plunger stroke and fuel injected. Engine speed is then controlled.

The engine is stopped by a separate lever that moves the control sleeve to open the spill port and prevent any injection.

Injection Advance System

As engine speed increases, the timing of injection must be made earlier, or advanced. [**Fig. 3-12**] An automatic advance device consists of a spring-loaded piston, whose other end is exposed to supply pump pressure. This pressure increases as speed rises, to a maximum that is controlled by a pressure regulating valve. [**Fig. 3-13**]

The piston is linked to the pivoted roller ring, so that piston movement moves the roller ring against shaft and cam rotation to advance injection timing.

Fuel Lines

The fuel lines from the pump to the spray valves must be of equal length. These lines carry full injection pressure and are made of heavy steel tubing. [Fig. 3-13]

No attempt should be made to substitute a copper line for a damaged feed line. Those tubes should be removed and stored in a safe place when the fuel pump is to be removed.

No service is required by these lines except that of ordinary care in removal and installation and of storage to protect them from dirt and distortion.

Spray Valves

Spray valves are known as nozzles or injectors. They are really pressure-operated spray valves and consist of a spring-loaded valve, which opens a passage to a spray tip when pressure from the fuel pump is high enough to over-

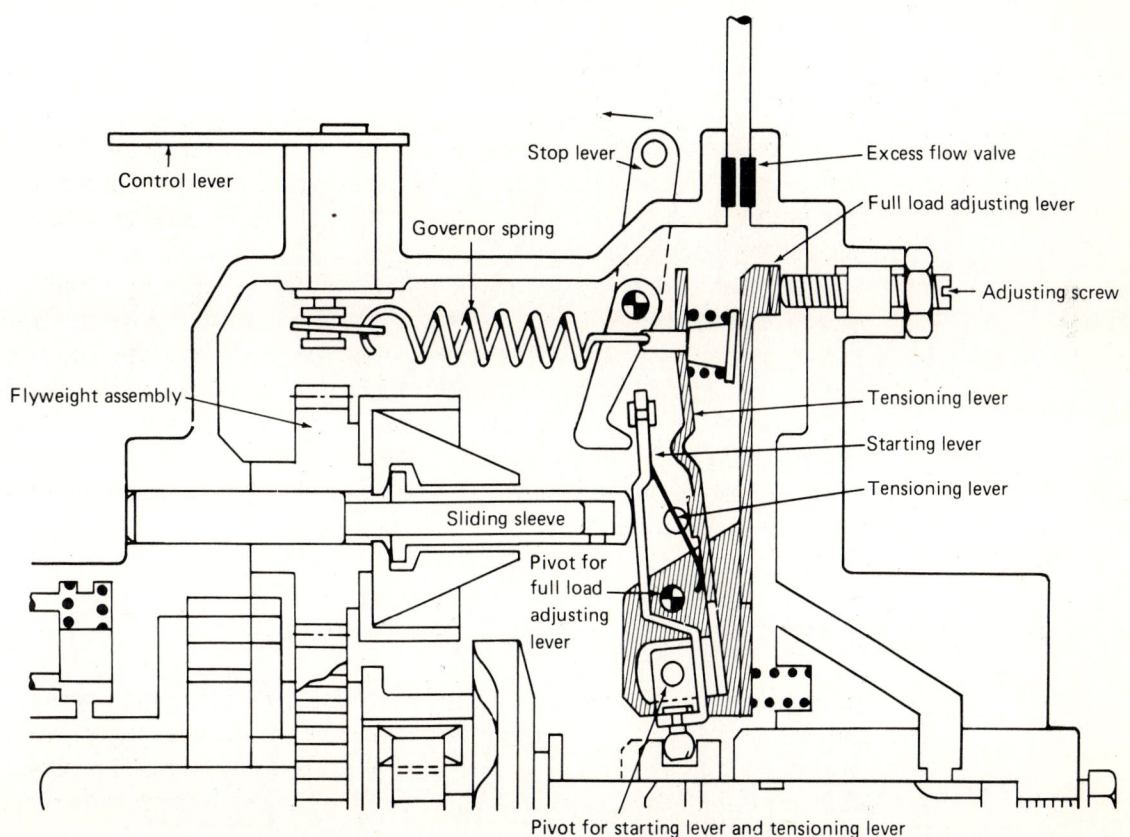

FIGURE 3-11(a). Robert Bosch Distributor Governor. (Courtesy of Robert Bosch Sales Corp.)

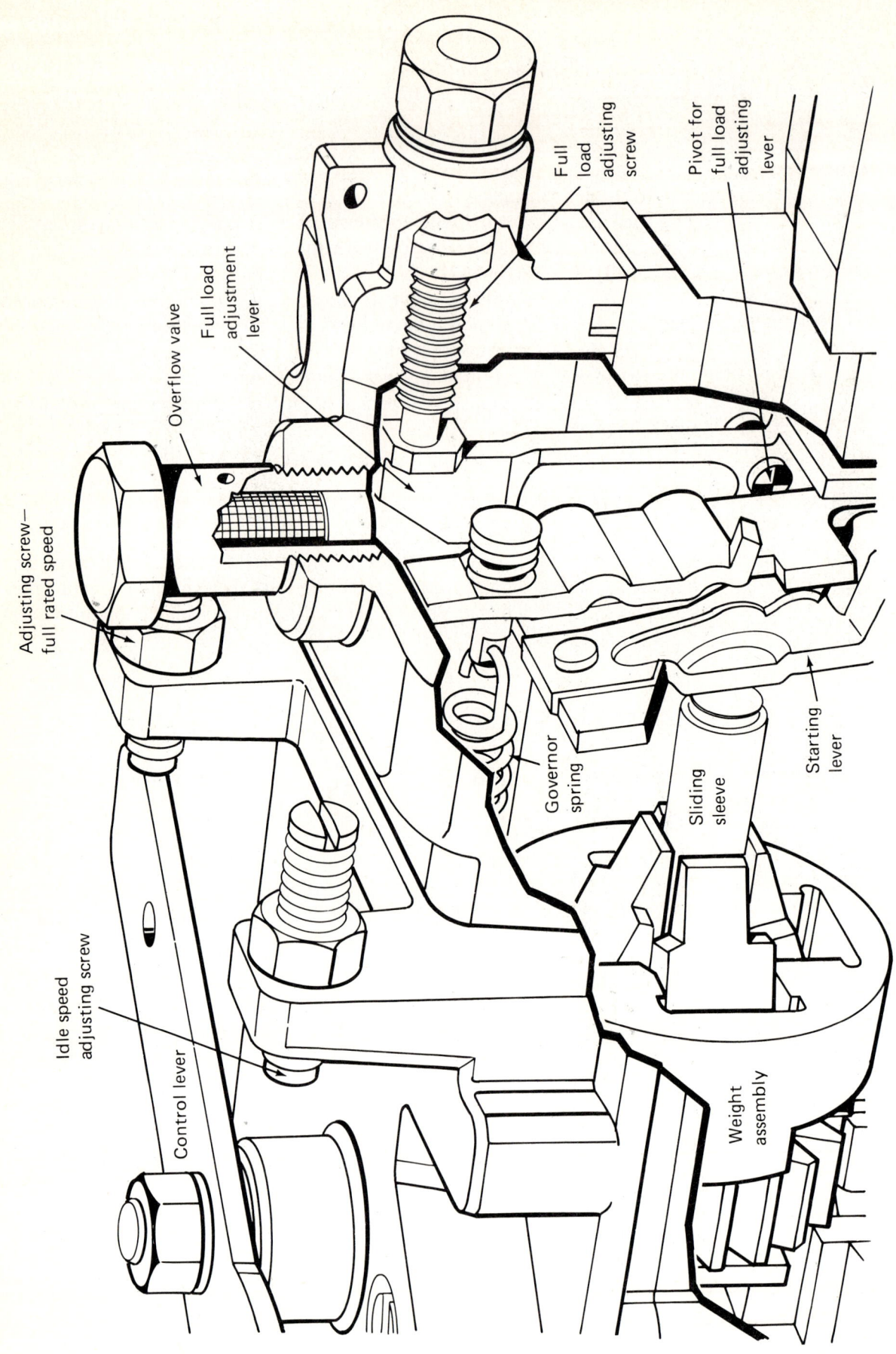

FIGURE 3-11(b). Speed Adjustments. *(Courtesy of Robert Bosch Sales Corp.)*

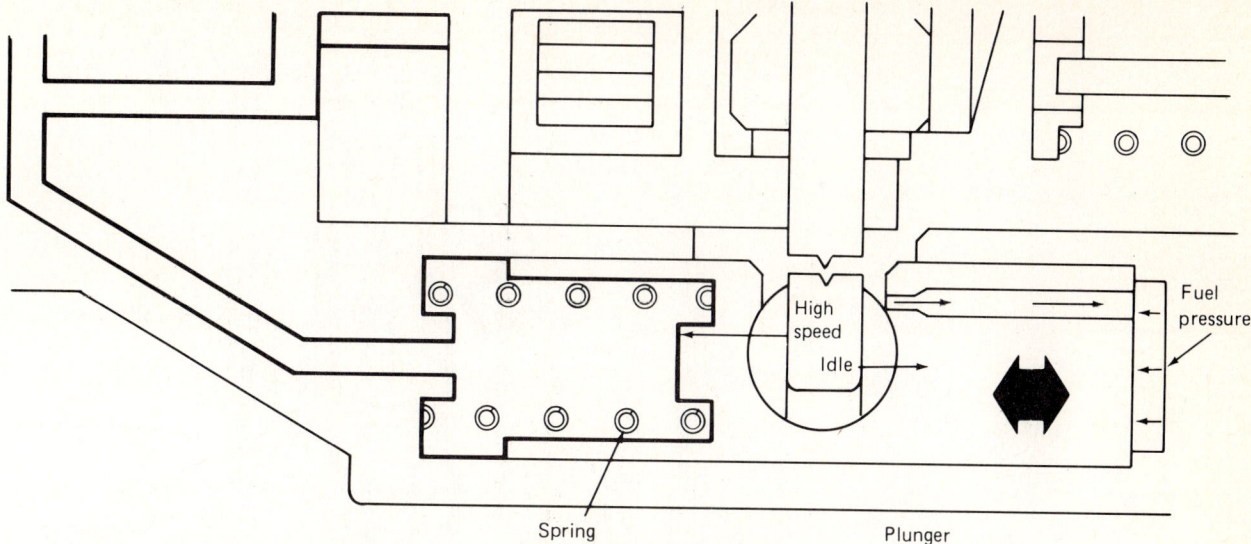

FIGURE 3-12. Injection Advance Device. *(Courtesy of Robert Bosch Sales Corp.)*

come spring tension. [**Fig. 3-14**] The valve itself is designed to close sharply and positively in order to prevent dribble after injection.

The valves are mounted in sockets bored in the cylinder head at each cylinder. Their spray is usually directed into a cavity called the *precombustion chamber*. [**Fig. 3-15**]

The design intention is to generate combustion in the small chamber, then direct its flame into the main combustion chamber over the piston to create high turbulence and complete burning of the fuel.

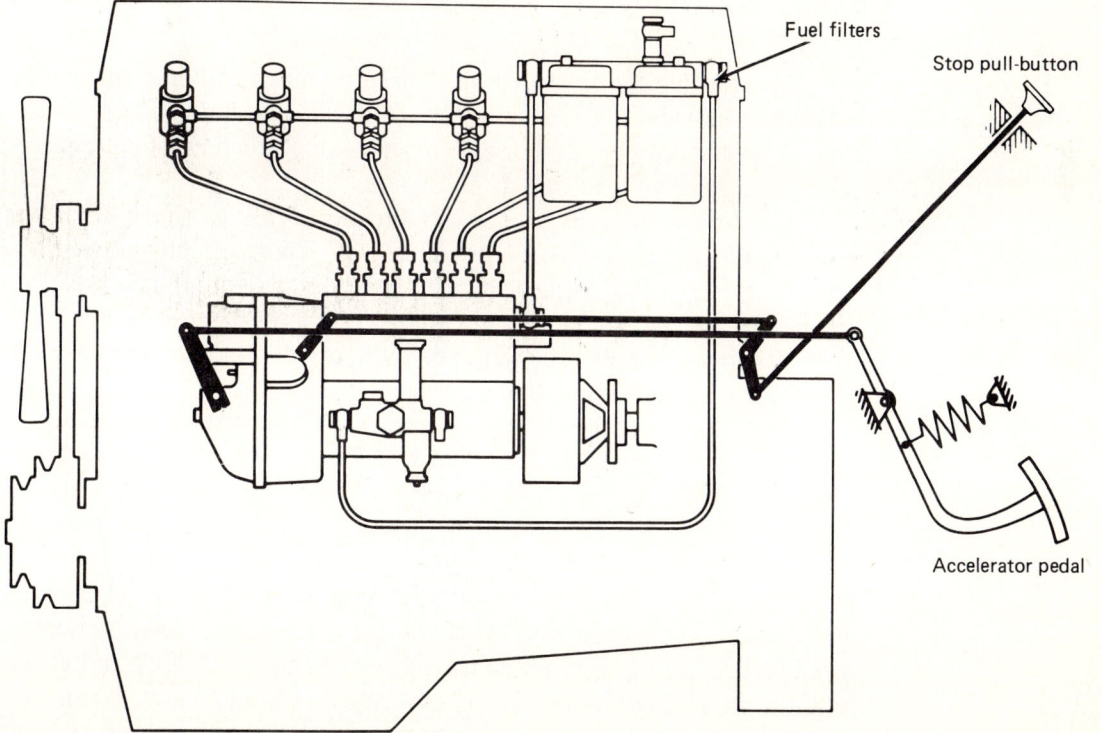

FIGURE 3-13. Typical Fuel System on Engine. *(Courtesy of Robert Bosch Sales Corp.)*

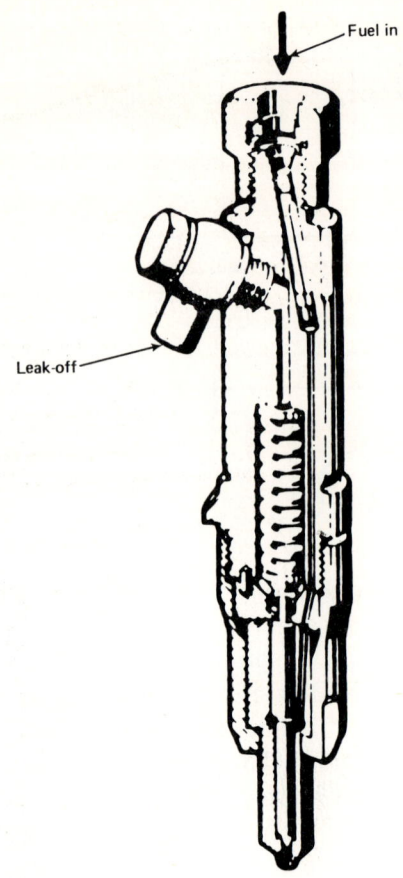

FIGURE 3-14. Spray Valve. *(Courtesy of Robert Bosch Sales Corp.)*

Spray valve sockets must be cleared of carbon when a spray valve is removed.

A gasket may be used under the valve tip, which must be removed when an exchange valve is installed.

Hold-down clamps must be tightened evenly to the specified tension. Usually 10 to 15 pound feet is specified. Overtightening will distort the spray valve and may crack the seat in the head. Undertightening will allow a compression leak.

A spray valve must be adjusted on a test stand. The valves are furnished as exchange and rebuilt units.

One should never attempt to adjust or repair these valves in the field. They are closely fitted and will not tolerate dirt or even fingerprints.

Starting Aids

Most engines equipped with Bosch fuel systems have resistance heaters in the head, called *glow plugs*. When the engine is cranked to start, or before cranking, the glow plugs are heated by battery current and warm the air in the precombustion chamber to supplement compression heat and assure prompt firing. [**Fig. 3-16**]

Diesel Engine Fuel Systems / 23

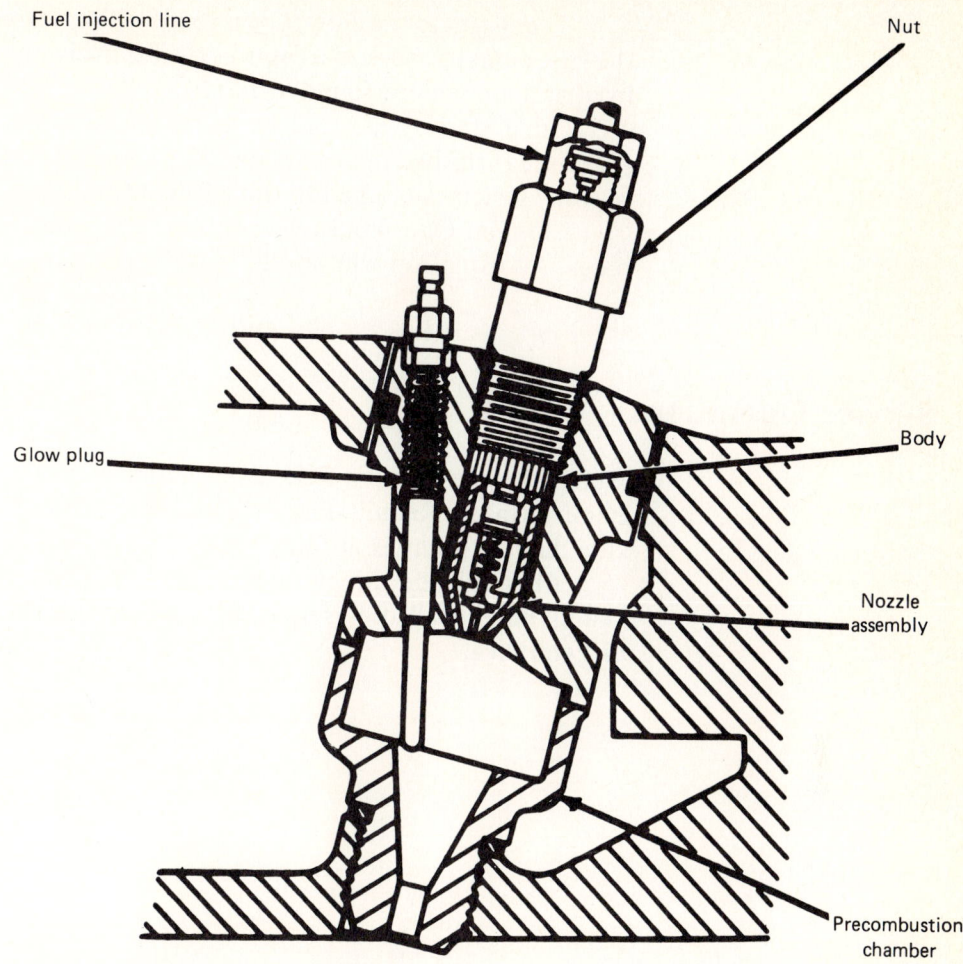

FIGURE 3-15. Spray Valve and Glow Plug Installed. *(Courtesy of Caterpillar Tractor Company)*

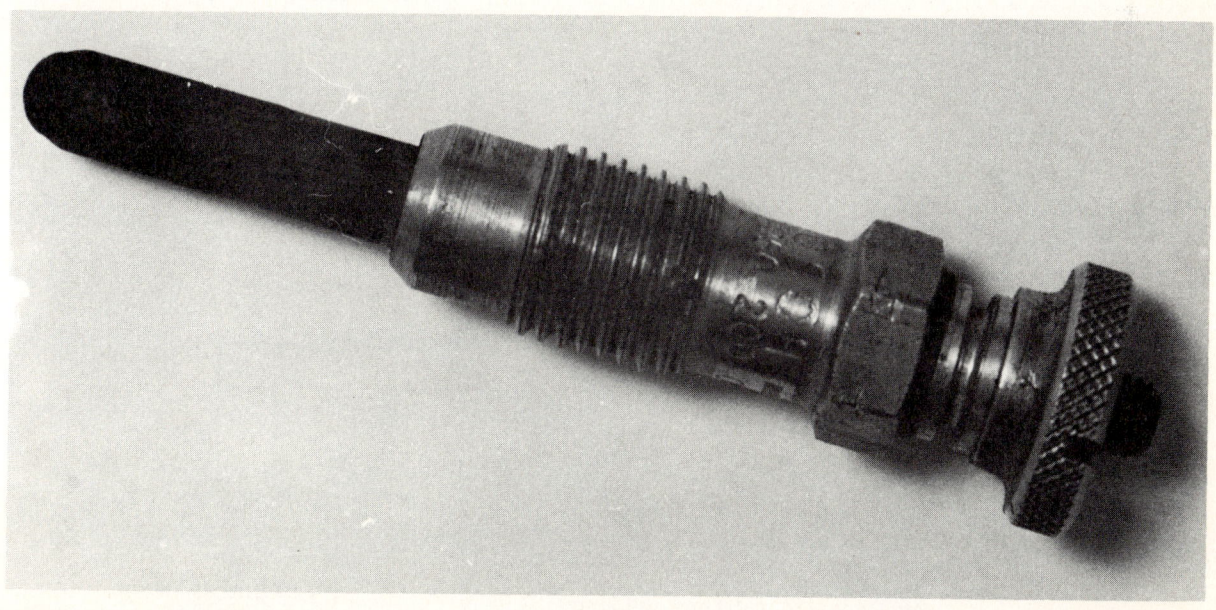

FIGURE 3-16. Glow Plug.

24 / Diesel Engine Service

The circuit to the glow tube is fused with at least a 50-ampere fuse. This is the first point to check in a failure-to-start problem. The glow plugs are long-lived and are seldom found at fault.

Other starting aids are used; the most popular is starting fluid. Such fluids are furnished in aerosol cans for spraying and in capsules for applying in spray systems attached to the air manifold. [**Fig. 3–17**]

Caution: One should *never* use starting fluid in combination with glow plugs. An explosion will result.

Service Operations

Note: In no case may a field disassembly and attempt to repair a Bosch fuel pump be made. Rebuilt units are available and must be used when checks indicate that the pump is at fault.

Although the side-mounted supply pump can be removed and an exchange unit installed, it is necessary first to clean the pump in order to prevent dirt entrance.

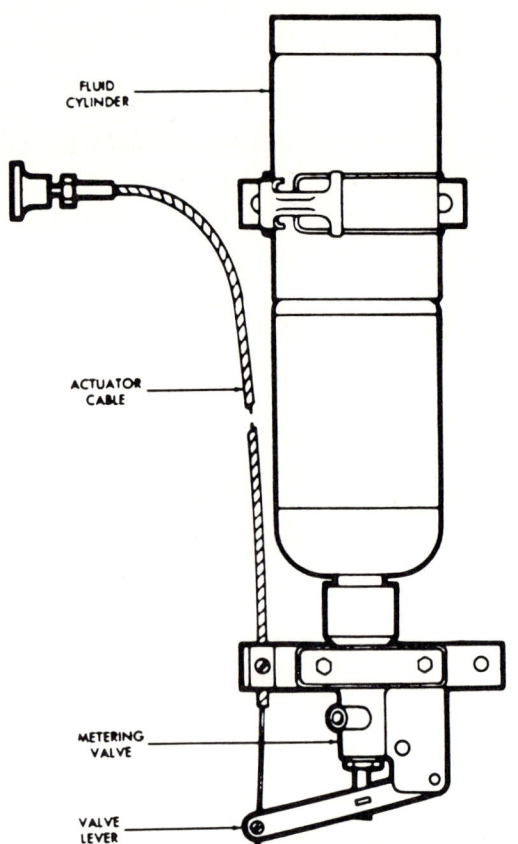

FIGURE 3-17. Starting Fluid System. *(Courtesy of General Motors Corp.)*

Diesel Engine Fuel Systems / 25

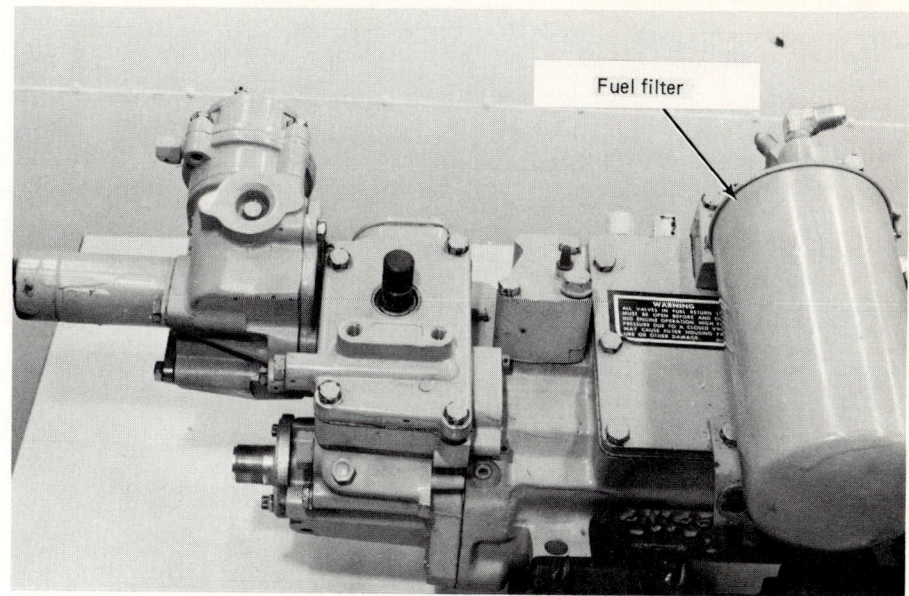

FIGURE 3-18. Fuel Filter Location. *(Courtesy of Caterpillar Tractor Company)*

Filter Service

The closely fitted parts in the pump will not tolerate water or dirt. Fuel filters must be serviced frequently enough to prevent contamination. [**Fig. 3-18**]

Each engine builder prescribes fuel filters. After-market filters can be of high quality but can also be of such poor quality as to be useless. Only filters made by old-line, reputable companies or those furnished by the engine manufacturer should be considered. Price is not the basis for selection, because a low-quality filter may allow dirt and water to enter the fuel pump, with expensive consequences.

Bosch Injection Pump Installation and Timing

All engines using these injection pumps have marked gears or drive chain links. Some have keyed gears, others drive by a splined sleeve. The engine builder has detailed manuals for the pumps on the engine models. [**Fig. 3-19**] Many pumps have slotted bolt holes in the mounting flange, allowing enough pump rotation to adjust timing exactly.

Bosch fuel pumps can give long service if good maintenance is practiced. Most engine faults that may be blamed on the pump are due to other causes. One should refer to the trouble-shooting section of this book. One should never replace a Bosch fuel pump without checking all other possible causes for the fault.

Priming an Exchange Pump

When an exchange fuel pump is installed, or when the engine has run out of fuel, priming of the system is necessary.

26 / Diesel Engine Service

FIGURE 3-19. Typical Gear Train. *(Courtesy of Caterpillar Tractor Company)*

1. See that the fuel tank contains some fuel.
2. Fill a new fuel filter and install it.
3. Squirt some clean engine oil into the fuel pump inlet before reconnecting it to the suction tube.

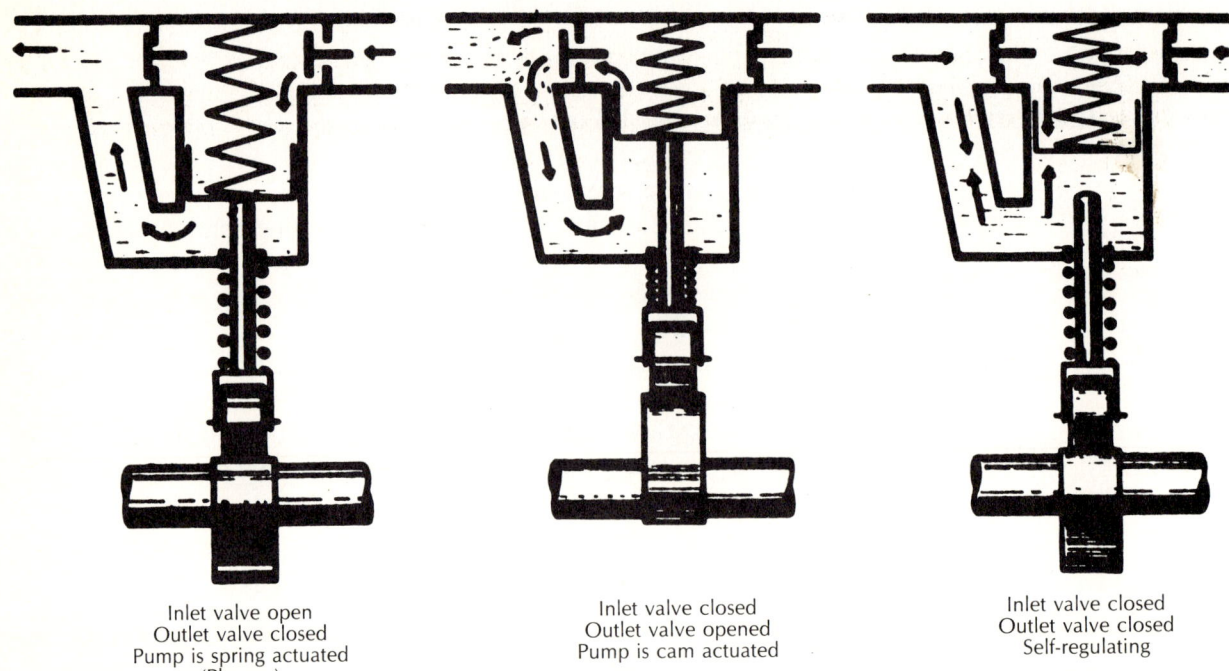

FIGURE 3-20. Fuel Supply Pump. *(Courtesy of American Bosch Division, United Technologies)*

Diesel Engine Fuel Systems / 27

4. Leave all fuel line nuts loose at the spray valve end.
5. Operate the hand priming pump until solid fuel is obtained. Pumping effort will increase.
6. Crank the engine with the throttle in "run" position. Continue to use the priming pump. Stop cranking when all fuel lines discharge solid fuel.
7. Tighten fuel line nuts. Operate the starting aid and start the engine.
8. It may be necessary to loosen some fuel lines to bleed air, with the engine running.

Fuel Supply Pump

In the Bosch distributor-type systems, the vane-type supply pump is located in the drive end of the fuel pump housing. [**Fig. 3-20**] Usually the supply pump pressure can be measured at a point on the fuel pump that is exposed to supply pump pressure. This measuring point varies with different models of fuel pump; one must refer to the manufacturer's manual for the correct point.

A judgment can be made of the need for replacement by comparing the supply pressure at full engine speed to that specified. Low supply pressure may indicate a worn supply pump or a faulty regulating valve. [**Fig. 3-21**]

AMERICAN BOSCH FUEL SYSTEM

The American Bosch Division of United Technologies produces fuel injection pumps and spray valves for many engine builders which are similar to those offered by Robert Bosch. On some engines the fuel pump of either maker may be used.

American Bosch in-line pumps use a variety of plunger scroll designs, depending on the engine builder's needs. The single-plunger distributor pumps use a sleeve-metering system similar to the ones used by other manufacturers.

In-line pumps are driven at engine camshaft speed. Distributor pumps are driven at engine speed.

All fuel-handling parts of both types of pump are lubricated by fuel. Fuel pump camshaft and bearings are oiled by an oil supply in the fuel pump base or by passages from the engine oil system. Self-lubricated fuel pumps have a breather opening. Those that do not are oiled from the engine. American Bosch distributor pumps are oiled from the engine. [**Fig. 3-22**]

The cam-driven metering plungers move through a constant stroke; the surrounding pinion is turned by the control rack. The helix on the plunger may be placed on the top or on the bottom, or there may be two helixes that are opposite in direction. The top helix gives a constant start and a variable end of injection. The bottom helix produces a variable start and a constant end of injection. [**Fig. 3-23**] The double-helix plungers give variable start and end of the injection. [**Fig. 3-23**]

28 / Diesel Engine Service

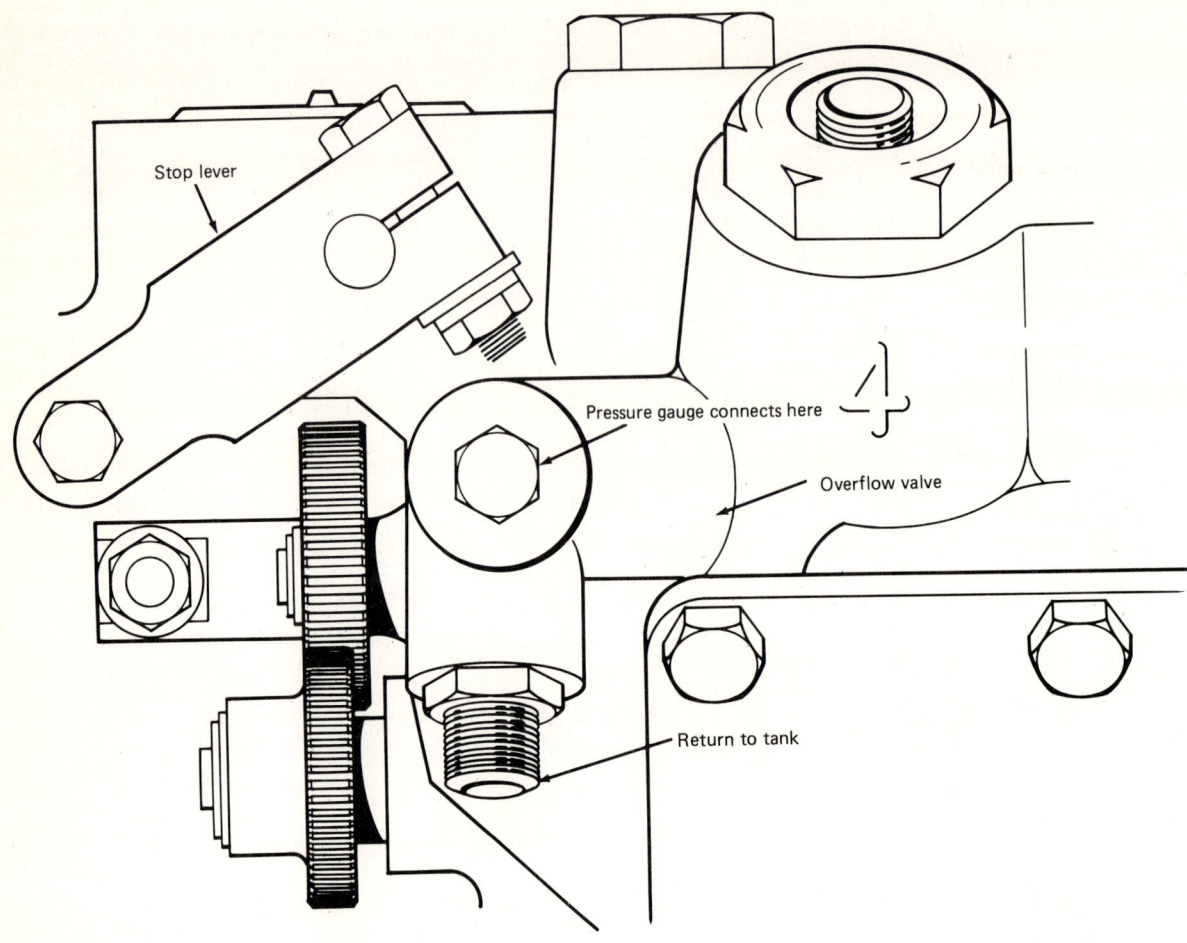

FIGURE 3-21. Supply Pressure Measuring Point. *(Courtesy of Robert Bosch Sales Corp.)*

American Bosch Governor

The centrifugal governor on these fuel pumps is nearly the same for both in-line and distributor type. In-line pumps turn at half engine speed and the governor has a step-up gear drive. Distributor pumps drive at engine speed.

Both governors transmit weight force through a sliding sleeve to a fulcrum lever that pivots on the sleeve. The lower end of the lever is pin-connected to the operating lever. The upper end is connected to the rack control shaft by a link. [Fig. 3-24]

An extension of the fulcrum lever carries a projection called a *torque cam,* which contacts the *stop plate* at full speed. The sloping contour of the stop plate causes an automatic movement of the control rack position, moving it toward increased fuel as long as the operating lever is held against its stop and engine speed is below the governor setting. [Fig. 3-24]

As engine speed increases, the sliding sleeve is pushed away by the weight

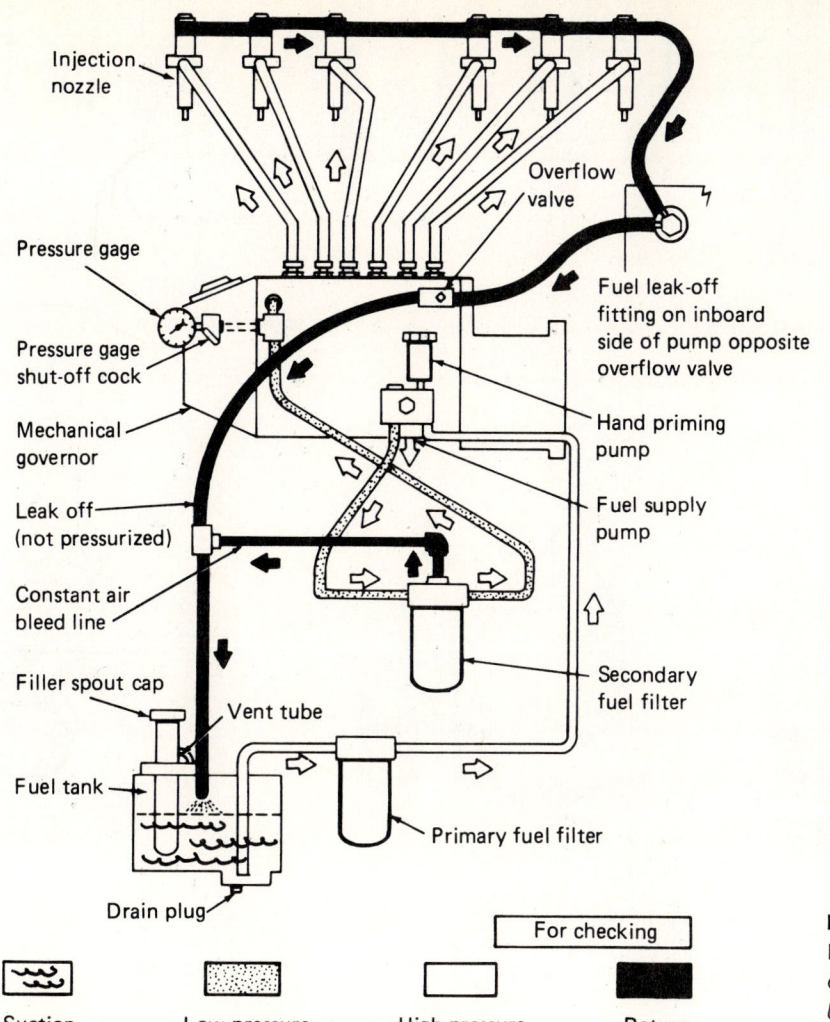

FIGURE 3-22(a). American Bosch Fuel Pump. (Courtesy of American Bosch Division, United Technologies)

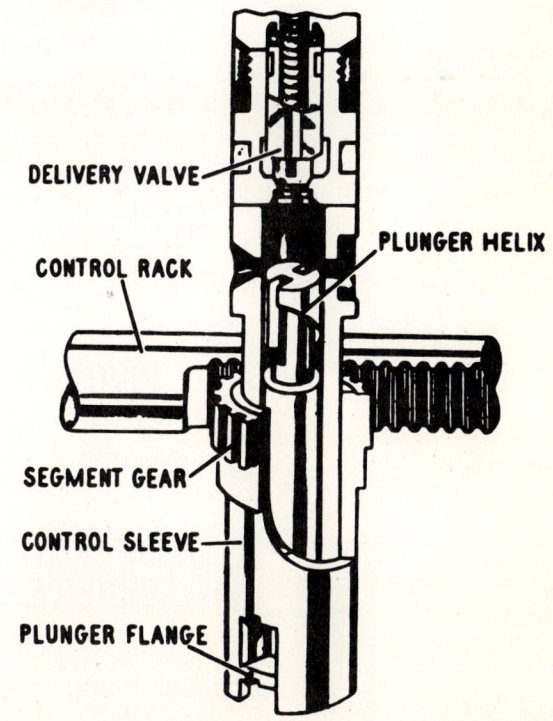

FIGURE 3-22(b). Metering System, In Line Pump. (Courtesy of American Bosch Division, United Technologies)

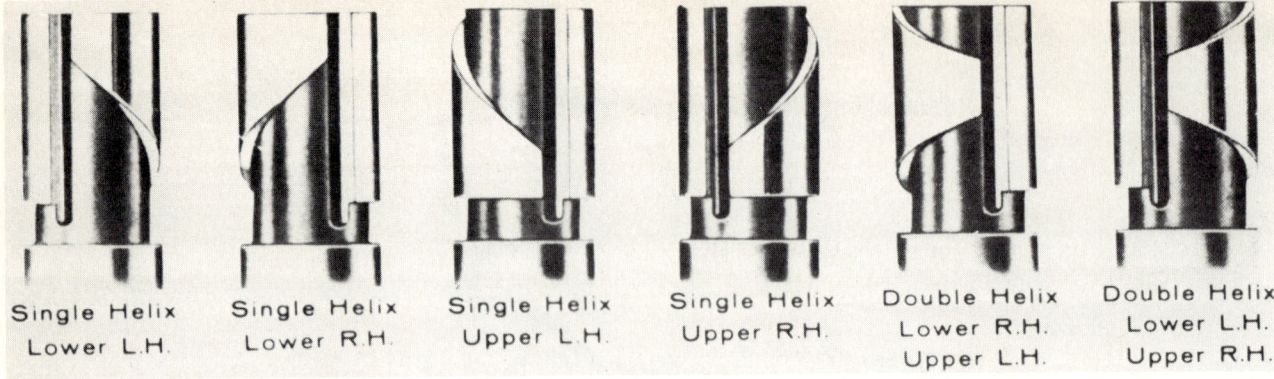

FIGURE 3-23(a). Pumping Plungers. *(Courtesy of American Bosch Division, United Technologies)*

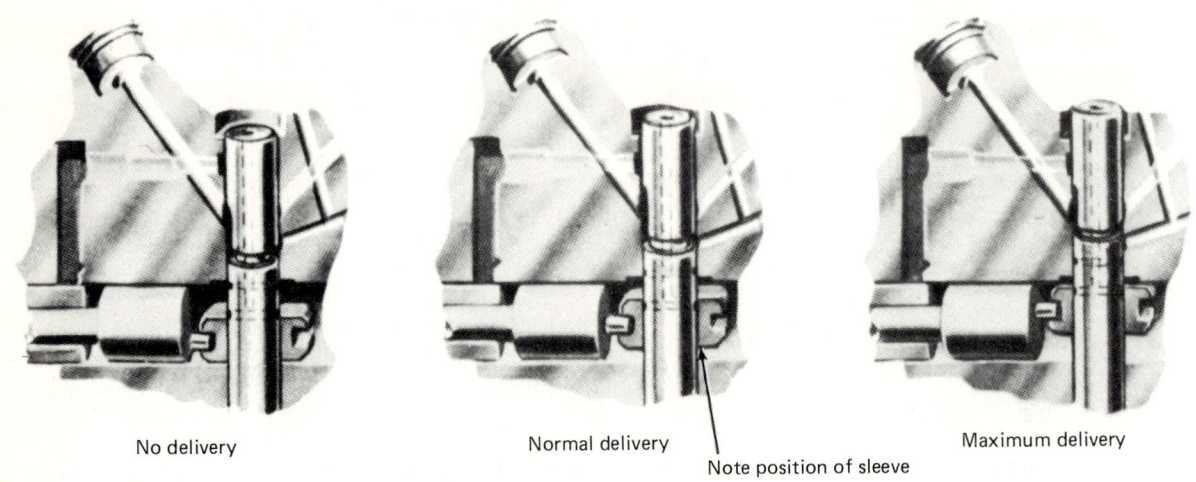

FIGURE 3-23(b). Sleeve Metering Distributor Pump. *(Courtesy of American Bosch Division, United Technologies)*

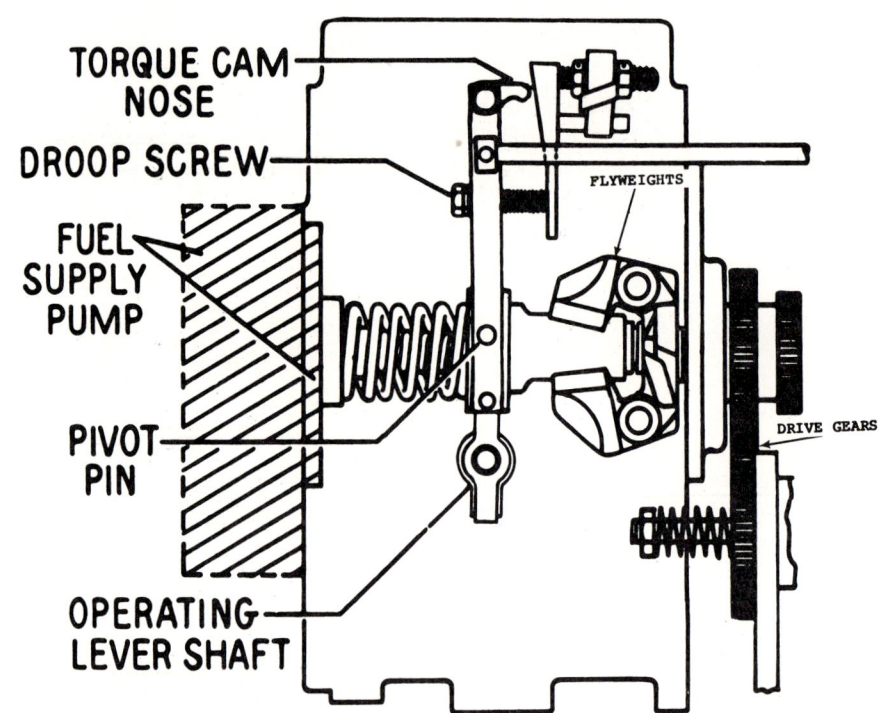

FIGURE 3-24(a). American Bosch Governor. *(Courtesy of American Bosch Division, United Technologies)*

Diesel Engine Fuel Systems / 31

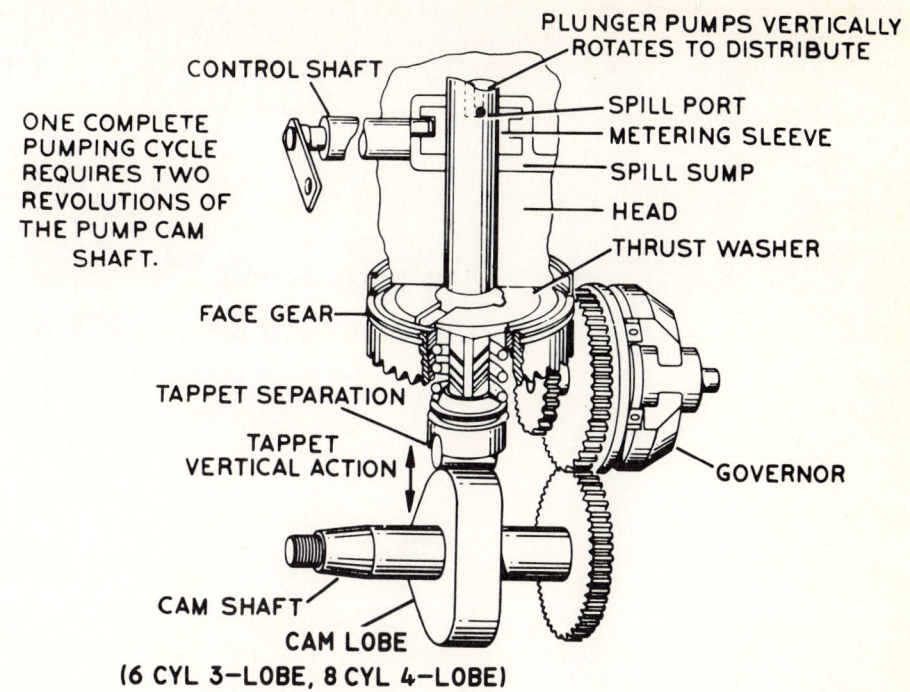

FIGURE 3-24(b). Distributor Pump Hydraulic Head and Governor. *(Courtesy of American Bosch Division, United Technologies)*

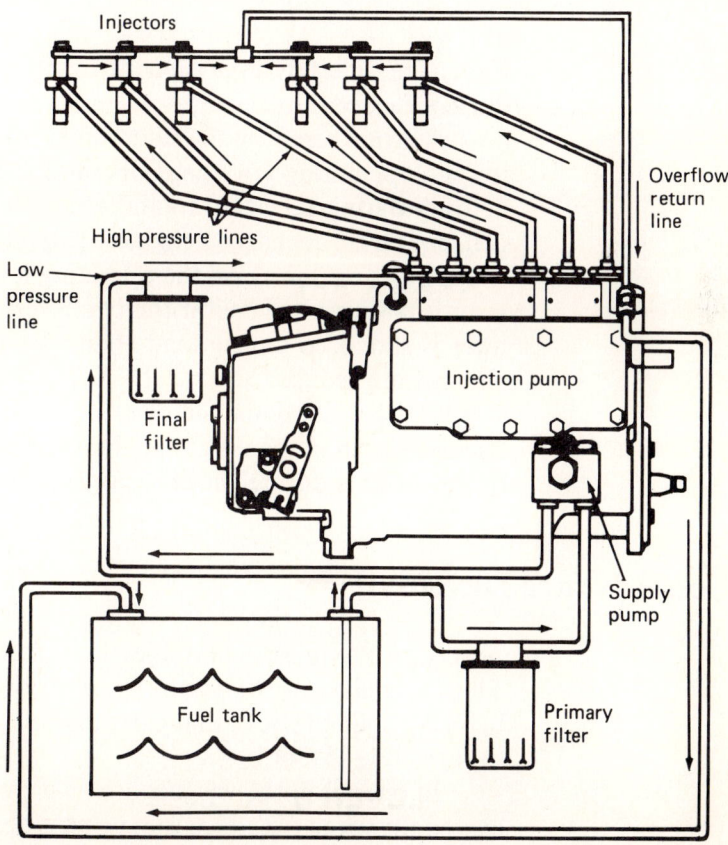

FIGURE 3-24(c). Typical Fuel System Installation. *(Courtesy of American Bosch Division, United Technologies)*

32 / Diesel Engine Service

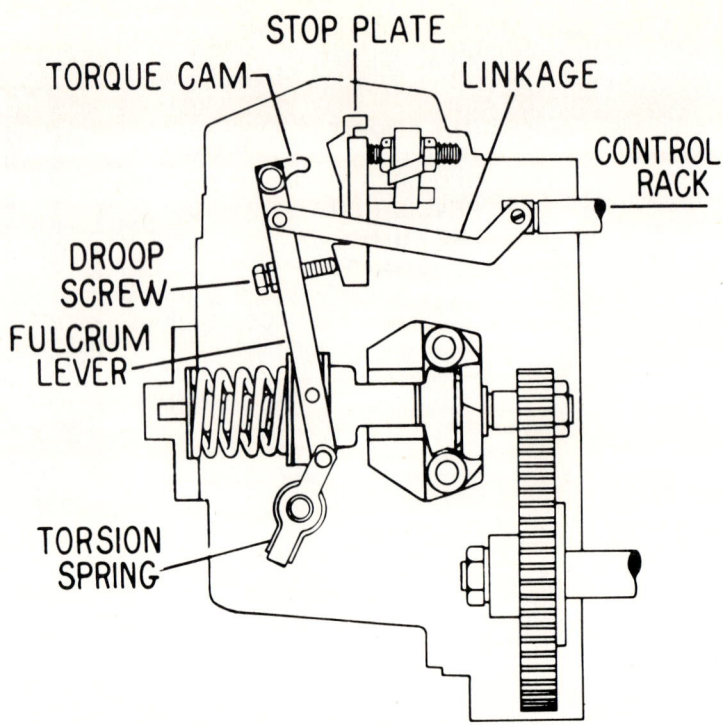

FIGURE 3-25. Idle Speed Position. *(Courtesy of American Bosch Division, United Technologies)*

force. The fulcrum lever is pushed with it and pulls the control rack toward the less fuel position.

A droop screw on the fulcrum lever is set to contact the stop plate on its lower slope during low-speed operation. This has the effect of reducing fuel delivery during acceleration and accommodates the need for controlled fuel delivery on turbocharged engines. [**Fig. 3-25**]

A torsion spring on the operating lever at its connection to the fulcrum lever is tensioned whenever the two levers are not in line. Thus the operating lever is stabilized during load changes, and shocks caused by sudden speed changes are reduced.

Shock loading on the parts affected by weight force is prevented by a friction clutch in the weight spider. The clutch provides momentary slippage of the weights during sudden speed changes. [**Fig. 3-26**]

Injection Advance Device

American Bosch offers an injection advance system called the Intravance®. This unit makes use of weight force to control pressure oil flow into the space behind a splined sleeve that can move endwise along the camshaft, causing it to turn. [**Fig. 3-30**]

In operation the weights are held by a spring. As engine speed starts to exceed the speed for which the spring is designed, weight force moves the control valve to the right. This small movement permits pressure oil to enter the passages to the space behind the splined sleeve. [**Fig. 3-30**]

Diesel Engine Fuel Systems / 33

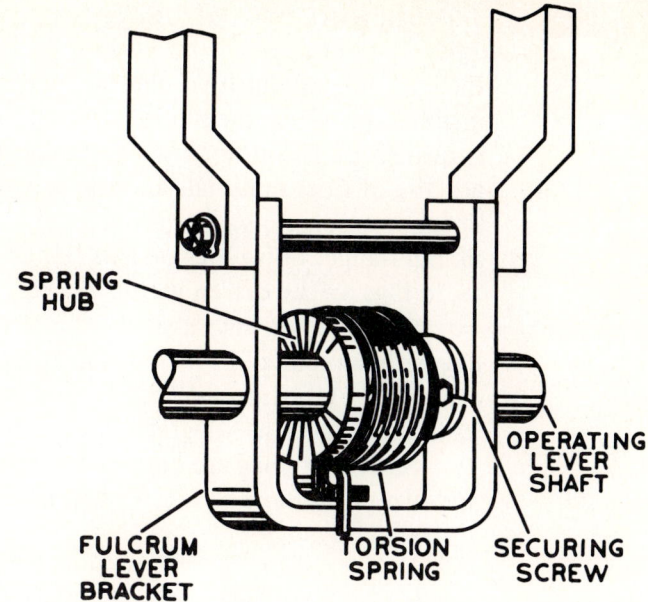

FIGURE 3-26. Torsion Spring Detail. *(Courtesy of American Bosch Division, United Technologies)*

Oil pressure then forces the sleeve to move; the splines act to turn the camshaft to advance. As the sleeve moves, it carries the follow-up rod, which pulls the spring against the control valve. When the valve edge just closes the oil passage, the sleeve stops, and advance action is held.

This unit can advance injection up to 20 degrees. [**Fig. 3-27**] When engine speed falls, weight force is reduced, and the spring pulls the follow-up rod and splined sleeve back to retard injection. The action of this device depends on engine speed, but it is independent of the main governor.

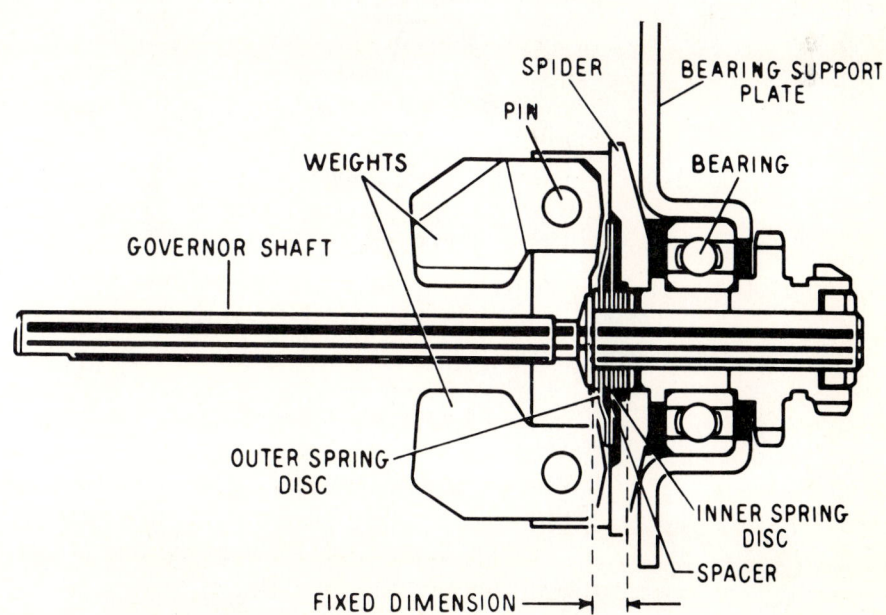

FIGURE 3-27. Friction Clutch. *(Courtesy of American Bosch Division, United Technologies)*

Timing and Adjustments

Each engine builder furnishes detailed timing procedures for injection pump installation. When the pump is driven by an accessory shaft, a drive coupling is furnished. This unit consists of an adjustable drive member, an intermediate disc of fiber material, and the drive hub. The drive hub has slotted holes through which capscrews are inserted to the tapped holes in the adjustable timing flange. Drive lugs on this flange engage slots in the intermediate disc, as do lugs on the driven member. [Fig. 3-28]

This unit is always assembled with the center line of the adjustable hub in line with the center line on the edge of the hub. The O marks are together. The center line of the hub is over the center of the keyway, and each line on the adjustable member is 3 degrees. Thus precise timing is made possible.

Such couplings are designed to tolerate slight misalignment of the drive and driven shafts. The less misalignment, the less wear there is on the coupl-

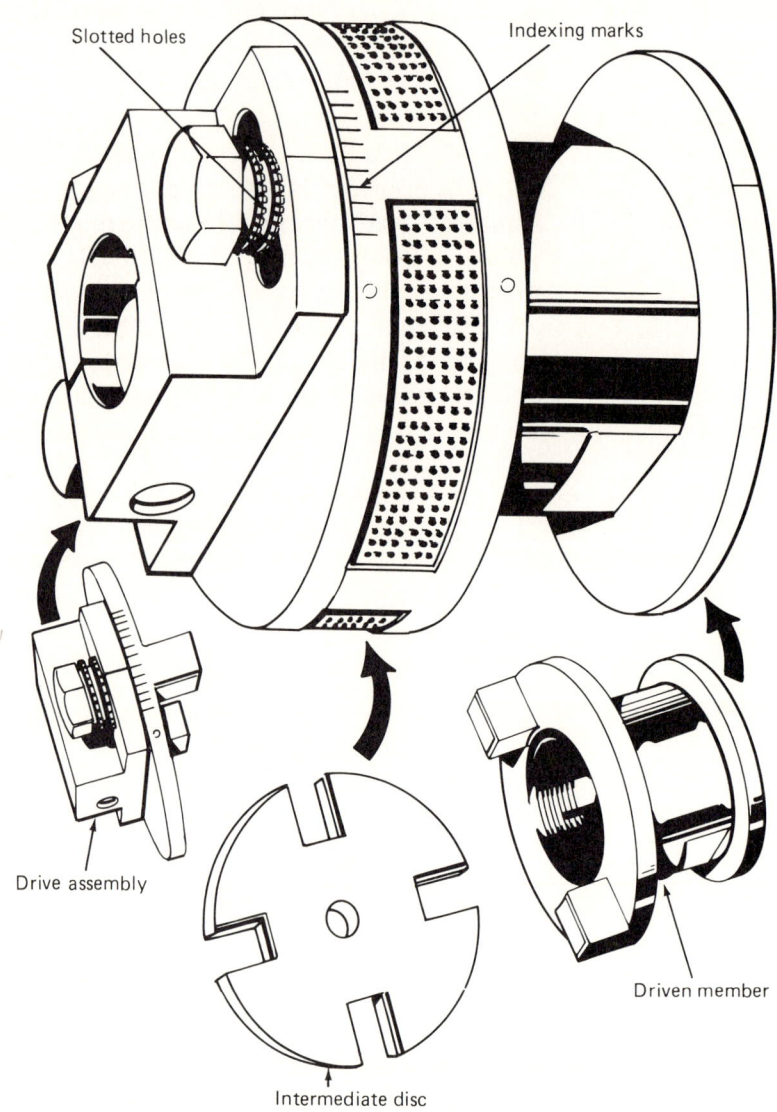

FIGURE 3-28. Drive Coupling. *(Courtesy of American Bosch Division, United Technologies)*

ing. Also, some endwise movement is tolerated, and the intermediate disc must have from 0.004 inch (0.1016 millimeter) to 0.014″ (0.355 millimeter) size/clearance).

Field Repairs and Adjustments

Field adjustments to all fuel pumps must be made only by experienced mechanics using proper tooling. Injection pumps are completely calibrated and adjusted on test stands; field adjustments to satisfy operator demands will result in voided warranties and possibly expensive damage.

Throttle Linkage

Linkage connecting the operating lever or pedal with the throttle lever on the fuel pump must be adjusted to match the throttle lever travel. Breakaway levers give some tolerance and permit take-up for wear. [**Fig. 3–29**]

Speed Adjustment

Although both idle and maximum speed are set on the test stand, a field adjustment of low idle may sometimes be needed. The idle-speed adjusting screw is located behind the throttle lever on in-line pumps. It is located on the lower rear of the pump housing on distributor pumps. [**Fig. 3–30**]

Speed adjustments are sealed after testing and should never be made unless necessary. To raise idle speed, one should loosen the locknut and turn the screw in, with the engine warm and running at idle. A slight adjustment is usually enough.

Engine idle speed is normally at 500 to 600 rpm. All engines should be set to idle fast enough to reduce vibration. [**Fig. 3–30**]

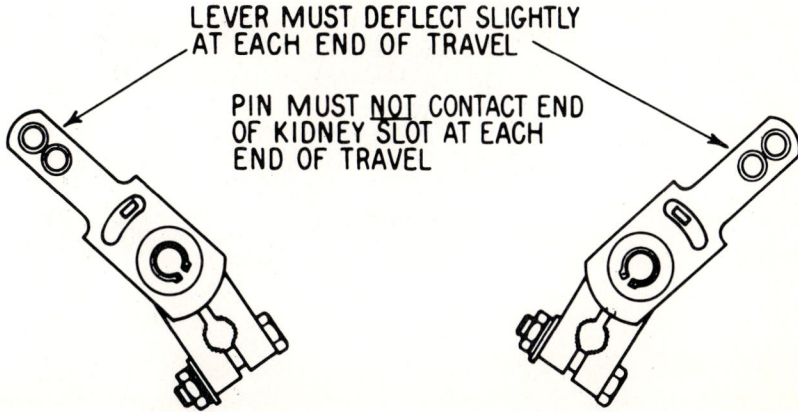

FIGURE 3–29. Breakaway Throttle Levers. *(Courtesy of American Bosch Division, United Technologies)*

36 / Diesel Engine Service

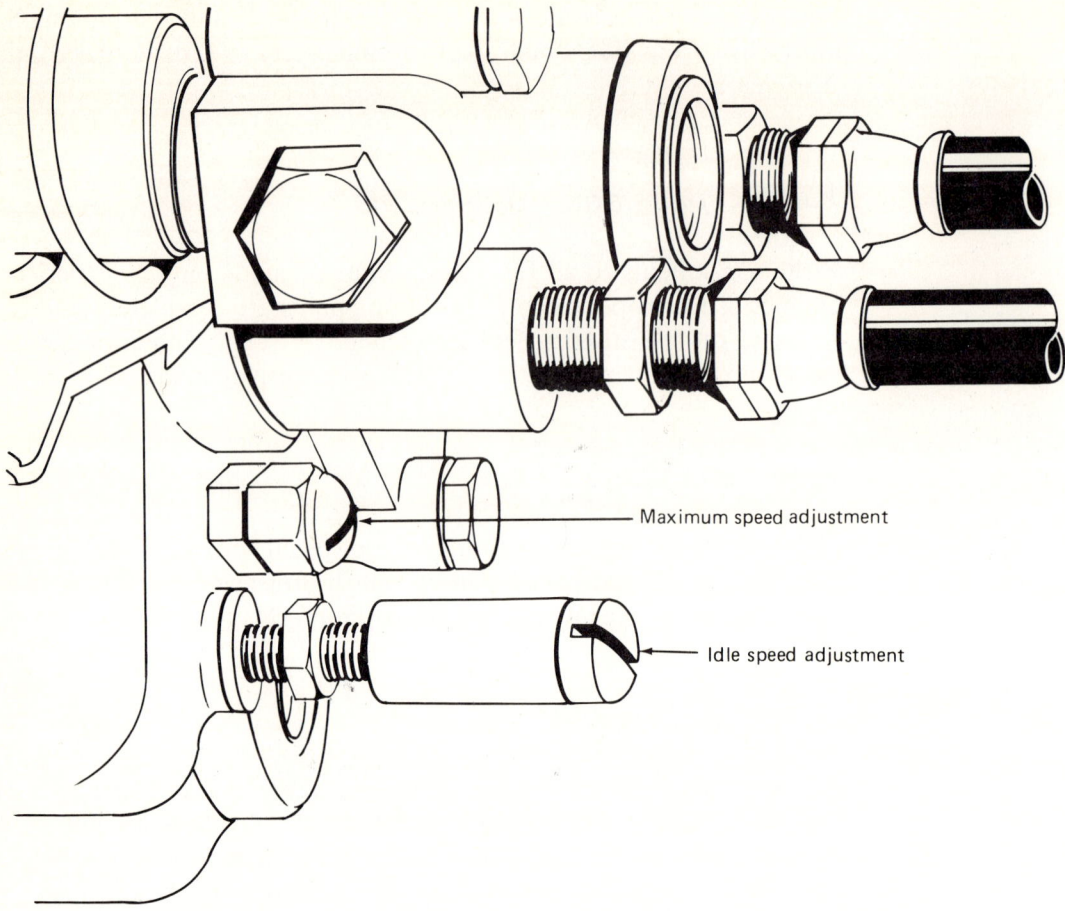

FIGURE 3-30(a). Speed Adjustments. *(Courtesy of American Bosch Division, United Technologies)*

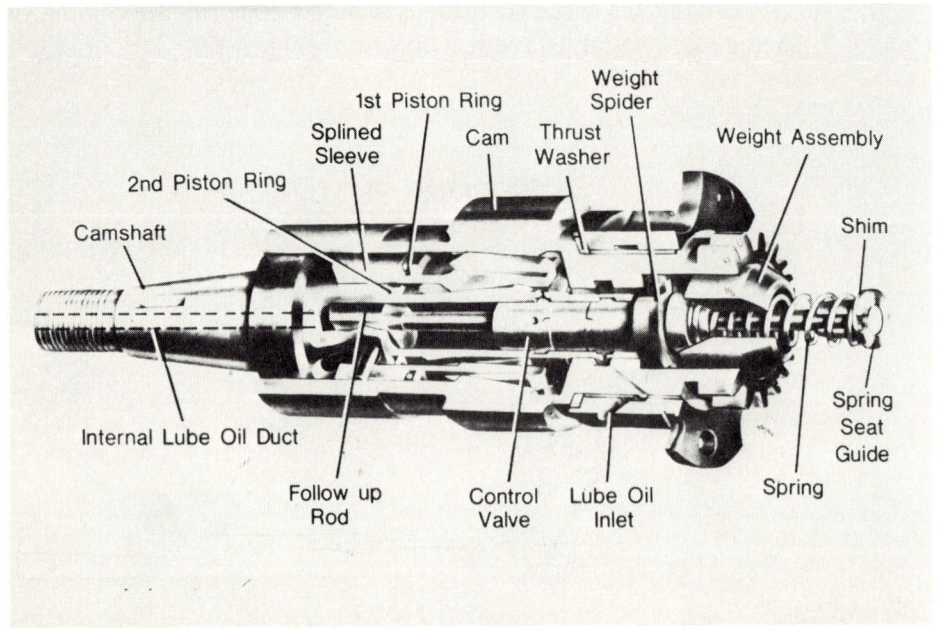

FIGURE 3-30(b). Intravance Device. *(Courtesy of American Bosch Division, United Technologies)*

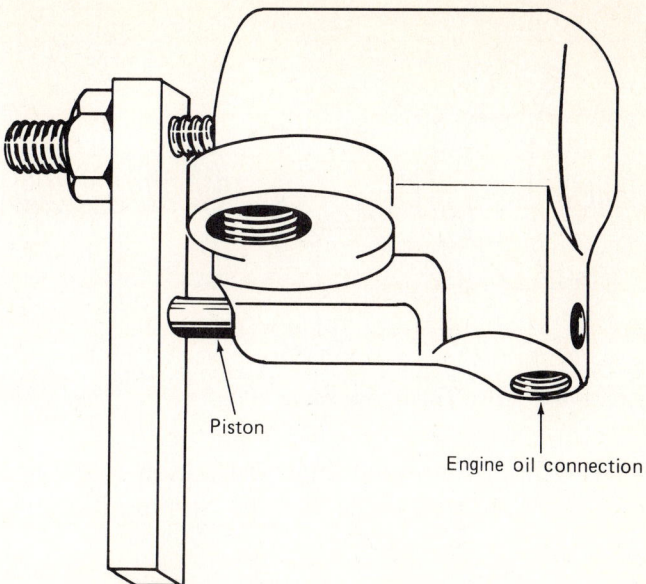

FIGURE 3-31. Starting Fuel Device. *(Courtesy of American Bosch Division, United Technologies)*

Starting Fuel System

This device consists of a small piston in a housing mounted under the governor cover behind the stop plate. The piston is subject to engine oil pressure routed to the housing. **[Fig. 3-31]**

At start, the governor spring holds the fulcrum lever, stop plate, and piston in, moving the fuel control rod to increase fuel delivery.

As soon as the engine starts, oil pressure pushes the piston out against the stop plate, forcing it into the normal, run position. The control rod moves with it, and this position is held until the engine is stopped. **[Fig. 3-31]**

Overflow Valves

Bosch fuel system pressure in the pump is set by a relief valve called the *overflow valve*. Fuel pressure in the supply passages is maintained by the valve spring and ranges from 10 psi (0.69 Bar) at idle to 50 psi (3.45 Bar) at speed.

The overflow valve is located under a cap nut at the drive end of the fuel passage on in-line pumps and on the side of the hydraulic head on distributor pumps. **[Fig. 3-32 and 3-33]**

Service to this valve consists of cleaning and inspection for damage.

A sticking or faulty overflow valve can cause hard starting, rough operation, and low power.

Shut-down Lever

In-line fuel pumps have a separate shutdown lever, which acts to move the rack control to no-fuel position. Distributor pumps may use a cable-

38 / Diesel Engine Service

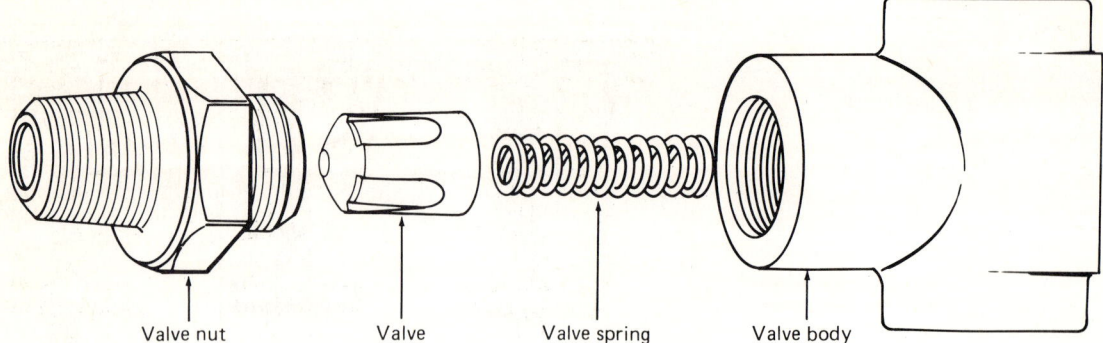

FIGURE 3-32. Overflow Valve for In-Line Pump. *(Courtesy of American Bosch Division, United Technologies)*

controlled shut-off position of the operating lever or may use a solenoid to move the lever to the stop position.

Fuel Supply Pump

The supply pump used on the in-line fuel pump is a simple plunger-type pump, mounted on the side of the pump housing, and operated by the main camshaft. When a hand priming pump is used, it is mounted on the supply pump. **[Fig. 3-34]**

Distributor pumps are equipped with a gear-type supply pump, mounted on the governor end of the pump. **[Fig. 3-35]** Because these are positive displacement pumps whose output is proportional to speed, a bypass valve is provided to limit the maximum pressure by allowing discharged fuel to return to the suction side when discharge pressure reaches the limit for which the valve is set. The hand priming pump can be used, mounted on the supply pump.

American Bosch fuel injection equipment is used on many engines from small to large. It has been very reliable, with a long service life.

TROUBLE-SHOOTING FUEL SYSTEMS

Most fuel pumps operate well over long service periods. One should never consider replacing the fuel pump until thoroughly checking every other possible cause.

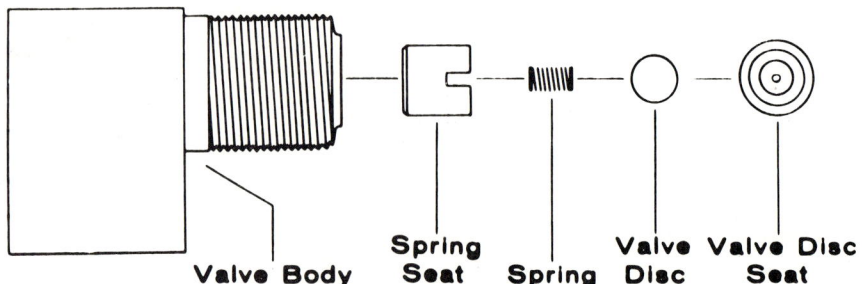

FIGURE 3-33. Overflow Valve on Distributor Pump. *(Courtesy of American Bosch Division, United Technologies)*

Diesel Engine Fuel Systems / 39

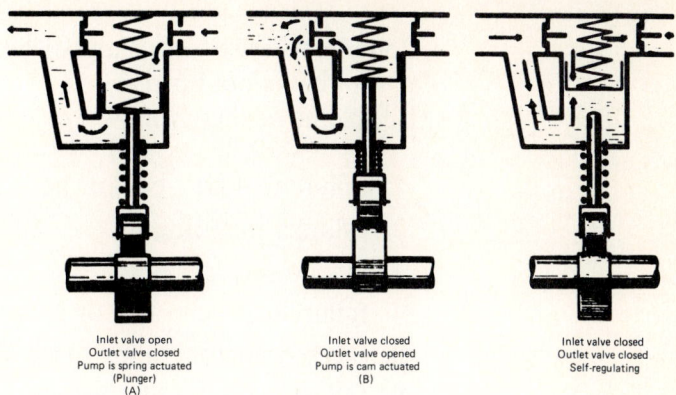

FIGURE 3-34. **Fuel Supply Pump.** *(Courtesy of American Bosch Division, United Technologies)*

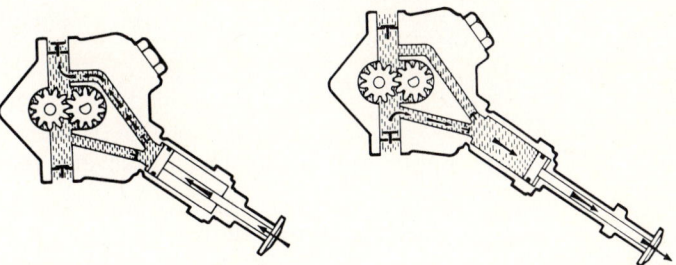

FIGURE 3-35(a). **Supply Pump and Priming Pump.** *(Courtesy of American Bosch Division, United Technologies)*

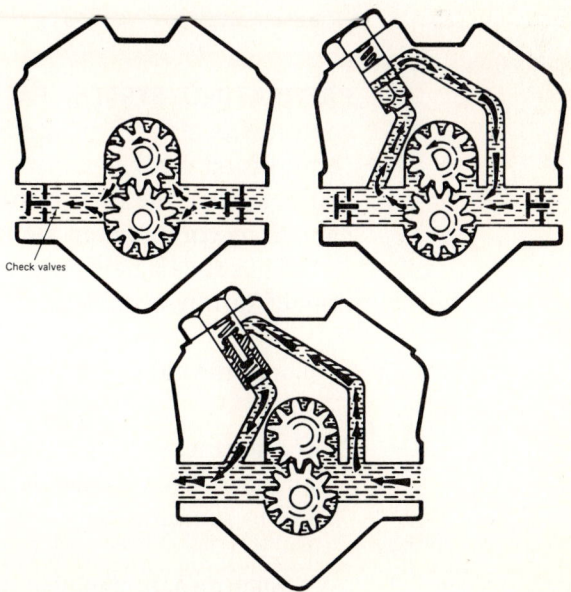

FIGURE 3-35(b). **Supply Pump on Distributor Pump.** *(Courtesy of American Bosch Division, United Technologies)*

Fuel System Complaints

Complaints arise from many different symptoms. A *symptom* is some engine action that is not normal.

Some of these follow:

1. failure to start—cold or hot
2. cylinder misfiring
3. low power
4. failure to reach rated or governed speed
5. excessive detonation—firing knock
6. failure to run at idle—stalling
7. engine stopping from normal speed
8. excessive oil consumption
9. excessive fuel consumption
10. black or dense exhaust smoke
11. power loss at random times during operation

Failure to Start

An engine may fail to start for several reasons:

1. preheating system failure
2. aerated fuel supply
3. battery low—low cranking speed
4. extreme temperature

PREHEATING SYSTEM FAILURE. A diesel engine burns fuel by the heat of compressed air. Air at atmospheric temperature is compressed during the compression stroke, and this compression heats the air. We need at least 700°F to burn the fuel; if the air is quite cold, we may not be able to heat it enough by compression alone.

Thus starting aids are necessary. Unless the starting aid system is operational, starting is difficult if not impossible.

1. Check the glow plug circuit. If there is no current at the glow plugs, check the fuse.
2. If a starting fluid system is used, check to be sure that it is working. An engine may fire on starting fluid, which has a low ignition temperature, but fail to run on fuel.

Such an action may indicate a fault in fuel supply. If white smoke is present in the exhaust during cranking, the fuel supply is adequate, but the starting aid is not functioning. No smoke indicates no fuel being injected.

AERATED FUEL SUPPLY. No diesel engine fuel system can furnish fuel for injection unless it is receiving solid fuel from the tank and filter system.

1. Check the first fuel filter to see if it is full. The fuel system may require priming. A leak may allow fuel to run back to the tank and empty the fuel system while the engine is parked. This will surely cause hard starting and extended cranking.
2. To prime any Bosch system, loosen the connection at each spray valve, and operate the hand priming pump located on the supply pump.

Note: Some systems have the priming pump in the fuel filter head.

Continue to pump while barring the engine. When fuel issues from each spray valve line, stop the priming pump and secure the lines.

It is always good to fill the filters and fuel pump with clean fuel before trying to prime it. If you cannot pick up fuel with the priming pump, look for a suction leak. A small vacuum gauge teed into the fuel suction line may help.

LOW BATTERY, SLOW CRANKING SPEED. One should never continue to crank a diesel engine until the batteries are discharged. Any engine in reasonable condition will start within 30 seconds. If the engine does not start, one must find what is wrong before the ability to crank is lost.

The mechanic should observe battery condition and evidence of corrosion at the terminals. A dirty battery may self-discharge by a leakage path through dirt on its top, and corrosion may hinder current flow from the terminals.

If heat is found on a cable or its connections, one may be sure of high resistance at that point; the remedy is to clean the battery and the cable ends. Soda water and a terminal cleaning brush will help get the system working.

EXTREME TEMPERATURES. Diesel engines require starting aids in cold climates. A mechanic must consider the conditions and make sure that the starting aid used is able to work before cranking the engine.

Several devices are available to heat the oil, coolant, and engine block during shutdown. One may have to resort to such a device or drain the oil after shutdown and keep it warm until it is put into the crankcase just before starting.

Cylinder Misfiring

This symptom is one for which the fuel pump may be blamed. However, several other faults can cause this problem.

In general a steady miss in one cylinder should lead to the spray valve or to an investigation of that cylinder's condition. Sticking valves, or even a scored cylinder, may cause a miss. One should always remove the oil filler cap or vent to observe for puffs as the engine idles.

In the Bosch system one should check for the miss by loosening each fuel line in turn, with the engine idling. The effect can be readily noted.

An intermittent miss, or one that appears to affect cylinders at random, indicates a suction air leak if it is generally accompanied by intermittent heavy detonation or knocking. This condition is often observed after priming and generally clears up in a few minutes.

In rare cases a discharge valve on the fuel pump may fail. This will give a steady miss.

Other causes are water in the fuel and too light a fuel grade.

Pump exchanges are often required after such problems are found.

Low-Power Complaints

These complaints occur as a result of many misunderstandings on the part of truck drivers and other operators. The only valid test of engine power is a dynamometer.

Some service items that can be checked are

1. dirty fuel filters,
2. observation of exhaust smoke,
3. restricted air supply,
4. worn throttle linkage,
5. poor fuel quality,
6. improperly adjusted valves or injectors (on some engines), and
7. dragging brakes or misalignment.

In general an engine that does not pull well will exhibit some sign or action. The exact circumstances under which the engine loses power must be discovered by questioning the operator. The simple things should always be checked first.

Failure to Reach Rated or Governed Speed

Governors do not usually change their setting. Thus, if the governor allowed the engine to perform before the complaint, some other fault may have occurred during operation.

All of the faults applicable to low-power problems apply to this complaint of failure to reach full speed.

Excessive Detonation—Firing Knock

Again, one must question the operator as to the circumstances. An engine that ran normally until this noise occurred may be burning a fuel that is too light. Such fuels as furnace oil, kerosene, or even No. 1 fuel will increase engine knock and reduce power. Of course, if gasoline is put into the tank, knock will increase and engine damage may follow.

When an exchange fuel pump is installed, it must be timed to the engine

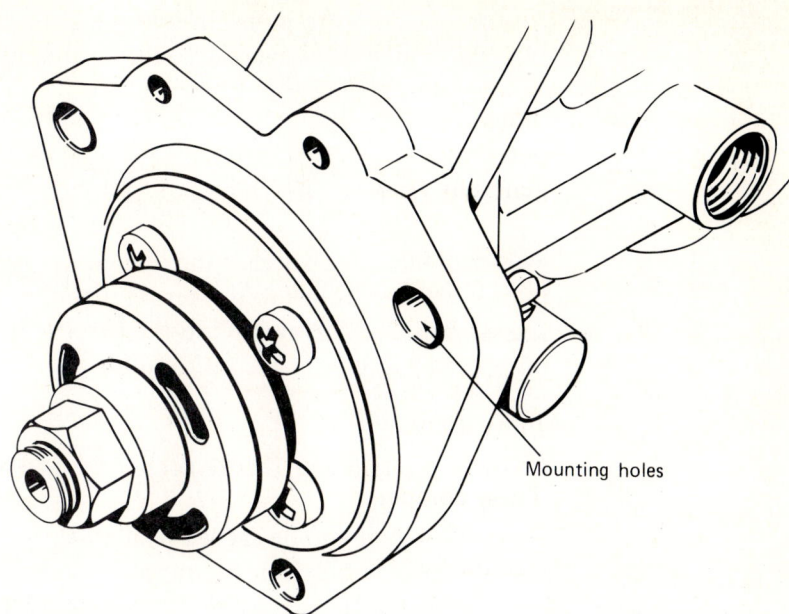

FIGURE 3-36(a). Pump Mounting Flange. *(Courtesy of American Bosch Division, United Technologies)*

camshaft or drive assembly. Bosch pumps are timed at the drive end; timing instructions are included in all engine builders' manuals.

Where the drive is through a splined coupling, a blank spline is used to make it impossible to install the pump out of time. The pump mounting capscrew holes are usually slotted, so that fine tuning adjustment can be made on the engine. [**Fig. 3-36**] In most cases, a small rotation of the pump after loosening the capscrews will correct a timing problem.

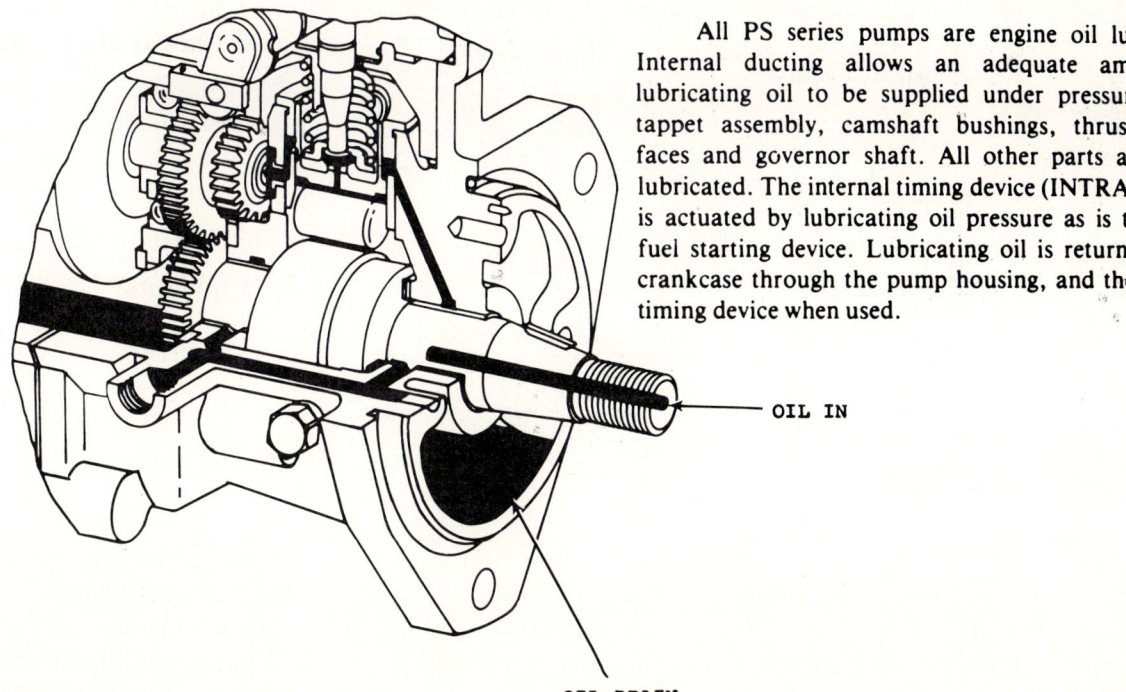

All PS series pumps are engine oil lubricated. Internal ducting allows an adequate amount of lubricating oil to be supplied under pressure to the tappet assembly, camshaft bushings, thrust washer faces and governor shaft. All other parts are splash lubricated. The internal timing device (INTRAVANCE) is actuated by lubricating oil pressure as is the excess fuel starting device. Lubricating oil is returned to the crankcase through the pump housing, and the external timing device when used.

FIGURE 3-36(b). Lubricating Oil Passages. *(Courtesy of American Bosch Division, United Technologies)*

Note: Turn the pump in operating direction to retard timing.

An engine with severe detonation should never be permitted to continue. Damage can result.

Failure to Run at Idle—Stalling

This problem suggests that the idle governor is set too low. Most Bosch fuel pumps have an adjustable external stop that can be set to obtain proper idle speed. [**Fig. 3-37**] Any bind in the throttle linkage or interference at the fuel pump can cause poor return to idle.

Engine Stopping from Normal Speed

Such a stoppage can be caused by a part failure. It can also be due to running out of fuel or a break in the supply system.

What was the nature of the stoppage? Was it sudden, like turning off a switch? Or did the engine falter, then die? Running out of fuel seldom stops an engine suddenly. A broken pump drive part may stop the engine. The

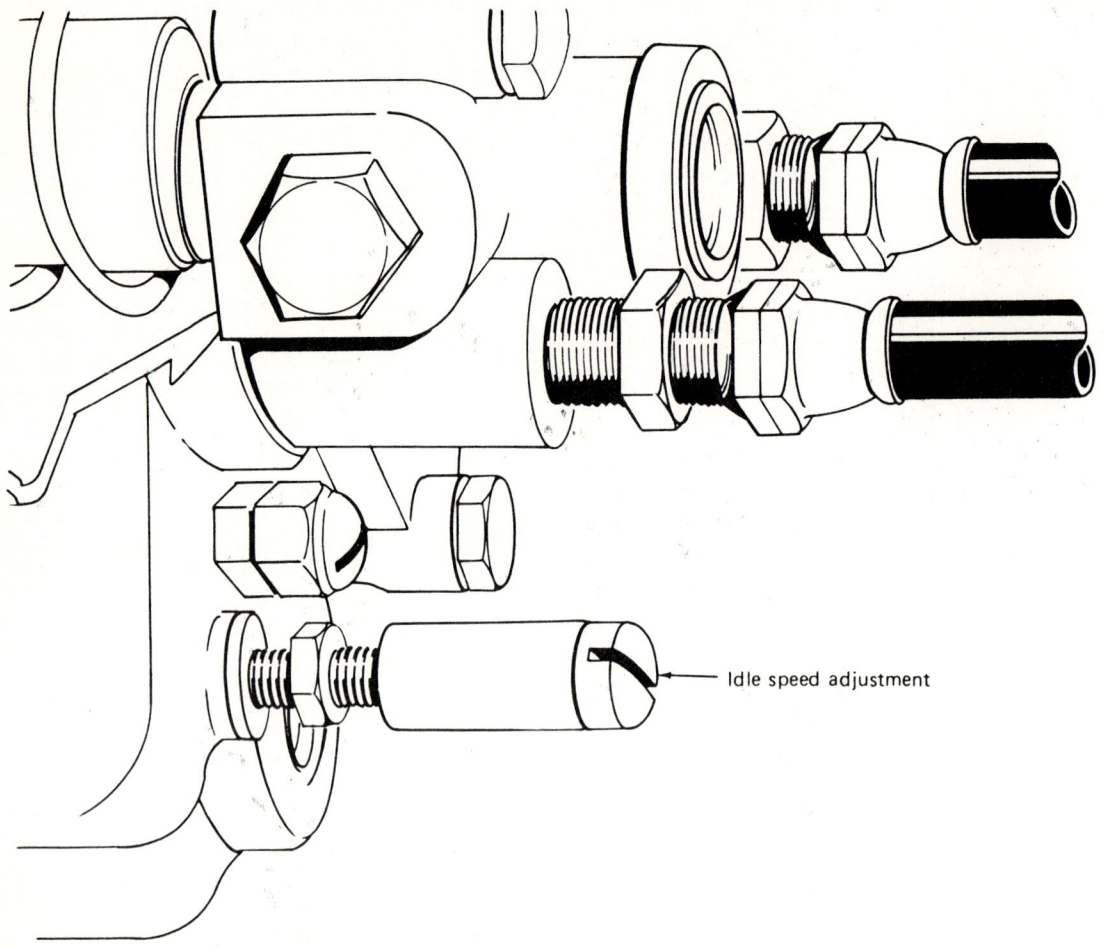

FIGURE 3-37. Idle Speed Adjustment. *(Courtesy of American Bosch Division, United Technologies)*

mechanic should loosen a fuel line and see whether fuel is flowing. If it is not, the pump may not be turning. If it is, one should look for a loss of fuel supply.

Some engines have an electrically operated shutdown valve. This circuit should be checked.

Excessive Fuel Consumption

Abnormal fuel consumption is seldom the fault of the fuel pump. Leaks, long idle periods, or other operating faults are much more frequent. Excessive exhaust smoke can mean a restricted air supply, wrong pump adjustment, sticking spray valves, or any combination of operating wear.

Dense Exhaust Smoke

We have already pointed out the main causes of heavy exhaust smoke. There are several cases in which an engine fault causes smoke. In general the fuel system will not cause smoke unless it has been tampered with to produce overfueling.

Intermittent puffs of black smoke can mean a sticking spray valve. Smoke throughout the speed range indicates a plugged air cleaner. Sudden increase in smoke density may indicate a turbocharger failure.

Smoke during acceleration on a turbocharged engine indicates that the smoke control in the fuel pump is inoperative.

Exhaust smoke is the residue of incompletely burned fuel. Lack of air or overfueling is indicated.

Power Loss at Random Times

This complaint can be caused by several simple faults:

1. floating debris in the fuel tank
2. faulty suction hose connections
3. loose suction connections

Many strange items are found in fuel tanks. Anything that can move can be drawn into the suction connection and restrict or stop the fuel flow. A strong light must be used in checking the tank.

When assembling fittings on the ends of hose, it is possible to damage the hose lining so that it can pull loose in service, causing a restriction. Also, the connection must match the mating part. Some fittings have a 37-degree flare, while Society of Automotive Engine fittings have a 45-degree flare. A leak can result from mismatching such fittings.

Suction connections must be secure. Burrs, poor machining, or failure to use thread sealer may allow a leak. Care must be exercised when tightening fittings in diecast filter heads. It is better to use a little sealing compound on the threads than to overtighten and crack the casting.

These trouble-shooting principles apply to any diesel engine fuel system.

46 / Diesel Engine Service

The Bosch system is no different than any other in its requirement for a solid clean fuel supply.

STANADYNE-ROOSA MASTER FUEL SYSTEM

This fuel pump differs in several ways from the Bosch system. It is a distributor-type pump. Control is afforded by a governor-controlled metering valve that is turned to regulate the fuel inlet to the pumping cylinder. Two cam-driven plungers move in this cylinder, each plunger driven by a roller in contact with the cam lobes inside the cam ring.

Because two opposite lobes bear on the two plungers, both plungers are driven together to pressurize the fuel. There are six lobes on a six-cylinder engine pump, giving three pressurizing plunger movements. The rotor is driven at engine speed from the engine gear train.

The cam ring is a free fit in the housing. An ingenious timing piston engages a ball-end screw in the bottom of the cam ring.

At engine start, the cam ring is positioned for retarded timing of injection. As speed increases, transfer pump pressure increases and bears on the timing piston to move the cam ring toward advance. [Fig. 3-38]

The transfer pump is a vane-type unit, mounted on the main fuel pump

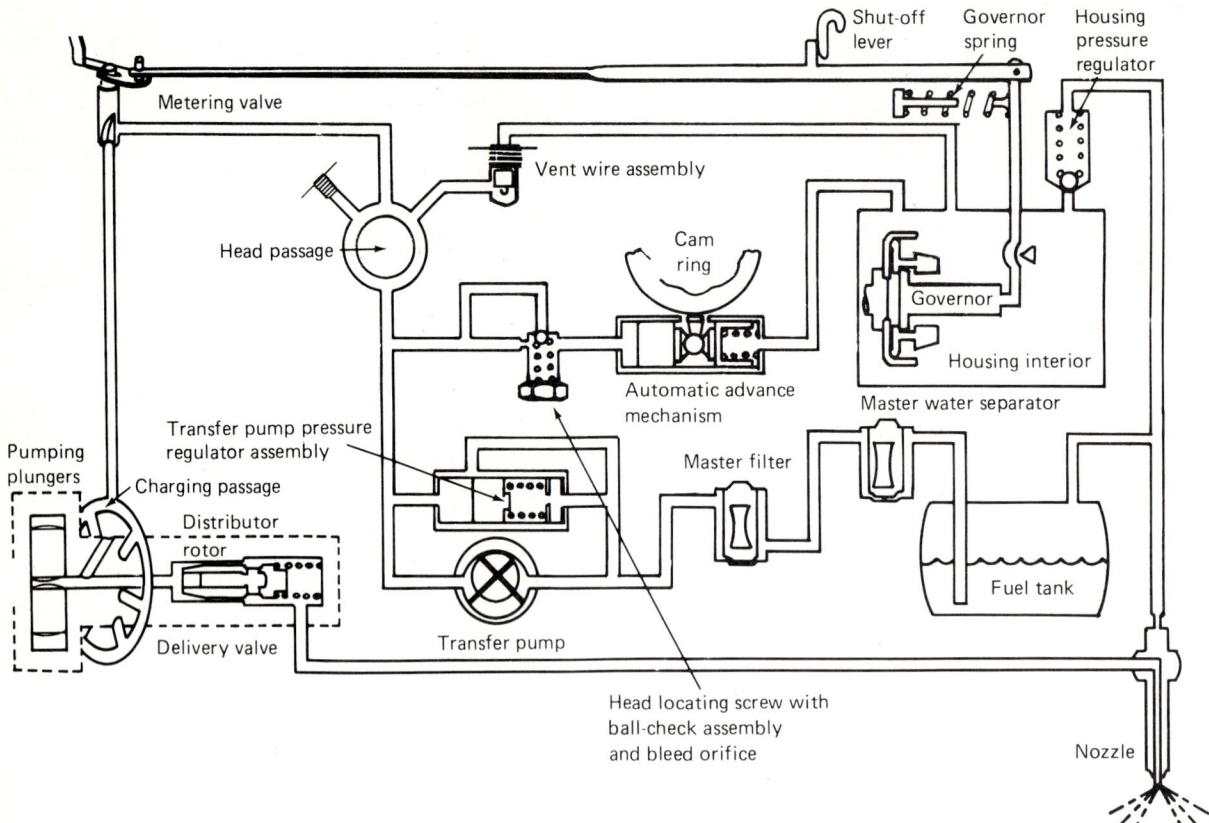

FIGURE 3-38. Roosa Master Fuel System. *(Courtesy of Stanadyne, Inc.)*

shaft. Fuel enters through an inlet screen and discharges to a passage that surrounds the rotor. [**Fig. 3-38**]

The fuel quantity entering the pumping chamber between the plungers is controlled by a metering valve, whose position is controlled by the governor. [**Fig. 3-39**] A leaf spring attached to the rotor limits the travel outward of the pumping pistons, and thus the maximum amount of fuel discharged to the injection nozzles.

The governor controls the metering valve position at all speeds. The manual throttle controls the tension of the governor spring, which resists weight force. Thus the metering valve position and fuel delivery are controlled by throttle setting, acting through the governor spring tension.

Because the two pumping plungers are moved apart by the amount of fuel passing the metering valve, the system varies the amount of injection by controlling the fuel supply to the pumping chamber, rather than controlling the fuel during high pressure injection.

Internal timing of the fuel pump is performed on a test stand. No field repairs or adjustments that require pump disassembly should be attempted.

Both idle and maximum speeds are set on the throttle stop screws. These are adjusted on the test stand and retain their settings for long periods. There is seldom a legitimate need to attempt field adjustment.

Timing of these fuel pumps to the engine is fully covered in engine builder's manuals.

Priming the Pump on the Engine

Stanadyne-Roosa Master fuel pumps are equipped with a vent plug, located under the governor cover. [**Fig. 3-38**] On some models, a J wire is contained in the plug to vibrate and keep the vent open. With this feature, the pumps are essentially self-priming.

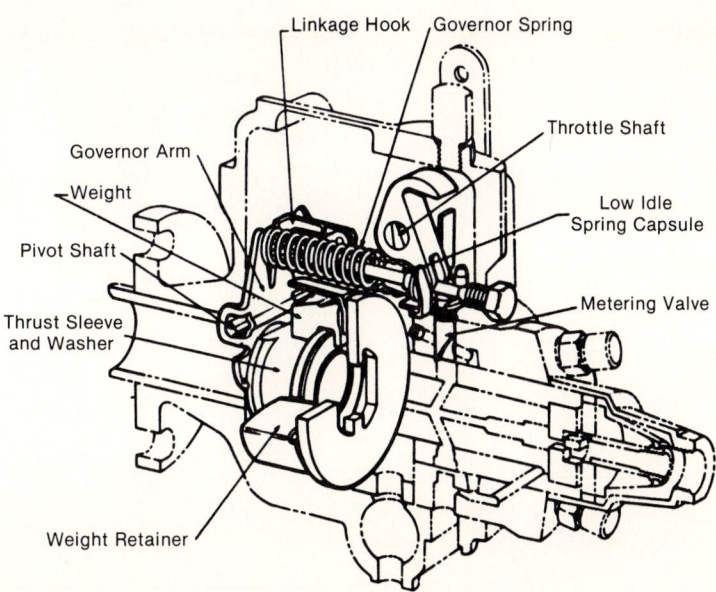

FIGURE 3-39. Roosa Master Governor. *(Courtesy of Stanadyne, Inc.)*

48 / Diesel Engine Service

When an exchange pump is installed, a small amount of clean fuel squirted into the fuel inlet will help the transfer pump during the first start.

All connections to the injection nozzles should be loosened, so that air can escape. As soon as solid fuel appears, they can be secured.

A properly installed pump, correctly timed and primed, will assure a prompt first start, provided the starting system is working properly.

Checks on Engine

Several checks can be made to determine the cause of complaints. [Fig. 3-40] Transfer pump pressure will never exceed the specification unless some abnormal restriction occurs.

Leaks from various seals are caused by either a worn seal or a sticking pressure regulator. Either of these faults will require internal pump repairs, which should never be attempted in the field. An exchange fuel pump should be installed.

Checks of transfer pump pressure and inlet vacuum can be made. The procedure follows:

1. Remove the lock plate screw from the top center of the head, between two discharge outlets. Leave the locking plate and seal in place.
2. Install the adapter, part 21900, in place of the locking screw.
3. Install a shut-off valve in the adapter, then connect tubing to a 0-to 160-psi gauge.

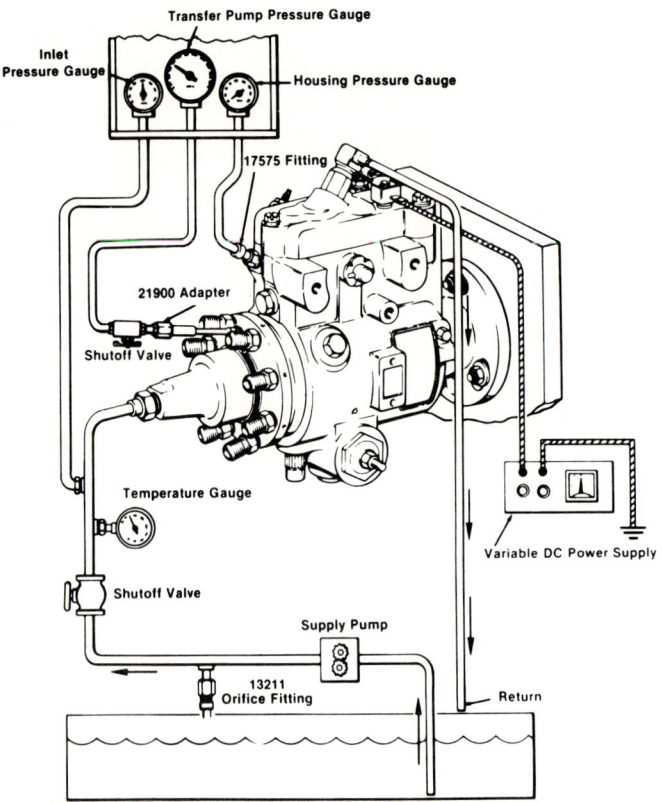

FIGURE 3-40. Gauges on Pump. *(Courtesy of Stanadyne, Inc.)*

4. Install a tee and a shut-off valve at the fuel pump inlet, then connect a good vacuum gauge.

Note: Gauges equipped with telltale pointers will be useful.

With the gauges connected as described, the engine should run to warm up (about 15 minutes). The pressure gauge valve should be shut during warm-up, and the vacuum gauge valve left open.

After operating temperature is reached, the pressure gauge valve and the throttle should be opened. The maximum pressure obtained should be noted. Pressure should be 100 to 120 psi. It must never exceed 130 psi. High pressure indicates a sticking pressure regulator. The pump should be repaired in a shop equipped for fuel pump rebuilding.

A new fuel filter should have been installed before making vacuum checks. Any vacuum reading higher than 10 inches of mercury indicates a plugged filter or a restricted supply line.

New filter vacuum should be about 2.5 inches of mercury. Low vacuum readings suggest suction leaks.

The Stanadyne-Roosa Master fuel system makers suggest the use of an electric fuel pump mounted close the fuel tank. This will reduce the possibility of suction leak and provide a positive fuel supply. Such a supply pump should be set to provide a slight pressure, not more than 5 psi, at the fuel pump inlet.

Injection Nozzles

The Stanadyne-Roosa Master system offers several types of injection nozzles. **[Fig. 3-41]** They are similar to other makes but offer a variety of spray tip designs and valve configuration.

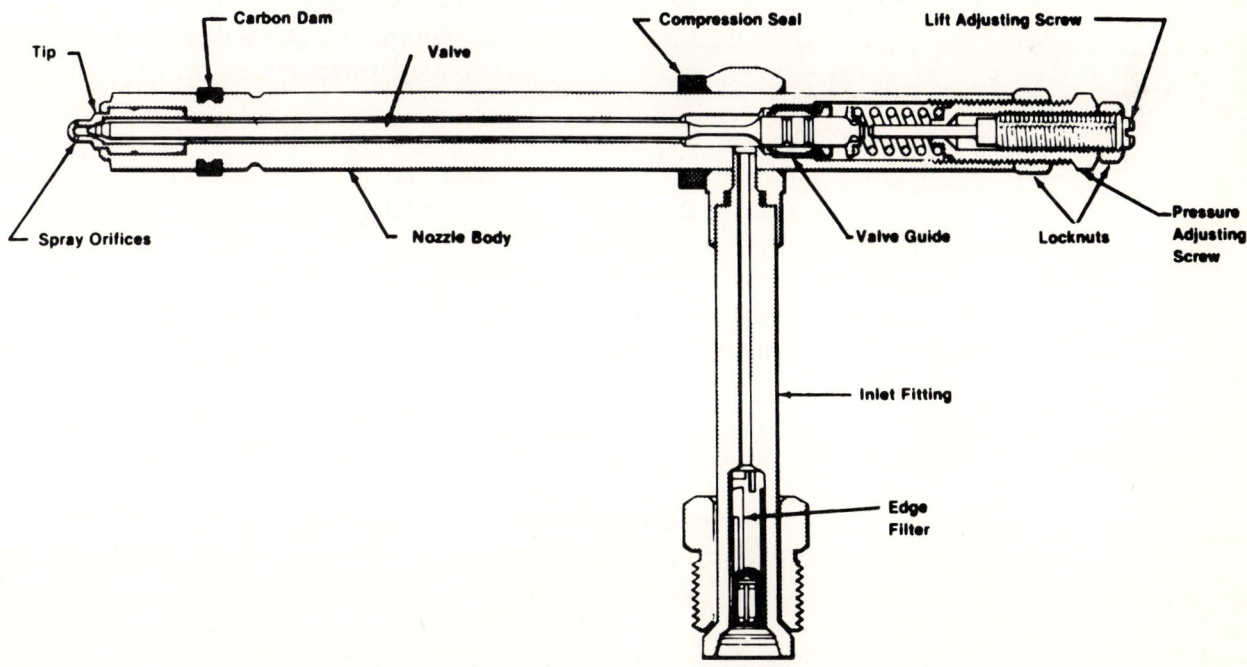

FIGURE 3-41. **Roosa Master Spray Valve.** *(Courtesy of Stanadyne, Inc.)*

50 / Diesel Engine Service

They all are designed to open against spring pressure when injection pressure is imposed, and to close sharply when pressure falls.

No timing or adjustment is provided for the field service mechanic. They can be exchanged for rebuilt injection nozzles.

Later designs are of small diameter; special brushes are needed to clean the sockets in the cylinder head.

The correct engine make and model must be used in ordering exchange units. This rule applies to all subassemblies as well as all injection equipment.

Although external carbon can be cleaned from the spray tip of any spray valve, no field test can be made. Thus, the use of exchange units is recommended.

New seat seals and carbon dams should always be used when spray valves are replaced.

The prescribed torque on the hold-down is 20 pound-feet (27–32 Newton-Meters).

CATERPILLAR FUEL SYSTEM

Description

This fuel pump is close to the original Bosch design. The pumping elements are cam-driven and deliver fuel at injection pressure to each cylinder in turn. The pump crankshaft is driven at half crankshaft speed through gears that are marked for timing.

Older fuel pump housings have no openings, but the individual pumping elements can be removed and exchange units installed. [**Fig. 3-42**]

Fuel from the tank is routed through the filters, then from the pump-mounted supply pump into the fuel passage surrounding the pumping elements. [**Fig. 3-43**]

Like the Bosch system, these elements contain a ported barrel and a hollow ported plunger. The plunger stroke is constant, but a moveable sleeve on the lower end of the plunger is moved by the control shaft lever to control the point in the plunger stroke at which a spill port is opened, releasing fuel pressure and ending the effective stroke.

The Caterpillar governor derives its force from rotating flyweights, mounted on the pump drive shaft.

Fuel Systems Service

The injection pumps can be replaced individually for exchange units. They are not identified by cylinder number. One should always clean the fuel pump area using a steam jet or solvent gun.

1. Clean area. Place the throttle in shut-off position.
2. Remove fuel lines and protect from dirt.

Diesel Engine Fuel Systems / 51

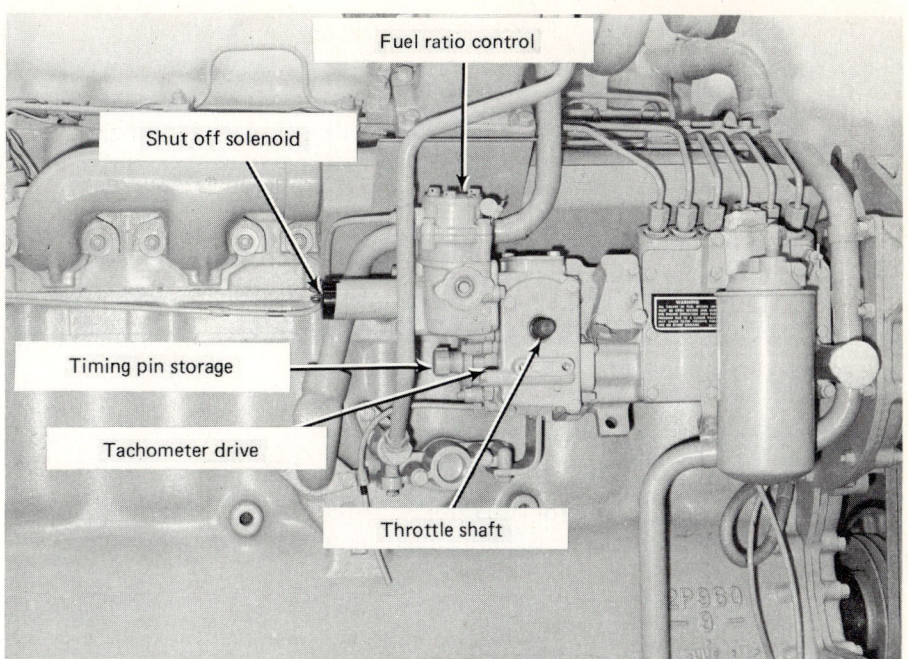

FIGURE 3-42. Caterpillar Fuel System. *(Courtesy of Caterpillar Tractor Company)*

3. Unscrew the body nut and carefully withdraw the pump assembly.
4. Be sure that no dirt enters the socket; use new gaskets to install the exchange unit.
5. Install the new pump, making sure that the sleeve engages the control lever.
6. Later pumps have a body cover plate whose removal allows the pumps to be installed more easily.

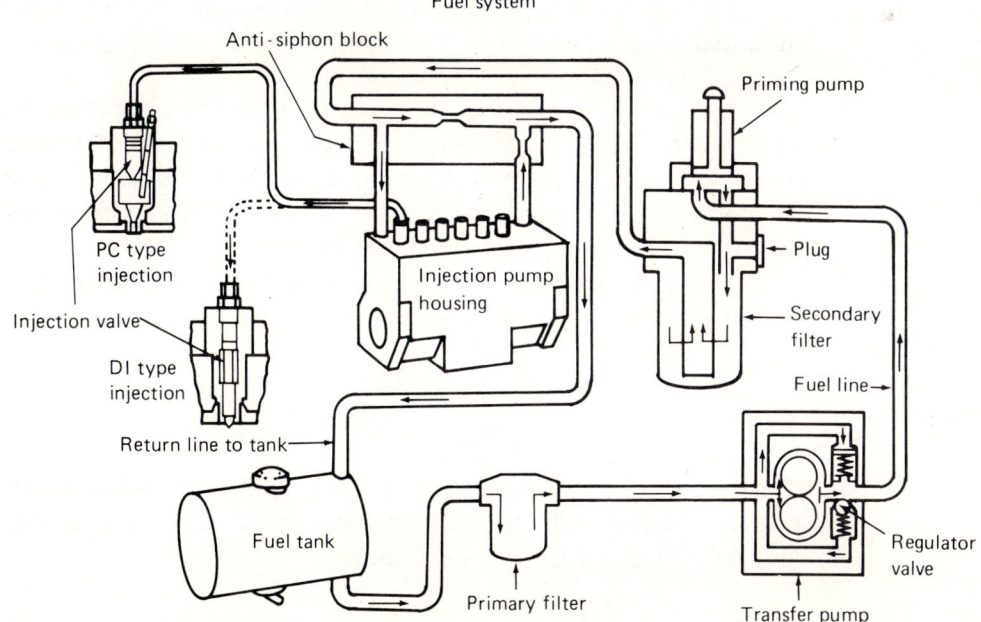

FIGURE 3-43. Caterpillar Fuel System. *(Courtesy of Caterpillar Tractor Company)*

7. Secure the body nut to 70 pound-foot (95 N-m) with a torque wrench and special socket. *Caution:* Be certain that the control levers are engaged in the sleeve grooves. Do not loosen the lever clamp screws but be careful when installing the pump. The engine can run out of control if the levers are not positioned in the groove. It is unlikely that injection pumps will need replacement, unless water, dirt, or bad fuel, such as gasoline, has caused damage.

Several other service operations can be performed on Caterpillar fuel pumps. All components should be checked before replacing the fuel pump.

Manual Priming Pump

Should the priming pump become worn and unable to pick up fuel, it can be removed and repaired. A new plunger seal can be installed, or an exchange pump put on. [Fig. 3-44]

Bleed Valve

This valve is located above the hand priming pump, in a tube fitting. Its purpose is to permit a small flow from the pump to the fuel tank in order to remove entrapped air and heat from the pump housing. Because the housing pressure is 25 to 32 psi, as regulated by the bypass valve in the supply pump, some fuel always flows through the bleed valve.

Dirt or wear can cause this valve to fail. Replacement requires a special fitting. A standard tube fitting should never be used in this location.

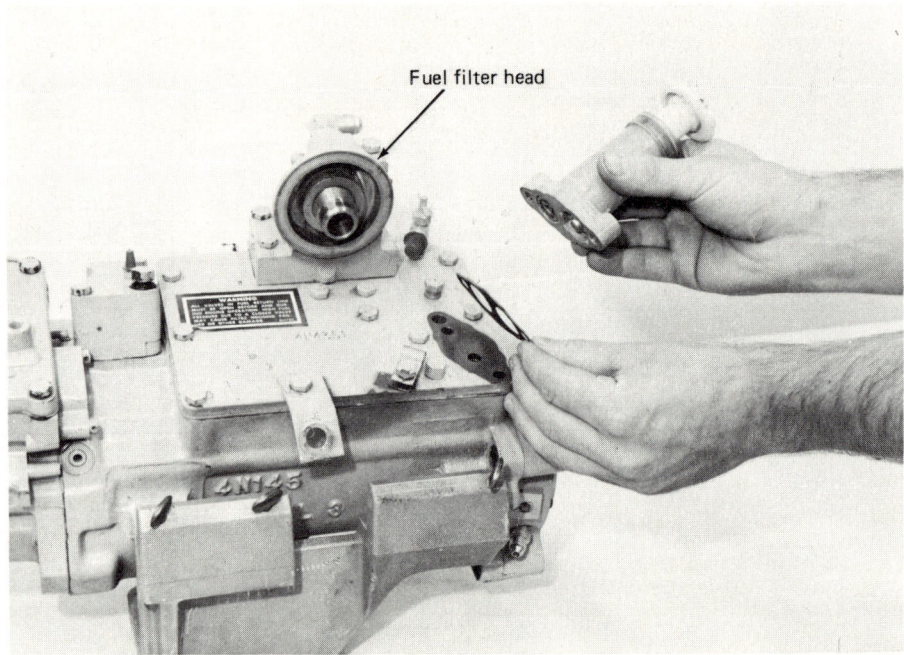

FIGURE 3-44. Manual Priming Pump. *(Courtesy of Caterpillar Tractor Company)*

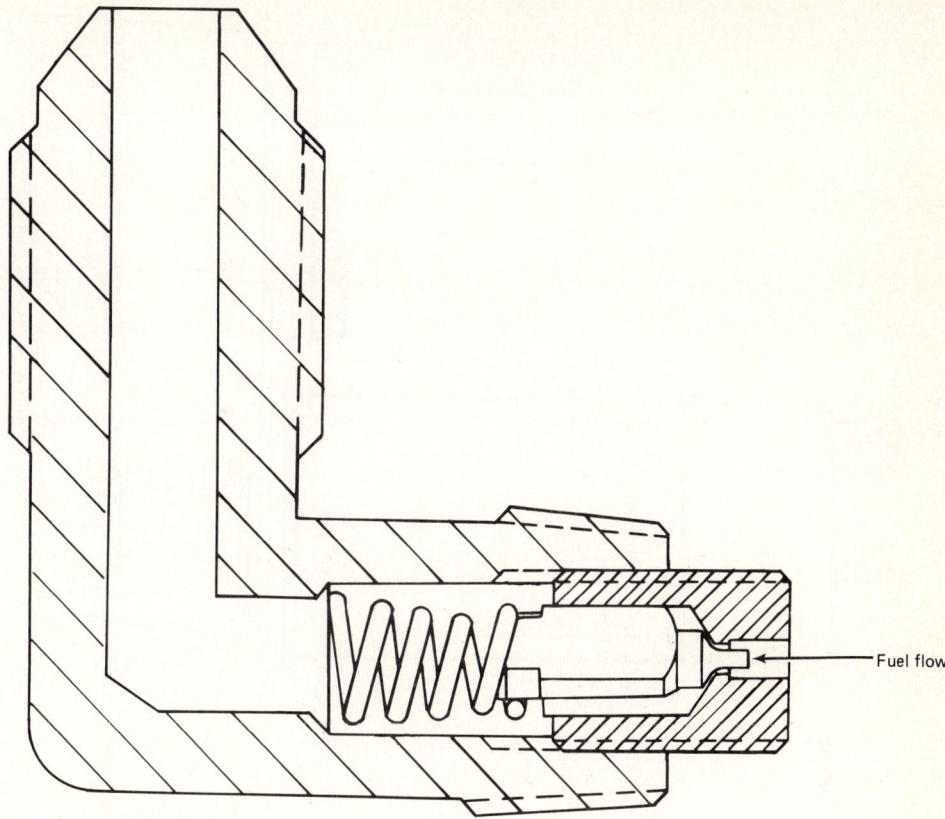

FIGURE 3-45. Bleed Valve. *(Courtesy of Caterpillar Tractor Company)*

This tube and valve can be useful when using the priming pump. When solid fuel comes out the bleed valve, the fuel pump housing is full. **[Fig. 3-45]** Further priming is done by loosening the fuel lines at the spray valves and by barring the engine over. When solid fuel comes from the lines, the engine should start.

Solenoid Shut-Off

A solenoid shut-off is mounted on the end of the governor housing. There are two openings for it, but one is covered by a plate.

The solenoid may act to run the engine, in which use it is mounted in the lower hole, or it may be used to stop the engine; it is then mounted in the upper hole. **[Fig. 3-46]**

It can be replaced if faulty. Current to its terminals should be applied when the switch is operated. The ground side as well as the hot side should be checked. The circuit must be complete for the solenoid to work.

Note: Although governor adjustments can be made on the engine or on a test stand, the special tools required make direct training necessary. It is not the purpose of this book to encourage makeshift methods for such important adjustments. Therefore, the use of exchange fuel pumps when adjustment of the governor is indicated, is recommended.

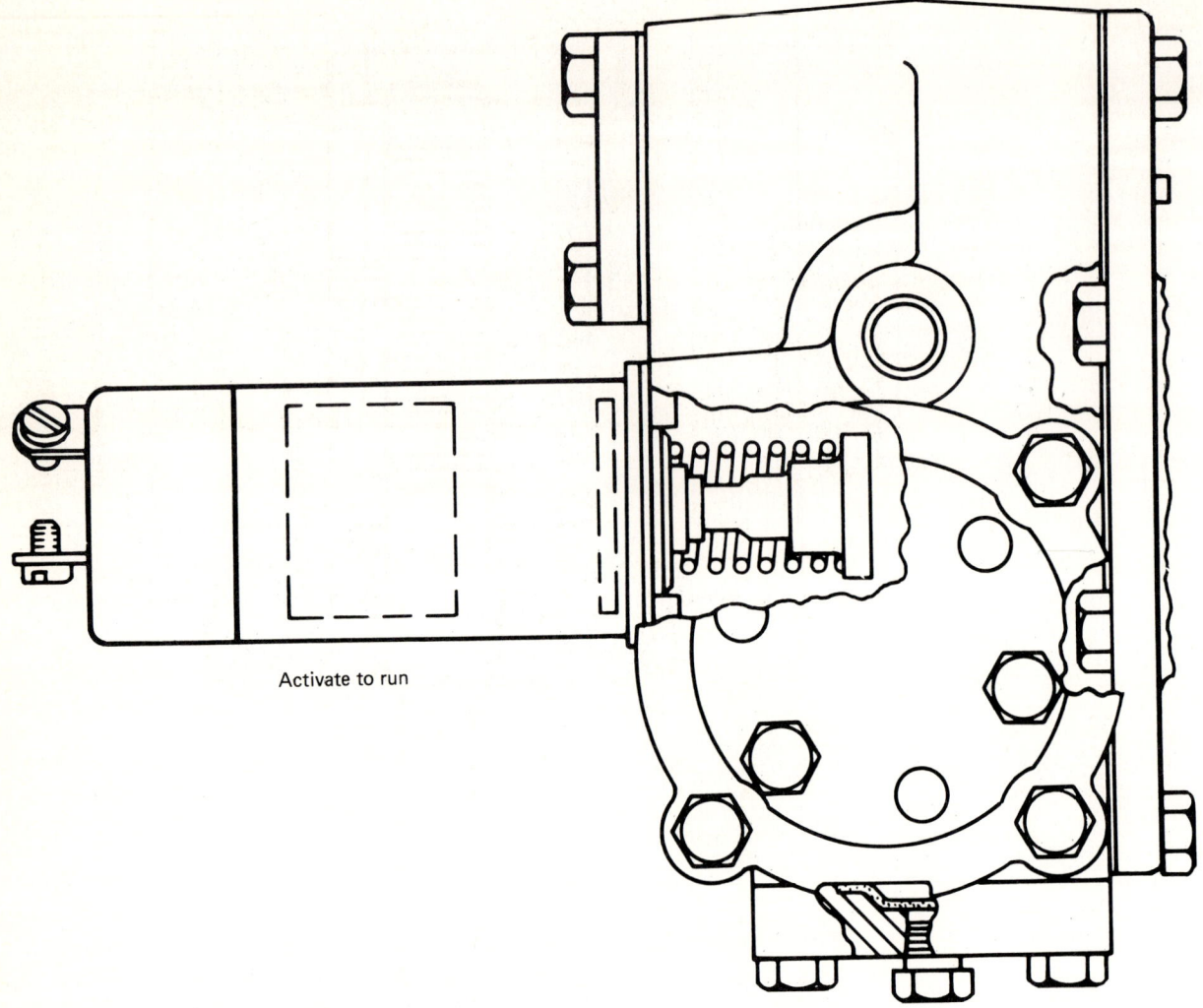

FIGURE 3-46. Solenoid Shut-off. *(Courtesy of Caterpillar Tractor Company)*

Fuel Ratio Control

This unit is mounted on the governor housing behind the governor control lever. It is used on turbocharged engines to reduce fuel delivery and exhaust smoke during acceleration only. Its operation is controlled by air pressure in the intake manifold; it responds to turbocharger speed increases during acceleration. [**Fig. 3-47**]

This unit can be rebuilt when exchanged but must be adjusted on the fuel pump using special tools and techniques. No field adjustment should be attempted without thorough training and the necessary tools.

Note: When an exchange fuel ratio control is being installed, the governor control lever must be in shut-off position. The operating rod of the control is pin-connected to the governor control lever. [**Fig. 3-48**]

Load Stop

A stop is provided to limit governor travel toward more fuel. This is adjusted with the same indicator as the rest of the governor adjustments. Again, this

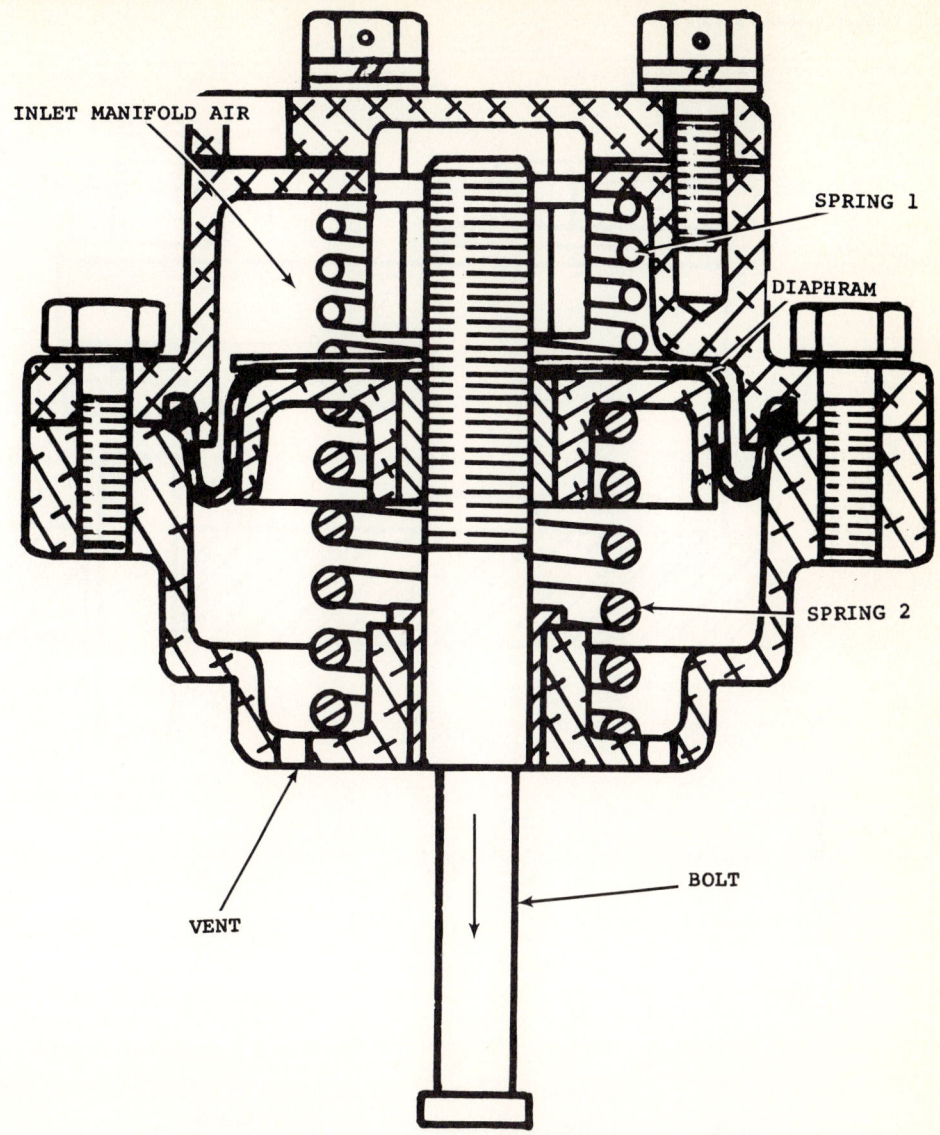

FIGURE 3-47. Fuel Ratio Control. *(Courtesy of Caterpillar Tractor Company)*

FIGURE 3-48. Install Fuel Ratio Control. *(Courtesy of Caterpillar Tractor Company)*

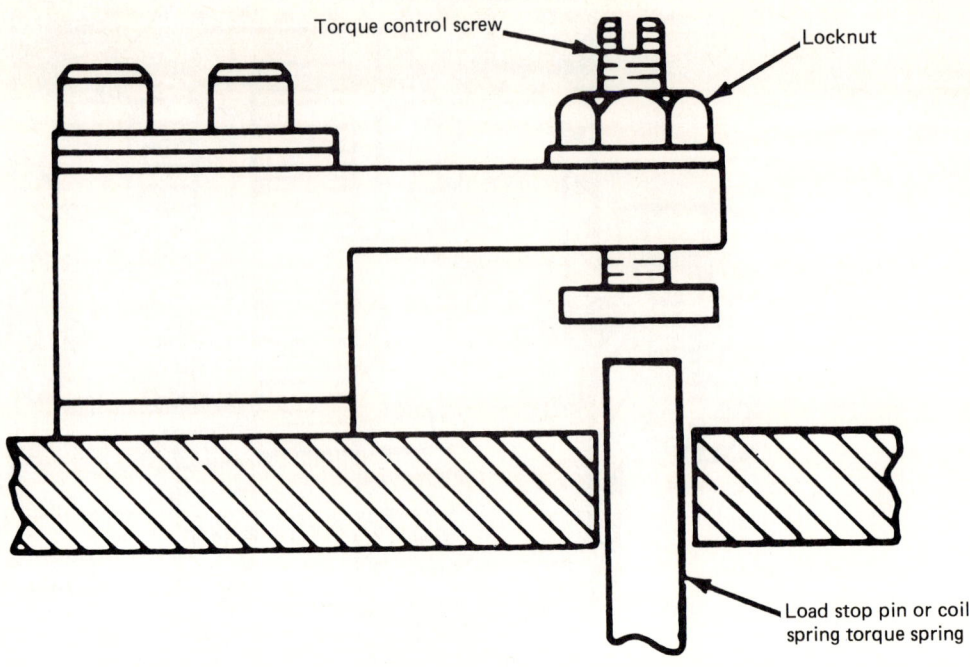

FIGURE 3-49. Load Stop. *(Courtesy of Caterpillar Tractor Company)*

stop should not be adjusted by mechanics who have not been trained on the Caterpillar fuel system. **[Fig. 3-49]**

Speed Adjusting Screw

A small cover on the pump end of the governor housing covers adjusting screws for idle and maximum speed setting.

1. After washing the pump thoroughly, remove the two screws and the cover.
2. The lower screw adjusts low idle speed. Turning it in, to the right, raises idle speed.
3. The engine must be warmed up and idling when making this adjustment.
4. Try the setting by moving the throttle manually to increase speed, then moving it smartly to idle position.
5. Idle speed should be stable, at the value stamped on the pump data plate. **[Fig. 3-50]**

Note: Use an accurate tachometer, and be sure that the adjustment meets the specification. A recess in the cover engages the screw head and keeps it from turning.

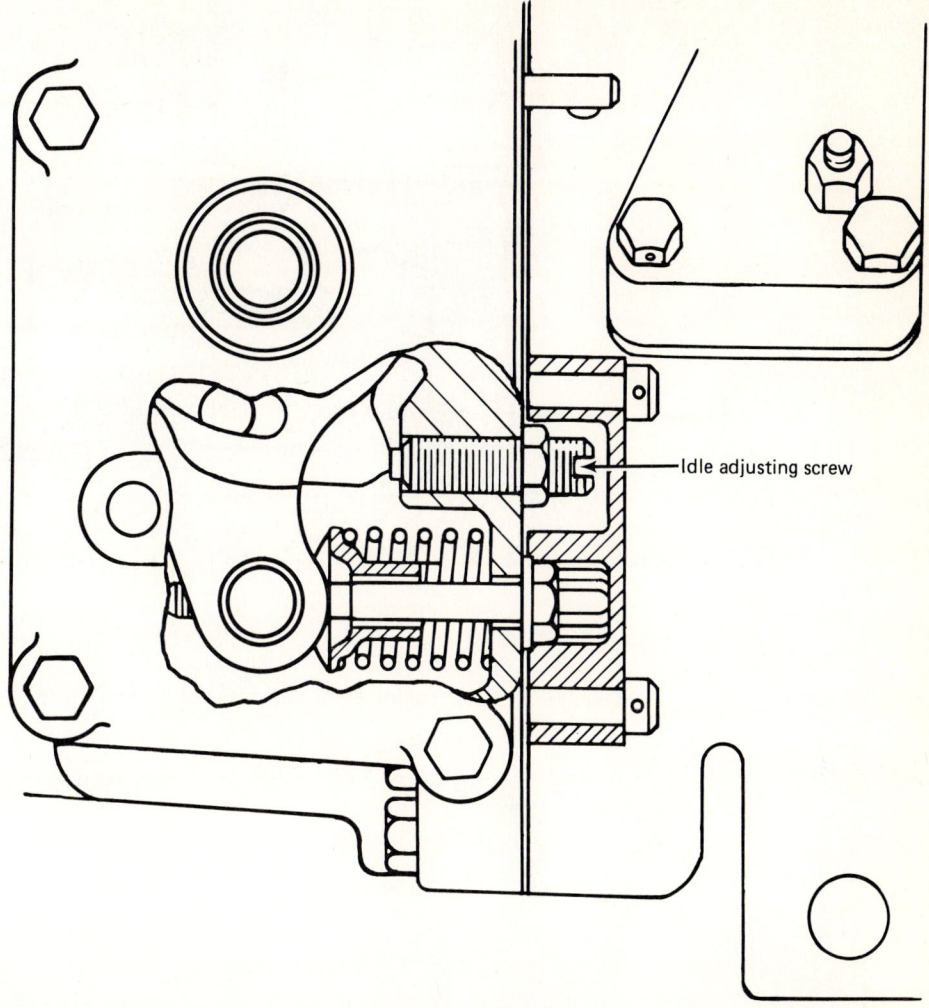

FIGURE 3-50. Speed Adjusting Screws. *(Courtesy of Caterpillar Tractor Company)*

The top screw adjusts maximum governed speed. Turning the screw in causes lower speed.

This screw should never be adjusted unless engine speed is outside the range stamped on the data plate. Increased speed is not a substitute for power. **[Fig. 3-51]**

Note: These adjustments do not require much movement and must be made with care.

Dynamometer Testing

If the engine can be loaded on a dynamometer, a quick check can be made of its performance.

1. Connect a good tachometer to the adapter on the pump.
2. Connect a continuity light to the brass terminal screw on the load stop cover. Connect the other lead to a current source or use a self-powered light.

FIGURE 3-51(a). **Engine Data Plate.** *(Courtesy of Caterpillar Tractor Company)*

FIGURE 3-51(b). **Engine Data Plate.** *(Courtesy of Cummins Engine Company)*

Diesel Engine Fuel Systems / 59

3. Start the engine and warm it up by applying load. After warm-up, remove all load and run at high-idle, speed control at full throttle. Record high-idle speed.

4. Increase the load gradually until the continuity light just comes on. Repeat this operation to make sure. Record the speed, which is full power and should be rated engine speed.

5. Compare the speeds found to those stamped on the data plate or with the engine manual.

6. If full-load speed is below standard, a small adjustment of high-idle speed may correct it, provided that the limit is not exceeded.

7. Be sure that the dynamometer is correctly calibrated and check all other causes of low power. If the fuel pump is at fault, remove it and install an exchange assembly of the same specifications. [Fig. 3-52]

Fuel Pump Installation

Fuel pump installation requires the use of special tools for checking timing. A rebuilt pump is timed internally during assembly. [Fig. 3-53]

1. Remove the starting motor and No. 1 spray valve.
2. Attach the turning device in the starter hole. [Fig. 3-54]

    Note: The engine turns counterclockwise viewed from the flywheel looking forward.

3. Turn the engine to find top center on compression on No. 1 piston. Both valves will be closed, rocker levers free.

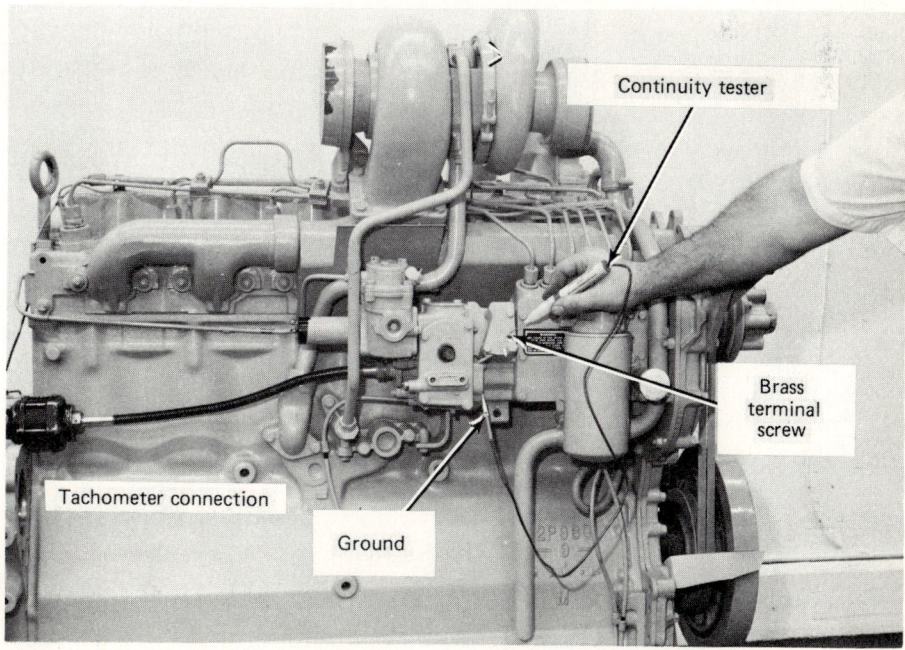

FIGURE 3-52. Check Balance Point. *(Courtesy of Caterpillar Tractor Company)*

60 / Diesel Engine Service

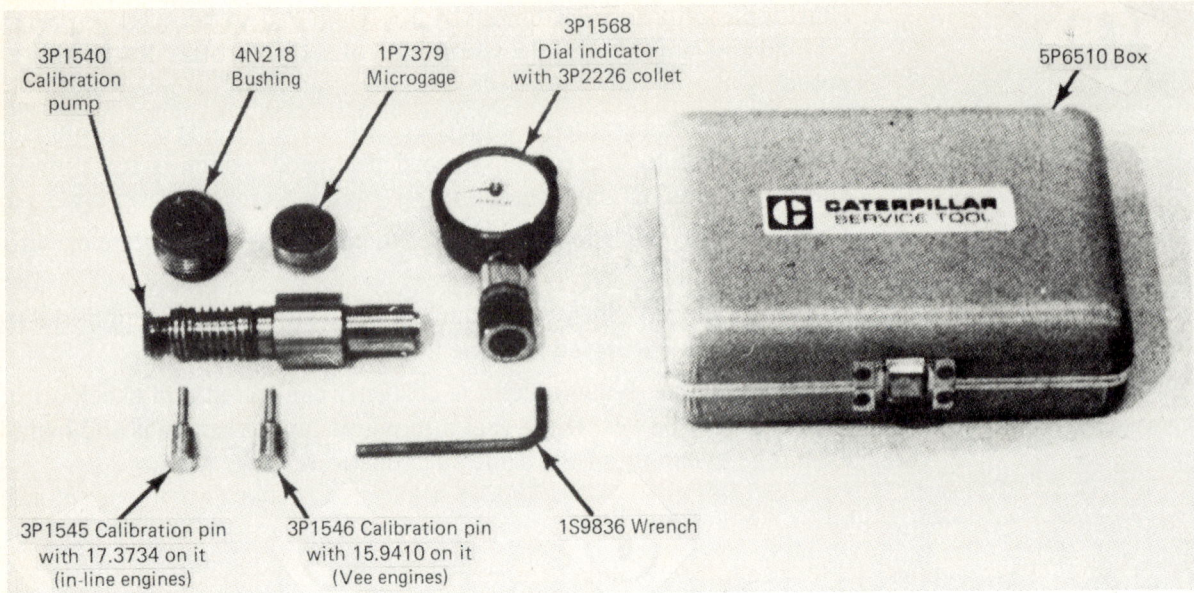

FIGURE 3-53. Timing Check Tools. *(Courtesy of Caterpillar Tractor Company)*

4. Install the timing pin in the lower front hole of the cover in Figure 3-55 after removing the screw.
5. Some engines have a timing pointer on the flywheel. The pointer must register with the flywheel marks at 1½°. **[Fig. 3-56]**
6. Some engines have a bolt hole below the starter. When the crankshaft is in the right position, a ⅜-inch-20 by 1½-inch capscrew will enter the flywheel through this hole. **[Fig. 3-57]**

FIGURE 3-54. Turning Device Installed. *(Courtesy of Caterpillar Tractor Company)*

Diesel Engine Fuel Systems / 61

FIGURE 3-55. Timing Pin in Pump Cover. *(Courtesy of Caterpillar Tractor Company)*

7. Always approach top center in the direction of rotation. Turn the engine back at least 30 degrees, then forward to the mark.

   Note: Check the valve rockers on No. 1. They must both be free, valves closed. If they are tight, the engine must be turned one revolution and top center No. 1 reset.

Do not turn the engine with the timing bolt in place.

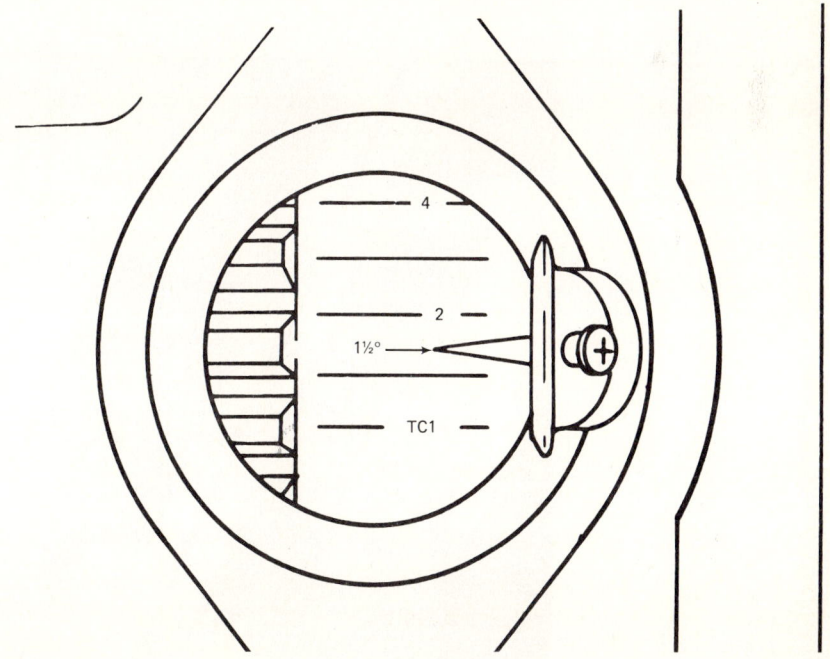

FIGURE 3-56. Timing Marks on Flywheel. *(Courtesy of Caterpillar Tractor Company)*

FIGURE 3-57. Timing Bolt in Place. *(Courtesy of Caterpillar Tractor Company)*

8. With the timing pin in the pump camshaft slot and with the engine spotted as described, remove the cover over the fuel pump drive gear. Mark the gears before removal.

9. Loosen the capscrew holding the fuel pump drive gear to the tapered camshaft. Use the puller plate and two long capscrews to pull the gear.

    Note: This puller plate can be improvised from scrap steel at least ½-inch thick. [Fig. 3-58]

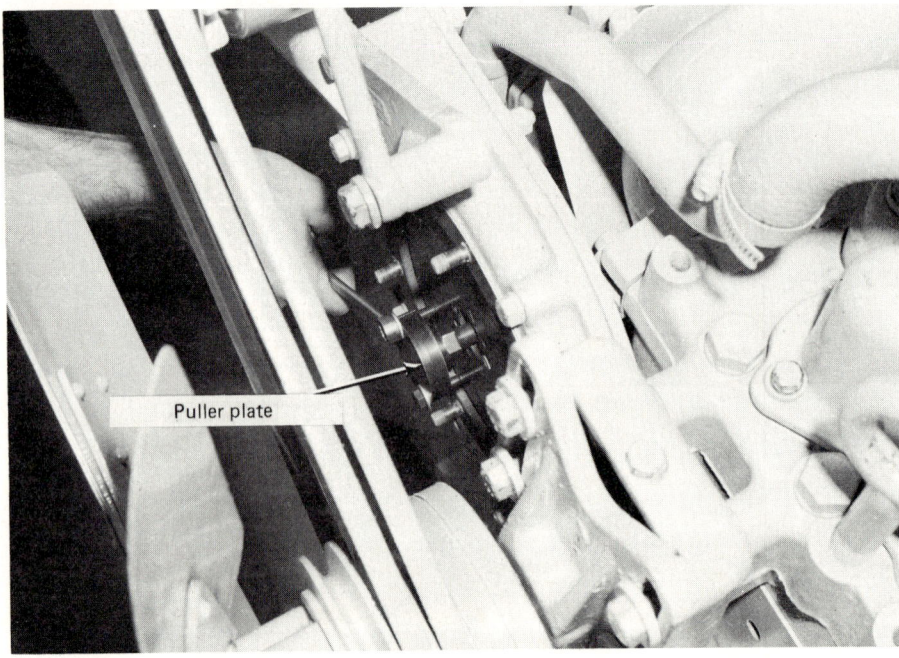

FIGURE 3-58(a). Pull Fuel Pump Drive Gear. *(Courtesy of Caterpillar Tractor Company)*

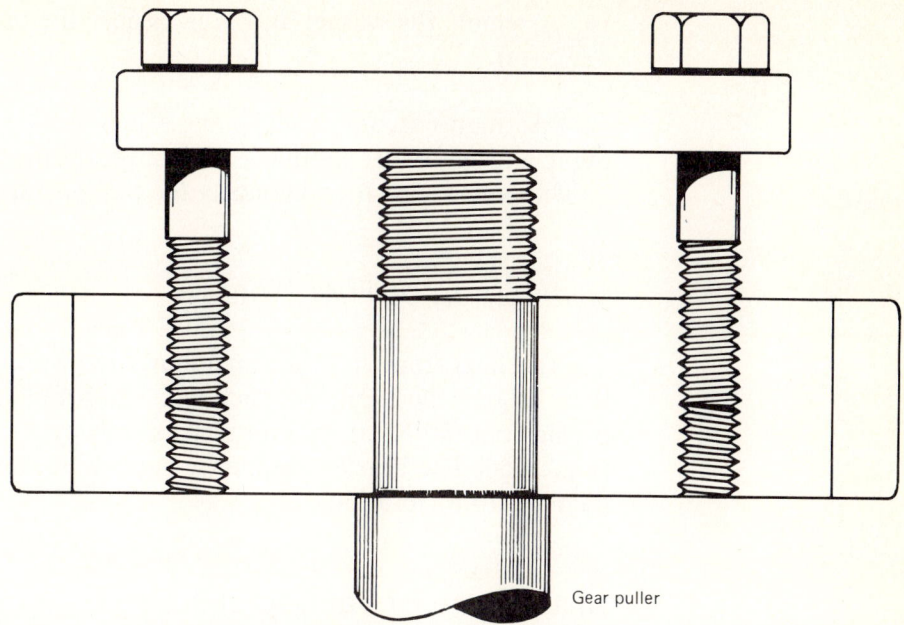

FIGURE 3-58(b). **Improvised Puller Plate.** *(From a Sketch by P. M. Uhl)*

10. The fuel pump is attached with three studs that go through the gear plate behind the drive gear.
11. With the engine spotted as described, the old fuel pump can be disconnected and removed and the new one installed. Put clean oil on the front seal and use a new gasket.
12. Secure the three nuts evenly.
13. Put the gear and automatic timing device back on the hub. Bring the marked teeth on the fuel pump drive gear and idler gear together. [**Fig. 3-59**]

FIGURE 3-59. **Align Marked Teeth.** *(Courtesy of Caterpillar Tractor Company)*

64 / Diesel Engine Service

14. Assemble the washer and bolt. Secure the bolt to 110 pound-foot (149 N-m).

At this point, one should make a final check of fuel pump timing. If both pump and engine are spotted as described, the mechanic should put the cover on with a new gasket and connect the pump to spray valve lines.

Installation of Fuel Lines

The fuel lines from the injection pumps to the spray valves must be installed in their original position. All fuel connection nuts should be tightened to 30 pound-foot (40 N-m) with a special socket and a torque-measuring handle. [**Fig. 3-60**] The bracket should be applied and secured in order to prevent damage from vibration.

Installation of Tachometer Drive

The tachometer drive assembly on the back of the fuel pump is usually furnished on an exchange pump housing. If it is not, the drive from the old pump should be used. [**Fig. 3-61**]

To remove the drive,

1. Remove the cover capscrews and the cover.
2. Remove the adapter assembly. Discard the seal ring. [**Fig. 3-62**]
3. Check the coupling for wear. It is usually good for further use.

FIGURE 3-60. Install Fuel Lines. *(Courtesy of Caterpillar Tractor Company)*

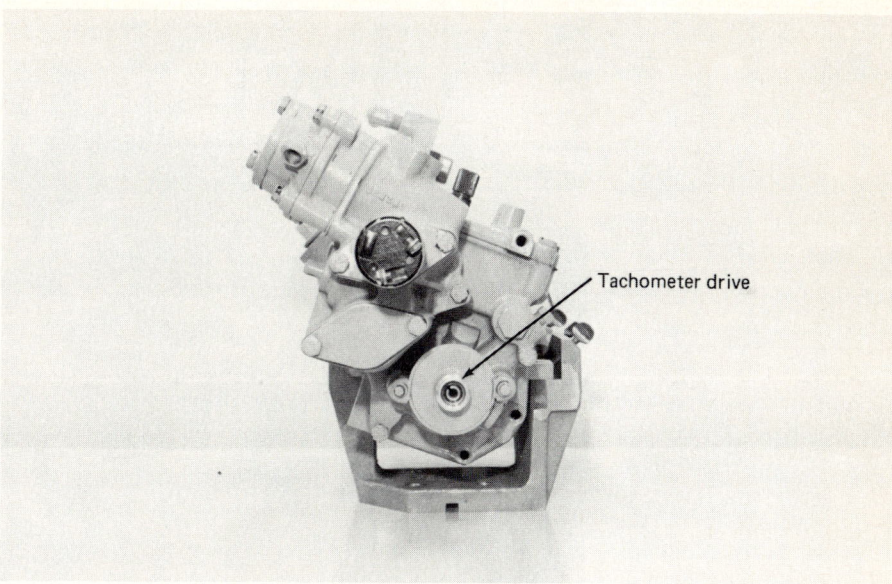

FIGURE 3-61. Tachometer Drive. *(Courtesy of Caterpillar Tractor Company)*

4. Place a new seal ring on the adapter, and wet it with clean fuel.
5. Install the coupling, cover, and clamp screws. Tighten the screw evenly to 20 pound-foot. (14.8 N-m)
6. Connect the tachometer cable.

Installation of Fuel Filter

A new fuel filter should be installed on the exchange pump. It should be filled with clean fuel and the seal wet. It should be hand tightened. [**Fig. 3-63**]

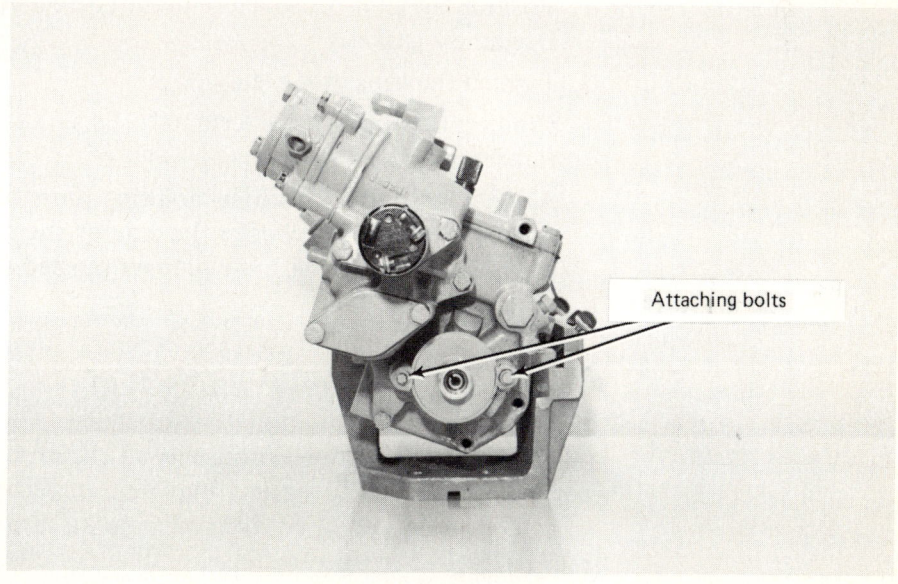

FIGURE 3-62. Remove Tachometer Drive. *(Courtesy of Caterpillar Tractor Company)*

FIGURE 3-63. Install Fuel Filter. *(Courtesy of Caterpillar Tractor Company)*

Installation of Fuel Shut-off Solenoid

There are two types of shutoff solenoid. One is designed to operate to hold the control shaft in the run position. It has a spring on the plunger. The other is used to stop the engine only and does not have a spring. **[Fig. 3-64]**

The right type must be on the exchange pump, or one can install the one from the old pump.

1. Position the shut-off solenoid in the correct bore in the back of the pump. Use a new gasket.
2. Tighten the capscrew evenly.
3. Connect the wires to the terminals. Be sure to connect the hot wire to the right terminal.

   Note: The solenoid without a spring must have a space of 1.397 inches (35.48 millimeters) between the side of the end cap and the face of the solenoid end plate, with the plunger pushed all the way in. **[Fig. 3-65]**

Installation of Fuel-Priming Pump

This hand pump may be furnished on the exchange fuel pump. If not, the pump from the old fuel pump may be transferred. **[Fig. 3-66]**

The pump should be checked to make sure that it works well. It should pick up fuel when operated in a container of clean fuel. If it does not pump a complete stroke, an exchange priming pump should be installed.

Diesel Engine Fuel Systems / 67

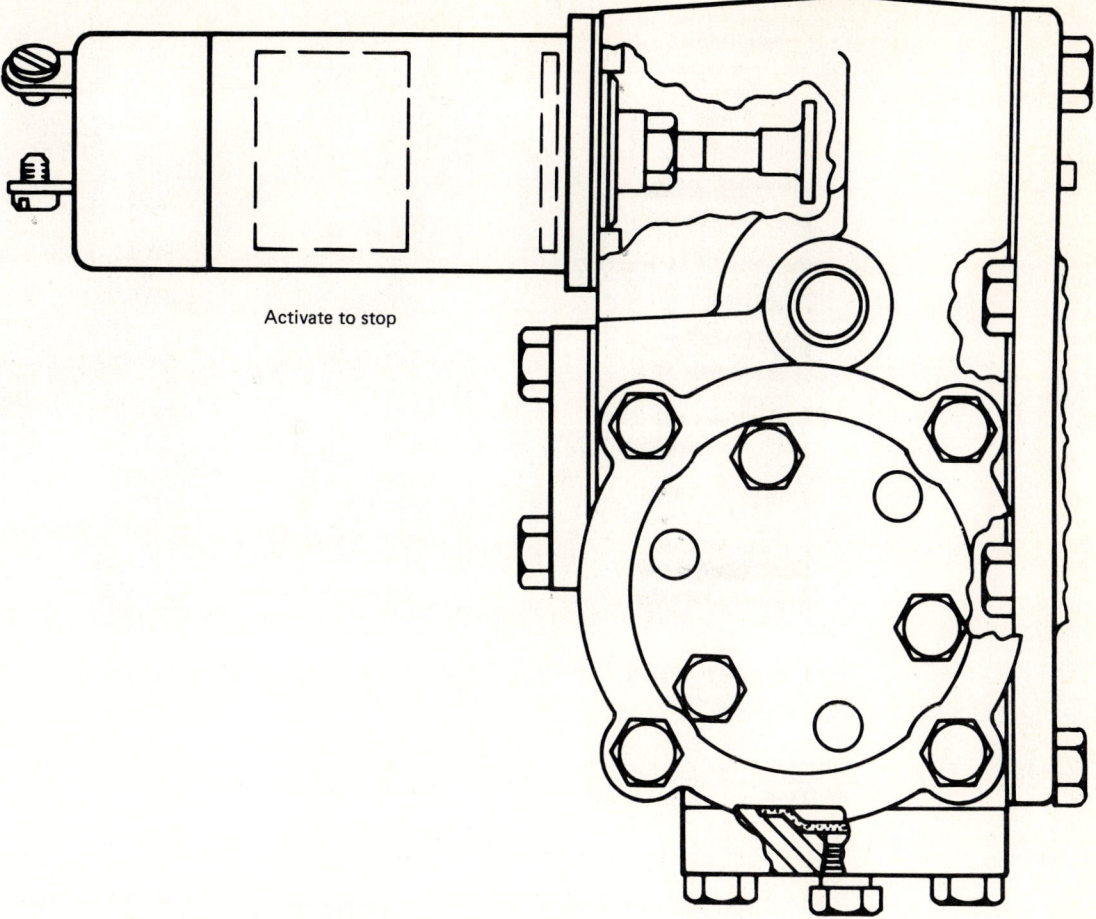

FIGURE 3-64. Shut-off Solenoids. *(Courtesy of Caterpillar Tractor Company)*

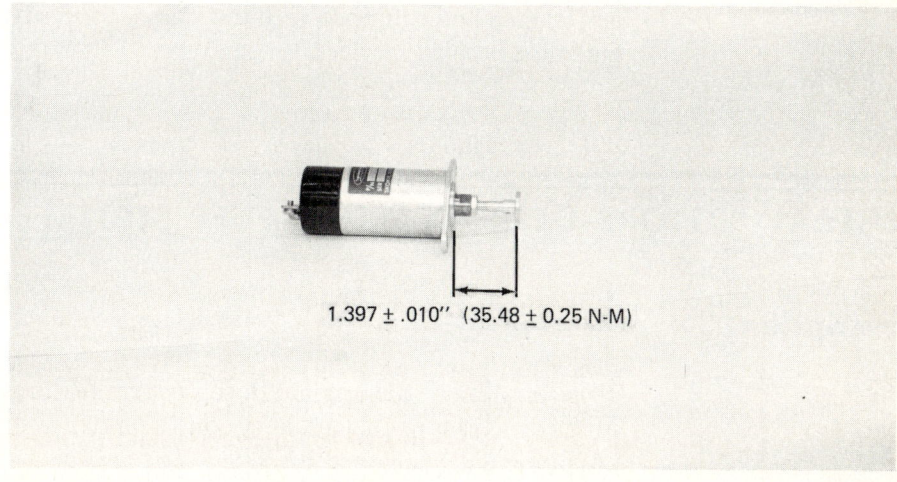

1.397 ± .010″ (35.48 ± 0.25 N-M)

FIGURE 3-65. Check Plunger Extension. *(Courtesy of Caterpillar Tractor Company)*

FIGURE 3-66. Install Priming Pump. (Courtesy of Caterpillar Tractor Company)

Water Separator

Some engines are equipped with a water separator connected in the fuel suction line, ahead of all filters and the fuel pump. Because water in the fuel will cause immediate damage, it must be kept from going through the fuel system. It does no good to use chemical treatment of the fuel.

This separator has a glass bowl so that the water is visible. The unit should be checked daily and any water drained out by opening the valve on the bottom and the vent valve at the top. [Fig. 3-67]

The Caterpillar fuel system is used on all Caterpillar models. It is designed to give long service life with ordinary and proper maintenance.

The line and field mechanic should not attempt to rebuild these units unless trained and equipped with the special tools needed. A manual for the engine model being serviced is necessary.

GENERAL MOTORS-DETROIT DIESEL FUEL SYSTEM

Description

This variation of the original Bosch system features injectors that perform all of the functions of fuel quantity control and injection except those of the supply pump and the governor.

Each injector contains the variable volume metering pump and the spray

Diesel Engine Fuel Systems / 69

FIGURE 3-67. Water Separator.

valve. Because there are no high pressure lines, injection pressure is very high, up to 25,000 psi. The injected fuel is thus composed of fine droplets; no pre-combustion chamber is necessary.

Fuel is supplied to the common fuel manifold and to each injector by a simple gear-type pump. The injectors are cam-driven from the engine camshaft through push rods whose length is adjustable.

Injector

Control of fuel quantity is furnished by turning the injector plunger by the pinion that encloses the plunger and by the toothed control rack. In turn, the rack is moved in and out of the injector by the lever that is carried on the control shaft or tube. As the rack is moved into the injector, more fuel is injected. [Fig. 3-68]

The outer limit of rack travel turns the plunger to a no-fuel position and stops all injection.

Governor

The control tube is turned in its needle-bearing mounting brackets by a link from the governor. The governor is driven by the blower shaft and is a pair of flyweights whose force is exerted on a tube called the *riser,* which is grooved on its outer end to receive a fork that is attached to the vertical control shaft. [Fig. 3-69]

The upper end of the vertical control shaft is keyed into a two-armed

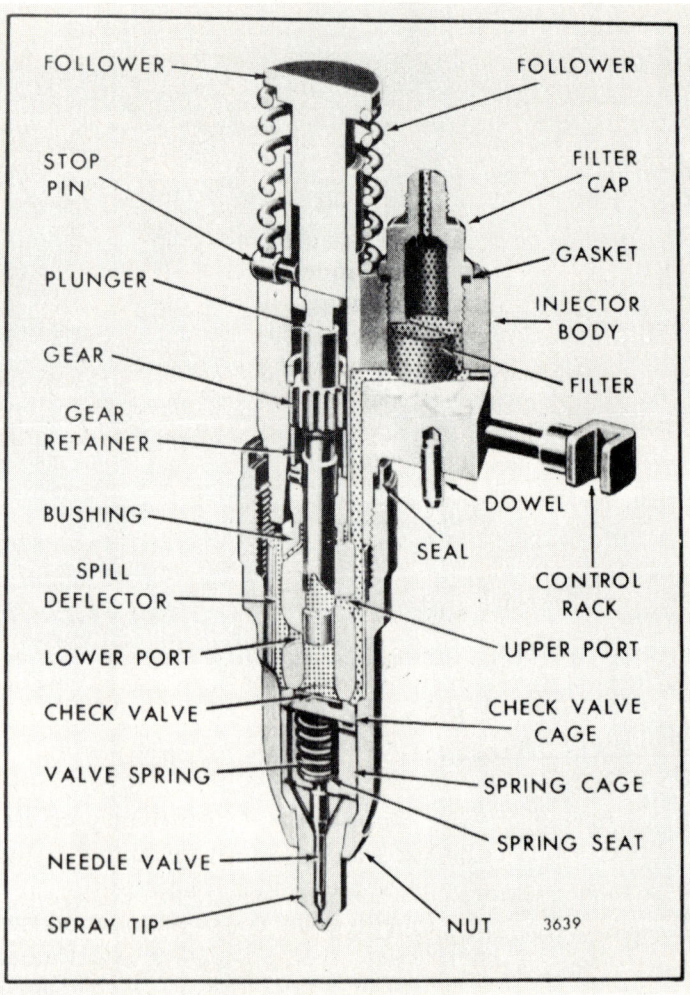

FIGURE 3-68. General Motors Injector. *(Courtesy of General Motors Corp.)*

Diesel Engine Fuel Systems / 71

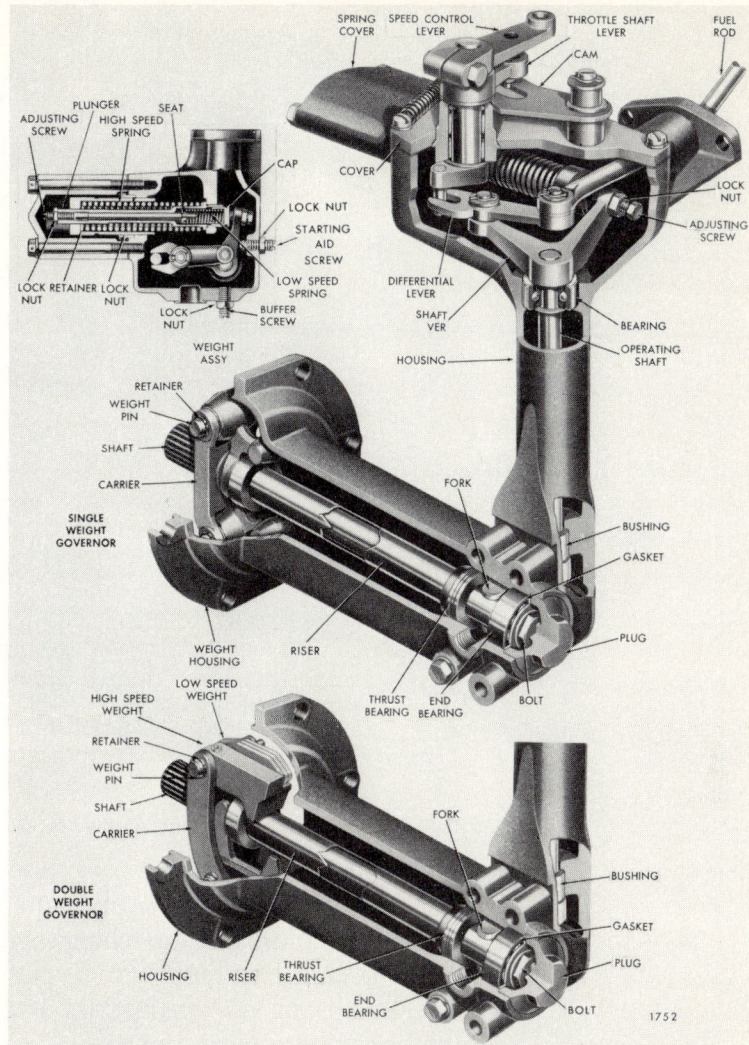

FIGURE 3-69. General Motors Governor. *(Courtesy of General Motors Corp.)*

lever. One side carries an adjusting screw and locknut, while the other arm is pin-connected to the differential lever.

One end of the differential lever is forked to engage the pen end of the external speed control lever. The other end connects to the fuel control link that is pin-connected to the control rack lever.

An internally shaped cam engages a pin on the throttle lever, which is pinned to the throttle shaft below the speed control lever. Its function is to limit the movement of the throttle shaft, preventing strain on the internal parts. [Fig. 3–69]

Fuel Pump

The fuel supply pump is usually driven by the lower blower rotor shaft through a forked coupling on the pump shaft and a drive disc on the rotor shaft. [Fig. 3–70] On some engines the pump is mounted on the rear of the gear cover and driven by a coupling to the balance shaft.

72 / Diesel Engine Service

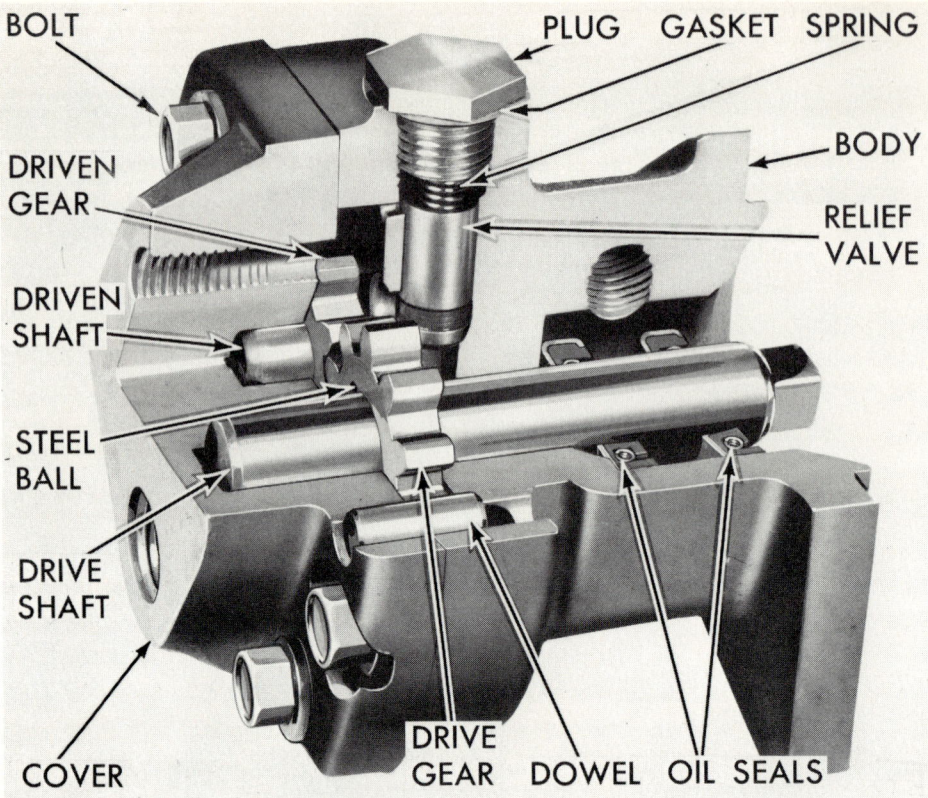

FIGURE 3-70. General Motors Fuel Pump. *(Courtesy of General Motors Corp.)*

Fuel pumps for vertical engines are furnished for right-hand and left-hand rotation and are *not* interchangeable. The original pump must be replaced by the same part number and rotation.

Fuel pumps on V-71 engines are driven by the right-hand blower rotor shaft and always rotate the same way.

There are two different sizes of fuel pumps. The standard pump has ¼-inch wide gears. The pump for some applications has ⅜-inch wide gears. One must exchange with the right pump number.

Fuel pump discharge pressure is limited by a relief valve to 65 psi. Normally this relief valve remains closed and opens only should the filter restriction exceed that pressure.

Some fuel leakage always appears at the tapped drain openings. This is normal, because the seals must be lubricated by fuel. If leakage exceeds 1 drop per minute, the fuel pump should be removed and rebuilt or an exchange pump installed.

If fuel pump wear is suspected, a pressure gauge installed in the fuel inlet manifold should read 50 to 70 psi at high-idle speed. A pressure lower than 50 psi can reduce engine power. This test must be performed after fuel filters have been replaced.

Tuneup

Although Detroit Diesel engines are capable of long service life, over time some adjustments are necessary to maintain performance. In general the steps to be performed are

Diesel Engine Fuel Systems / 73

1. Adjust valve clearance and injector timing (engine cold).
2. Check governed speed (high idle).
3. Check idle speed.
4. Check speed control cable and fittings.
5. Check for free injector rack travel.

Unless the engine speeds are found out of specification, no adjustment to the governor will be required. It is characteristic of all governors to maintain correct speeds; all other possible causes should be checked before adjusting the governor.

Checking Valve Clearance

Generally the valve clearance should be adjusted to the cold values. It is not normally possible to maintain engine operating temperature during the time required to make the adjustment. The specified clearance for the engine model must be followed; the settings must be exact.

INJECTOR TIMING GAUGES. Injector timing is done while the valve clearance is being adjusted. [**Fig. 3–71**]

ROCKER LEVER POSITIONS. When adjusting valve clearance, the injector rocker lever must be on cam, noticeably higher than the valve rockers.

When injectors are adjusted, the valve rocker levers are on cam, valves open. Both adjustments can be made on all cylinders during one crankshaft revolution.

The gauge specified for the engine model and injector number must be used [**Fig. 3–72**]

These steps should be followed in order to make the adjustment:

1. Clean the top of the engine with steam or solvent.
2. Place the throttle in shutdown position.
3. Remove the rocker lever covers.
4. Provide a means of turning the crankshaft by hand. A simple bridge can be made to fit any gear-driven pulley. [**Fig. 3–73**]

TURNING DEVICE ON PULLEY. If the capscrew in the front of the crankshaft is used, do not try to turn the engine backwards.

5. Turn the engine until one cylinder's injector is on cam.
6. Using the correct feeler gauge, check the clearance between the rocker lever nose and valve stem or bridge.
7. To correct this clearance, loosen the locknut on the push rod (½ inch) and use a $\frac{5}{16}$-inch end wrench to turn the push rod—to the right to increase clearance and to left to reduce it.

74 / Diesel Engine Service

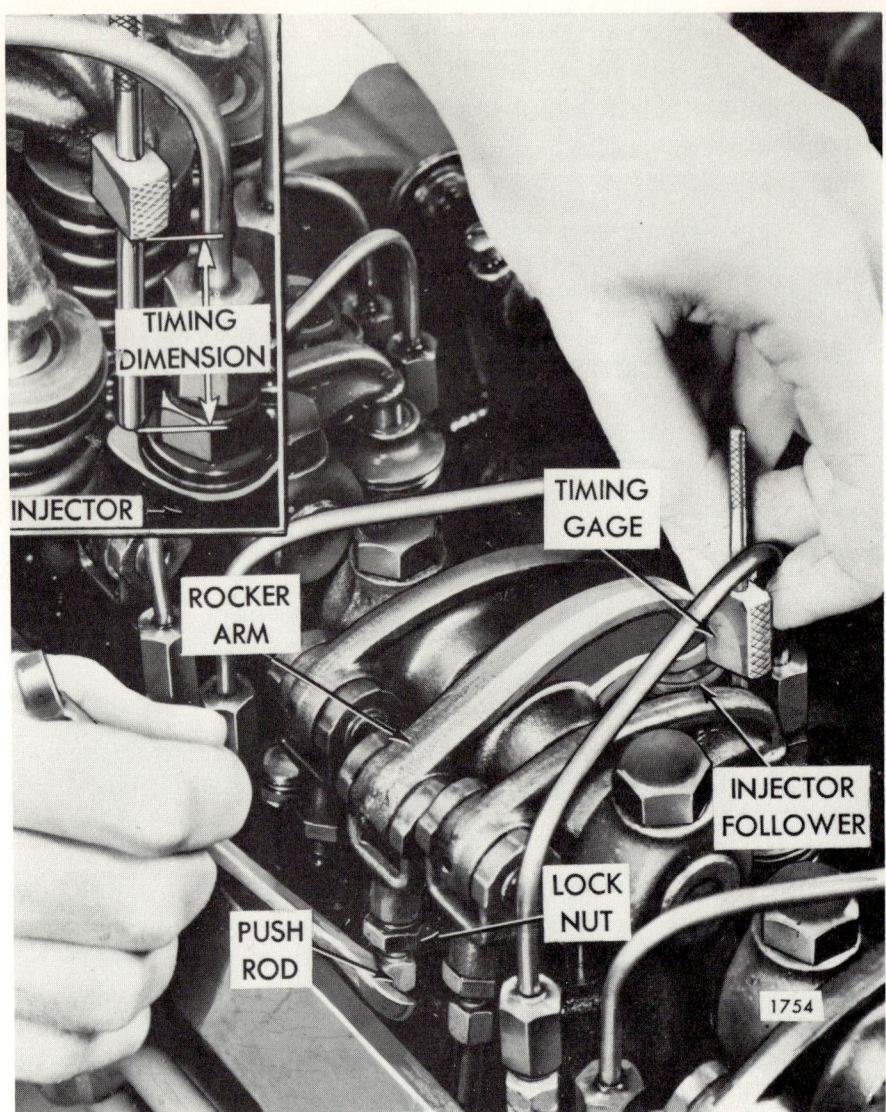

FIGURE 3-71. Using Timing Gauge. *(Courtesy of General Motors Corp.)*

Note: It is normal for clearance to reduce after some operating time from valve repair. If clearance is erratic, valve repairs may be required. Four-valve engines may need readjustment of the valve bridges. These bridges should retain adjustment for long periods, unless the valves under them start to fail. Thus, erratic bridge settings for equal contact with the valve stems are a clue to the need for valve service. One must recheck the clearance after securing the push rod locknut.

TIME INJECTORS

1. After selecting the correct timing gauge, bar the crankshaft to bring the valve rockers on one cylinder to on-cam position, valves open.
2. Insert the small tip of the gauge into the hole in the injector body top. Turn the gauge so that the flat side faces the plunger spring follower edge.

FIGURE 3-72. Rocker Lever Positions.

3. Loosen the push rod locknut, and turn the push rod until the gauge will pass just over the follower edge. Be sure to hold the gauge vertical.
4. Secure the push rod locknut and recheck the gauge setting. When the gauge barely clears the follower edge, the setting is correct.
5. As you adjust each injector, check the rack control levers for correct setting. [Fig. 3-74]

ADJUST RACK CONTROL LEVER SCREWS

1. Disconnect any rod or link from the stop lever.
2. Loosen the locknut enough to free the buffer screw, and back the screw out about ⅝ inch from the face of the nut.
3. Loosen all rack control lever screws. Be sure that the levers are free on the shaft.
4. Place the speed control in full speed position.

76 / Diesel Engine Service

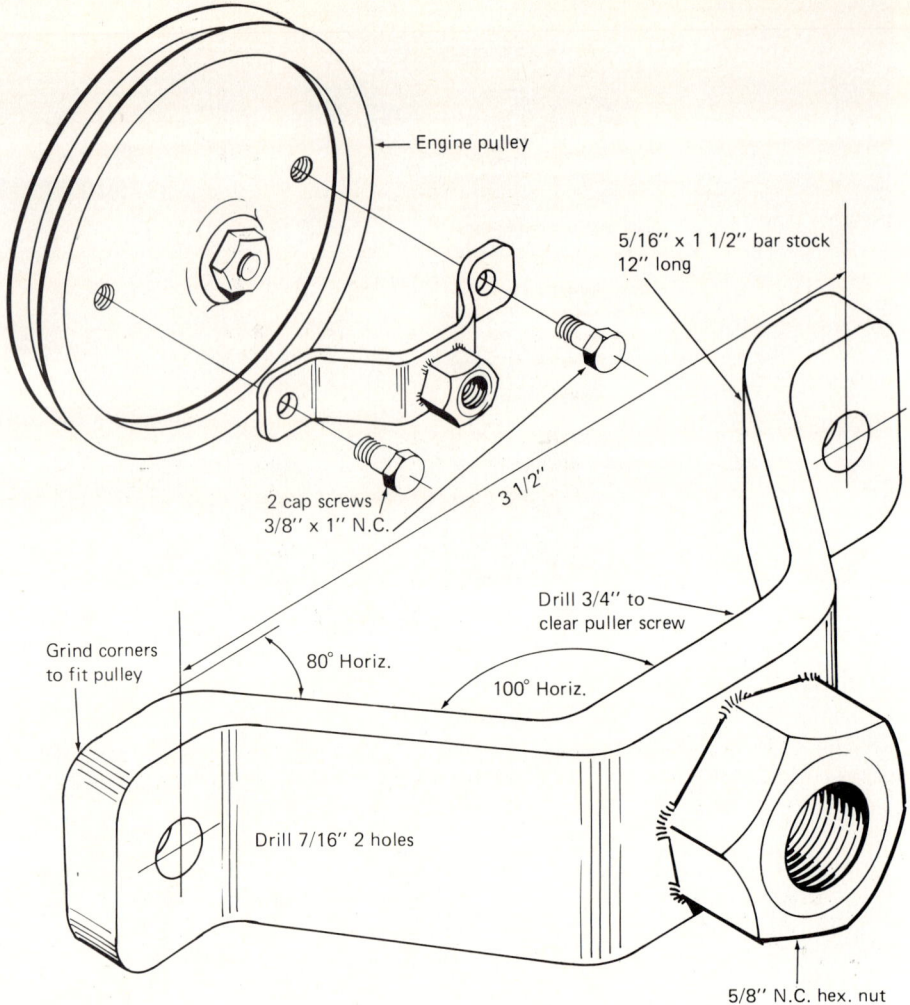

FIGURE 3-73. **Turning Device on Pulley.** *(From a sketch by P.M. Uhl)*

5. Attach a light spring to the stop lever in order to hold it in the run position.
6. Turn the inner lever adjusting screw on No. 1 injector (the one closest to the governor) in until you feel an increase in resistance. This is a light adjustment. [**Fig. 3-75**]
7. Turn the outer screw down to light contact.

CHECKING RACK FREEDOM

1. Place a screwdriver tip on the side of the rack clevis, [**Fig. 3-76**] and turn the rack a little. Let it spring back. It should return freely.
2. Adjust the screws by backing the inner screw off and tightening the outer screw until the stop lever just reaches its stop before the rack becomes tight.
3. When you are sure of this adjustment, secure the screws alternately to 30 pound-inch. Do not overtighten. Check the adjustment after tightening the screws.

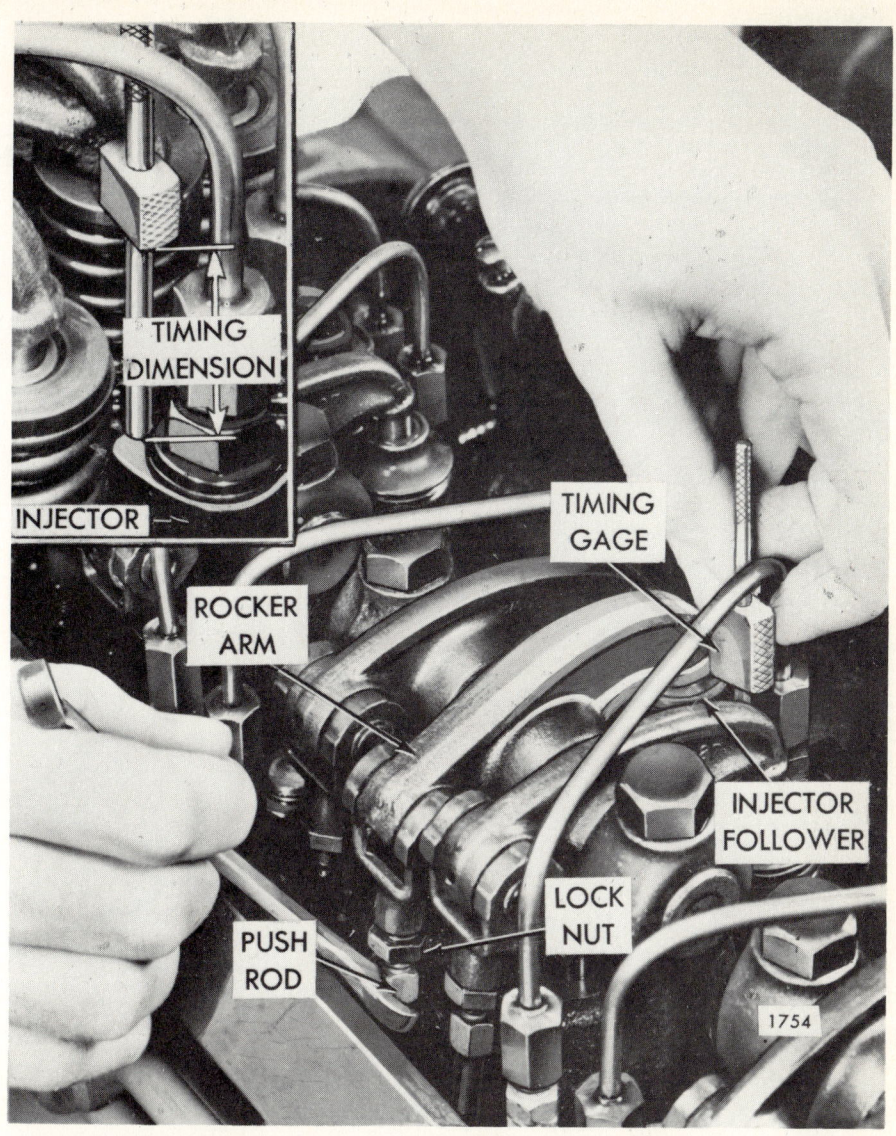

FIGURE 3-74. Adjust Push Rod. *(Courtesy of General Motors Corp.)*

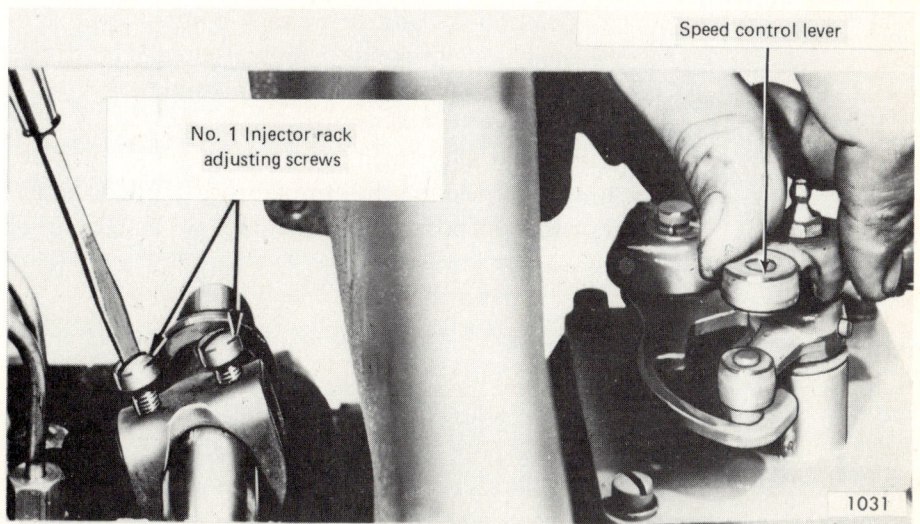

FIGURE 3-75. Adjust Rack Control Lever Screws. *(Courtesy of General Motors Corp.)*

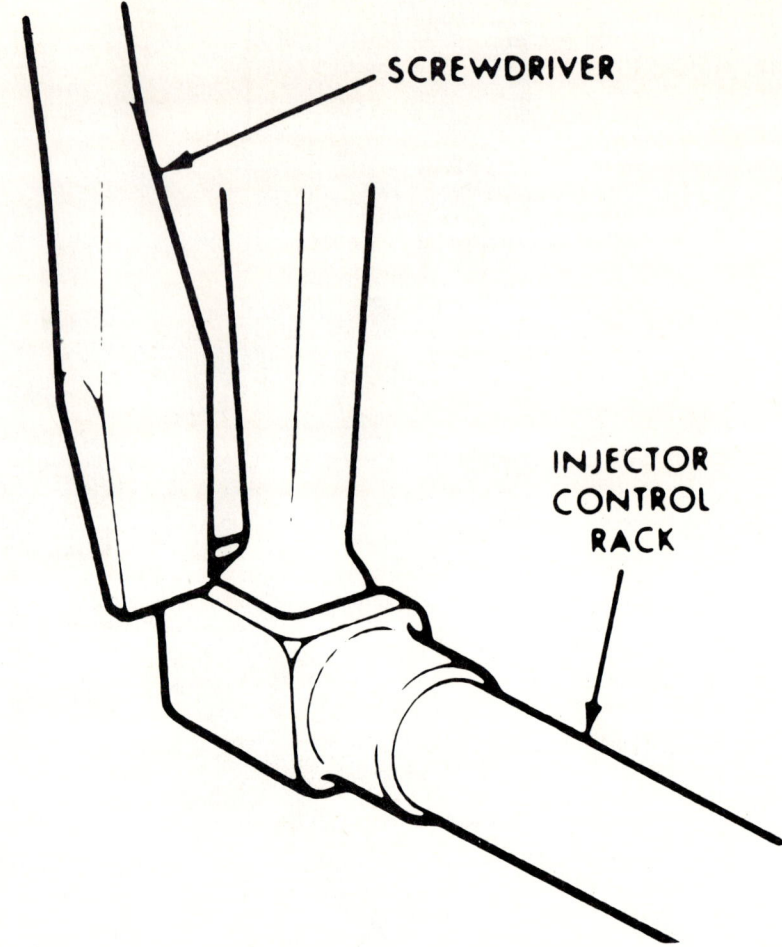

FIGURE 3-76. Adjust Rack Freedom. *(Courtesy of General Motors Corp.)*

4. Adjust the No. 2 injector rack control lever in the same way, using the No. 1 lever to establish full-fuel position.
5. Continue to adjust the rack control lever screws on the rest of the cylinders, and check No. 1 rack after each is adjusted. It is important to adjust all levers as nearly alike as possible.

Final adjustment of the lever screws will be made with the engine running, by checking the temperature of the exhaust manifold at each cylinder. For this purpose, a good contact pyrometer is valuable. [**Fig. 3-77**]

Adjusting the Governor

The foregoing instruction applies to an engine in day-to-day operation, with nothing more than minor complaints. In general, if the engine idles smoothly and will reach specified maximum governed speed, no load, the governor does not need adjustment. One should look for other causes of power loss, as described in the trouble-shooting section.

Diesel Engine Fuel Systems / 79

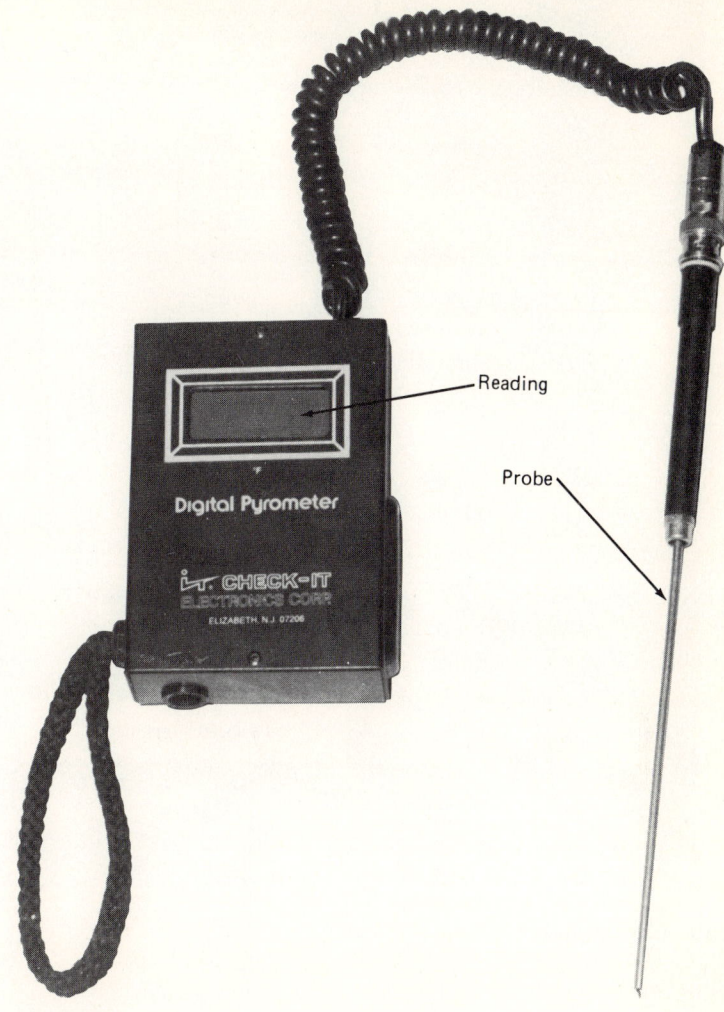

FIGURE 3-77. Contact Pyrometer.

After major repairs, such as cylinder head or blower replacement, a complete step-by-step adjustment of the governor will be necessary [**Fig. 3–78**]. Injectors and valves should be adjusted first; no bind should exist in the rack control shaft or linkage.

SINGLE-WEIGHT GOVERNOR

1. Be sure that the buffer screw is backed out ⅝ inch.
2. Remove the spring cover.
3. Remove the governor cover.
4. Remove the governor link from the differential lever and the injector control shaft lever.
5. Check the gap between the low speed spring cap and the high speed plunger. The special gauge for this check is 0.170 inch. Adjust if necessary.

80 / Diesel Engine Service

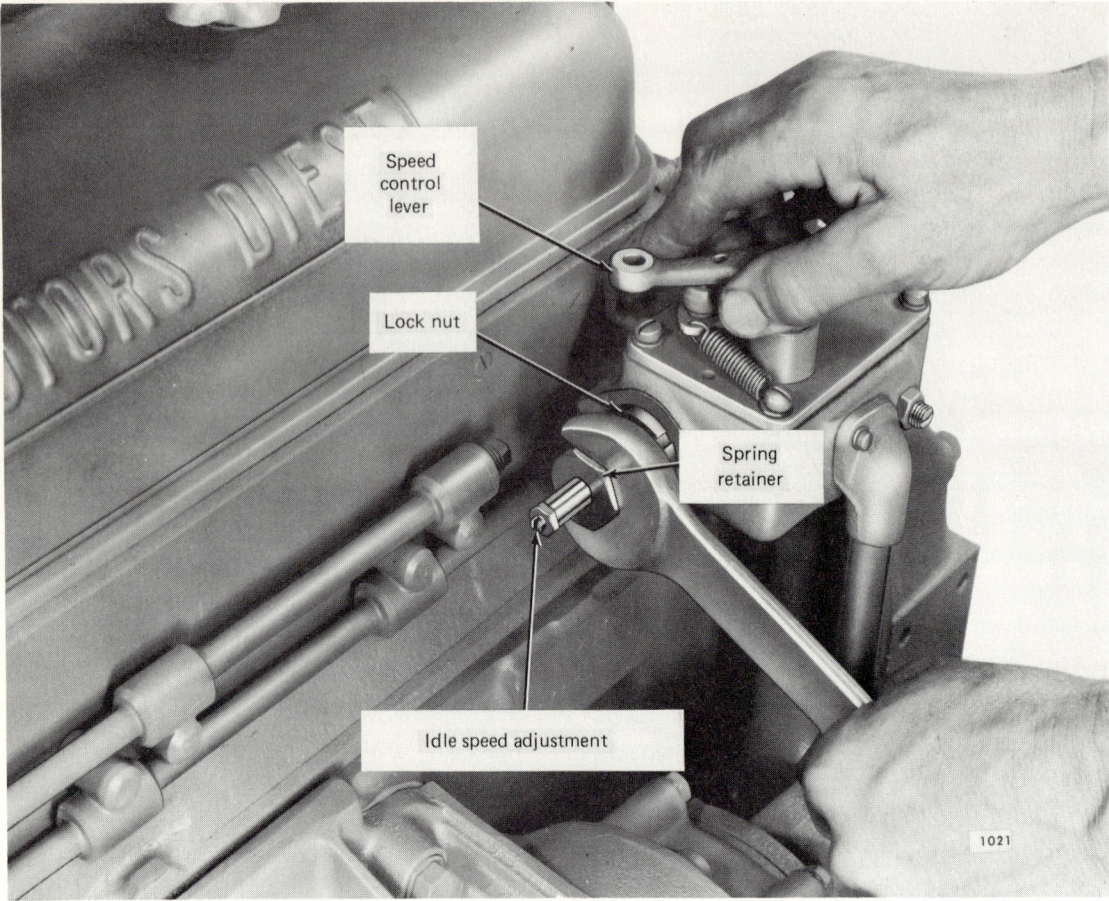

FIGURE 3-78. Adjust Governor. *(Courtesy of General Motors Corp.)*

6. Leave the gauge in and tighten the lock nut. Be sure that the gap is correct. **[Fig. 3-78]**
7. Install the governor link rod. Install the cover with a new gasket. The foregoing is for a single-weight governor.

DOUBLE-WEIGHT GOVERNOR. One should follow steps 1 through 4 of instructions for the single-weight governor. Then the mechanic should set the gap by the following method:

1. With the governor link rod removed, start the engine and control its speed at 1,100 to 1,300 rpm by manual control of the rack control shaft. Set the gap to 0.002 inch to 0.004 inch. Secure the locknut.
2. Place the speed control in full speed position. Turn the spring retainer in to bring engine speed to the specified no-load speed. Use a good tachometer to read the speed and do not set it above that recommended.
3. Secure the locknut on the spring retainer.

Adjusting Idle Speed

1. Adjust the idle speed to 485 rpm and secure the screw locknut.
2. Turn the buffer screw in carefully to pick up about 15 rpm. Lock the nut.
    Note: The buffer screw or starting aid is the last thing adjusted. If this screw is set too far in, the engine will overspeed.
3. Replace the governor spring cover with a new gasket.
4. Stop the engine; reconnect the link rod to the differential lever and rack control lever. Secure the pin with cotter pins.
5. Install the governor cover with a new gasket. Check to be sure that the stop lever brings the rack control to full-out position.

All governors are set to specified limits for the engine model at the factory. It is unlikely that maximum speed will change in service unless major disassembly has occurred or some fault in the injectors or fuel supply causes a loss of power. All other adjustments must be correct with no failure of fuel supply reducing engine speed.

Caution: Governor adjustments must be approached with care. No diesel engine will run at higher speeds than those for which it is designed, without damage.

Replacing Injector

After checking to be sure that an injector is faulty, it may be necessary to replace it with an exchange unit of the same number.

1. Remove the two capscrews and lift the rocker lever assembly for that cylinder back.
2. Remove the fuel lines from the injector and manifold connections. Store them in a clean place.
3. Loosen the rack lever adjusting screws and slide the lever aside to clear the rack clevis.
4. Remove the injector clamp bolt, washer, and clamp.
5. Lift the injector out carefully to avoid striking the tip.
6. Clean all carbon from the copper tube. A special reamer is available for this. If used, the reamer should be filled with grease to trap particles. Be sure not to cut the copper seat. Wire brushes are made for this purpose and are less expensive than the reamer.
7. Insert the exchange injector into the tube, with its dowel in the locating hole in the cylinder head.
8. Install the clamp washer with curved side down, and capscrew. Tighten the capscrew to 20 to 25 pound-foot (27 to 34 N-m). See that the clamp does not interfere with any other parts. Do not overtighten.
9. Remove the shipping caps from the fuel connections, and put them on the old injector.

10. Install the fuel tubes, starting the nuts by hand. Tighten to 12 to 15 pound-foot (16 to 20 N-m), using a torque wrench and special socket.

    Note: These fuel tubes can be twisted if overtightened and can allow fuel leakage. In case dilution of the engine oil is suspected, observe for any signs of washing near the fuel tubes. Correct as needed.

11. Replace the rocker levers, making sure that the valve bridges are in proper position. Secure the capscrews to 90 to 100 pound-foot (122 to 136 N-m) unless a load limit bracket is used. In these cases, tighten the capscrews to 75 to 85 pound-foot. (102 to 115 N-m).

12. Slide the control lever into engagement with the rack clevis. Adjust the screws as described.

13. Replace the rocker cover.

    Note: The foregoing governor description applies to common limiting-speed governors used on truck and bus engines. Many other applications use variations of the basic governor, but they present no mystery if the basic system is understood.

Other Governor Descriptions

When the speed control lever acts to vary the main spring tension, the governor is called a *variable speed governor.*

A *dual-range* action allows the governor to control the engine at full speed during acceleration, with a lower governed speed once cruise speed is reached.

A variation to permit operation at a lower speed for unloading, while providing full speed for highway use, is called a *variable low-speed* governor. The spring tension change, and therefore the governed speed change, is controlled either by a manual cable or by air pressure from the service air system.

A *fast-idle cylinder* may be installed in place of the buffer screw, with an air-operated stop cylinder mounted on the governor cover.

In addition to these variations, an engine driving the load through a torque converter can have a separate governor driven from the converter output shaft and linked to the main governor speed control lever.

Variations in output shaft speed cause the torque converter governor to vary the engine speed control to maintain output shaft speed under varying load.

Hydraulic Governors

Engines driving generators of electric power are usually equipped with hydraulic governors, which allow more precise speed regulation.

All of these governors sense engine speed by the effect of rotating weights. On some systems, weight force is modified by trapped air, which acts like a spring. In hydraulic governors weight force moves a pilot valve against a spring, and the pilot valve controls oil flow to a power piston, which turns the terminal lever to move the fuel control toward full fuel. Return of the lever is by a spring.

One some engines, engine oil is circulated through the hydraulic governor. This system has the advantage of automatic shutdown if the engine runs out of oil. The disadvantage is that engine oil is often dirty, and dirt and carbon can cause governor failure. [**Fig. 3-79**]

Several engines use a separate sump or reservoir for governor oil; there is an oil pump in the governor to provide pressure for governor action. These systems are often provided with automatic shutdown devices that operate should low oil pressure or high coolant temperature occur.

Hydraulic governors can be adjusted to maintain engine speed at a constant, preset value as long as load variation is within engine capacity. Response time of such governors can be as low as one-hundredth of a second. Such governors are called *isochronous*.

All governors require a speed change in order to control fuel injected. The amount of change is called *speed droop*. Mechanical governors usually need an 8 to 12 percent speed droop as load is increased from no-load to full rated load. That change must be made by load application in a controlled way, as by a dynamometer. In cases when a dynamometer cannot be used, governor action must be judged by reading maximum no-load speed.

Governor Modifications

Governors on Detroit Diesel engines may have several modifying attachments; some assist starting and some modify governed speed to achieve greater economy or improve torque performance. As described, a separate governor may be used to control engine speed to achieve control of torque converter output shaft speed.

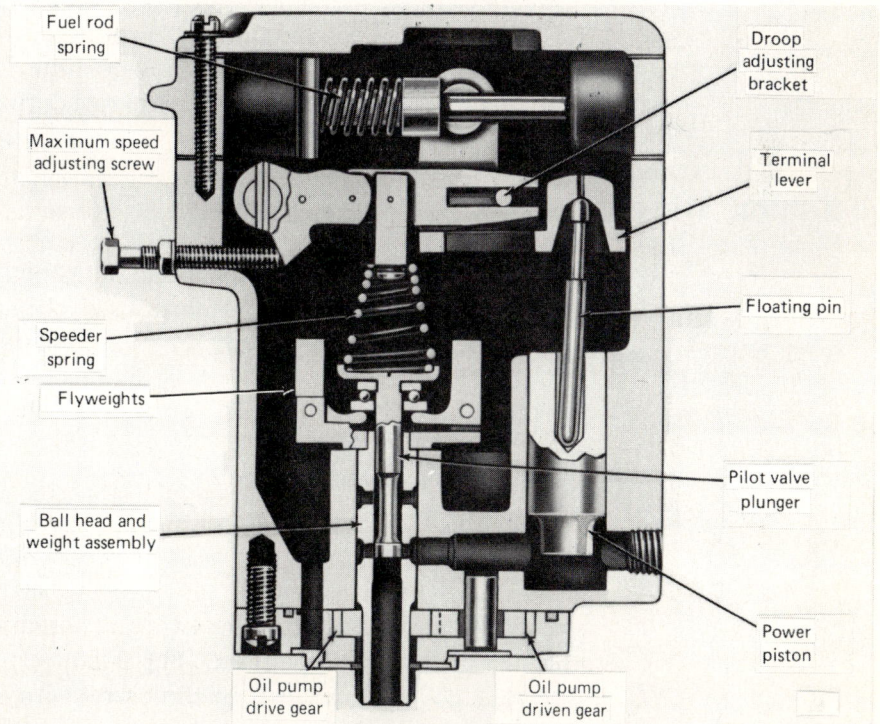

FIGURE 3-79. Hydraulic Governor. *(Courtesy of General Motors Corp.)*

None of these governors present any obscure service problems. The field mechanic should have manuals covering the service details and should not attempt adjustment without such references.

Injector timing, valve lash, and rack control lever must be correctly adjusted; the rack control shaft must be free.

CUMMINS FUEL SYSTEM

Description

This system, which is used on current engines, makes use of a law of hydraulics that has been known for many years. The volume of liquid flow from any opening of a pressurized passage to a space that is at lower or atmospheric pressure is governed by three factors:

1. The length of time that the hole is open
2. The size of the opening
3. The pressure on the fluid

Because Cummins injectors are cam-driven, and the moving plunger acts as a valve, the length of opening time depends on engine speed. The size of the opening is fixed during manufacture, so that its effect on liquid volume is fixed.

Pressure on the fluid can be regulated over a wide range, so that this regulation is used for all control functions.

This system is called *PT* for pressure time. It is the only diesel fuel system that functions by controlling pressure rather than volume.

Fuel Pump Description

The fuel pump consists of four basic subassemblies. Although each of these units would function if separately housed, they are contained in one housing to provide a compact assembly. The standard pump weighs about 13 pounds.

In order of fuel flow, these component assemblies are

1. fuel supply pump,
2. governor and pressure regulator,
3. throttle, and
4. shutdown valve.

The supply pump is a gear-type unit and draws fuel from the supply tank through the suction-type filter. [Fig. 3-80]

The governor consists of the barrel with its fuel passages; the weight-driven plunger, which is moved to control the barrel passages; the idle and maximum speed springs, which resist weight force and plunger movement; the

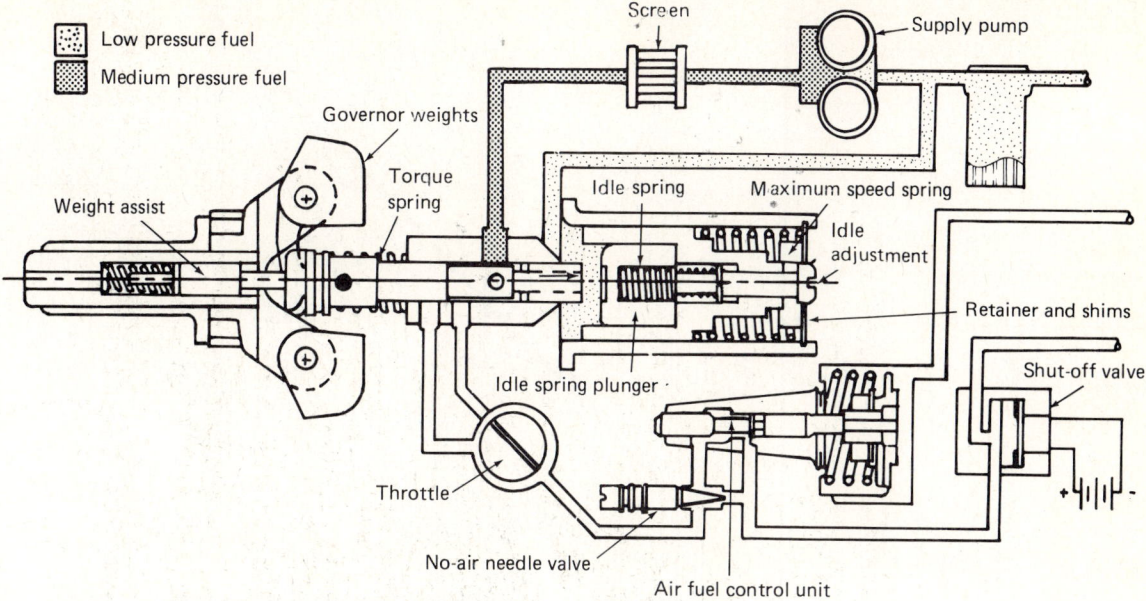

FIGURE 3-80. Cummins Fuel System. *(Courtesy of Cummins Engine Co., Inc.)*

idle spring plunger, which acts as a bypass valve to regulate pressure; and modifying springs called *weight assist* and *torque control springs,* which shape the fuel delivery curve over the engine speed range. [**Fig. 3-81**]

The throttle is a spigot-type rod in which is drilled a hole that indexes with the passage from the governor. As the throttle is turned, the hole is opened, allowing more fuel to flow. Throttle travel is fixed by two stop screws. The inner end of the throttle rod contains a restricting plunger, which is used for fine calibration.

The shutdown valve is located on top of the pump and is usually solenoid-operated. Its function is to stop fuel flow to the injectors, thereby stopping the engine.

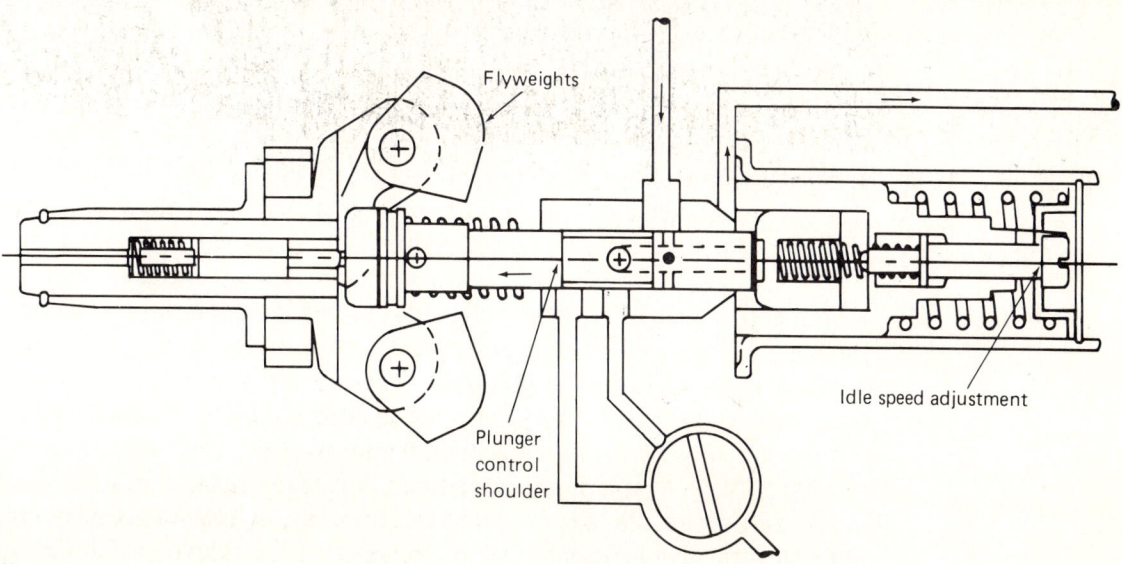

FIGURE 3-81. Cummins Governor. *(Courtesy of Cummins Engine Co., Inc.)*

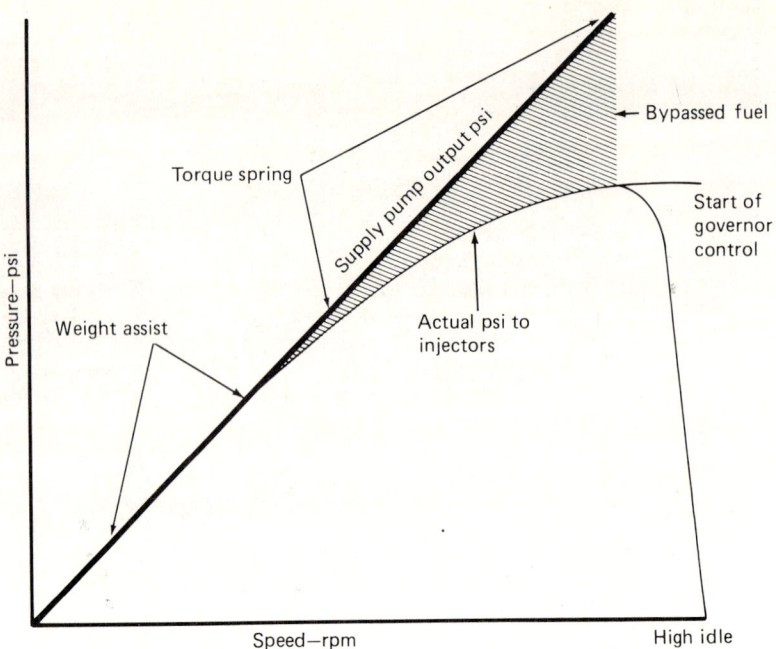

FIGURE 3-82. Fuel Delivery Curve. *(From a sketch by P. M. Uhl)*

Looking at the pressure-speed curve, one can see the supply pump output as a straight line, because it is a positive-displacement pump where output varies with speed. [**Fig. 3-82**] There is no way to get higher pressure at any point in the speed range, so that any pressure regulation must be accomplished by a controlled bypasss of fuel.

The capacity of the engine to burn fuel is limited by the air supply and by the strength of the parts to withstand high cylinder pressure. Thus, we must reduce fuel pressure by allowing some fuel to escape from the pump output line back to the pump inlet. This feature is called a *bypass*.

The hollow governor plunger is closed by the idle spring plunger, whose recessed end is acted on by supply pump pressure. This idle spring plunger recess varies in size, and the plunger is selected by number to match the calibration requirements of various engine models.

Fuel pressure pushes the plunger back against spring pressure, allowing fuel to escape to the bypass passage to the pump inlet. Some fuel is bypassed at all speeds; the volume of bypass increases with engine speed. The plunger recess diameter determines the bypass volume. [**Fig. 3-83**] Thus a high-rated engine uses a smaller plunger recess diameter than a lower-rated engine.

There are three passages in the governor barrel. The first hole on the weight end is an idle fuel port and leads around the main throttle passage. As the weights revolve, their force is exerted to move the plunger endwise and bring the plunger control shoulder to cover the idle port partially. The engine then runs at idle speed. Idle speed can be adjusted by turning the small screw to add or remove tension from the idle spring.

From idle to maximum speed, the governor plunger control shoulder moves across the idle port in the barrel, then through the blank bore until the main throttle port is reached. When the control shoulder is just at the edge of the main port, the engine is running at full-load, rated speed. As the load de-

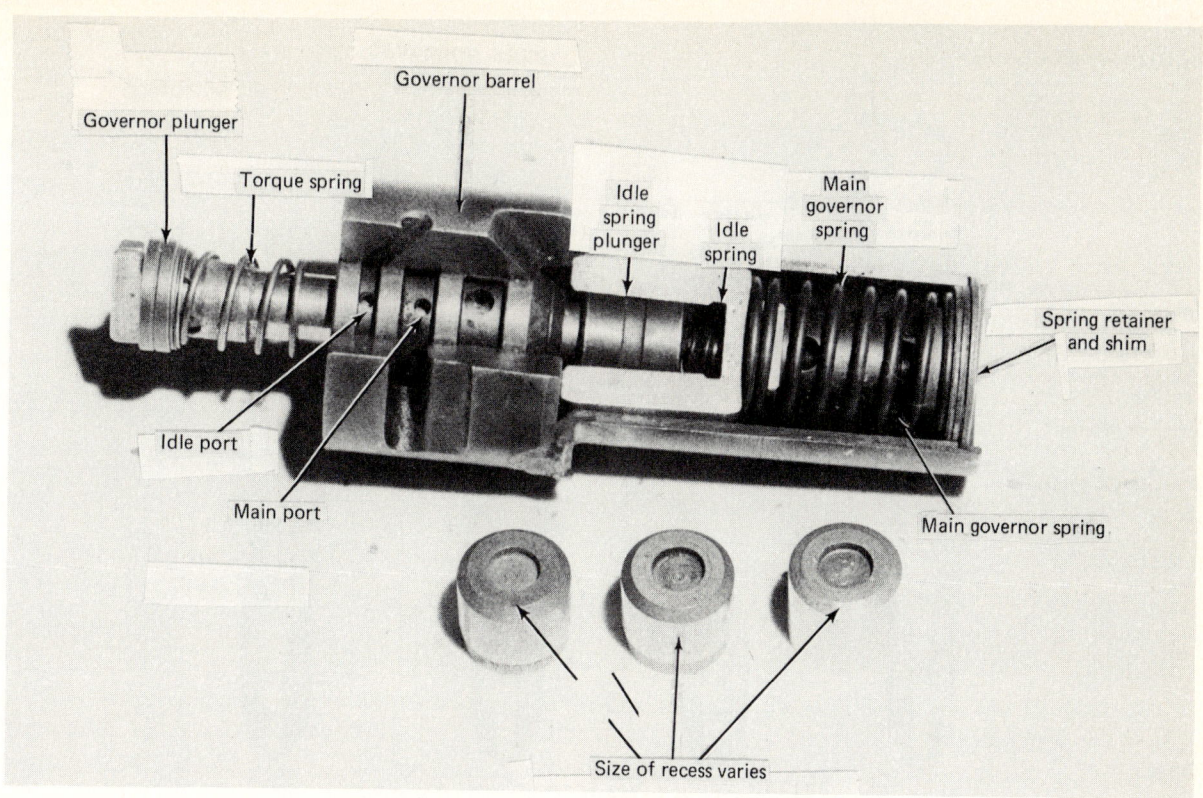

FIGURE 3-83. Variety of Idle Spring Plungers. *(Courtesy of Cummins Engine Co., Inc.)*

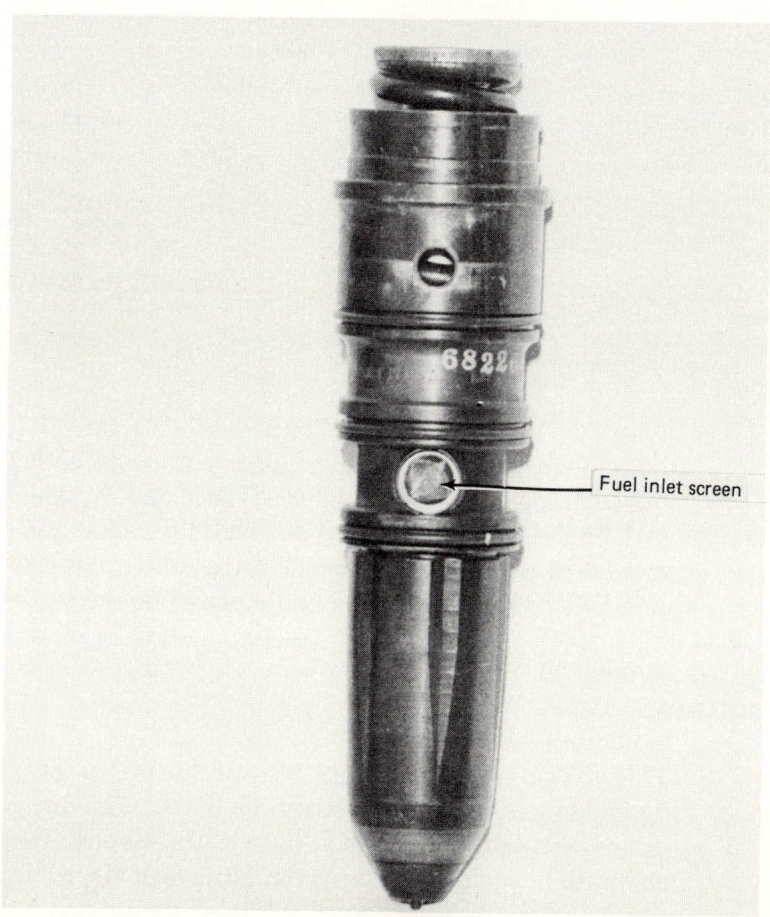

FIGURE 3-84(a). PTD Injector. *(Courtesy of Cummins Engine Co., Inc.)*

88 / Diesel Engine Service

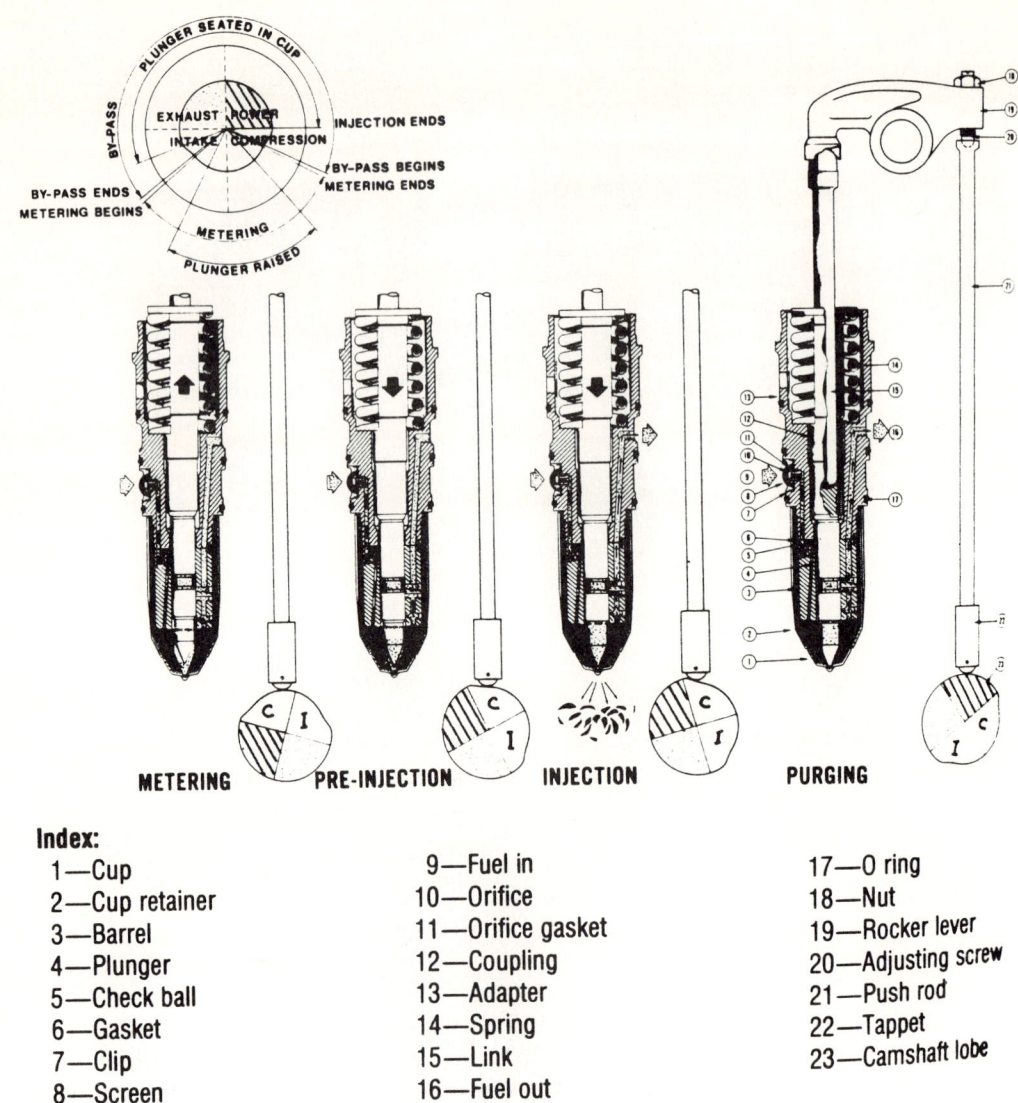

FIGURE 3-84(b). PTD Injector. *(Courtesy of Cummins Engine Co., Inc.)*

creases, engine speed increases until there is just enough fuel flow to run the engine at maximum governed speed, no load, or high idle.

This is the highest speed at which the engine can pull itself. Any higher speed must be induced by load. Thus, a truck going downhill can force the engine over its governed speed and damage can result.

Injector Description

This version of the Cummins PT injector is current, although the Top Stop design is increasing in popularity. The two designs are alike, except for the addition of a plunger stop nut at the top of the body. The plunger travel is set in calibration, and no further service adjustment is required to maintain plunger travel as wear occurs in service. [**Fig. 3-84**]

Fuel flow through the injector starts at the lower groove on the body and

Diesel Engine Fuel Systems / 89

enters through a small screen. Fuel flows through an orifice plug whose opening is adjusted during a test to match the specification for that injector number.

A drilled passage leads fuel past a ball check valve, which prevents back flow. This is its only function. From the ball check, fuel flows down through a passage to the end of the barrel, where it flows around a groove in the top of the cup to the feed hole and to the return passage above.

The plunger acts as a valve as it moves in response to the cam and spring. The lower end of the plunger, at the start of its tapered end, uncovers the feel hole during its upward movement. The amount of fuel entering the injector cup depends on the fuel pressure and the speed of plunger movement.

As the plunger reaches the top of its stroke and starts down by cam action, the volume of the cup cavity is reduced, and some fuel would be forced back if it were not for the ball check. [**Fig. 3-85**] Continued downward movement of the plunger closes the fuel hole. All fuel in the cup is driven into the cylinder through the spray holes.

Injection pressure may be 18,000 to 20,000 psi. Distribution of the fuel over the combustion space area is due to the number and size of the spray holes.

As the plunger continues downward, the return passage is opened by the

Start upstroke (fuel circulates)

Fuel at low pressure enters the injector at (A) and flows through the inlet orifice (B), internal drillings, around the annular-groove in the injector cup and up passage (D) to return to the fuel tank. The amount of fuel flowing through the injector is determined by the fuel pressure before the inlet orifice (B). Fuel pressure in turn is determined by engine speed, governor and throttle.

Upstroke complete (fuel enters injector cup)

As the injector plunger moves upward, metering orifice (C) is uncovered and fuel enters the injector cup. The amount is determined by the fuel pressure. Passage (D) is blocked, momentarily stopping circulation of fuel and isolating the metering orifice from pressure pulsations.

Downstroke (fuel injection)

As the plunger moves down and closes the metering orifice, fuel entry into the cup is cut off. As the plunger continues down, it forces fuel out of the cup through tiny holes at high pressure as a fine spray. This assures complete combustion of fuel in the cylinder. When fuel passage (D) is uncovered by the plunger undercut, fuel again begins to flow through return passage (E) to the fuel tank.

Downstroke complete (fuel circulates)

After injection, the plunger remains seated until the next metering and injection cycle. Although no fuel is reaching the injector cup, it does flow freely through the injector and is returned to the fuel tank through passage (E). This provides cooling of the injector and also warms the fuel in the tank.

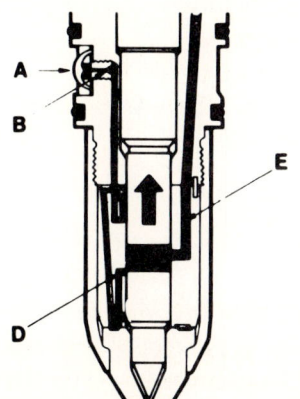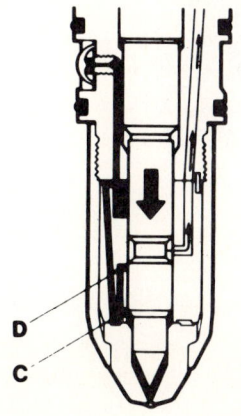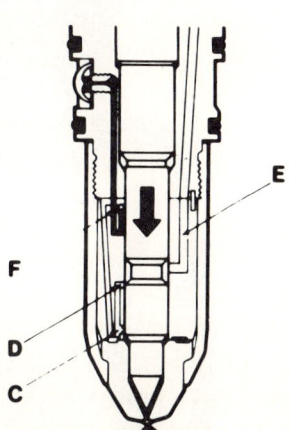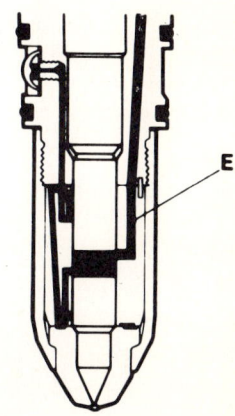

FIGURE 3-85. Injector Sequence. *(Courtesy of Cummins Engine Co., Inc.)*

90 / Diesel Engine Service

plunger groove, and fuel returns to the tank through drillings in the heads and the return line. The plunger remains seated until the next cycle.

Injector Service

Service to the injectors consists of proper rocker lever adjustment to assure correct plunger travel. The injectors can be replaced with rebuilt units. If kept adjusted and protected from contamination, they have a long service life.

Because individual injectors cannot be temporarily "cut out" by loosening a fuel connection, their condition must be judged by other means. General engine performances, smoke level, and the standard trouble-shooting methods help to determine the need for replacement.

A good contact pyrometer applied to each exhaust connection is a useful method of determining which cylinders are out-of-step. [**Fig. 3-86**] To make this test, the engine should run at idle until temperatures stabilize. The probe of the pyrometer should be applied to each exhaust outlet, at the same distance from the cylinder head on each one.

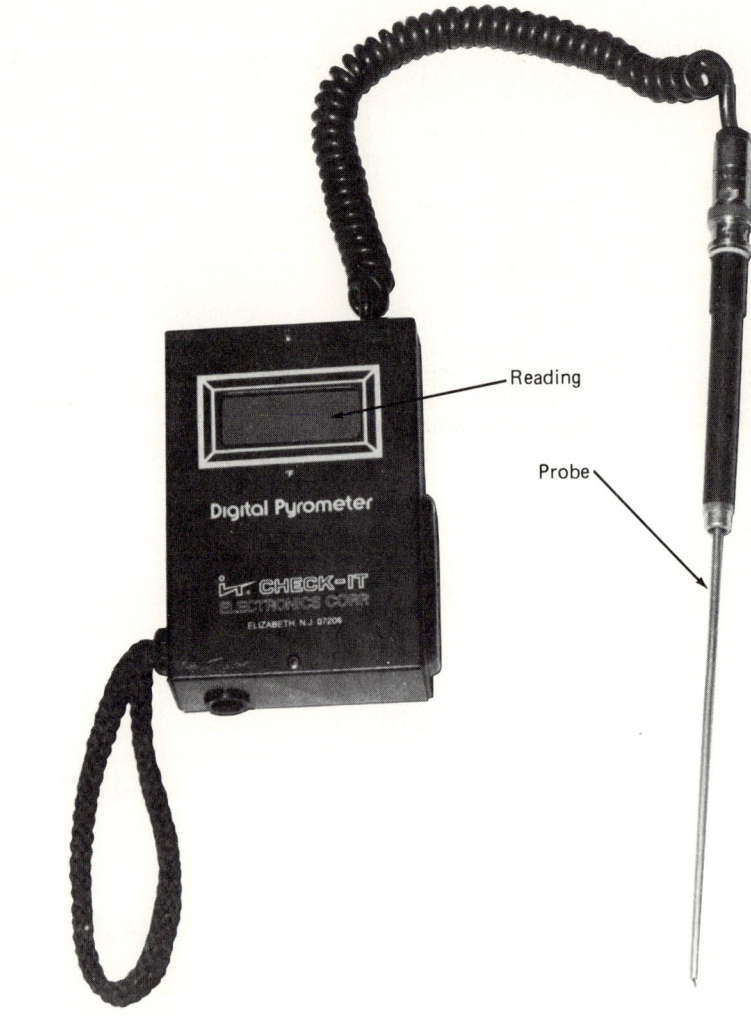

FIGURE 3-86. Contact Pyrometer.

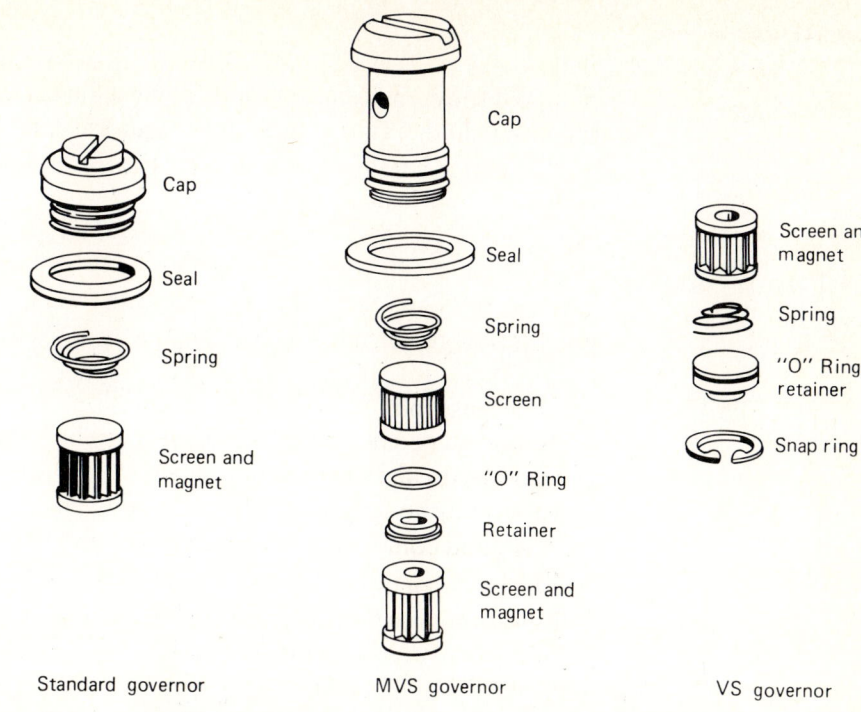

FIGURE 3-87. Fuel Pump Screens. *(Courtesy of Cummins Engine Co., Inc.)*

FIGURE 3-88. Remove Supply Pump.

92 / Diesel Engine Service

With the temperatures stable, the exhaust will read from 140°F to 170°F. Any cylinder that reads more than 15 degrees F high or low from the average of the others should be checked for a fault. This test is useful on both Cummins and Detroit engines.

Fuel Pump Service

No adjustments to the governor or fuel rate can be made successfully in the field without a test stand. Several service operations can be performed.

1. The small screen under the slotted cap on the top of the pump should be checked during tuneup. This screen has a small magnet inside; metal pieces found on the magnet can mean supply pump wear. [**Fig. 3-87**] Lint from the filter can be removed by an air jet directed to the outside of the screen.
2. The complete supply pump can be replaced. Remove the small tube and the pulsation damper. Remove the outside two capscrews and slide out the pump. Be sure to replace it with the same part number exchange. Use a new gasket. [**Fig. 3-88**]

FIGURE 3-89. Fuel Pressure Measuring Point. *(Courtesy of Cummins Engine Co., Inc.)*

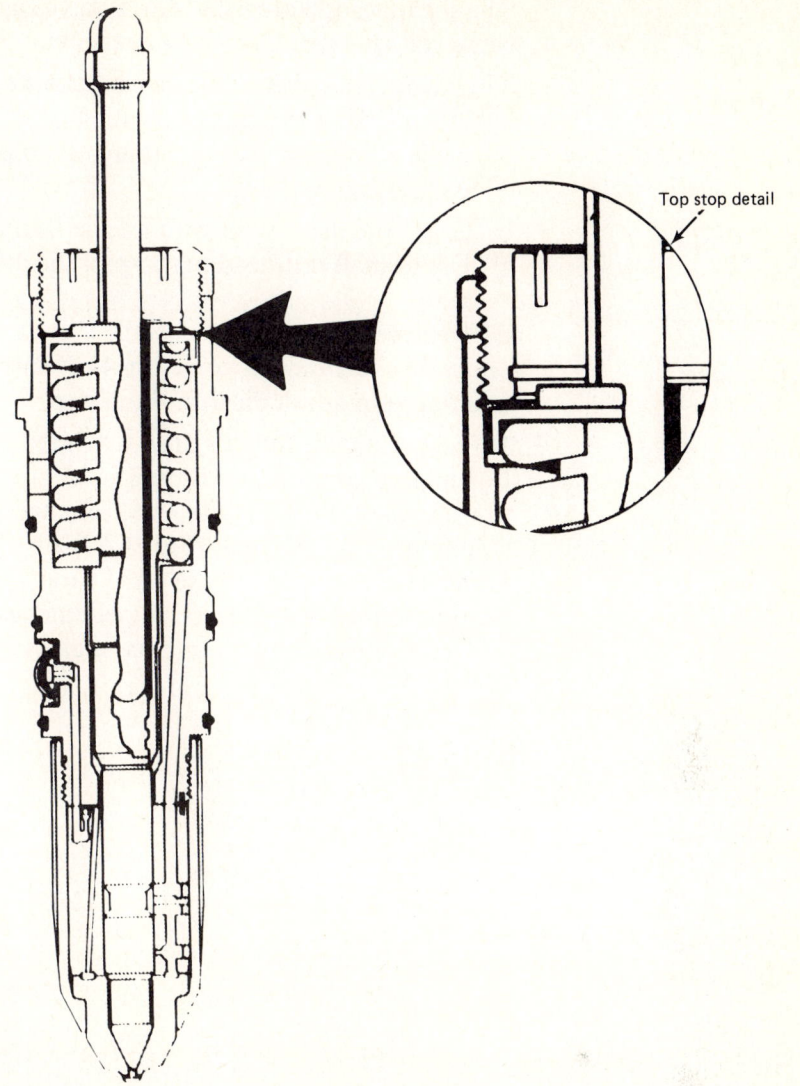

FIGURE 3-90(a). "Top Stop" Injector. *(Courtesy of Cummins Engine Co., Inc.)*

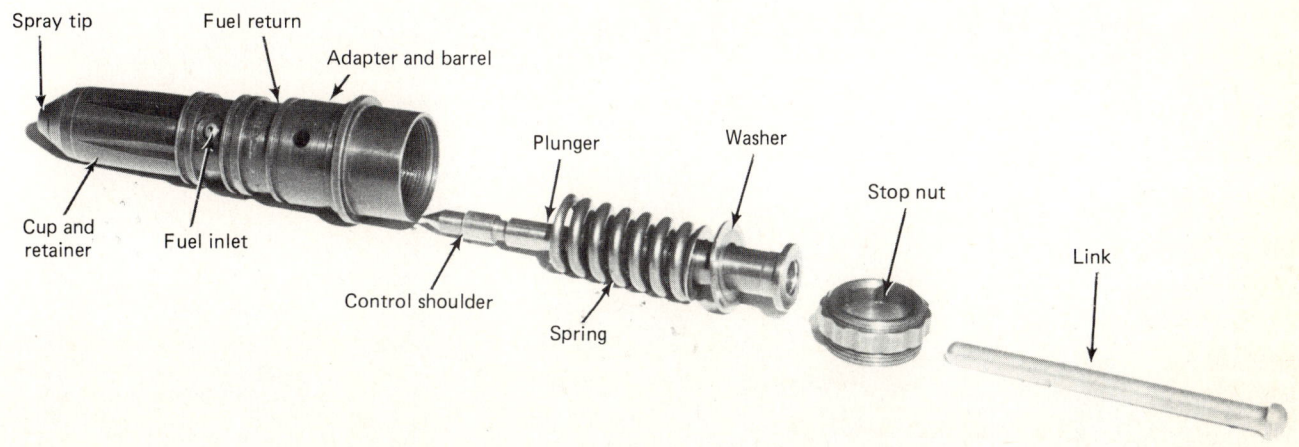

FIGURE 3-90(b). "Top Stop" Injector Exploded View. *(Courtesy of Cummins Engine Co., Inc.)*

3. The solenoid-operated shutdown valve can be replaced. Be sure to use the same voltage unit. Use a new seal ring.
4. The tachometer drive seal can be checked for leakage by removing the cable and filling the housing with fuel. If the fuel is drawn in while the engine is idling, the seal is leaking. To replace the seal:
    a. Remove the housing.
    b. Grasp the drive stem with a tube flaring clamp.
    c. Use a small roll-head bar to pull the drive assembly. It is held in by the seal.
    d. Replace the seal and reinstall. Wet the seal with clean fuel.
    Note: The tachometer drive on AFC current pumps is in the drive cover and does not admit air into the pump.
5. If fuel pressure is low, or after extended service, the entire fuel pump can be replaced. Because no timing is required, removal and replacement are not complex. Be sure that the exchange pump is calibrated to the same standard as the original. [**Fig. 3-89**]
    Note: These engines use the same variety of governor variations that have been described for Detroit Diesel engines. No unusual service problems will be presented by Cummins engines using different governors. [**Fig. 3-90**]

AIR INDUCTION SYSTEMS

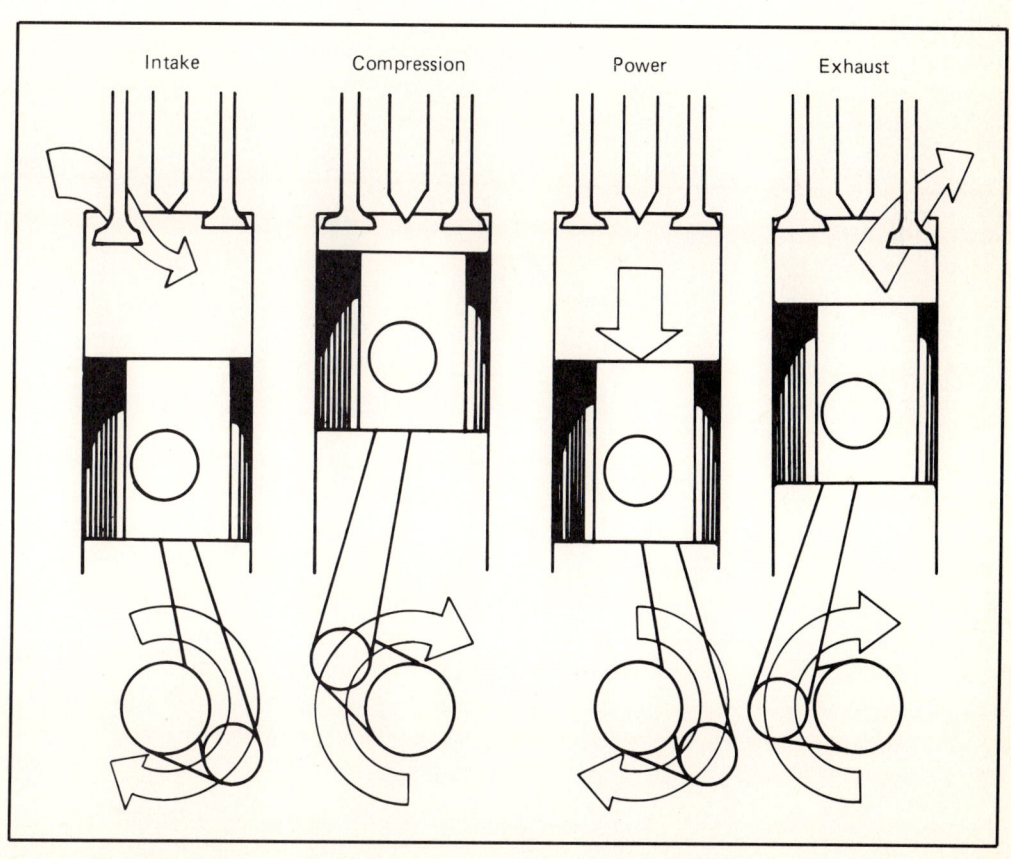

FUNCTIONAL DESCRIPTION

Because most commercial diesel engines are of the four-stroke cycle design, the engine pistons act as pumps to draw in air and expel exhaust gas. [**Fig. 4–1**] The exception is the Detroit Diesel, which must drive an air pump called the *blower* in order to move the air. [**Fig. 4–2**]

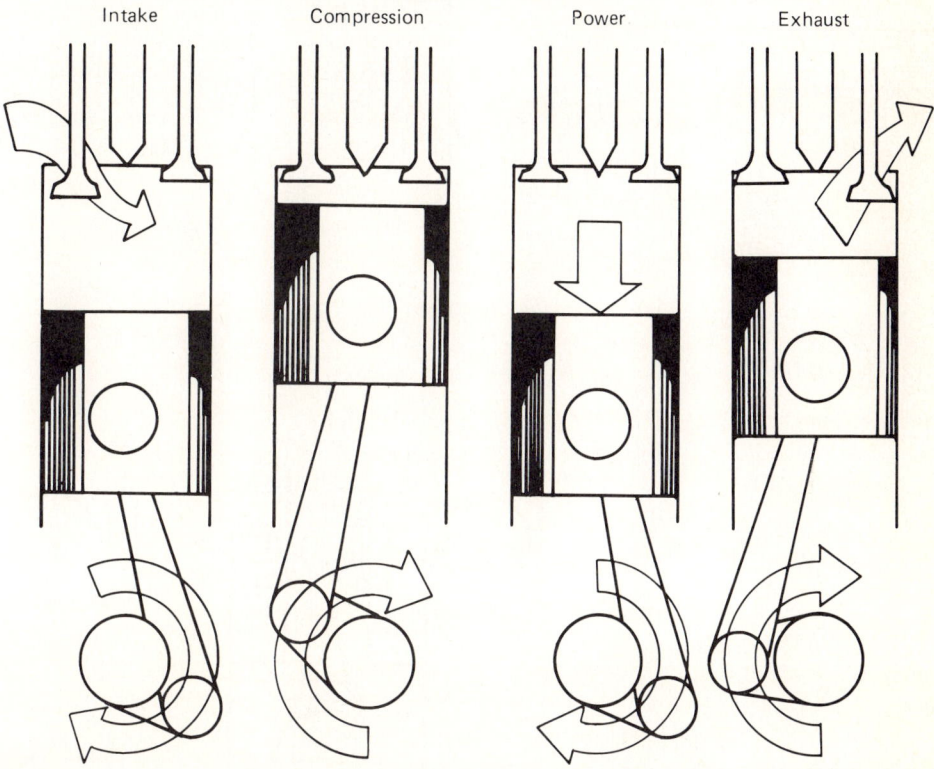

FIGURE 4–1. Four Stroke Cycle. *(Courtesy of Cummins Engine Co., Inc.)*

TURBOCHARGERS

The exhaust-driven turbo-supercharger is rapidly replacing the engine-driven supercharger as a means of increasing the air supply. The energy in exhaust gas, which is wasted in unblown engines, is used to drive a bladed rotor that is mounted on the same shaft as an air impeller rotor.

Although it is commonly thought that pressure and velocity of the exhaust gas drive the turbo charger, the heat of the gas really furnishes most of the driving energy. There is about 300° F difference between the incoming exhaust and the outlet temperature. [**Fig. 4–3**]

Because no mechanical drive is applied, no load is imposed on the engine. However, the turbine imposes a restriction on the exhaust, and backpressure is almost as high as the air pressure created by the air impeller.

The air is compressed by the impeller and is heated. This in turn lowers

98 / Diesel Engine Service

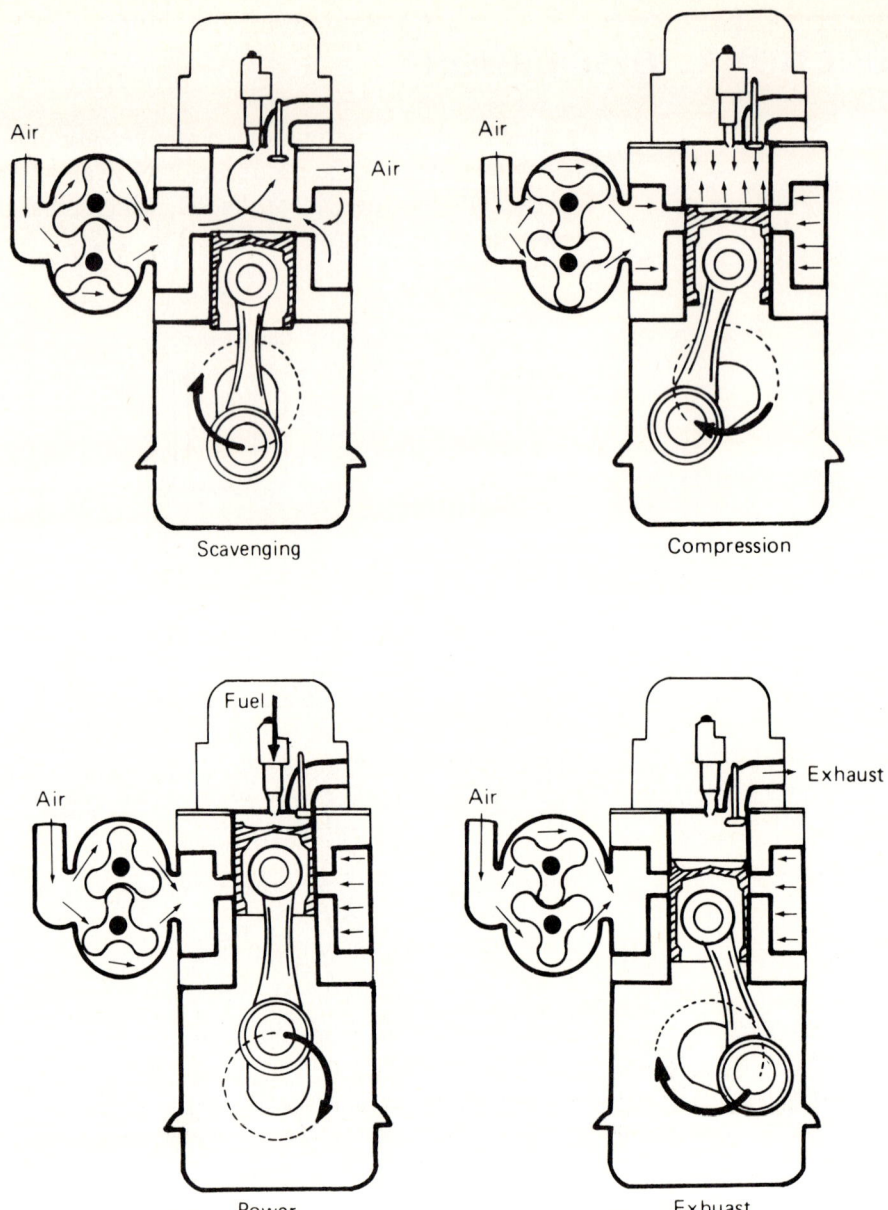

FIGURE 4-2. Two Stroke Cycle. *(Courtesy of General Motors Corp.)*

air density, so that a heat exchanger, called an *intercooler* or *aftercooler,* is installed in the intake manifold in order to transfer heat to the coolant, which circulates through the intercooler. Thus, some engines to which a turbocharger is applied have an intercooler added to increase air density and engine efficiency. [**Fig. 4-4**]

Turbochargers turn at extreme speeds. The shaft must float on a film of oil, and the supply of clean oil must not be interrupted. The oil drain back to the crankshaft must be kept open in order to avoid excess oil retention in the turbocharger bearing housing.

Oil filters have been used on the oil feed, but the turbocharger will fail if the filter plugs. Adequate filtration of engine oil is the only way to avoid shortened turbocharger life.

Air Induction Systems / 99

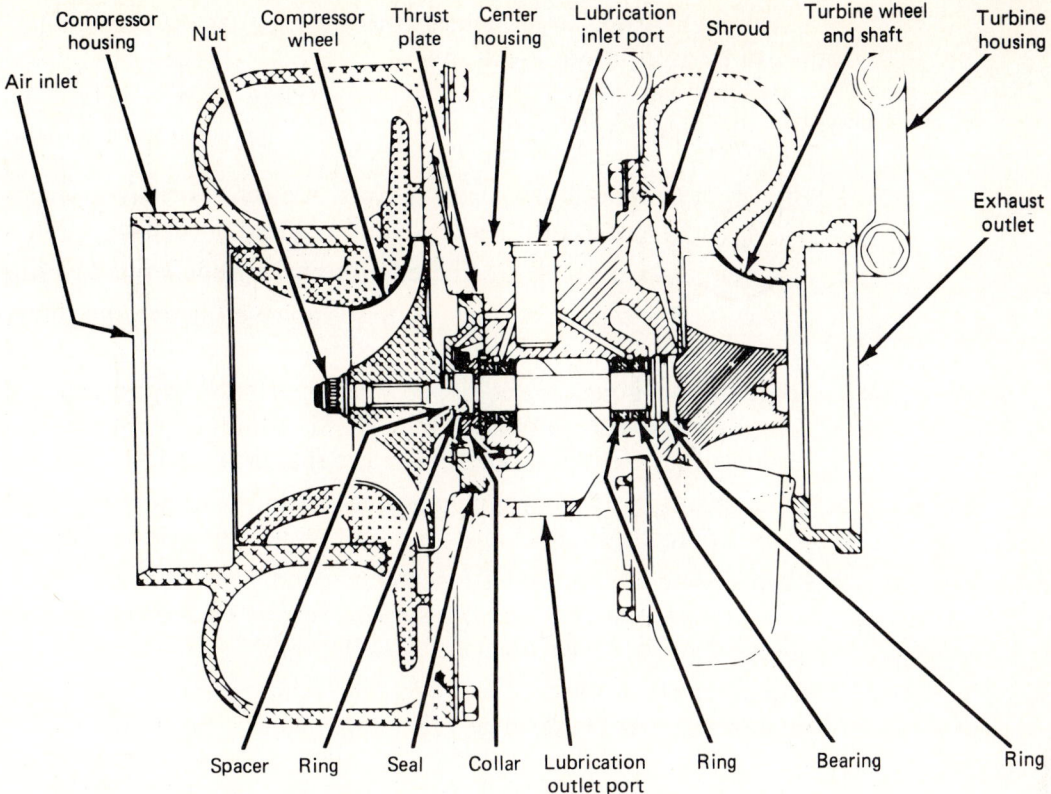

FIGURE 4-3. Turbocharger Cross-Section. *(Courtesy of Caterpillar Tractor Co.)*

FIGURE 4-4. Engine with Aftercooler. *(Courtesy of Caterpillar Tractor Co.)*

Although turbocharger failures are rare, they can be spectacular. A burst rotor can be as injurious as shrapnel.

Failures can be avoided in several ways, all of which are available to maintenance-minded operators:

1. Service air cleaners at frequent intervals. A dirty, restrictive element allows the turbocharger to turn faster, due to reduced air density.
2. Maintain a good supply of filtered oil to the turbocharger bearings.
3. Avoid overfueling the engine. To do so can result in overspeeding the turbine and may bring on a failure.
4. Some turbochargers are equipped with an exhaust bypass called a *wastegate*. Its purpose is to limit turbine speed. Although such devices are adjustable, they must never be adjusted so that the turbine is driven too fast.
5. Train operators. Turbocharged engines respond differently than unblown ones. Driving technique must be modified to suit the turbocharger.
6. Keep air-operated fuel controls in good condition. They act to limit throttle opening to match the air pressure in the intake manifold during acceleration only. They have no effect on full power.
7. Replace the turbocharger with a rebuilt exchange unit at reasonable service intervals. A yearly exchange will assure higher efficiency.

THE COOLING SYSTEM

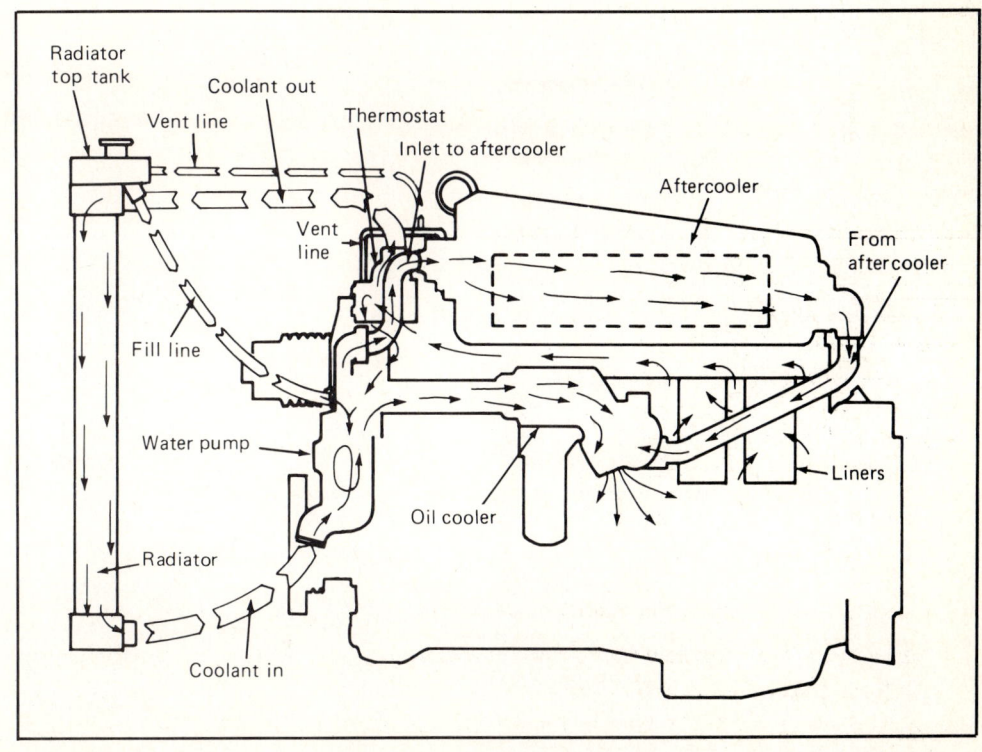

FUNCTIONAL DESCRIPTION

As the engine burns fuel, all the heat produced cannot be used for expansion and power. Some of it is expelled in the exhaust, and some is carried away by the coolant. No part of the engine should be heated to a point that causes damage.

Engine cooling systems provide a path for heat flow away from working parts. The cooling medium may be water or air, depending on engine design. Because water conducts heat efficiently and is not costly, it is used for cooling in most engines.

This is not to say that successful cooling systems cannot be made using air as the cooling medium. Quite large engines that circulate air around the cylinders to carry heat away have been made.

The first requirement is that adequate circulation must be designed into the system in order to assure that heated coolant flows away from the engine and that fresh coolant can always reach the heated parts.

Air-Cooled Engines

Air-cooled engines require ducting to direct cooling air and finned cylinders and head surfaces to increase the area that contacts the air flow. Efficient engine-driven fans are used to force the air through the ducting. Lubricating oil circulation is high in order to help cool the working parts. Fins are used on oil reservoirs, and the entire engine is designed to aid heat travel out of the metal masses.

Water-Cooled Engines

Liquid-cooled engines are designed with passages and walled areas called *jackets* surrounding the cylinders and valves and with centrifugal engine-driven pumps to circulate coolant through the system and move it to a heat-exchanger called the *radiator*. There the coolant flow divides into many small passages, which allow heat to be transferred into the surrounding air.

Because water is a corrosive liquid and will dissolve many materials if not chemically treated, several chemicals are used to modify its action. The liquid result of dissolving chemicals into the water is called *coolant*. Chemicals are also used to lower the freezing point of the liquid; the combination of anti-freeze and the resulting fluid may be only half water. Most service on cooling systems consists of maintaining the chemical combination and keeping the system full of solid coolant.

OPERATION

Cooling system design has progressed considerably over the years. Much of the design effort has been directed toward preventing air from entering the

104 / Diesel Engine Service

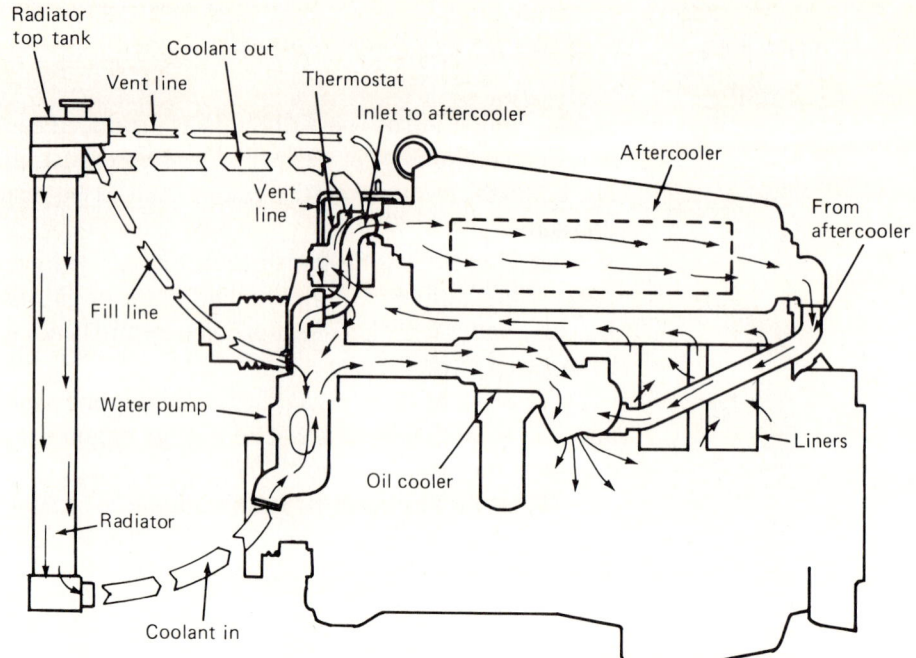

FIGURE 5-1. Cooling System Schematic. *(Courtesy of Caterpillar Tractor Co.)*

coolant, preventing air from displacing water, and reducing corrosion. [**Fig. 5-1**]

The first major improvement was the use of radiator caps containing a pressure-sensing valve between the top tank and the overflow opening. The purpose was to increase pressure on the coolant, raising its boiling temperature. [**Fig. 5-2**]

Because the boiling point is raised 3° F for each pound of pressure, a pressure cap that imposed 10 psi on the coolant would raise the boiling point 30° F. Originally this feature allowed the use of alcohol-based antifreeze compounds, because alcohol boils at 180° F at sea level.

The later use of glycol-based antifreezes reduced the need for this effect, but other advantages of pressure caps were recognized.

1. Pressure bearing on the coolant in the radiator encourages flow and reduces the chance for high vacuum at the water pump inlet.
2. As coolant expands with heat, the pressure cap reduces loss from the overflow.
3. Placing the coolant under pressure allows higher operating temperatures, increasing engine efficiency.

The next advance in cooling system design was the arrangement of piping to permit filling the engine from the bottom. A top vent line was added to allow air to escape to the top tank. The tank itself was made with a baffle, so that coolant from the engine contacted the air over a small area. [**Fig. 5-3**]

A fill line connected the top tank with the water pump inlet, so that coolant added could fill the block passages from the bottom, displacing any air to the vent, which is connected on the engine side of the thermostat.

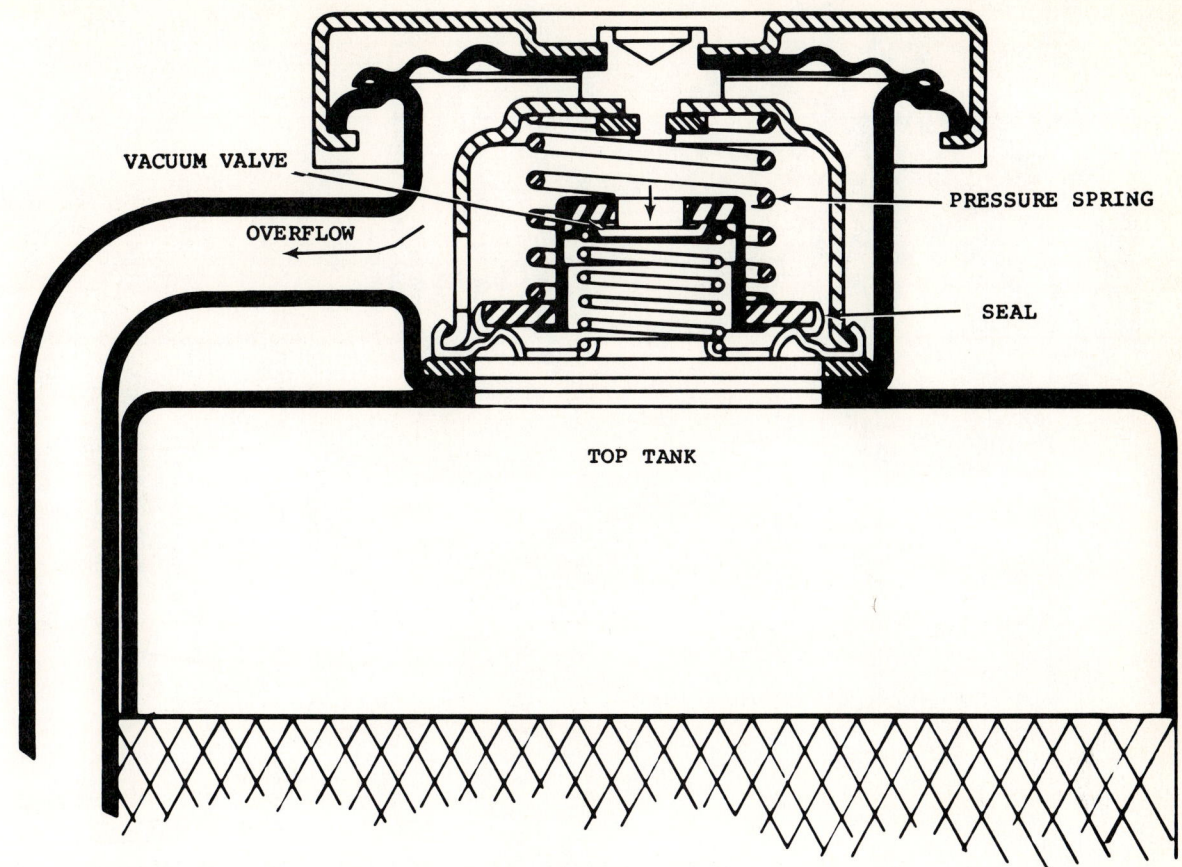

FIGURE 5-2. **Pressure Cap.** *(Courtesy of Caterpillar Tractor Co.)*

This system allows continuous venting and reduces the chance for the water pump to draw air. [**Fig. 5-3**]

Cross-Flow Radiators

Many modern vehicles have radiators in which coolant flow is across the core rather than downward. The core tubes are supported horizontally. The main advantage is a lower hood line. [**Fig. 5-4**]

Because entrained air must be vented, some components of cross-flow systems are special:

1. an aspirator, which serves to drive air out of the coolant
2. a U-tube vent pipe installed inside the filler-end tank

Coolant under engine block pressure flows to the aspirator through a vent tube from the engine side of the thermostat. Thus flow is established as soon as the engine starts, regardless of thermostat position.

Flow into the aspirator causes air to be drawn out of the engine, radiator, and piping, and directs it into the vent tube. This action keeps the entire coolant system except the supply chamber full of coolant.

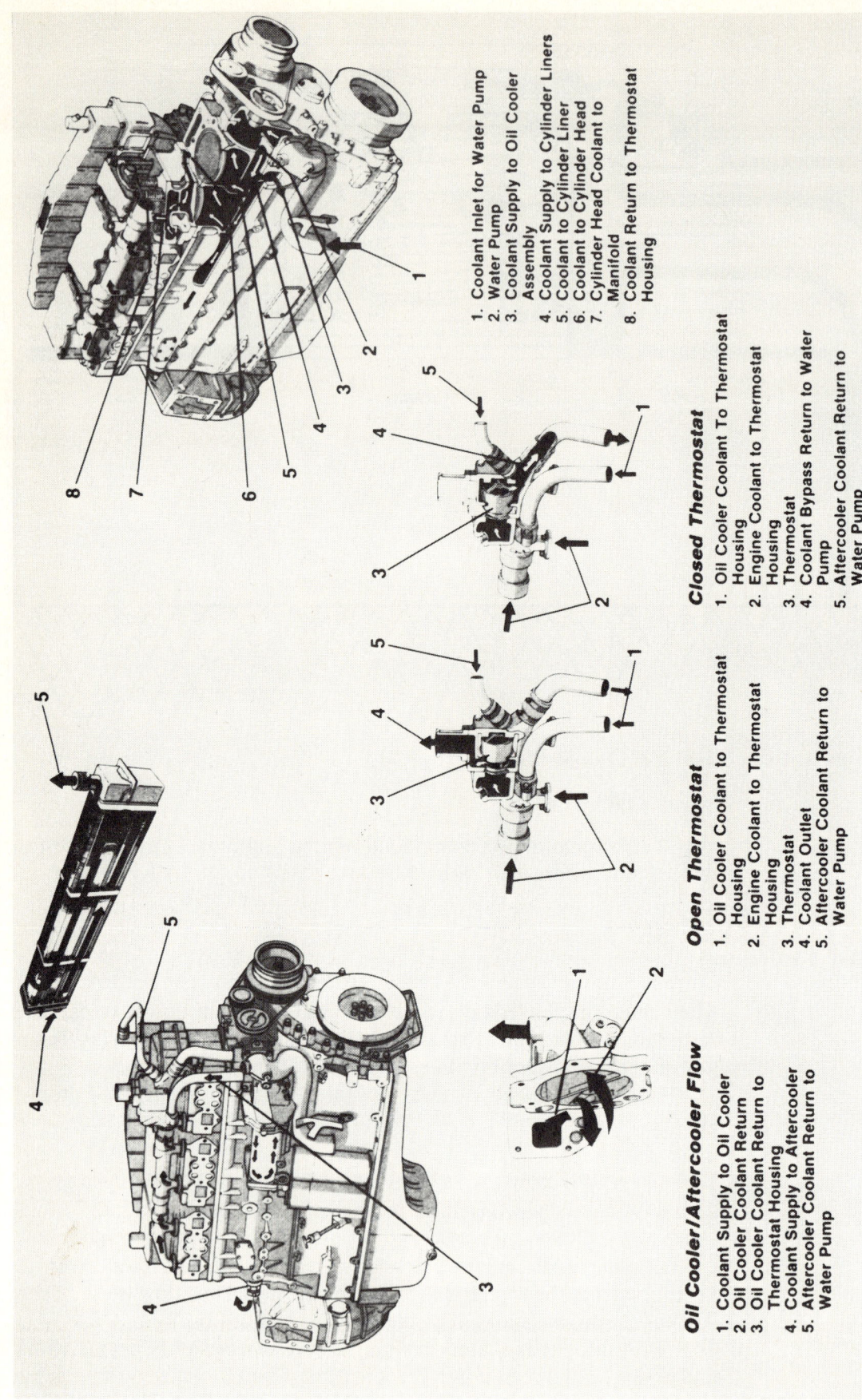

FIGURE 5-3. Full Flow Cooling System. (Courtesy of Cummins Engine Co., Inc.)

The Cooling System / 107

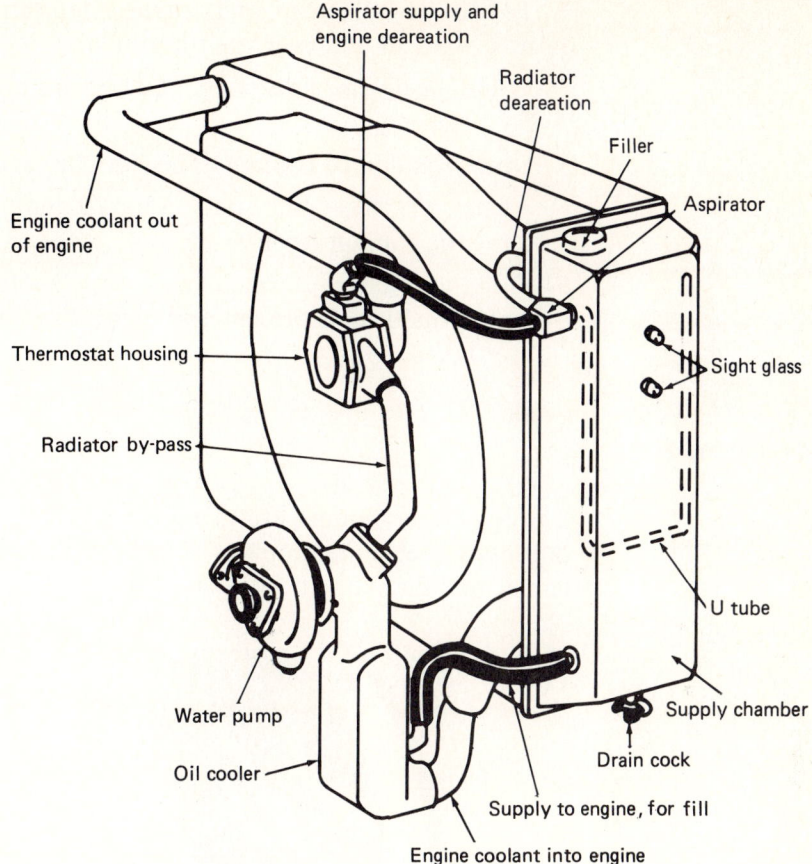

FIGURE 5-4. Cross-Flow Radiator. *(Courtesy of General Motors Corp.)*

When a receptacle is provided for overflow, the overflow tube carries coolant to the receptacle when the warm coolant expands. The tube ends below the level in the receptacle, and coolant is drawn back into the system as it cools down.

A small fill line provides a path for coolant to fill the water pump inlet and engine. The U-tube acts as a trap, to keep the coolant from seeking a common level, and reduces the chance for low coolant level and overheating at start-up. There are some service points to be observed on cross-flow systems. [Fig. 5-4]

1. Always drain the system from the drain at the bottom of the supply chamber.
2. After filling the system, run the engine for at least 10 minutes at 1,200 to 1,400 rpm. This allows the aspirator to remove all air from the system.
3. Add coolant to raise the level to be visible in the aspirator on the outside of the supply tank.

Leaks may occur between the radiator core and the supply chamber. Such leaks allow air to replace coolant, with overheating and failure the possible result.

One should test for leaks by the following method:

108 / Diesel Engine Service

1. Take off the filler cap, and run the engine at 1,200 to 1,400 rpm for 10 minutes. This should deaerate the system.
2. While the engine is running, add coolant up to the filler opening bottom. Stop the engine.
3. Drain 1 gallon of coolant. This may be saved.
4. Again start the engine and run for 10 minutes at 1,200 to 1,400 rpm. Observe the coolant level.
5. Stop the engine and watch to see if the coolant level rises much. If it does, an internal leak is present and repair is necessary. If it does not, the radiator is OK.

Water Pumps

Cooling water pumps are usually belt-driven, although gear-driven water pumps are becoming popular. They are mounted on the radiator end of the engine and discharge directly into block coolant passages. [**Fig. 5-5**]

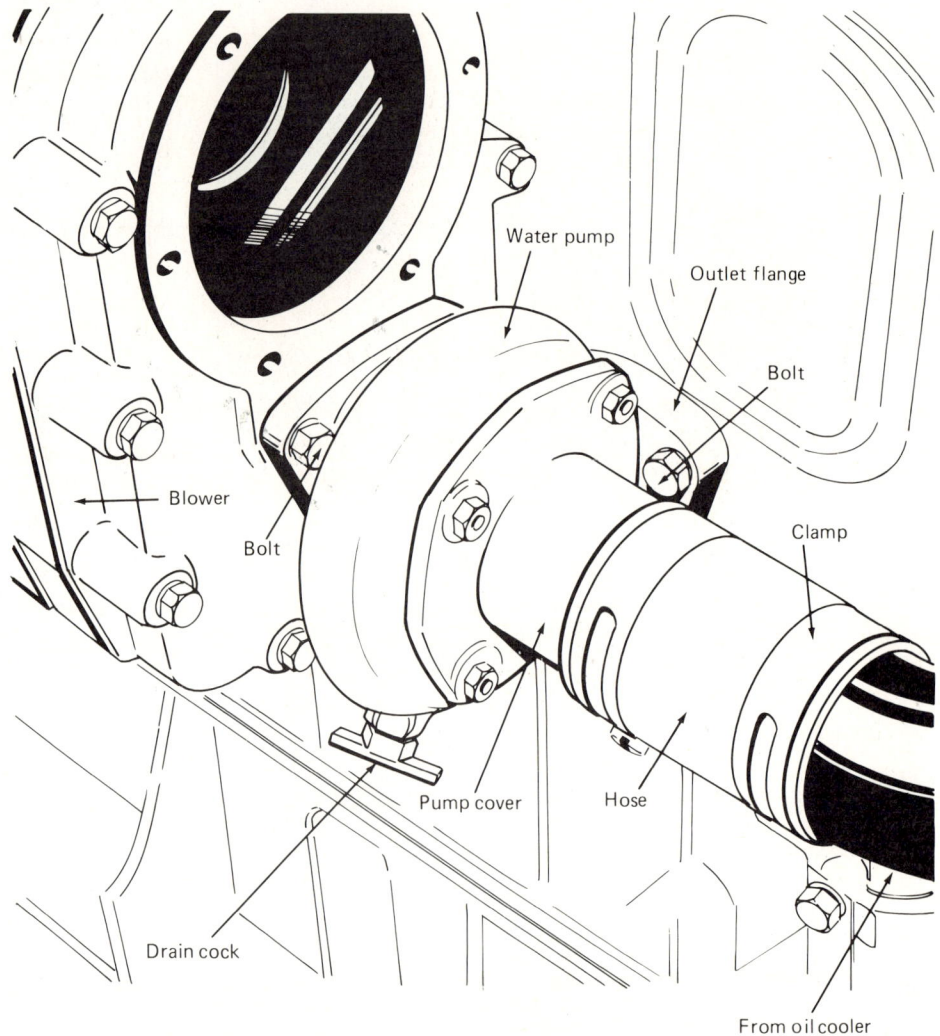

FIGURE 5-5. Typical Water Pump. *(Courtesy of General Motors Corp.)*

The Cooling System / 109

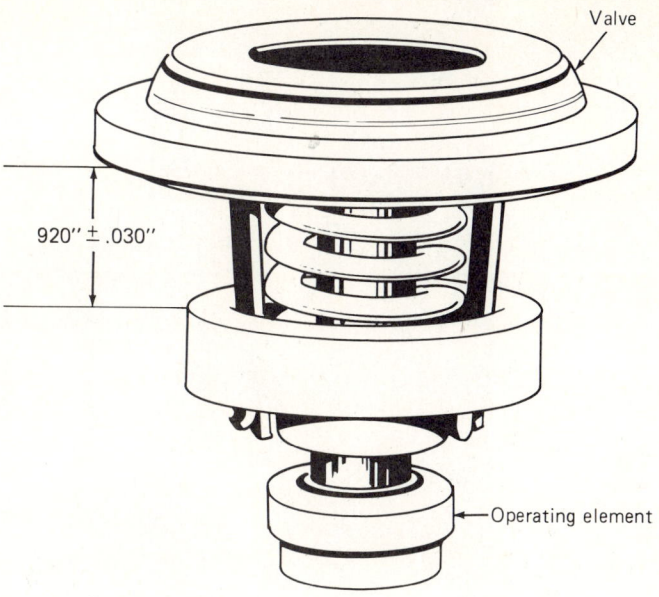

FIGURE 5-6(a). Thermostat. *(Courtesy of General Motors Corp.)*

Water pumps must move coolant through the engine. Resistance to flow is provided by block passage friction, but the chief restriction is the temperature-controlled valve called the *thermostat*. **[Fig. 5-6(a)]**

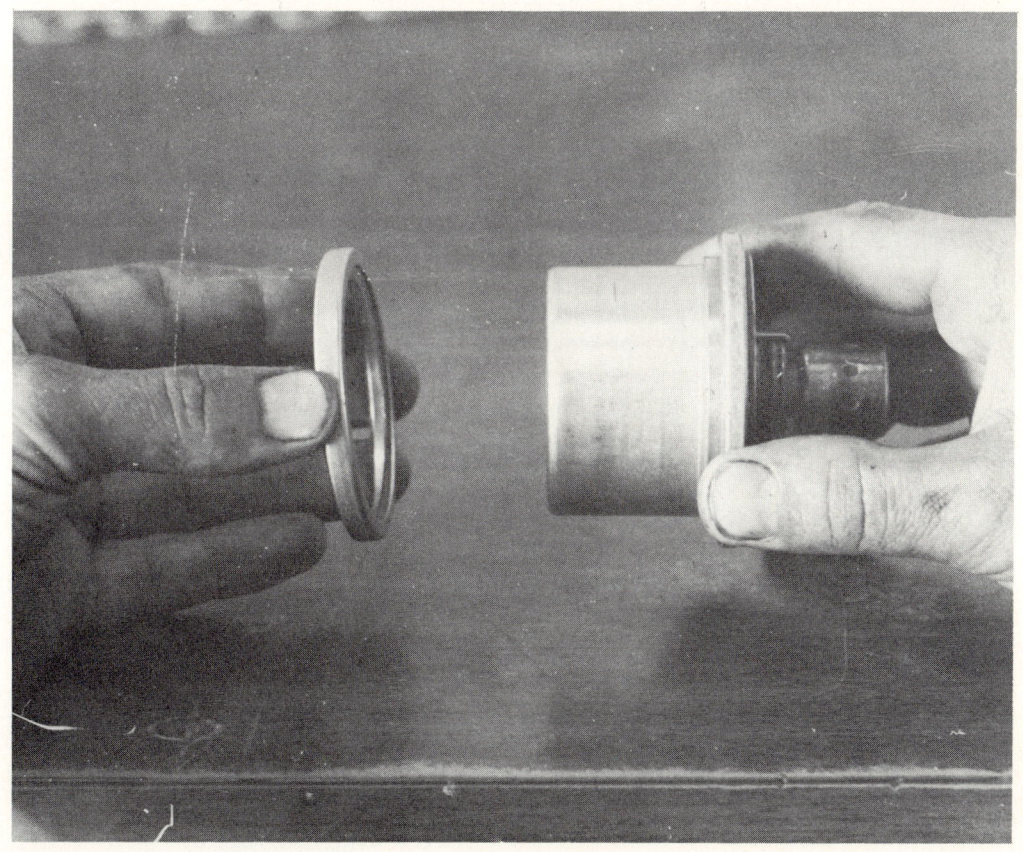

FIGURE 5-6(b). Thermostat and Seal. *(Courtesy of Cummins Engine Co., Inc.)*

FIGURE 5-7. **Engine Side with Aftercooler.** *(Courtesy of Caterpillar Tractor Co.)*

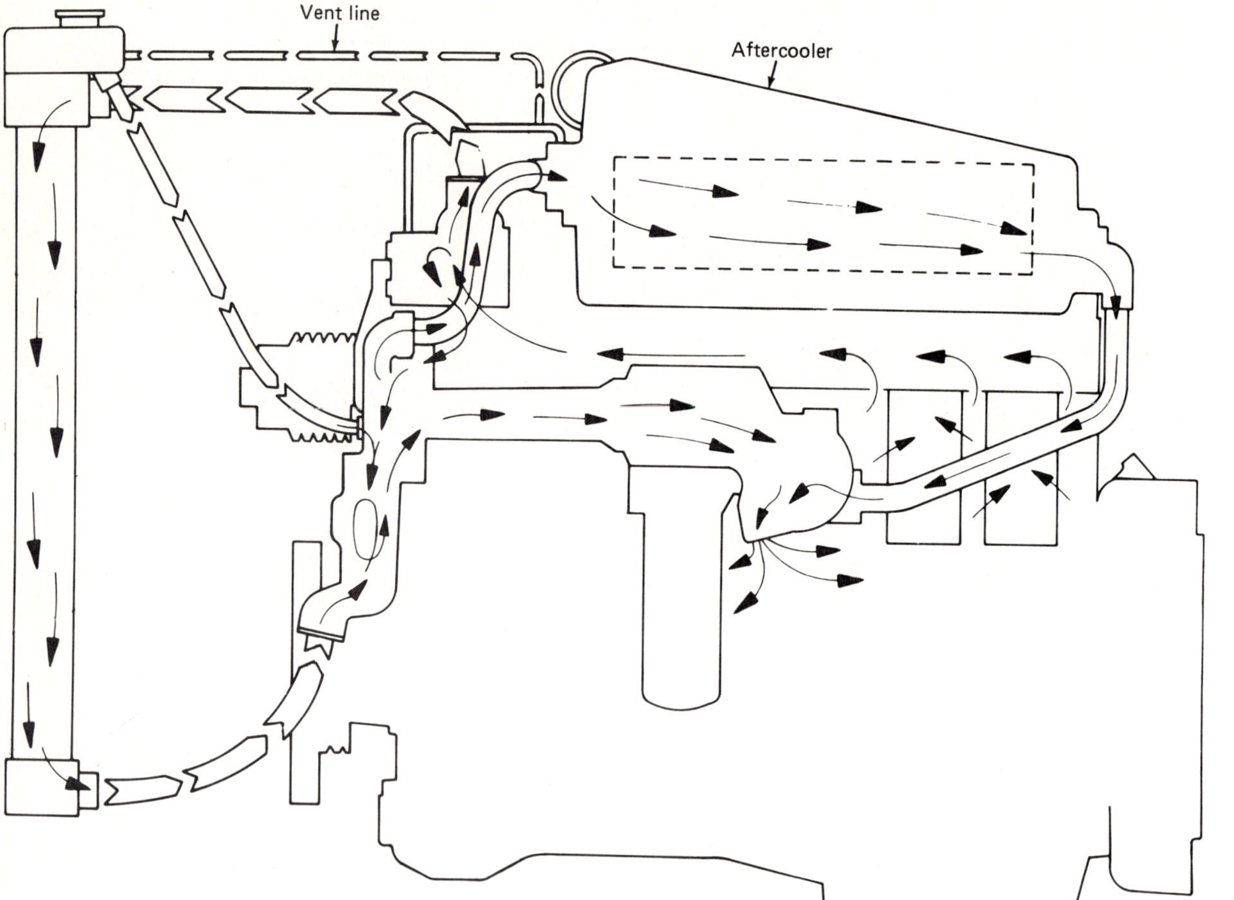

FIGURE 5-8. **Complete Coolant Flow.** *(Courtesy of Caterpillar Tractor Co.)*

Thermostat

The thermostat acts to direct coolant back to the water pump inlet while the engine is cold, thus bypassing the radiator. As the engine warms, the thermostat opens the passage to the radiator and closes the one to the bypass.

The thermostat operates over a 15° F range. That is, a thermostat marked to be fully open at 180°F will start to open at 165° F.

The restriction of the thermostat causes pressure on the coolant in the block. This pressure can be as high as 40 psi and bears on all coolant from the water pump to the thermostat. Because water under pressure conducts heat faster than it can without pressure, heat transfer is improved.

Coolant under pressure will not boil readily, and corrosion is reduced by pressure. Thus, the thermostat must be left in, even though one may consider it useless in hot weather. In fact, some makers offer optional orifice plates to use if the thermostat is left out.

FIGURE 5-9. Aftercooler Connections. *(Courtesy of Cummins Engine Co., Inc.)*

112 / Diesel Engine Service

Coolant Flow

Note: Engines with aftercoolers vent from the thermostat housing to the aftercooler and from the aftercooler to the radiator top tank. This serves to vent air from the aftercooler. [**Fig. 5-7**]

The coolant flow through the entire cooling system is as follows [**Fig. 5-8**]:

1. radiator outlet at bottom
2. water pump inlet
3. water pump to oil cooler
4. oil cooler to engine block
5. block to cylinder head
6. head to thermostat housing
7. back to radiator top tank or bypass

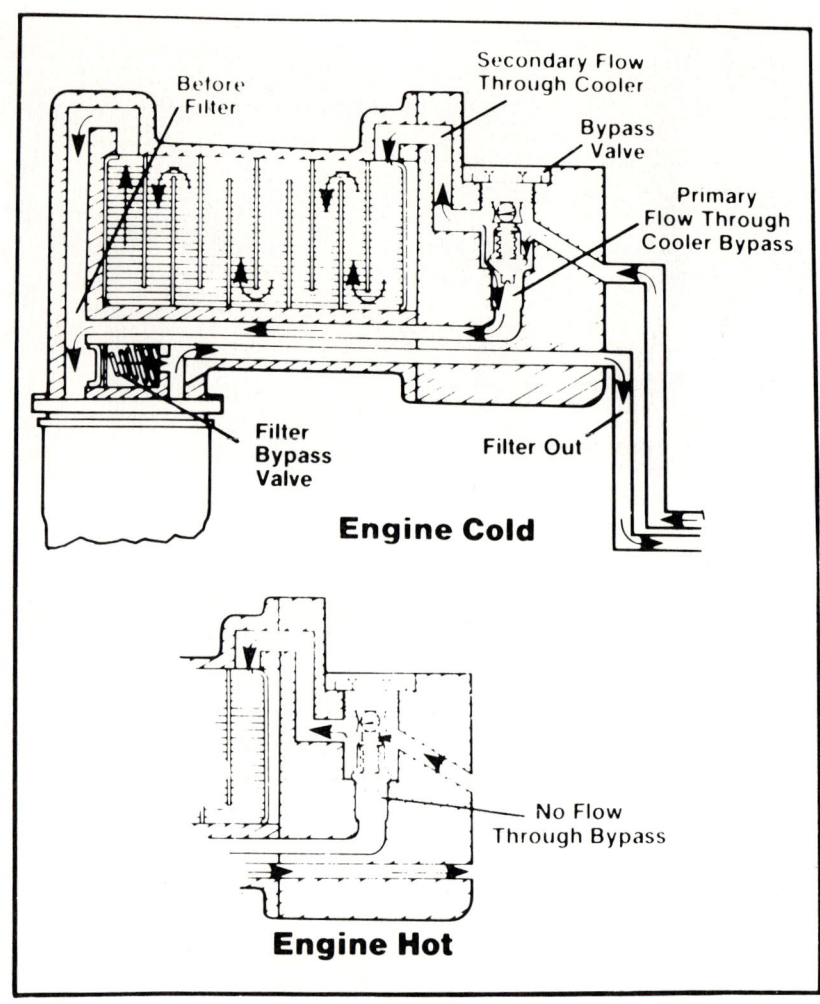

FIGURE 5-10(a). Oil Cooler. *(Courtesy of Cummins Engine Co., Inc.)*

The Cooling System / 113

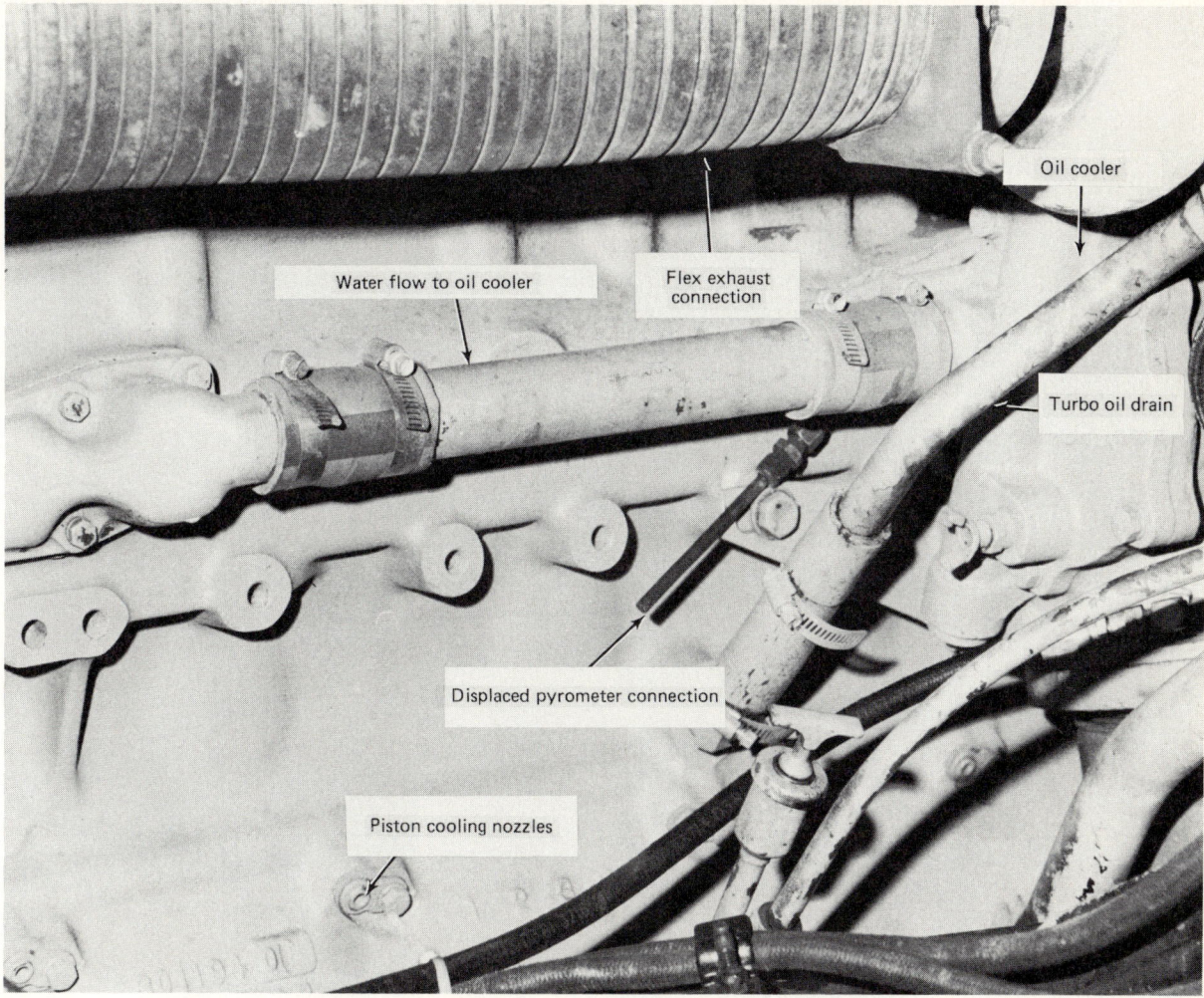

FIGURE 5-10(b). Water Connection to Oil Cooler. *(Courtesy of Cummins Engine Co., Inc.)*

Aftercooler

As described, the air from the turbocharger discharge is heated by being compressed in the air impeller of the turbocharger. In fact the temperature of the compressed air can be as high as 300° F.

Engine efficiency can be increased by cooling the intake air, thus increasing its density. To do this, an air-to-coolant heat exchanger, called an *aftercooler* or *intercooler,* is used. These two terms mean the same thing. **[Fig. 5-9]**

Coolant at 180° to 190° F is directed over a radiatorlike core, through which the air passes as it flows from the turbocharger to the engine intake. The heat of the air is thus transferred to the cooler coolant, because heat always flows from hot to cold.

Coolant piping may lead from the water pump to the aftercooler or from the rear of the block back to the aftercooler, then to the block. In either case the coolant is cooler than the air, and heat is transferred.

OIL COOLER

Engine coolant is also directed through a heat exchanger called the *oil cooler*. Oil picks up heat from the working parts as it circulates and is considerably hotter than the coolant. [**Fig. 5-10**]

Oil coolers usually consist of a tube assembly through which coolant flows, while the oil flows over the tubes. The tube assembly is contained in a housing, and coolant from the water pump is piped to it. Coolant then flows into the block passages. [**Fig. 5-10**]

Fan and Drive Belts

The use of air flow through the radiator to remove heat from the coolant makes the fan necessary. The fan and its drive are considered a part of the cooling system. [**Fig. 5-11**]

FIGURE 5-11. Fan and Belts. *(Courtesy of General Motors Corp.)*

The Cooling System / 115

The fan is most necessary on all stationary engines and most applications except on highway vehicles. Designers have recognized that the air flow created as the vehicle moves down the road is usually adequate for cooling and may overcool if weather is cold.

Thus, various controls are used on vehicles to modify air flow and to suit the cooling rate to the engine temperature. All of these are controlled by some form of temperature-sensing system, which may be either electrically operated or service air operated.

Shutters

Air flow through the radiator can be controlled by shutters, which are usually air-operated from a service air line controlled by a temperature-sensing valve installed in the thermostat housing. There are two basic types of shutters [**Fig. 5-12**]:

1. nonmodulating, which means that they are either open or closed
2. modulating, which means that they open enough to admit some air but restrict the air flow to control cooling

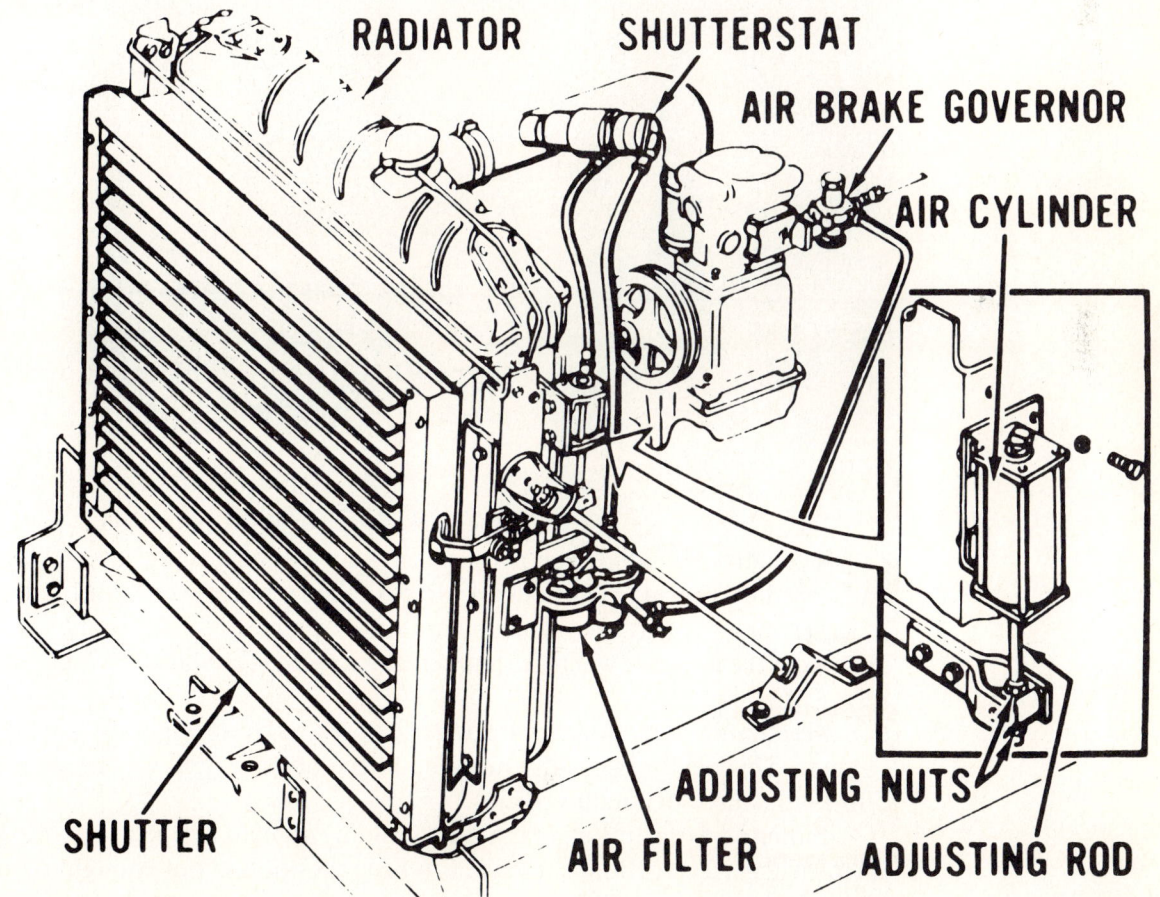

FIGURE 5-12(a). Shutters and Controls Schematic. *(Courtesy of General Motors Corp.)*

116 / Diesel Engine Service

FIGURE 5-12(b). Shutterstat.

The nonmodulated type is opened by air pressure acting on a cylinder and closed by a spring. The control valve is called a *shutterstat*. [Fig. 5-12(b)] It senses the temperature of the coolant on the engine side of the thermostat and opens the air valve when the temperature reaches the value at which its adjustment is set.

Modulating-type shutters are directly operated by a temperature-sending element. This device is usually installed in the bottom tank of the radiator. This kind is used on small vehicles.

Either type of shutter is capable of holding coolant temperature within a narrow range. The action is intended to keep the coolant hot enough to open the engine thermostat fully.

The Cooling System / 117

Fan Drive Clutches

Many vehicles are equipped with clutches on the fan hub, which disengage when cold and engage to drive the fan as the temperature increases.

Passenger cars have a viscous fluid sealed in a finned housing. These units sense the temperature of the air after it has picked up heat from the radiator. [Fig. 5-13]

Other systems have a magnetic clutch, which is activated to drive the fan when its controlling thermostatic switch is closed by coolant temperature.

Later designs have fan clutches engaged by air pressure. These are controlled either by heat sensors or by manual control.

Both types drive only when needed. Because the fan is the greatest accessory load on the engine, the action of these drives saves power and fuel.

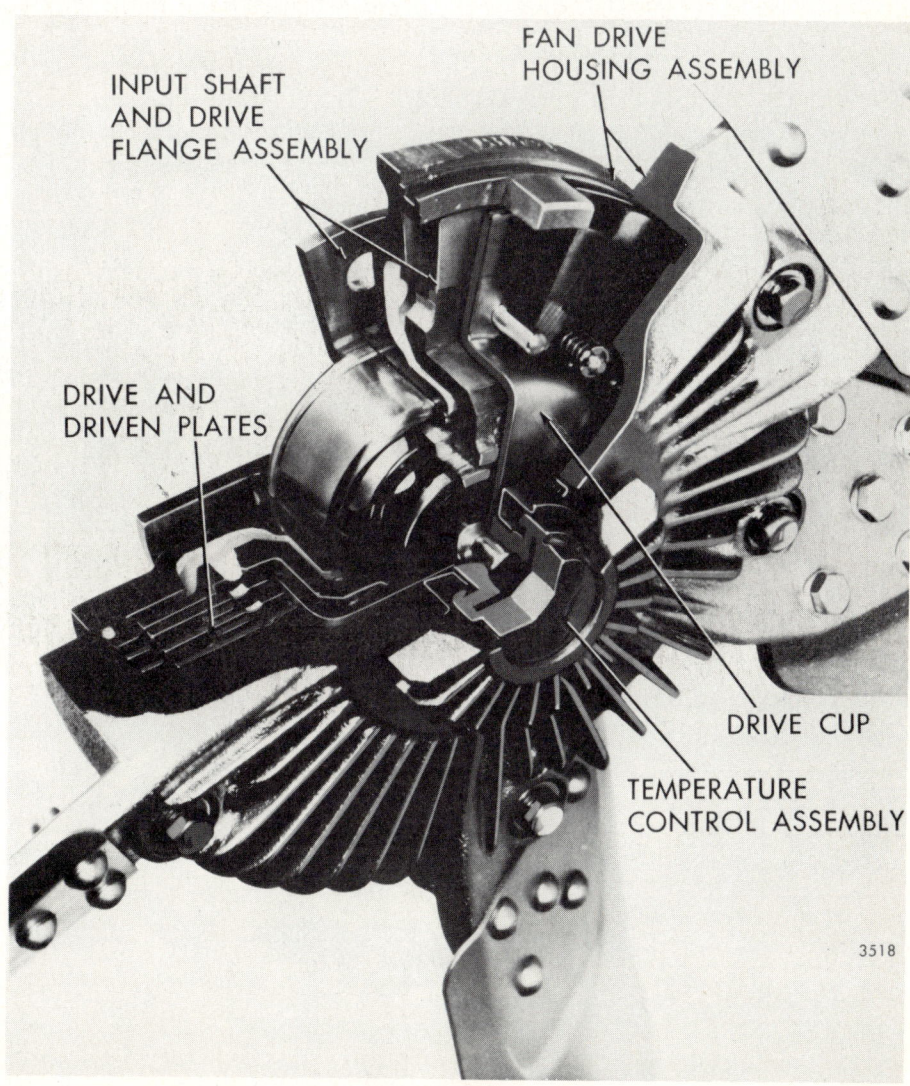

FIGURE 5-13. Viscous Fan Clutch. *(Courtesy of General Motors Corp.)*

118 / Diesel Engine Service

Motor Driven Fans

Most European vehicles and some American makes drive the fan with a separate motor, which is controlled by a temperature-sensing switch in some part of the coolant system. These fans run only when needed and impose no load on the engine. [Fig. 5-14]

COOLING SYSTEM SERVICE

All of the components of cooling systems can be replaced. Water pumps are exchangeable, as are shutterstats and fan clutches. Thermostats wear and should be replaced with new units at overhaul periods. Thermostatically controlled switches can be replaced as needed or at engine overhaul.

One of the primary causes for water pump replacement is leakage from

FIGURE 5-14. Motor Driven Fan. *(Courtesy of Ford Motor Co.)*

the internal pump seals. Such leakage should be repaired promptly, because air can be drawn in during operation, accelerating corrosion in the engine. Water pump impellers can be eaten away, either by air drawn in or by restricted water inlet passages.

Drive Belts

All drive belts must be checked during a tuneup for wear, cracking, and looseness. Pulley sheaves can be worn if a dusty atmosphere prevails. [**Fig. 5-15**]

The replacement of drive belts is common and is usually not complicated. The mechanic should always loosen the adjusting device fully and guide new belts on with a rag, rather than prying them on with a screwdriver.

Standard belt tensions are given in all manuals. Deflection of the belt one thickness for each foot of pulley separation is a good practical rule.

Any belt will lose tension after 10 minutes or more of operation. It should then be readjusted. The adjusting device should be tightened evenly to realign the pulleys. Adjustment of belt tension will pull the pulley to the drive side; the fasteners must be tightened starting at the side away from the drive.

Multigroove Belts

Later engines are furnished with belts that have several small V grooves, rather than one V shape. These are good belts and allow power to be transmitted with little loss. [**Fig. 5-16**]

Multigroove belts are vulnerable to wear from dust. Reasonable inspection frequency is necessary. Adjustment standards are the same as for other V belts.

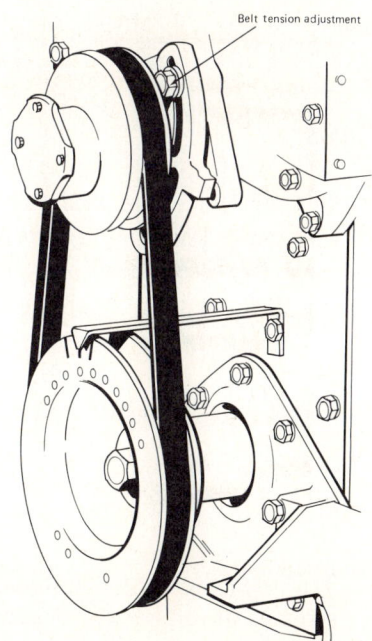

FIGURE 5-15. Drive Belts. *(Courtesy of General Motors Corp.)*

120 / Diesel Engine Service

FIGURE 5-16. Multigroove Belts. *(Courtesy of Cummins Engine Co., Inc.)*

Testing Pressure Caps

Several devices are available for testing the opening pressure of radiator pressure caps. The opening pressure is usually marked on the cap. Over a service time, the spring gradually loses its force, and the seal wears. Caps that are found more than 2 psi below the marked value should be replaced.

The smaller the radiator, the higher its operating pressure is. Passenger car pressure caps may be rated as high as 15 psi. Large truck caps are seldom above 7 psi, and most are 4 psi. [**Fig. 5-17**]

Aftercooler

This heat exchanger seldom gives trouble. There are no moving parts, and the stresses of temperature changes are the only forces that cause problems.

A leak of coolant into the intake air will cause white to gray smoke in the exhaust; a severe leak may allow coolant to fill a cylinder.

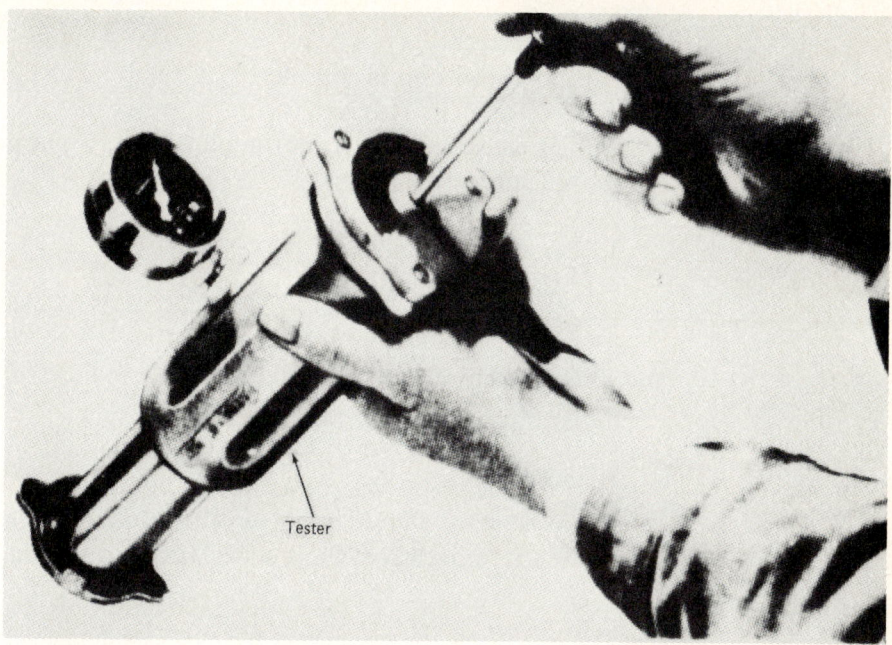

FIGURE 5-17. Test Pressure Cap. *(Courtesy of Cummins Engine Co., Inc.)*

Engine overheating may follow a coolant loss over an operating period. Any system that requires coolant addition at frequent intervals should be checked for leaks.

Aftercooler elements can be repaired in the same manner as radiators.

External leaks from gaskets or coolant connections can be checked visually.

The use of about 15 psi of air pressure applied to the radiator at the filler opening will aid in disclosing any leakage. Higher pressure should not be applied.

Cab Heaters

Leaks in cab heaters usually show up as coolant on the cab floor or external leaks from hose connections. These points will show up when the system is pressurized or by visual checks.

Hoses

All coolant hoses should be changed at least yearly. A hose that looks good on the outside may be softened and shedding particles on the inside. Not only are coolant leaks possible, but any particles may lodge in such small passages as in the radiator, aftercooler, and oil cooler.

Oil Cooler

Although this unit is considered as a part of the lubricating system, coolant flows through it. This is the primary source of coolant in the oil. [**Fig. 5-18(a)**]

Because leakage from the oil cooler is internal, the cooler must be removed for testing.

1. Drain the coolant from the engine.
2. Disconnect all water connections.
3. Remove the oil filter, if located on the oil cooler.
4. Remove the cooler housing.
5. Remove the cooler tube element.

At this point, we can use an old cooler housing to apply low air pressure to the element. Should only one or two tubes that leak be found, the mechanic can block both ends of those tubes with solder. [**Fig. 5-18(b)**] If more than two

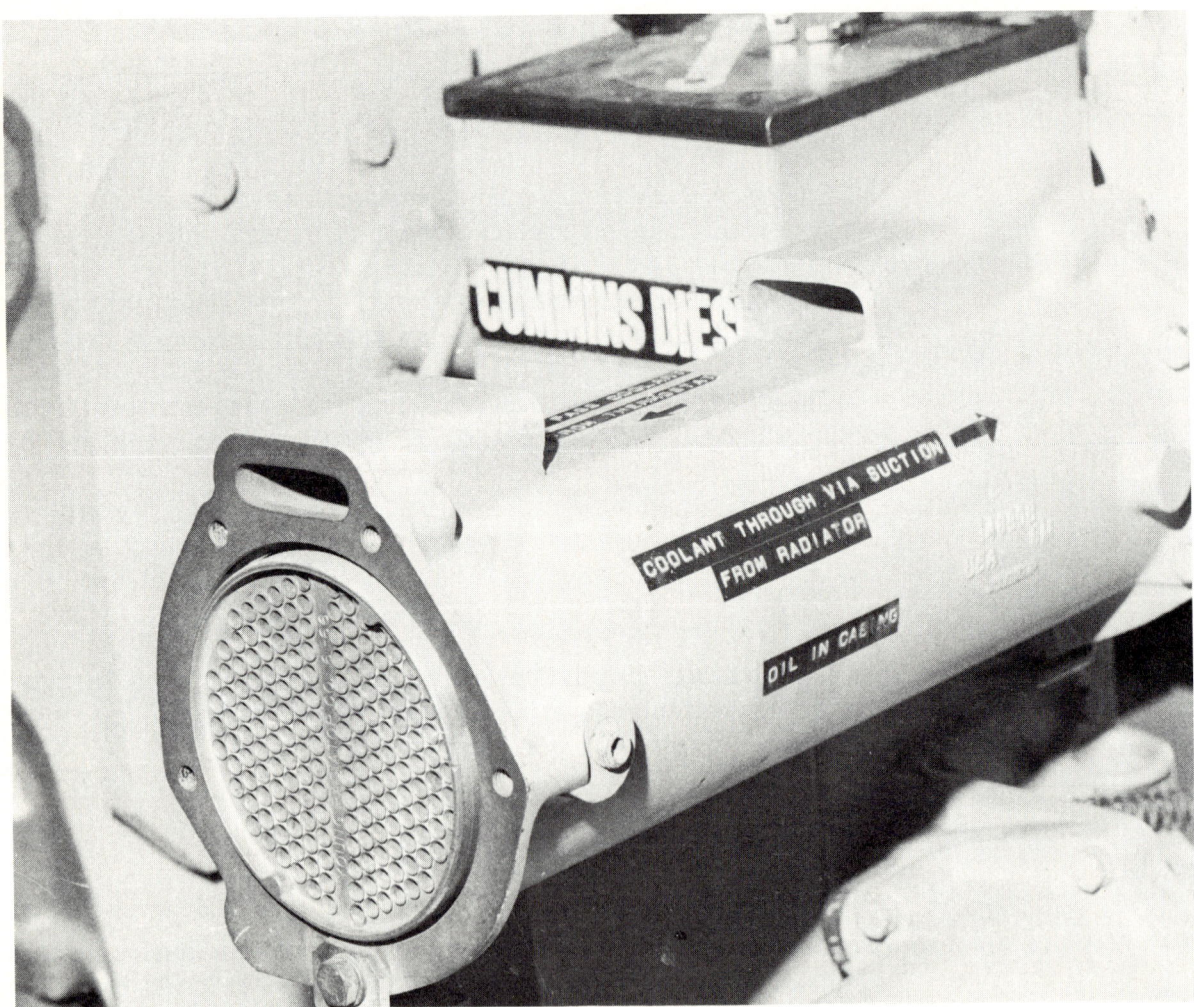

FIGURE 5-18(a). Oil Cooler. *(Courtesy of Cummins Engine Co., Inc.)*

FIGURE 5-18(b). Repair Oil Cooler. *(Courtesy of Cummins Engine Co., Inc.)*

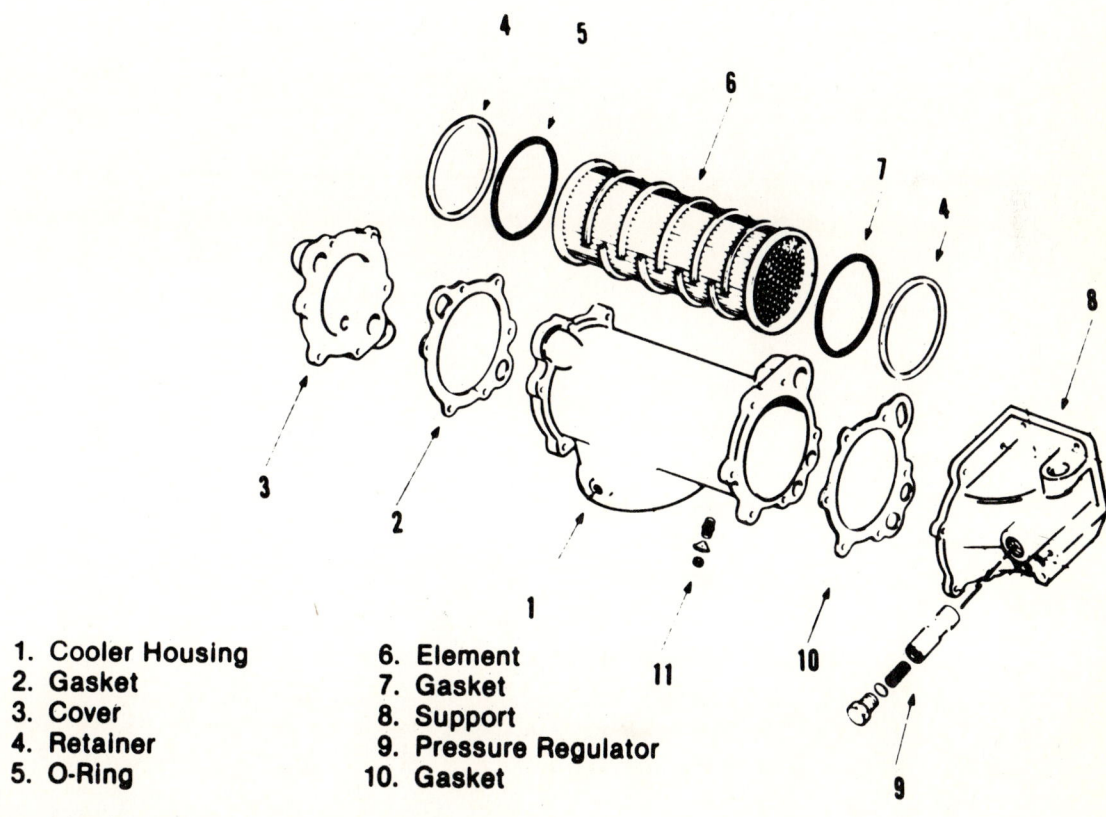

1. Cooler Housing
2. Gasket
3. Cover
4. Retainer
5. O-Ring
6. Element
7. Gasket
8. Support
9. Pressure Regulator
10. Gasket

FIGURE 5-19. Oil Cooler. *(Courtesy of Cummins Engine Co., Inc.)*

124 / Diesel Engine Service

tubes leak, the element must be replaced. The oil cooler should be reassembled with new seals and gaskets and then installed. [**Fig. 5–19**]

Cooling System Service

Objectives in servicing the cooling system are to keep it clean, full of coolant, functioning to protect the engine from overheating, and operating at the proper temperature. A daily check of coolant level, prompt repair of any leaks, and general inspection of all components, will allow continued service and reduce down-time.

The mechanic should refer to the trouble-shooting section for other service needs.

THE LUBRICATING SYSTEM

6

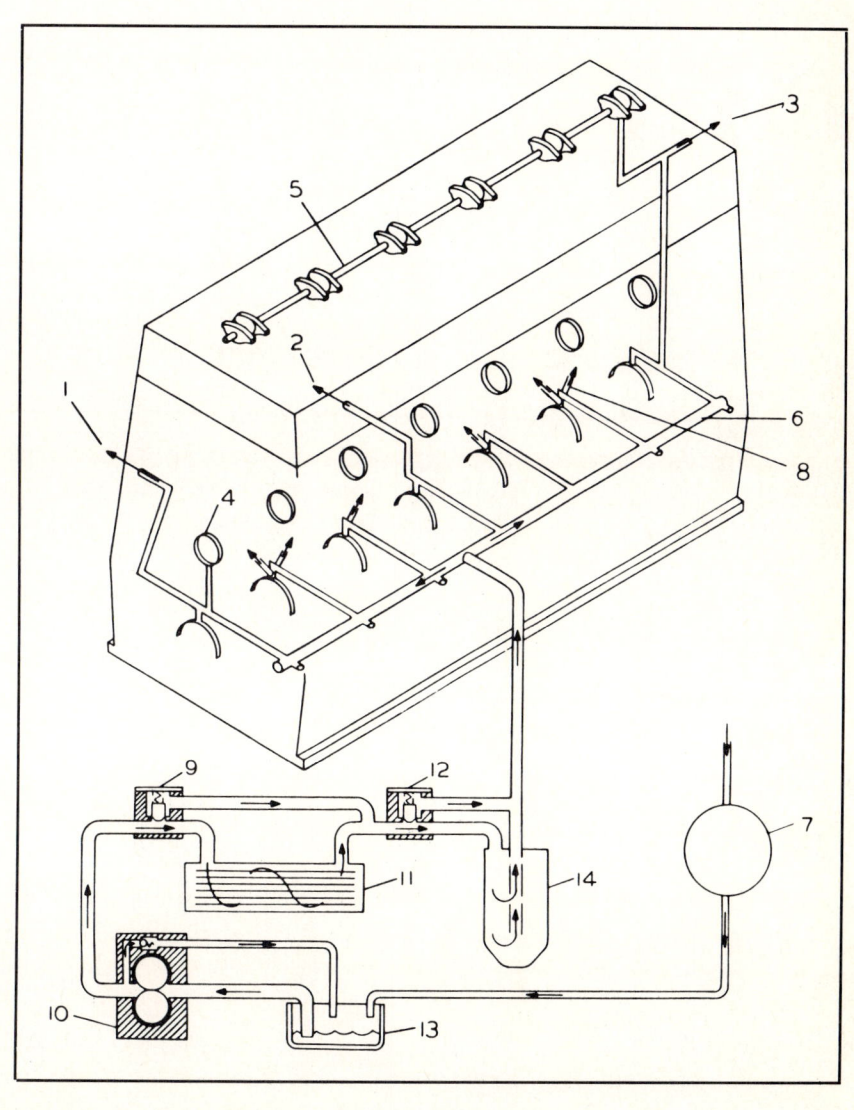

FUNCTIONAL DESCRIPTION

Aside from periodic oil and filter changes, the lubricating system is usually trouble-free. Many symptoms that require organized trouble-shooting methods have been pointed out. This chapter traces the oil flow and points out the causes of various complaints.

A typical oil flow diagram shows the general routing of lubricating oil through the engine. Because all commercial diesel engines are designed to supply oil under pressure to most working parts, many passages convey oil to the working surfaces. **[Fig. 6-1]**

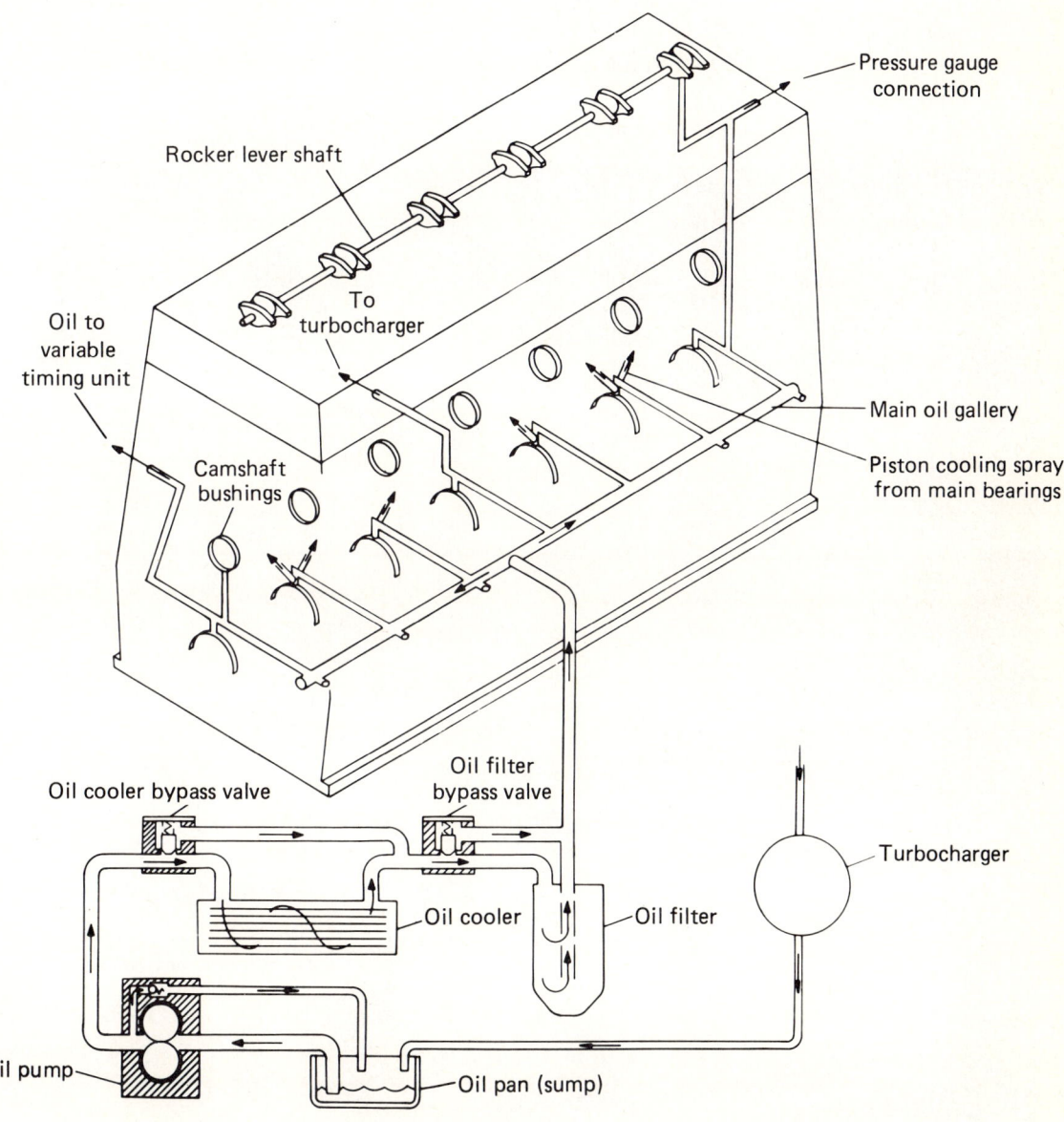

FIGURE 6-1. Typical Oil Flow Diagram. *(Courtesy of Caterpillar Tractor Co.)*

USES OF OIL PRESSURE

Although many people think of oil pressure as supporting the engine bearings in order to prevent failure, the oil pressure serves to charge the bearings with oil during the unloaded part of their cycle. Load pressure in the loaded part of the cycle prevents any oil from entering the working area.

Thus oil pressure provides a supply to the bearing clearance during the exhaust and intake strokes; that supply must resist the squeezing action of the load during compression and power strokes. It is significant that connecting rod journal wear is always located on the exhaust side of the journal. This phenomenon is caused because the oil that was injected under pressure during exhaust and intake has been squeezed out during compression and power, and the oil film is thus thinned.

Other lubricated wear points act in a similar way. Thus, oil pressure must be maintained.

FIGURE 6-2. Typical Oil Pan *(Courtesy of Caterpillar Tractor Co. and Empire Machinery Co.)*

Oil Sump or Pan

A *sump* is a reservoir. In the case of an engine the bottom of the open crankcase is used to support a pan in which the oil is contained. A screen is applied over the oil pump suction point; baffles may be present to prevent excessive oil movement. [**Fig. 6-2**]

Oil level in the sump is checked by a marked dipstick. There must always be enough oil to supply the oil pump under all operating conditions.

Because the oil expands when heated, one should always check oil level before starting the engine. It should never be checked just after shutdown, because a fair quantity can still be in the passages and surfaces, not drained down into the sump. [**Fig. 6-3**]

The mechanic must avoid overfilling the sump, because the connecting rods must not come within ½ inch of the oil as they pass over it at bottom center. Excessive oil consumption will surely result if the oil level is too high.

Nothing is gained by trying to keep the oil level at the high mark on the dipstick. It is safe as long as it is above half the distance between high and low. [**Fig. 6-3**]

Oil Pump

The main oil pump is usually a gear-type positive displacement unit. It is driven from the gears in the gear end of the engine and may be mounted on a main bearing cap or mounted externally on the gear housing. [**Fig. 6-4**]

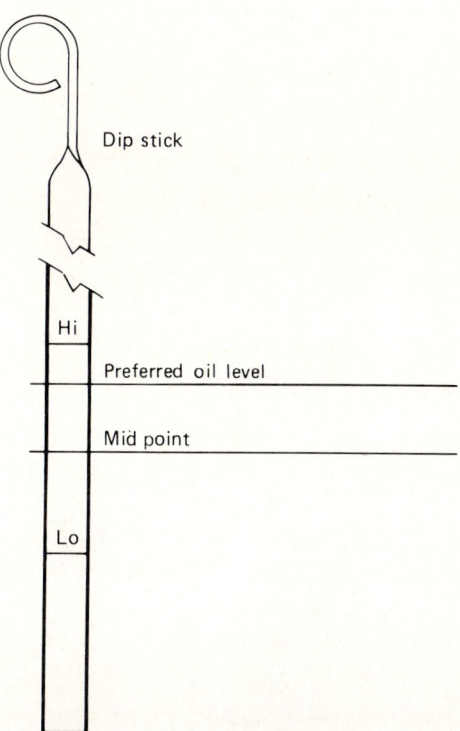

FIGURE 6-3. Dip Stick Markings. *(From a sketch by P.M. Uhl)*

130 / Diesel Engine Service

FIGURE 6-4. Typical Oil Pump. *(Courtesy of General Motors Corp.)*

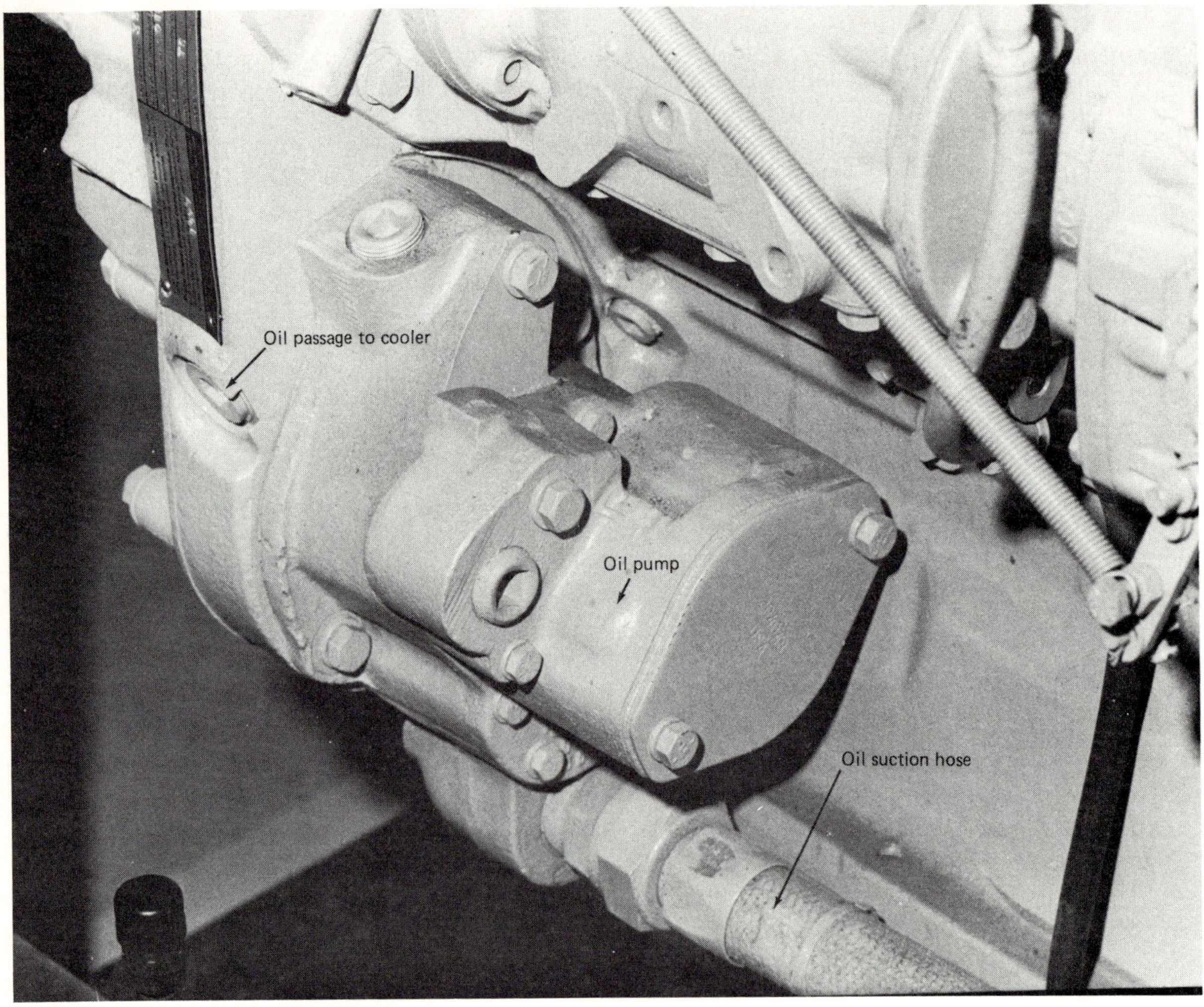

FIGURE 6-5. Externally Mounted Oil Pump. *(Courtesy of Cummins Engine Co., Inc.)*

A suction tube is mounted on the oil pump inlet and ends in the low part of the sump with a screen to bar coarse dirt from the pump. It is usually attached to the pumps by a flange.

Oil pumps are trouble-free, because they are well lubricated. Abrasives, such as sand and metal particles, must be kept out of the sump in order to reduce the chance of failure.

The pressure generated by the oil pump is controlled by a regulating bypass or relief valve, which may be in the pump housing or at another point in the oil system, close to the pump. When the oil pump is externally mounted, the suction tube is connected to a mounting pad on the side of the sump. A screen is used over the opening inside the sump. [**Fig. 6-5**]

Oil Suction Tubes

Various designs of tubing are used at the oil pump suction; some of them can leak, allowing air to enter the pump suction. This fault will result in loss of oil flow and engine failure.

All suction connections must be tight and any seals in good shape. Flexible suction hoses must be replaced at engine overhaul. The hose fittings must be assembled without damage to the hose lining.

From the oil pump, oil flows through passages or tubing to the oil cooler, then the oil filter. Some systems use a bypass valve to route the oil around these two units when it is cold. The oil is thicker when cold, and the cooler and filter restriction could reduce oil supply to the engine bearings during a cold start. The bypass closes when the oil warms up, and its pressure falls. [**Fig. 6-6**]

Oil Cooler

The oil cooler element is baffled to route oil flow around the water tubes, so that the oil contacts the tubes enough to give up its heat. Oil temperature is thus reduced about 30° F. [**Fig. 6-7**]

Oil can enter the cooling system under operating conditions because it has higher pressure than the coolant. When the engine is stopped, coolant may enter the oil because some pressure will remain in the cooling system, although there is no pressure in the oil system.

Oil Filters

Some engines filter the oil before it goes through the oil cooler, while some locate the filter after the cooler. All commercial engines have a full-flow oil filter, through which the full oil stream passes. Because these filters handle all of the oil, they are really more like strainers. They cannot keep the fine carbons from darkening the oil. [**Fig. 6-8**]

A bypass valve is necessary to keep oil flowing to the bearings should the filter become too restrictive. The more dirt the filter removes from the oil, the more restriction to flow it presents.

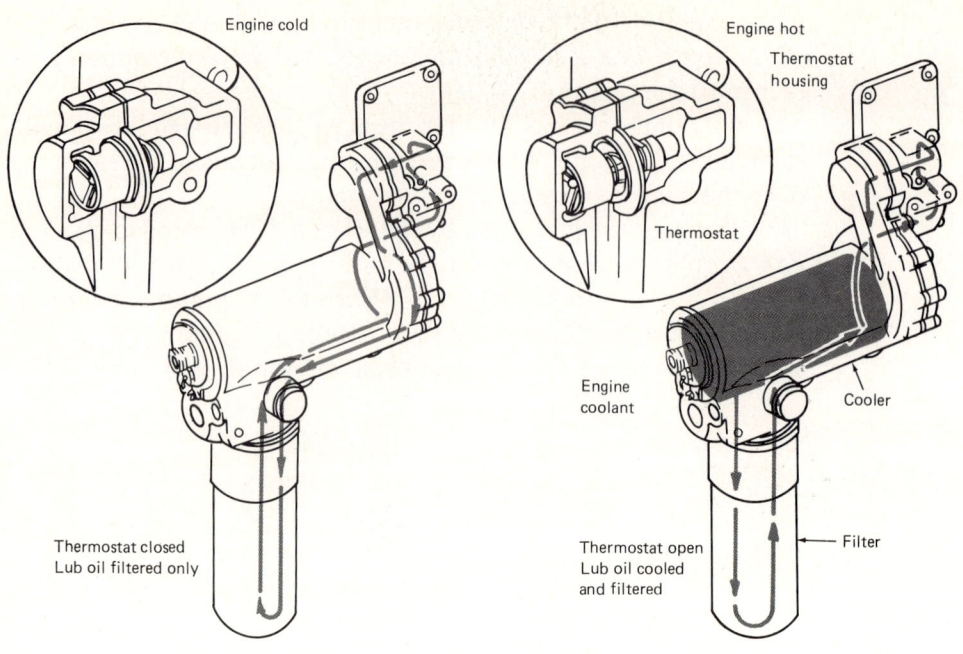

FIGURE 6-6. Oil Flow Through Cooler and Filter. *(Courtesy of Cummins Engine Co., Inc.)*

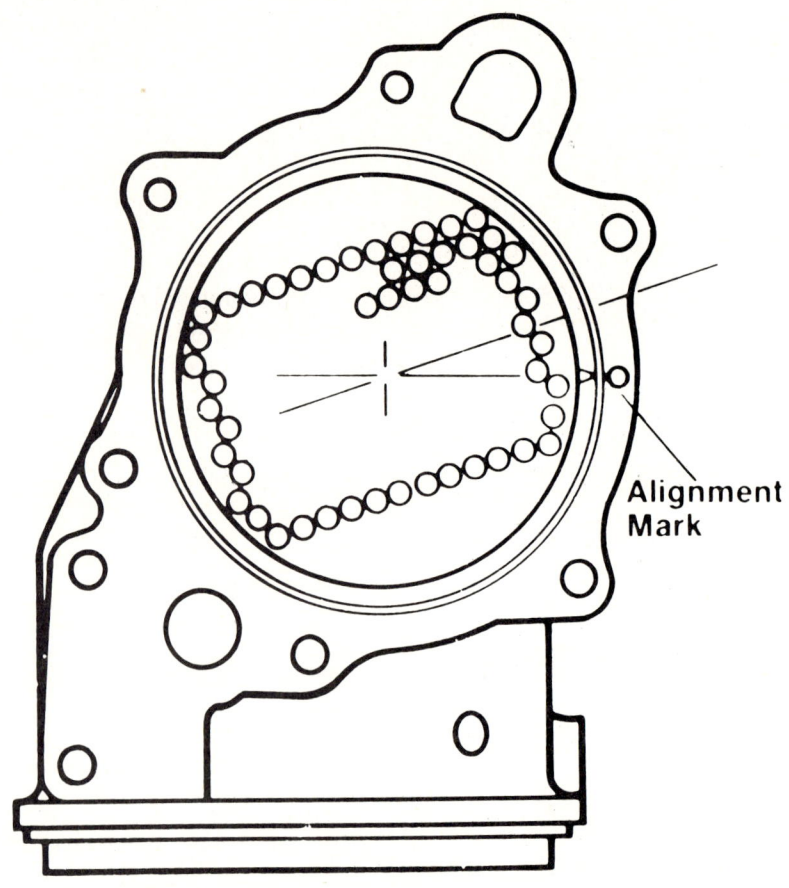

FIGURE 6-7. Oil Cooler Element Detail. *(Courtesy of Cummins Engine Co., Inc.)*

The Lubricating System / 133

FIGURE 6-8. Oil Cooler and Filter Mounting. *(Courtesy of Cummins Engine Co., Inc.)*

The bypass will open when the oil is cold and should close when the oil is warm enough to reduce its viscosity to normal. As the filter loads with dirt, the bypass stays open longer, and less oil passes through the filter.

For this reason, the filter must be changed at oil change periods; a filter change at midpoint will do no harm. It is also important to check the bypass valve when changing the filter. These valves are not heavily spring loaded, and a sticking valve can allow too much bypass. There are various designs of these valves, all of which do the same task. [**Fig. 6-9**]

Bypass or relief valves may be used at several points in the lubricating system. Such a valve is often used to bypass the oil cooler and to regulate oil pressure in the oil manifold that feeds the bearings.

Usual practice is to use a valve that opens at 15 psi for the oil filter, one of a 40-psi setting at the oil cooler, and a 60-psi pressure regulator valve. These valves are pressure *differences*. That is, the oil filter bypass opens when there is a difference of 15 psi between oil pressure at the filter inlet and that at the exit. [**Fig. 6-9**]

An oil pressure regulating valve allows oil to escape to the sump when its

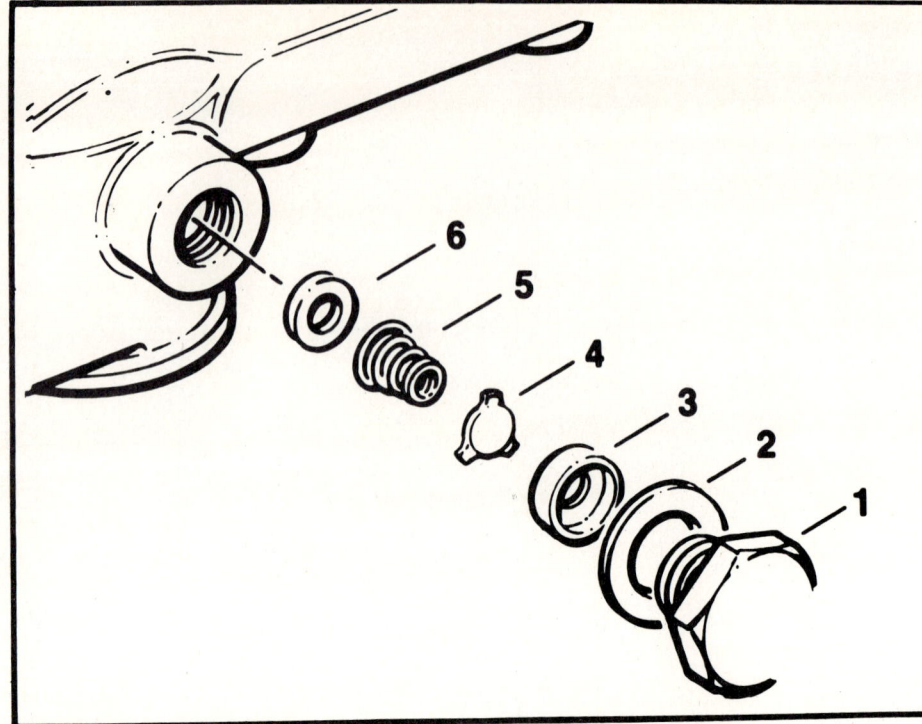

1 Plug
2 Gasket
3 Valve Seat
4 Valve
5 Valve Spring
6 Washer

FIGURE 6-9. Filter Bypass Valve. *(Courtesy of Cummins Engine Co., Inc.)*

opening pressure is reached. This keeps the oil pressure at the valve opening pressure.

Filters that receive oil from the system, filter it, and discharge to the sump without going to any working parts, are called *bypass filters*. [Fig. 6-10] These can be mounted on any part of the machine and are connected to the oil system by flexible hose. Rate of flow through these filters is low, about 1½ gallons per minute (gpm). These are depth-type filters, in comparison with the barrier-type full-flow filter. The oil must migrate through several inches of filtering material, which traps dirt mechanically.

Bypass filters can clean the oil visually clear. However, they are usually not large enough to remove all carbon completely, and the oil remains dark, although not as black as it would without such a filter.

SCAVENGING OIL PUMPS

This system is used but has been supplanted by oil pans that are designed to prevent isolation of the main oil pump suction tube. When engines must operate at steep angles fore and aft, they are usually equipped with an auxiliary

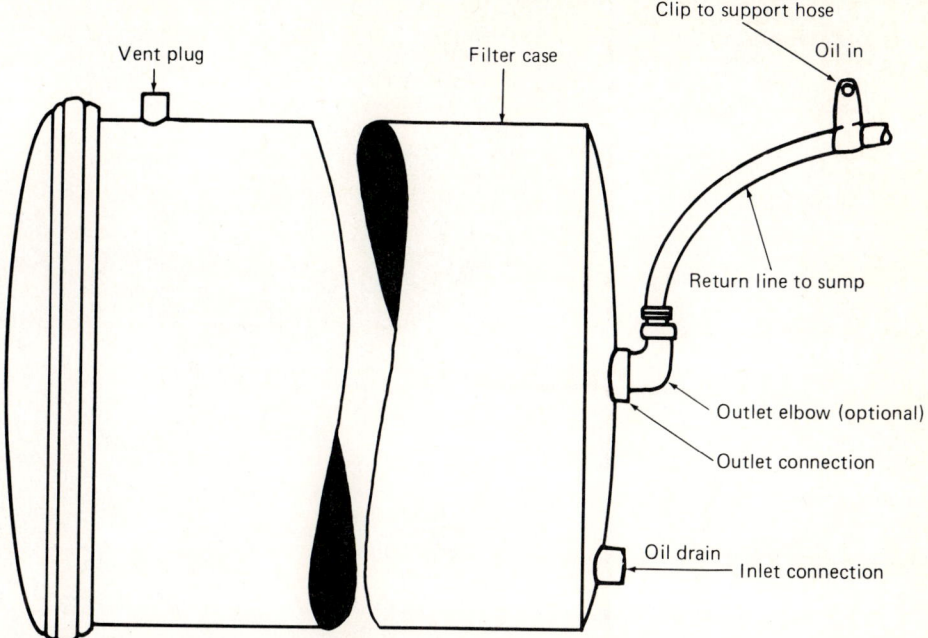

FIGURE 6-10. Bypass Oil Filter. *(Courtesy of General Motors Corp.)*

pump and a small sump at the front of the oil reservoir. The purpose of this system is to maintain the oil supply in the main sump, so that the main oil pump will not run out of oil.

The scavenging pump draws oil from the front sump and moves it back to the main sump. These pumps are usually driven from the crankshaft gear and are supplied with oil so that they do not run dry while the engine tilt does not drain oil into their sump.

TURBOCHARGER OIL SUPPLY AND DRAIN

Oil for the turbocharger is taken from a point just after the oil filter and cooler and routed to the turbocharger bearing housing by a flexible line. While the supply volume is not great, flow must be maintained. The turbocharger rotor will not tolerate any loss of lubrication. [Fig. 6-11]

Oil must drain back to the sump freely. The oil drain is a larger tube and must connect to the turbocharger close to the bottom of the bearing housing. Its opening into the crankcase should not be directly in line with a connecting rod in order to prevent any restriction to oil flow.

OIL SPECIFICATIONS

Engine builders supply complete lubricating oil grades and characteristics. Many makes and grades are available, but a general description of good lubricating and flow qualities follows.

136 / Diesel Engine Service

FIGURE 6-11. **Air Compressor Oil Flow.** *(Courtesy of General Motors Corp.)*

1. Oil grades should range from SAE10 to SAE40, depending on the climate to be encountered.
2. Multigrade oils can be used if they are produced by a recognized company. These oils are basically light oils, with chemical viscosity improvers to give good lubrication when temperatures are high.
3. Oils for diesel engines are compounded with chemicals that allow them to withstand the heat and carbon blowby that is characteristic of diesel engine use. They may be as much as 30 percent chemical.
4. Because sulfur is present in most crude petroleum, specifications for both lubricating oil and fuel requires no more than 0.55 percent by weight of free sulfur. High sulfur fuels and oil can form corrosive acids when exposed to the heat of combustion in an engine. Severe etching of machined surfaces may result.
5. Synthetic lubricating oils are offered. These oils are not based on petroleum but are chemical compounds designed to meet all lubricating require-

ments, with improved resistance to breakdown. They are expensive, depending for their sales on increased life and improvement performance. Their main advantage is the fact that they are designed to meet all engine needs and are not compromised by dependence on petroleum.

6. Branded additives of many makes have been marketed for some time. Laboratory tests have revealed that nearly all of these compounds are about 98 percent light mineral oil, with some coloring and aromatic chemicals. They have no value that could not be supplied by a good solvent, such as fuel.

Other Lubricated Parts

Other lubricated assemblies are

1. the air compressor,
2. some governors,
3. use of oil for piston cooling, and
4. working parts not pressure lubricated.

Service air compressors are lubricated by a tube connection to the oil manifold and return oil to the gear case through a passage in the mounting flange or through a drain tube to the engine crankcase. [**Fig. 6–11**]

Some governors are lubricated by engine oil, when they are a separate assembly from the fuel pump. Some hydraulic governors use engine oil as the operating medium.

Oil for piston cooling flows through a special gallery, from which nozzles direct the oil in a spray into the under side of the piston crown. A special oil pump or a separately driven pump may be used for this purpose. [**Fig. 6–12** and **Fig. 6–13(a)**]

The gear train, cylinder walls, and valve stems are lubricated by oil runoff from bearings and the rocker lever bushings, respectively.

This generalized treatment of the lubricating oil system covers the commercial engines being produced. In general all engines require the same working parts to be lubricated. Oil does several distinct functions:

1. separates working surfaces (lubricates) to prevent wear
2. cools by flowing through heated parts
3. cleans by removing carbons and other combustion products from the working parts
4. seals by covering rings, valve stems, and pressurized parts with a film that is renewed constantly

The circulation of lubricating oil protects the engine and allows it to operate for long periods before excessive wear makes overhaul necessary.

Lubricating systems are not troublesome. Only when maintenance is neglected does a failure occur.

Operating conditions account for most failures. Bearing failures may follow when low oil level is combined with turns that make the oil run away from the suction screen.

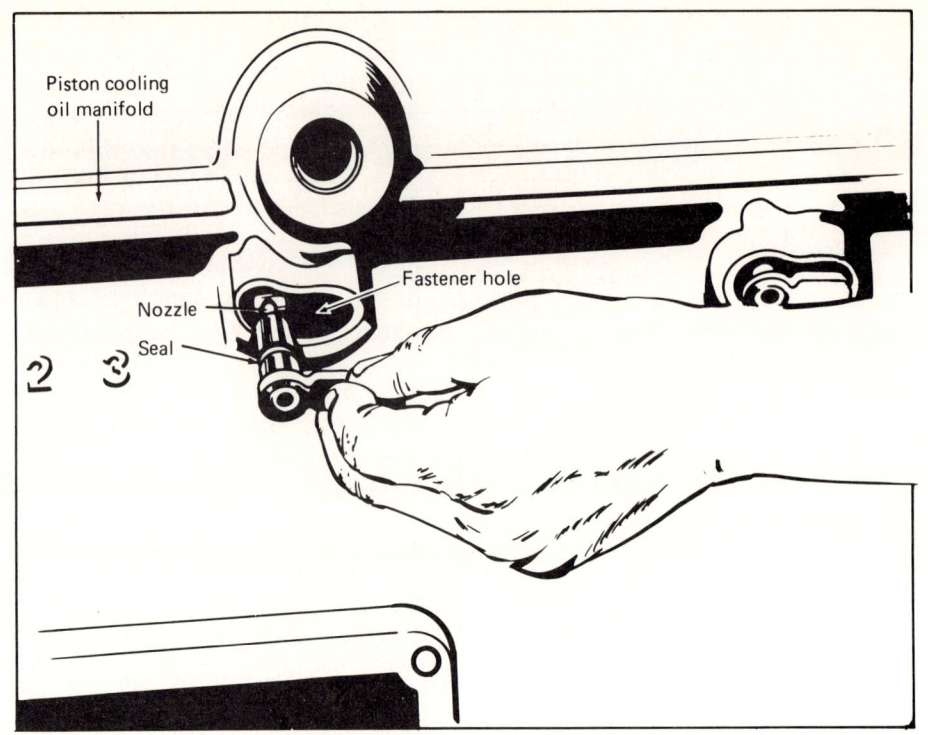

FIGURE 6-12(a). Piston Cooling Nozzle. *(Courtesy of Cummins Engine Co., Inc.)*

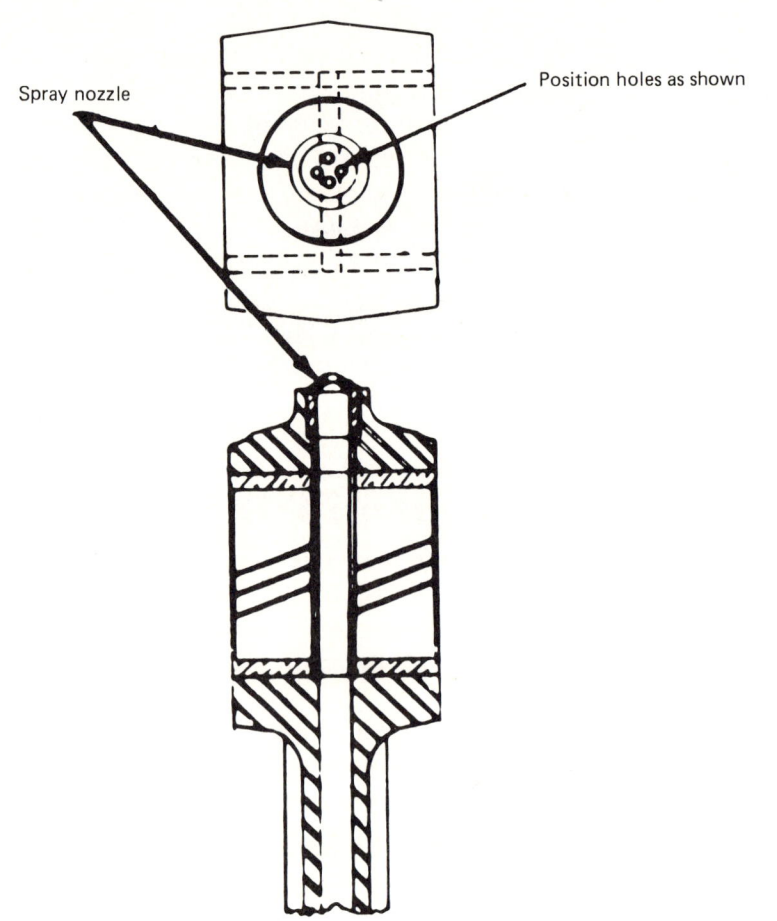

FIGURE 6-12(b). Piston Cooling Spray in Rod. *(Courtesy of General Motors Corp.)*

Engines operated on steep grades must have a scavenging system to move the oil to the suction point.

Bearings and crankshafts must match in size. The size of crankshaft journals should never be assumed. They must be measured, and the correct bearings then installed.

These rules apply to all commercial engines, regardless of size or application.

7
THE ENGINE BLOCK

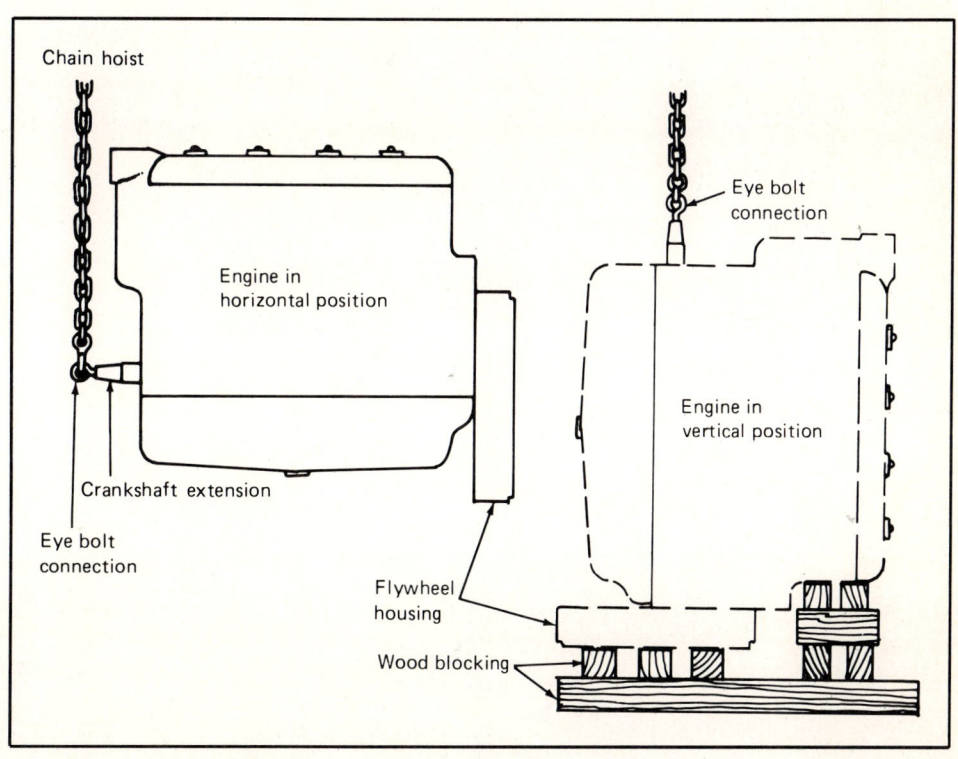

FUNCTIONAL DESCRIPTION

The block is the frame of the engine, to which all working parts are attached. Usually it is a cast-iron unit, although engines have been built from welded steel plates.

The block must be rigid to stand the high forces imposed and be machined to precise tolerances in many critical areas. The block is the most expensive part of an engine; any repairs done on it must preserve its integrity.

Each engine manufacturer has designed the blocks to suit his version of overall engine design. We point out various features, using typical designs for the class of engines that we are treating.

BLOCK FEATURES

Figure 7-1 shows a block used by Cummins and shows some of the typical mounting pads, water and oil passages, and the cylinder arrangement.

FIGURE 7-1. Typical Cylinder Block. *(Courtesy of Cummins Engine Co., Inc.)*

144 / Diesel Engine Service

Several cylinder arrangements are possible, depending on whether a vertical engine is concerned or whether it is a V-type. In the latter type the angle between banks is established by the block design.

Most commercial engines have replaceable cylinders called *cylinder liners* or simply *liners*. Some engines are being built with cylinders cast as part of the block. Four-stroke cycle engines usually are made with coolant spaces around the cylinders, called *jackets*.

Two-Stroke Cycle Blocks

Two-stroke cycle engines may feature a double-walled liner, with the space between the walls acting as a coolant jacket, or the liners may be thin sleeves that fit in cooled bores. [Fig. 7–2]

Notice the many threaded holes for attaching various components. It is important to check the threads in all such holes and attach parts with the right thread and lengths of fasteners. For this reason fasteners should be stored in

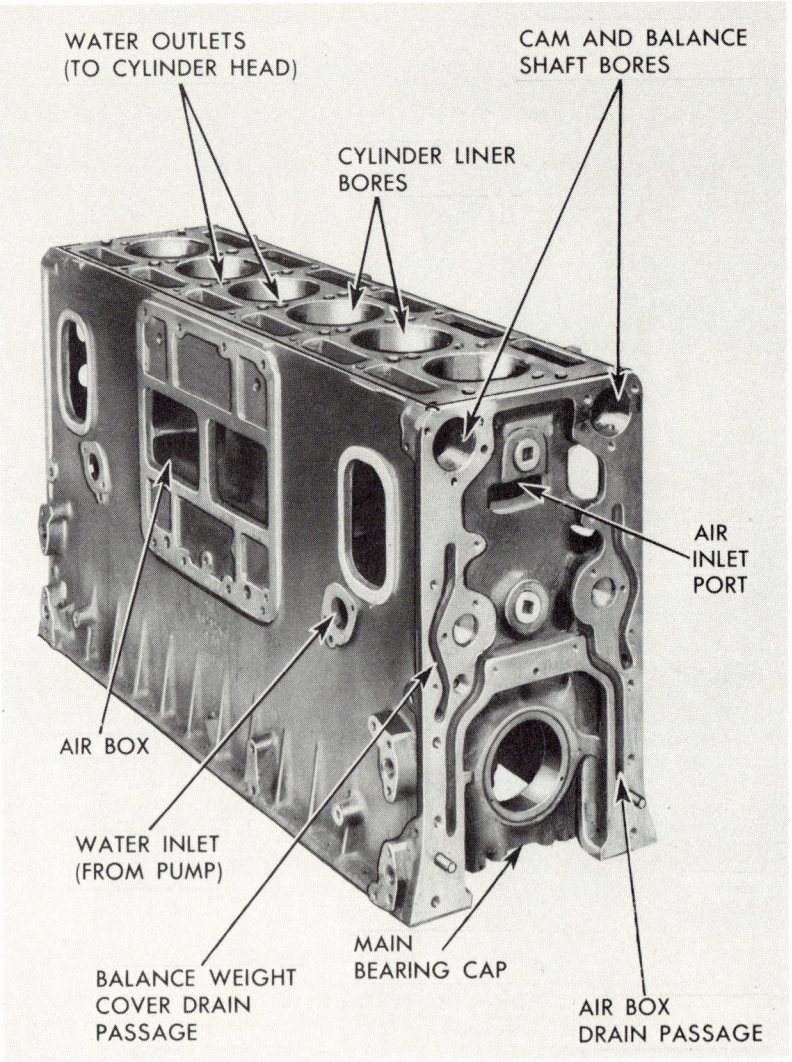

FIGURE 7-2. Two Stroke Cycle Block. *(Courtesy of General Motors Corp.)*

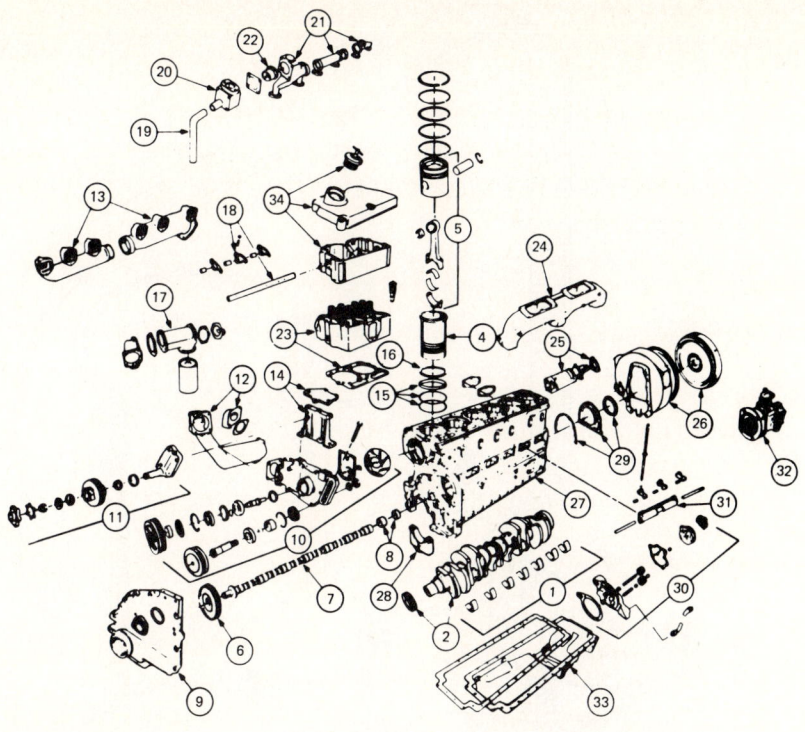

1. Crankshaft and bearings
2. Crankshaft gear
4. Cylinder liner
5. Piston and rod assembly
6. Camshaft gear
7. Camshaft
8. Camshaft bushings
9. Gear case cover
10. Water pump and drive assembly
11. Fan hub assembly
12. Water inlet connections
13. Exhaust manifold
14. Fan hub mounting bracket and support
15. Cylinder liner packing
16. Cylinder liner shim (optional)
17. Oil filter-cooler assembly
18. Rocker levers and shaft
19. Water bypass tube
20. Thermostat housing
21. Water mainfold
22. Thermostat
23. Cylinder head and gasket
24. Inlet manifold
25. Cranking motor and spacer
26. Flywheel and housing
27. Cylinder block
28. Main bearing cap
29. Rear seal assembly
30. Oil pump assembly
31. Cam follower assembly
32. Fuel pump
33. Oil pan and gasket
34. Rocker lever covers and housing

FIGURE 7-3. Exploded View of Block. *(Courtesy of Cummins Engine Co., Inc.)*

separate containers for each assembly to save the time consumed in sorting through a pile of different fasteners. [**Fig. 7–3**] A simple tray of scrap cans can be used to keep the fasteners separated. [**Fig. 7–4**]

ENGINE BLOCK SERVICE

An engine cannot be rebuilt while installed. It can be overhauled because most of the working parts are accessible. The entire engine must be removed from its working position in order to replace the crankshaft or do machine repairs. The order of block service follows:

1. Strip all attached parts from the block after removal.

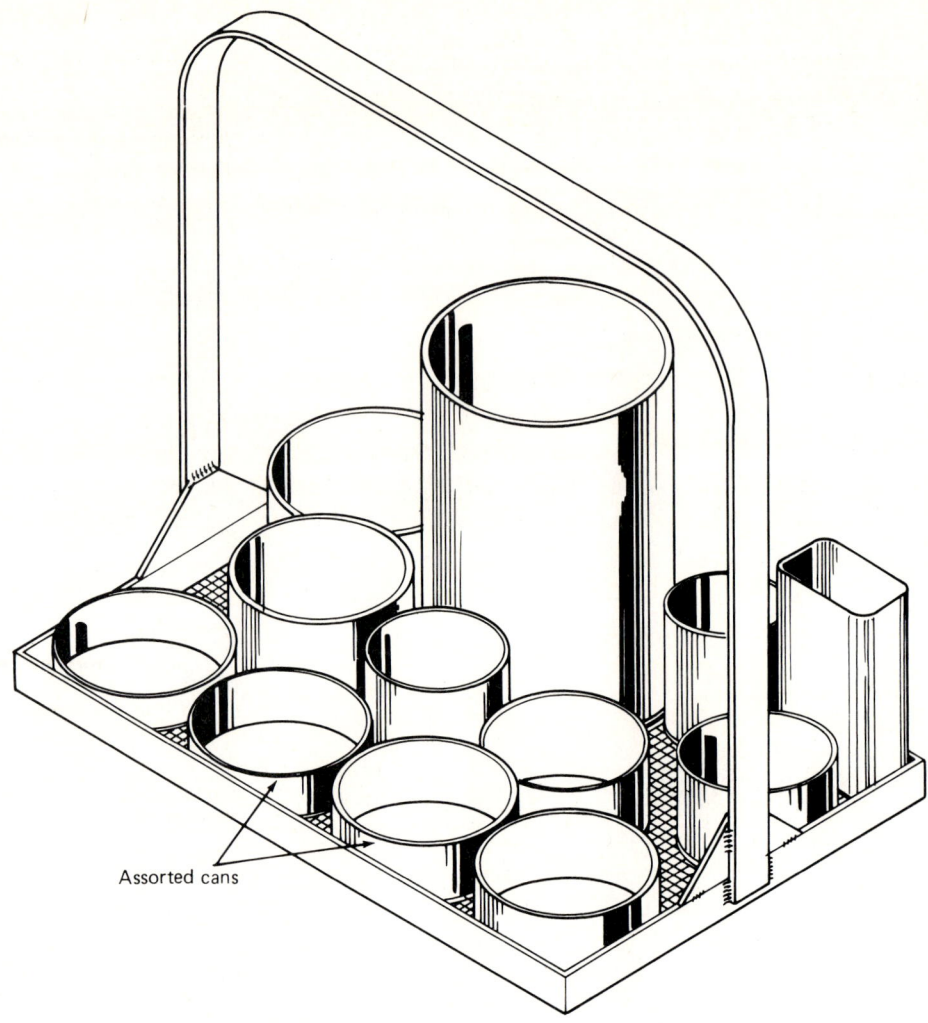

FIGURE 7-4. Tray for Fasteners. *(From a sketch by P.M. Uhl)*

Block Cleaning

2. Clean the block by immersing it in a tank of cleaning solution.
3. Mount the cleaned block on an engine rebuild stand or in a clean area with hoisting equipment provided.

Block Inspection

4. Check the cylinder bores for residue not removed in the cleaning tank.
5. Examine the surfaces for cracks or other damage. Magnetic inspection can be used.
6. Mark for machine repair such things as
   a. Cylinder counterbores.
   b. Main bearing saddles and caps.
   c. Head surfaces for wear or grooving.

d. Any distorted threads in holes.
e. Any other stressed areas for damage. Remove plugs from the oil gallery and clean with brushes.
f. Using the proper gauges, measure the following and compare readings with the specified dimension:
   (1) liner support ledges
   (2) main bearing bores
   (3) camshaft bore sizes [**Fig. 7-5**]

Machining Repairs to Blocks

Cylinder block repairs have been made to almost all machined surfaces. Most have been successful. Some have failed due to poor accuracy in machining or attempts to make do when real precision is required.

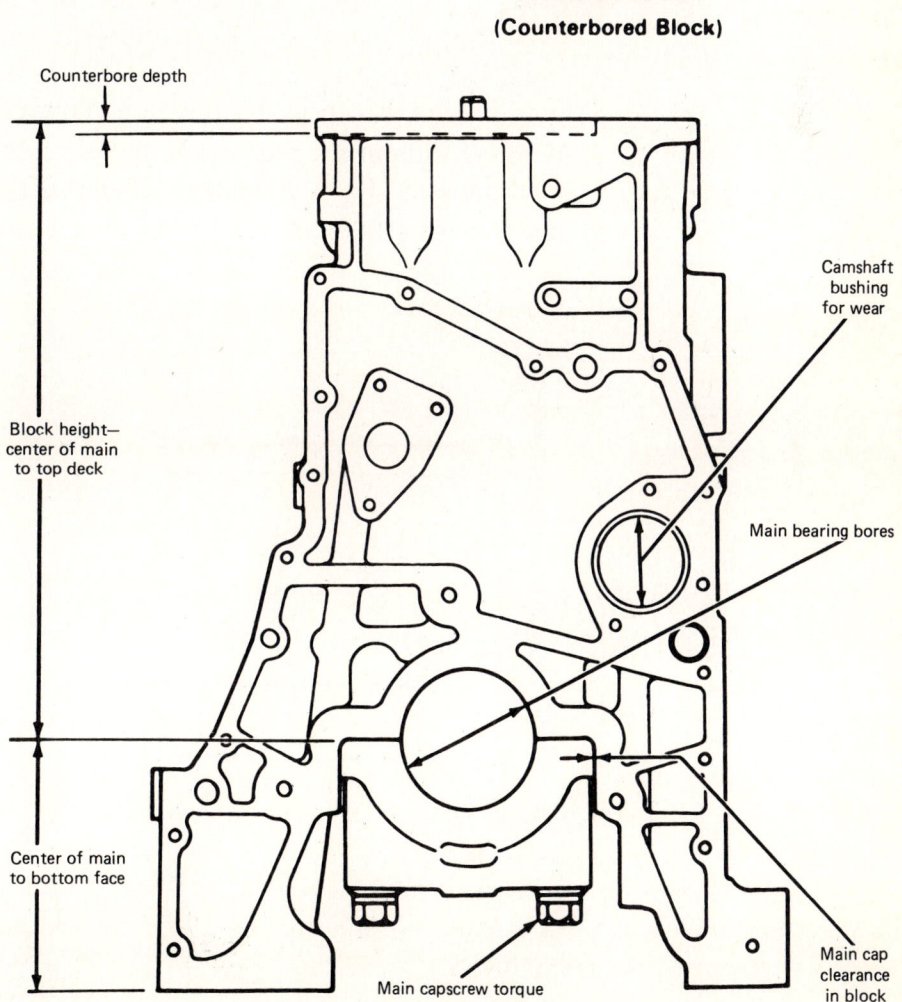

FIGURE 7-5. Block Inspection Points. *(Courtesy of Caterpillar Tractor Co.)*

A good machine shop can salvage blocks that would be scrapped from damage. Some things that can be done follow.

1. New liner counterbores can be installed in both top deck and bottom seal area.
2. When main bearings fail, the bearing saddle is usually warped. A new cap can be fitted, and a sleeve made to fit the remachined bore to recover new dimensions. This is a precision job for a shop that is equipped for it.
3. The head surface can be ground smooth, but no more than 0.010 inch can be removed.
4. Helicoil inserts can be used to restore damaged threads.
5. Special tools are sold by the engine manufacturer. If these are available the instructions for use, must be followed carefully. It is beyond the scope of this book to describe the use of the many special devices.

Beginning Engine Assembly

1. After all inspection is completed and repairs made, set the block on a rebuild stand, or place it on a clean shop floor on the head surface. **[Fig. 7-6]**
2. Install the crankshaft in new bearings. Secure the fasteners to specified tension. Be sure that thrust bearings are in place.
3. Working without a stand, install the rear crankshaft oil seal and the flywheel housing. Be sure to align the housing properly.
4. Install the camshaft bushings; install the camshaft and index the gear teeth.

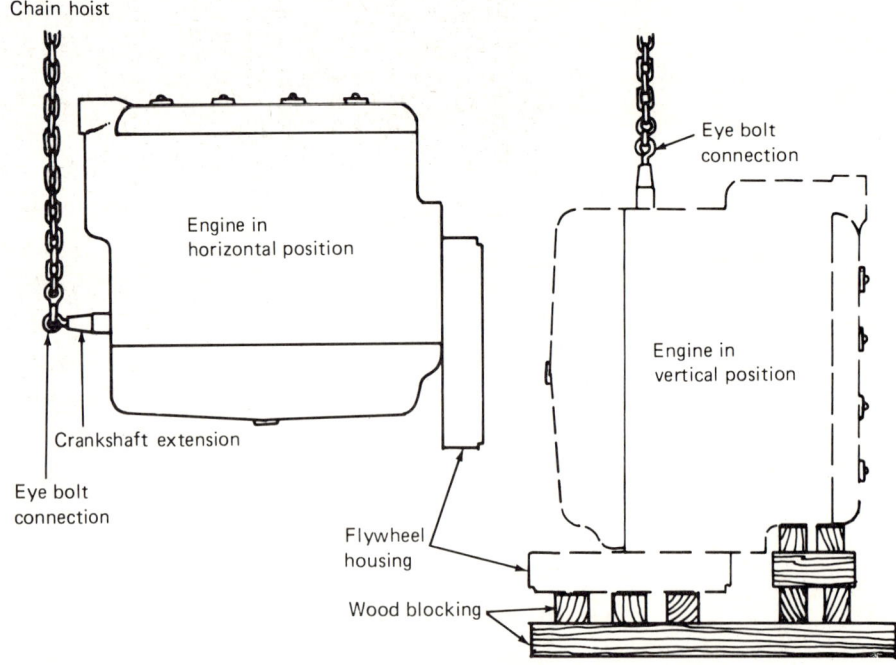

FIGURE 7-6. Engine Assembly Without a Stand. *(From a sketch by P.M. Uhl)*

5. Attach a heavy eyebolt to the front thread on the crankshaft. Be sure to engage all the threads possible.
6. Using a hoist, raise the front of the engine to a vertical position.
7. Arrange blocking under the flywheel housing as shown. A suitable thickness of blocks must be used to support the engine solidly.
8. Leave the hoist attached for safety, and continue assembly. All sides are now accessible.

This makeshift method will be useful only on vehicle-sized engines. Large engines are built in place, using cranes.

GENERAL BLOCK REPAIR PRINCIPLES

1. Remember that the block must align and support the crankshaft and cylinder heads. Working forces act to push them apart, and the block must not be warped by that force.
2. Keep the parts clean as assembly proceeds. Expensive repairs have been wasted by dirt entering during assembly.
3. Be sure that cylinder liners are correctly installed, with seals and prescribed lubricants and sealing compounds used. Check for distortion after installation.
4. Liners should project above the block surface from 0.003 inch to 0.006 inch. This height is established by the supporting ledge. The counterbore depth must be measured and any corrections made before assembly.
5. The main bearing bores must be aligned within 0.001 inch from end to end. The installed crankshaft must be hand-free.
6. New camshaft bushings must be installed, using the prescribed mandrel.
7. Follow tightening instructions for all fasteners, particularly for the head and bearing bolts. Tightening sequences should always start at the center and alternate outward. Several passes are required to tension the part evenly. Torque specifications are given for nearly all fasteners.

CYLINDER HEADS

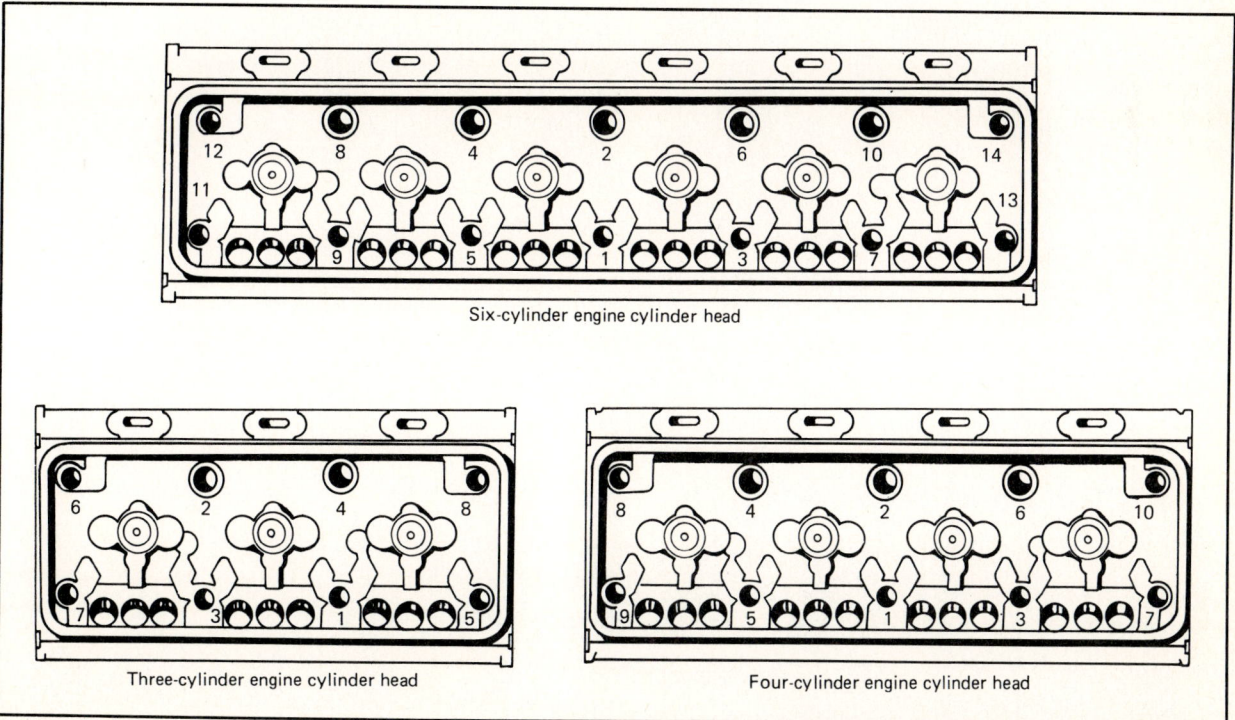

Six-cylinder engine cylinder head

Three-cylinder engine cylinder head

Four-cylinder engine cylinder head

FUNCTIONAL DESCRIPTION

There are many forms of cylinder heads. Some engines have one head for each cylinder. Others have one head over all cylinders. V-type engines have one or more heads on each bank.

In any case, cylinder heads are exchanged rather than rebuilt in the field. It is important to the line and field mechanic that the need for head replacement is known and that cylinder heads are not removed unless some system or failure makes it necessary.

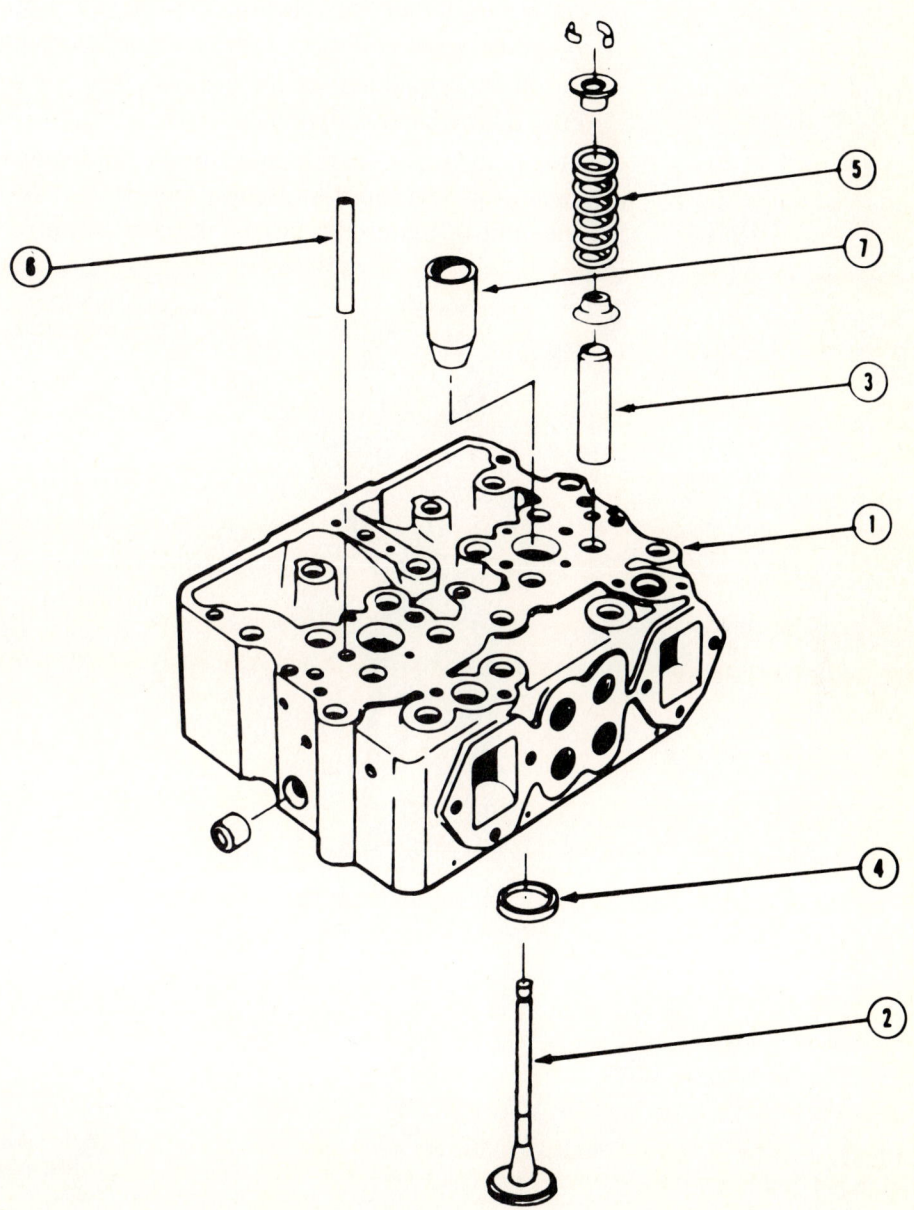

FIGURE 8-1. **Typical Cylinder Head.** *(Courtesy of Cummins Engine Co., Inc.)*

154 / Diesel Engine Service

CAUSES FOR HEAD REMOVAL

Some of the causes for head exchange follow. [Fig. 8-1]

1. Oil or coolant leaks from under the head. Such leaks may indicate a failed head gasket, or a warped head.
2. Serious valve failures.
3. Normal exchange during overhaul and rebuild.
4. Leaking injector seats. These are made of copper or brass on some engines and may leak compression or coolant if damaged or cracked.
5. Cast-iron construction of heads, possibly cracked by overheating or by improper tightening of injector hold-downs. Loss of coolant and excessive pressure in the cooling system are good indications.
6. A long misalignment of the head surface, called a *warp*, usually straightened as the head is tightened.
7. Head surfaces. These should be ground only when short irregularities, such as grooves and scratches, are present. Most engine builders specify the limit of material removed in resurfacing.

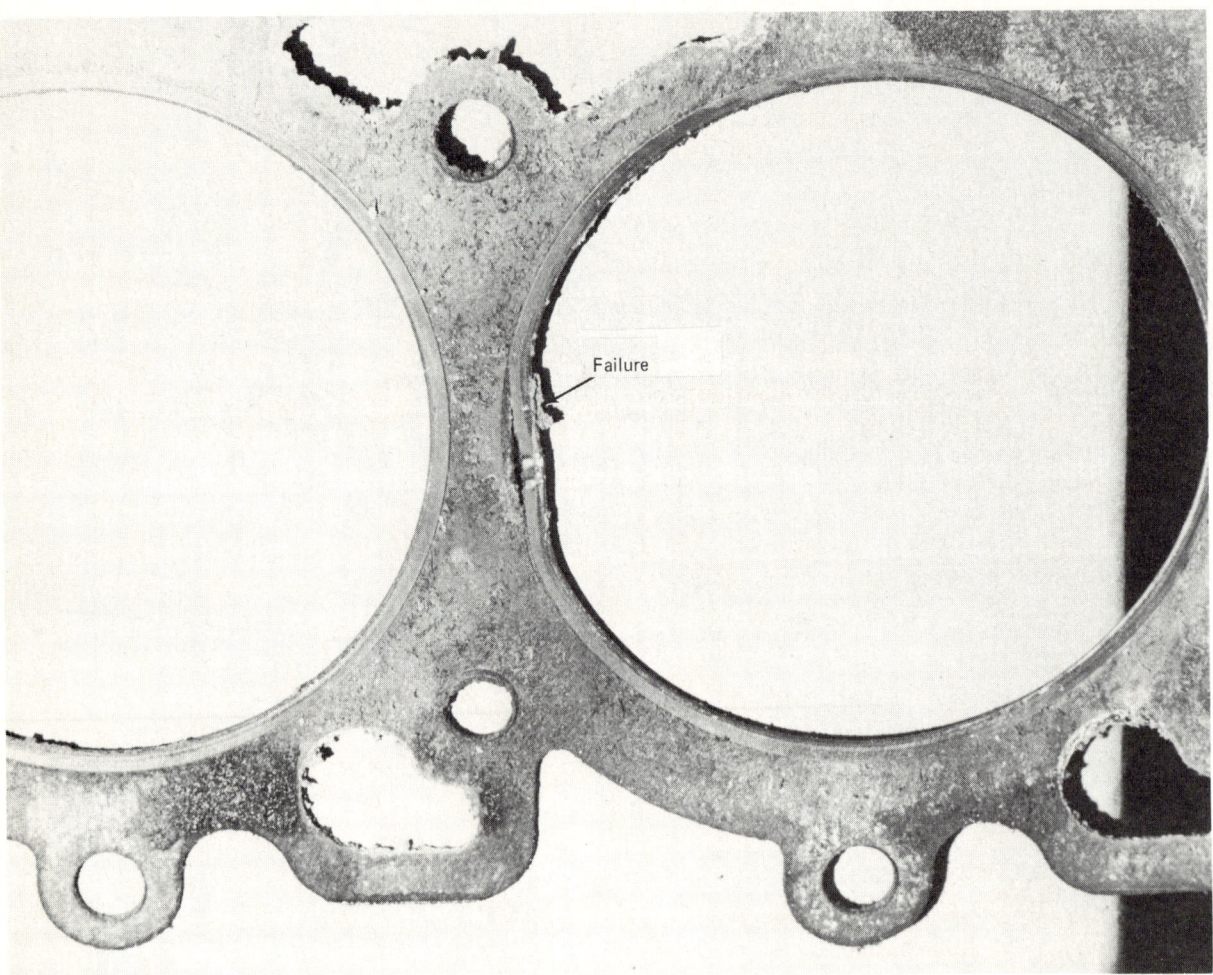

FIGURE 8-2. Failed Head Gasket. *(Courtesy of Empire Machinery Co.)*

Cylinder Heads / 155

Of course, heads are removed for reasons that do not involve head damage. Head gasket failures, or engine overhaul periods, require the heads to be removed but do not necessarily mean that head repairs are needed. As noted, exchange heads are usually installed at overhaul. [**Fig. 8-2**]

Cylinder Head Bench Repair

Given proper tooling and facilities, cylinder heads can usually be repaired to give a further service period. All such repairs are made after thorough inspection to assure that the casting is not ruined by cracks or other terminal damage.

1. Valve guides can be replaced and fitted to the valve stems. [**Fig. 8-3**]
2. Valve seats can be renewed by the use of hard inserts. Special tooling is required.

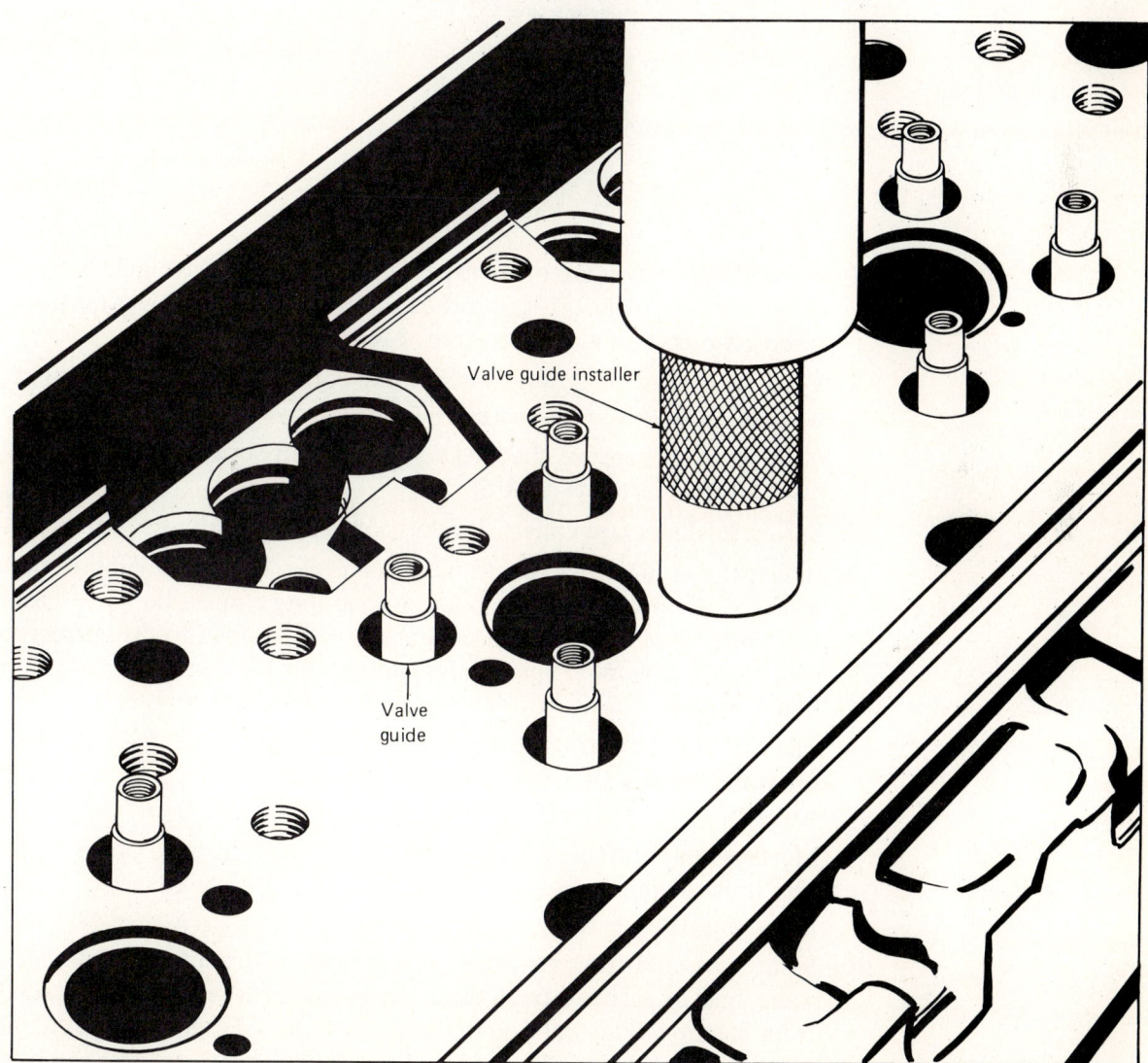

FIGURE 8-3. Replace Valve Guides. *(Courtesy of General Motors Corp.)*

156 / Diesel Engine Service

FIGURE 8-4. Valve Machine. *(Courtesy of General Motors Corp.)*

3. New exhaust valves are often used. Undamaged, used exhaust valves can be used in the inlet position, if the same size. All used valves must be refaced on precision machines. [Fig. 8-4]

4. Valve seats must be ground with special equipment. No finish lapping is needed if valves and seats are ground properly.

5. Used valve springs are checked for length in a spring tester. They must meet the standards. [Fig. 8-5]

6. Spring retainers and collets should be checked visually for wear.

7. Injector seats are replaced, using special tools. They must support the injector at the proper height; makeshift tooling cannot be used. Usually these seats or sleeves are rolled in place, then reamed to the correct support height. [Fig. 8-5] Spray valves are usually located in the precombustion chamber or head bore on replaceable seals and spacers.

8. The best way to finish seat valves after assembly is to strike the stem a smart mallet blow, letting the valve snap shut. Small irregularities are thus reduced.

9. Most rebuild shops use a surface grinder to renew the head gasket surface. The finished thickness of the head must not be less than the specified minimum. [Fig. 8-5(b)]

10. Some precombustion chambers are serviced without head removal, but some must be serviced from the cylinder side. Those that are threaded in from the top have a seal ring that must be renewed each time the chamber is removed. Minor cracks in precombustion chamber surfaces can be disregarded, but burned or broken areas require renewal.

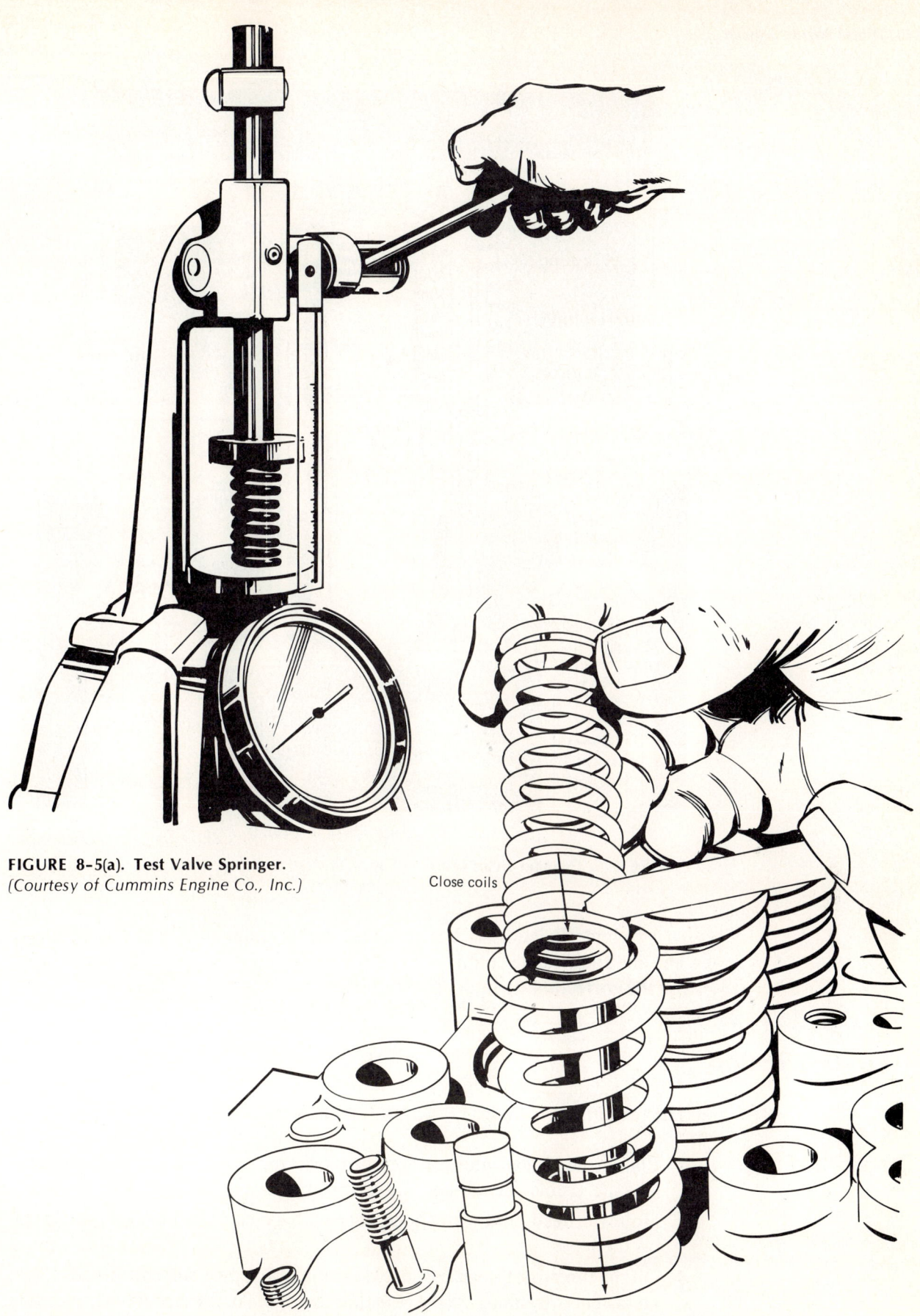

FIGURE 8-5(a). Test Valve Springer.
(Courtesy of Cummins Engine Co., Inc.)

Close coils

FIGURE 8-5(b). Valve Spring Installation. *(Courtesy of Caterpillar Tractor Co.)*

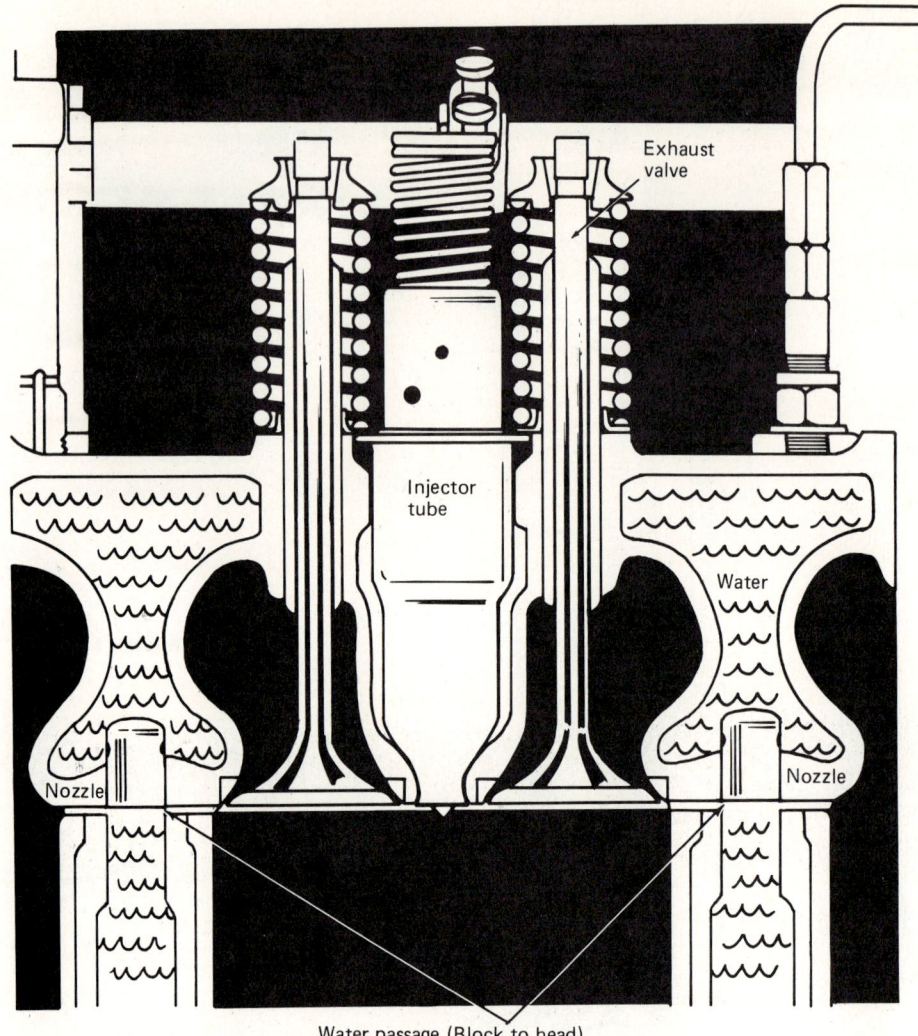

FIGURE 8-6. Cylinder Head Cross-section. *(Courtesy of General Motors Corp.)*

Cylinder head service is common and familiar to all rebuilding shops. The line and field mechanic does no rebuilding but removes and installs heads for the correction of faults as described.

Head Installation

The routine for cylinder head installation is as follows:

1. Place head gaskets and seals on the block, making sure that any top markings are visible. Install any water seals used. [**Fig. 8-7**]
2. Hoist the head, and wipe the surface with a clean rag just before setting it on the block.
3. Oil the threads of head capscrews or stud nuts. Just wet the threads. Start all fasteners by hand. Make sure that the right length capscrews are used in all holes. Several lengths are common.

Cylinder Heads / 159

FIGURE 8-7. **Head Gasket Markings.** *(Courtesy of Cummins Engine Co., Inc.)*

4. Make at least four passes over the sequence to bring the fasteners to final tension. The first pass should be one-fourth of the final torque. This method assures that the gasket is compressed evenly. **[Fig. 8-8]**
5. Never exceed specified torque. Damage due to overtightening is costly and inexcusable.
6. Some engines require retightening after a period of operation.

160 / Diesel Engine Service

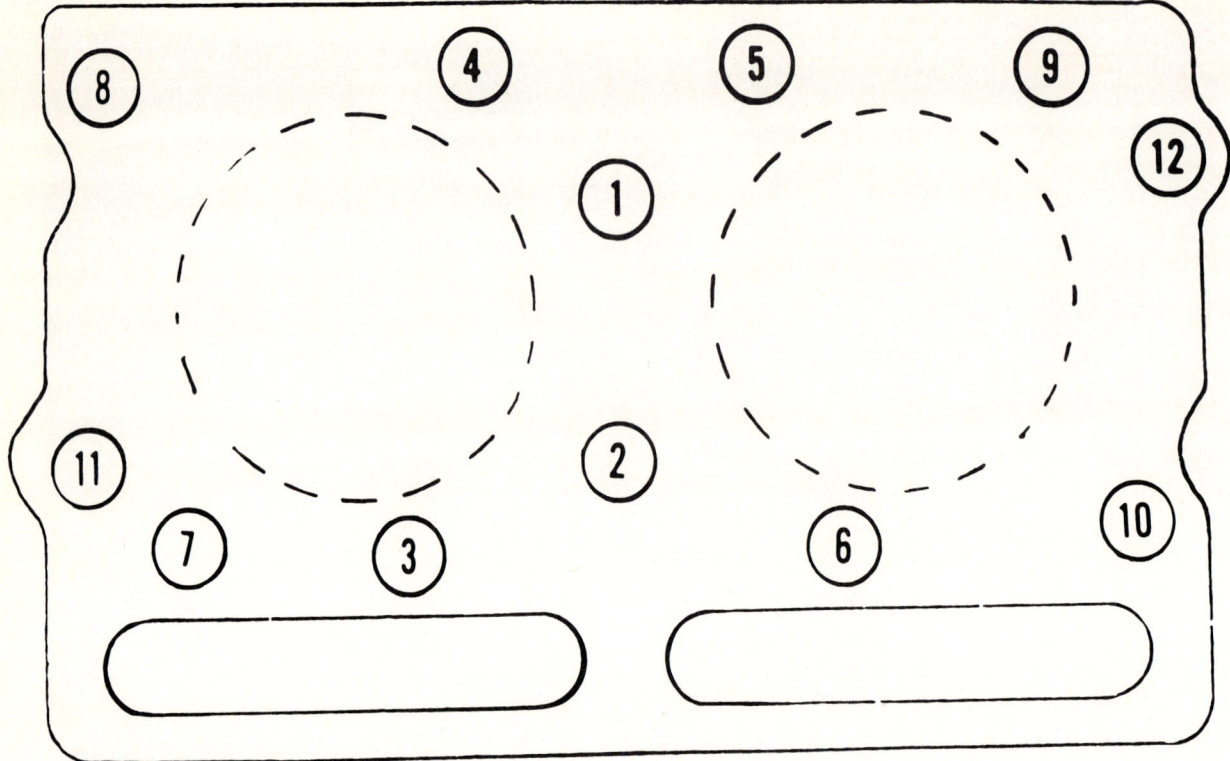

FIGURE 8-8(a). Typical Tightening Sequence. *(Courtesy of Cummins Engine Co., Inc.)*

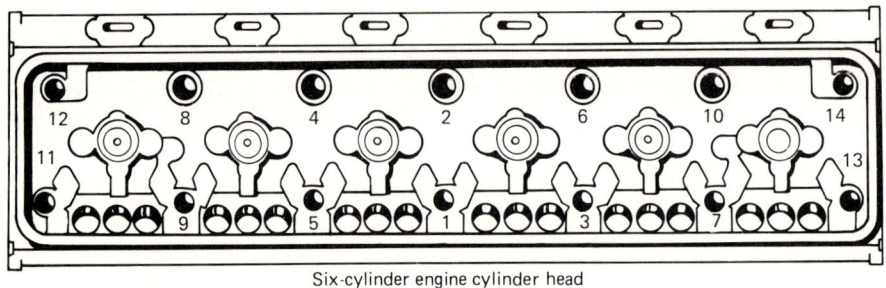

Six-cylinder engine cylinder head

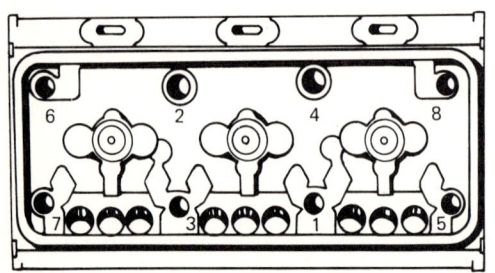

Three-cylinder engine cylinder head

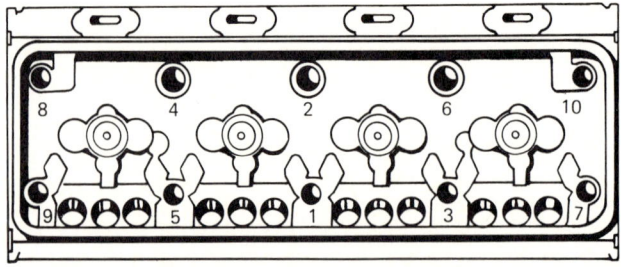

Four-cylinder engine cylinder head

FIGURE 8-8(b). Head Bolt Torque Sequence. *(Courtesy of General Motors Corp.)*

Cylinder Heads / 161

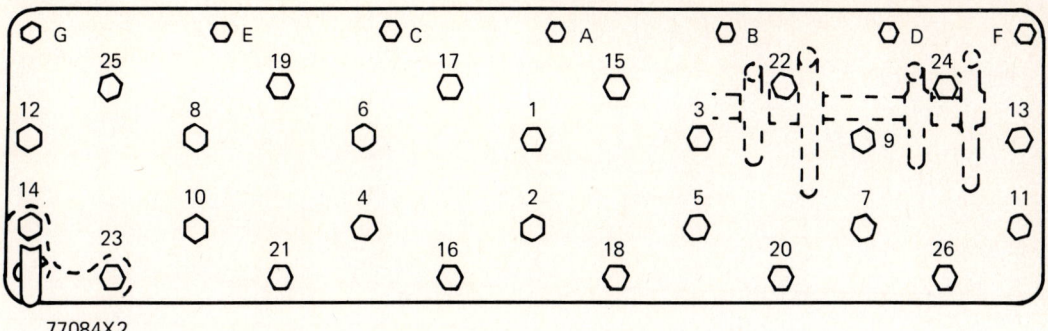

FIGURE 8-8(c). Caterpillar Head Tightening Sequence. *(Courtesy of Caterpillar Tractor Co.)*

CRANKSHAFT SERVICE

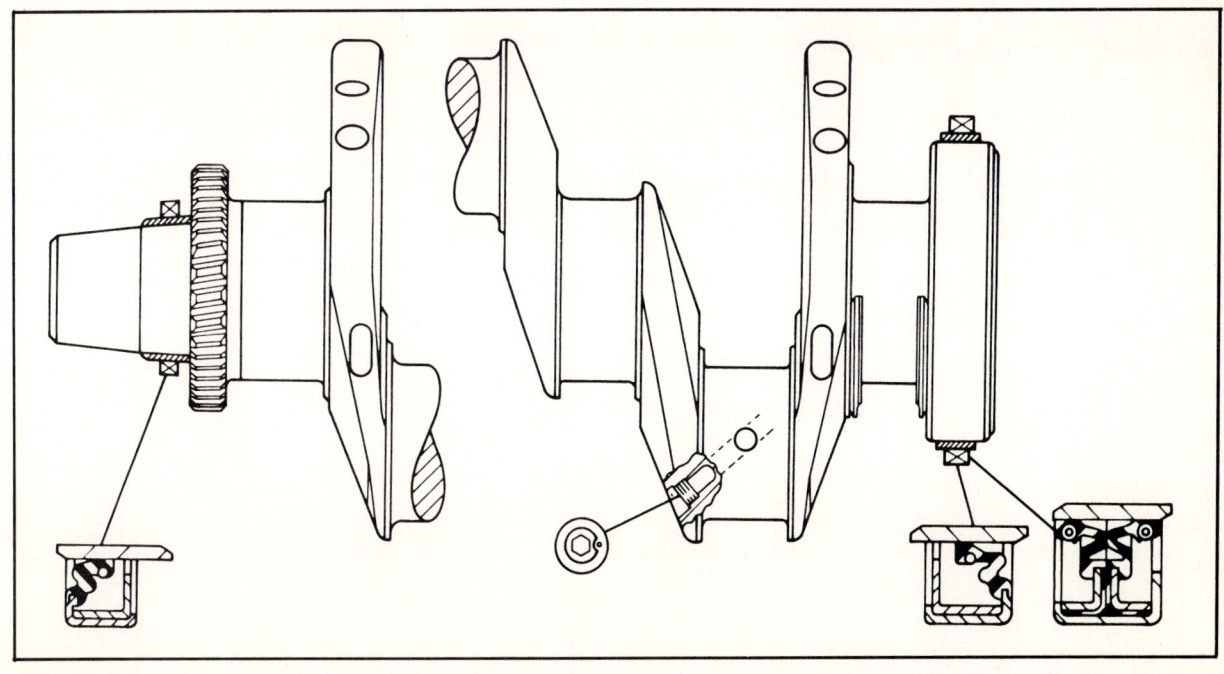

FUNCTIONAL DESCRIPTION

The crankshaft is a precision-machined forged steel member, whose journal surfaces are hardened by the induction method. It must translate the cylinder forces into rotary motion; it undergoes heavy stress during the firing strokes, although stress is relieved after such cylinder fires.

Thus the crankshaft is twisted slightly during the power stroke and springs back as load is relieved. A circular vibration is set up; it would cause breakage if not controlled. This is called *torsional vibration*. It cannot be seen or felt, but it is always present on any crankshaft during operation.

BEARING DESCRIPTION

Crankshafts are supported in main bearings, which are part of the block casting. Removable main bearing caps allow service of the bearing lining halves, which are made of various metals with a relatively soft surface and a steel backing. There is usually one more main bearing than the number of connecting rod journals. Thus, a 4-cylinder engine has 5 main bearings, while a 6-cylinder engine has 7. A V-type engine may have 5 for an 8-cylinder engine and 7 for a 12-cylinder unit.

The bearing halves are kept from turning with the shaft by dowel rings or by small portions of the bearing bent outward to form a stop. Grooves in the cap mating surfaces receive these *tangs* and require care to be sure that the bearing seats in the grooves.

When dowel rings are used, they are contained in a groove around the fastener hole, and the bearings have mating grooves to engage the ring. [**Fig. 9-1**]

Only one side of the bearing is doweled. Thus incorrect installation is discouraged. [**Fig. 9-2**]

Bearing Lubrication

Oil is fed to the main bearings by holes from the main oil gallery. Oil holes in the bearing halves index with these passages, and the main bearings are thus assured of a continuous supply of oil.

Crankshafts are drilled to feed oil to the connecting rod bearings. All journals must receive oil under pump pressure. [**Fig. 9-3**]

Thrust Bearings

Thrust bearings are radial bearing surfaces that confine the endwise movement of the crankshaft. They may be separate from the main bearings, or they may be on a flange formed on one main set, usually on the rear bearing. [**Fig. 9-4**]

166 / Diesel Engine Service

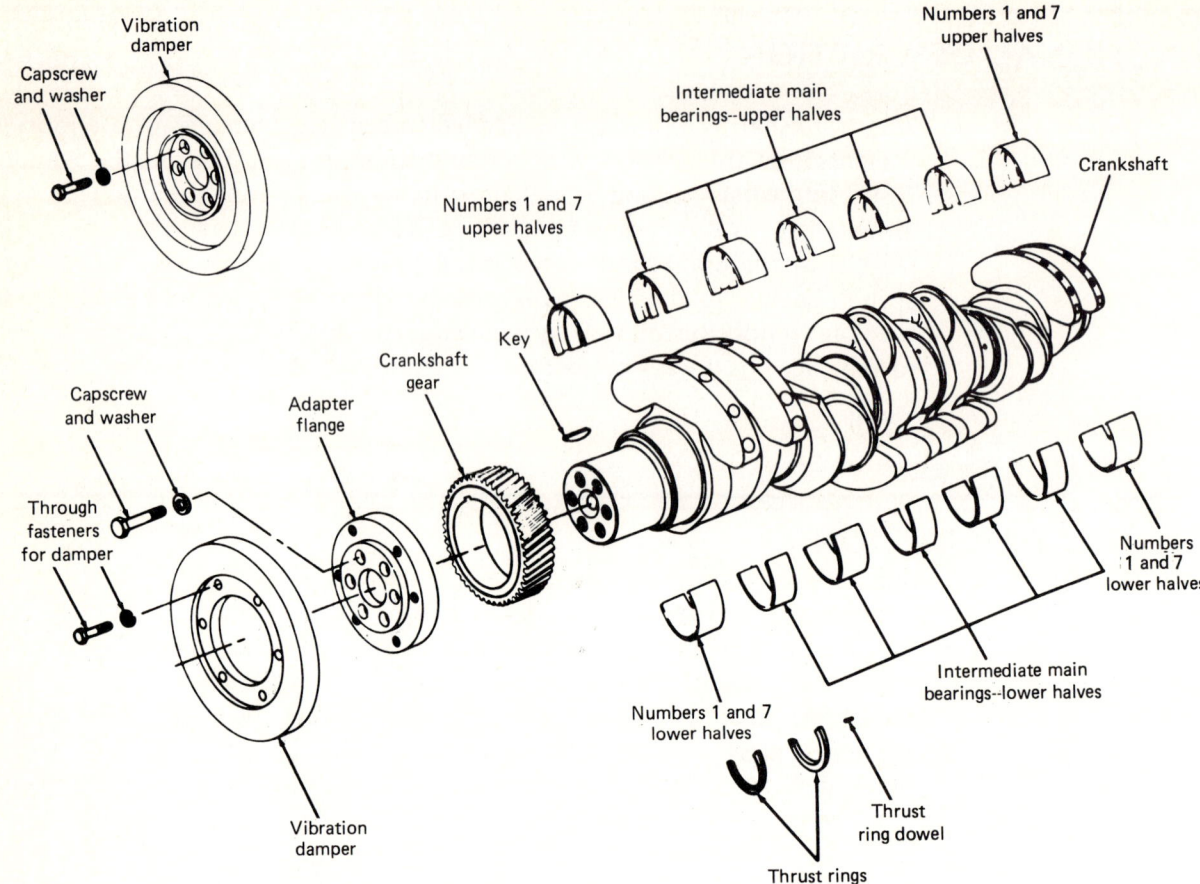

FIGURE 9-1. Typical Crankshaft and Bearings. *(Courtesy of General Motors Corp.)*

Thrust bearing halves are usually retained by dowel pins in the main bearing cap. The thrust bearings are heavily loaded by clutch throw-out action as the clutch is held disengaged. Wear rate thus depends on operator practice.

CRANKSHAFT AND BEARING SERVICE

Although the crankshaft cannot be removed unless the engine is taken out of its operating position, bearing service can be performed in place.

Crankshaft Inspection

1. Check to see crankshaft size is in the engine. Crankshafts are reground to 0.010-inch, 0.020-inch, and 0.030-inch undersize; undersize bearings are available.
2. Drain the oil and clean the lower surfaces of the engine.
3. Remove the oil sump (pan), being sure that *all* capscrews are removed.

Crankshaft Service / 167

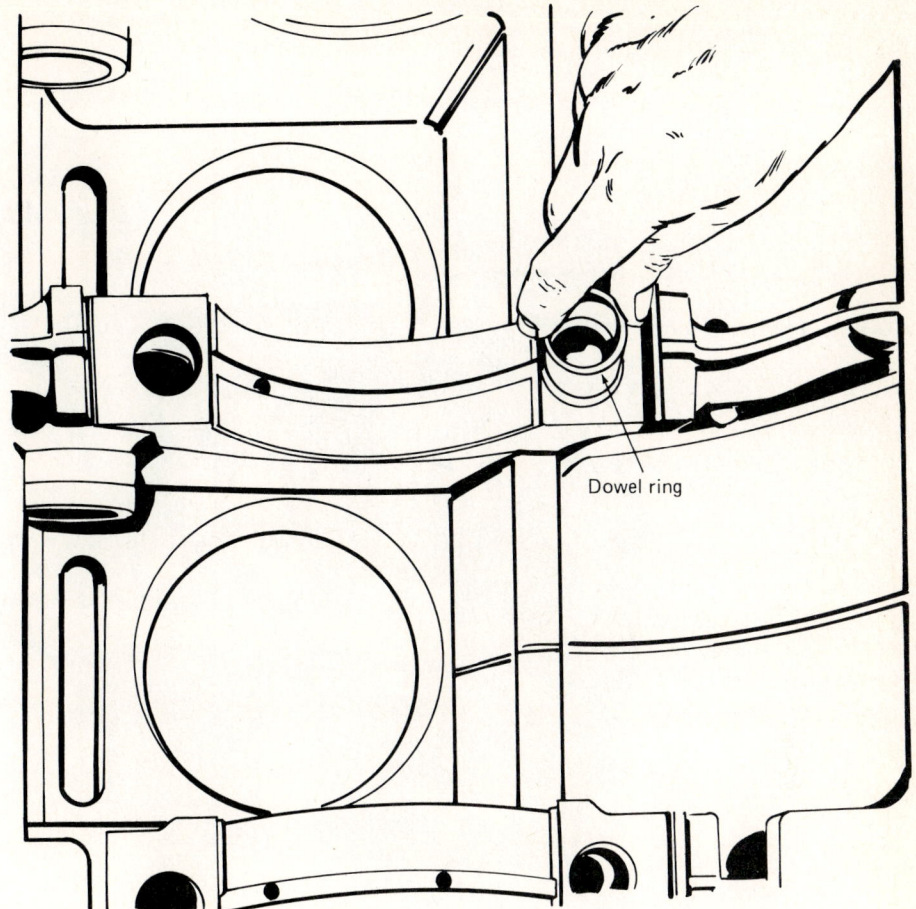

FIGURE 9-2. Main Bearing Dowel Ring. *(Courtesy of Cummins Engine Co., Inc.)*

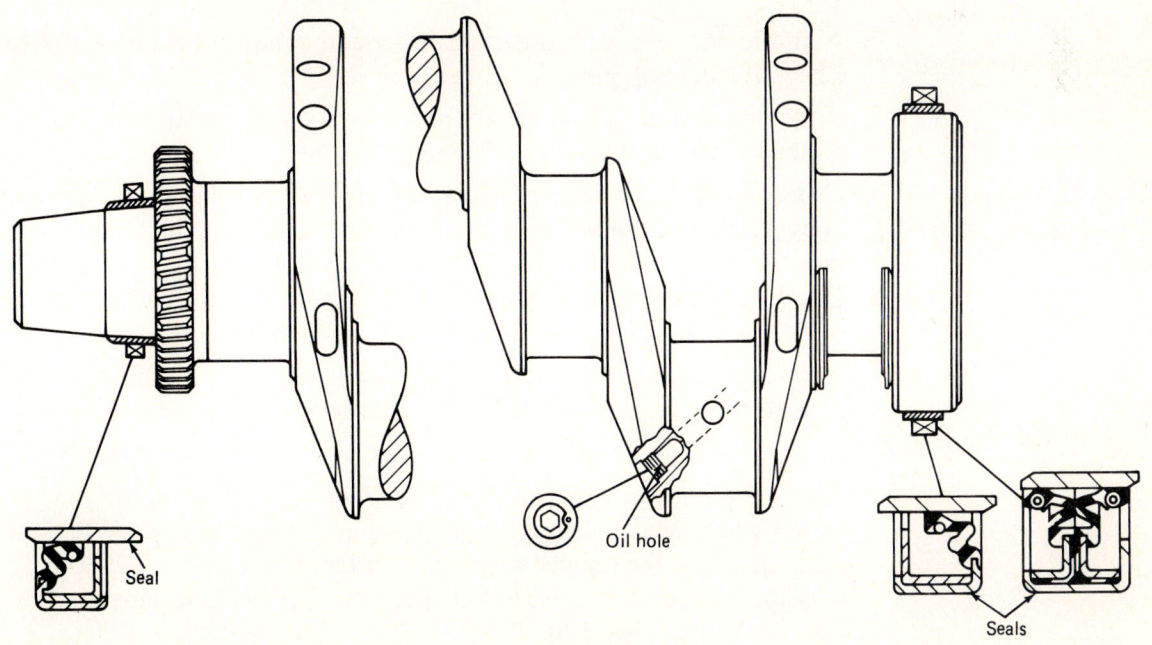

FIGURE 9-3. Crankshaft Oil Holes. *(Courtesy of Caterpillar Tractor Co.)*

168 / Diesel Engine Service

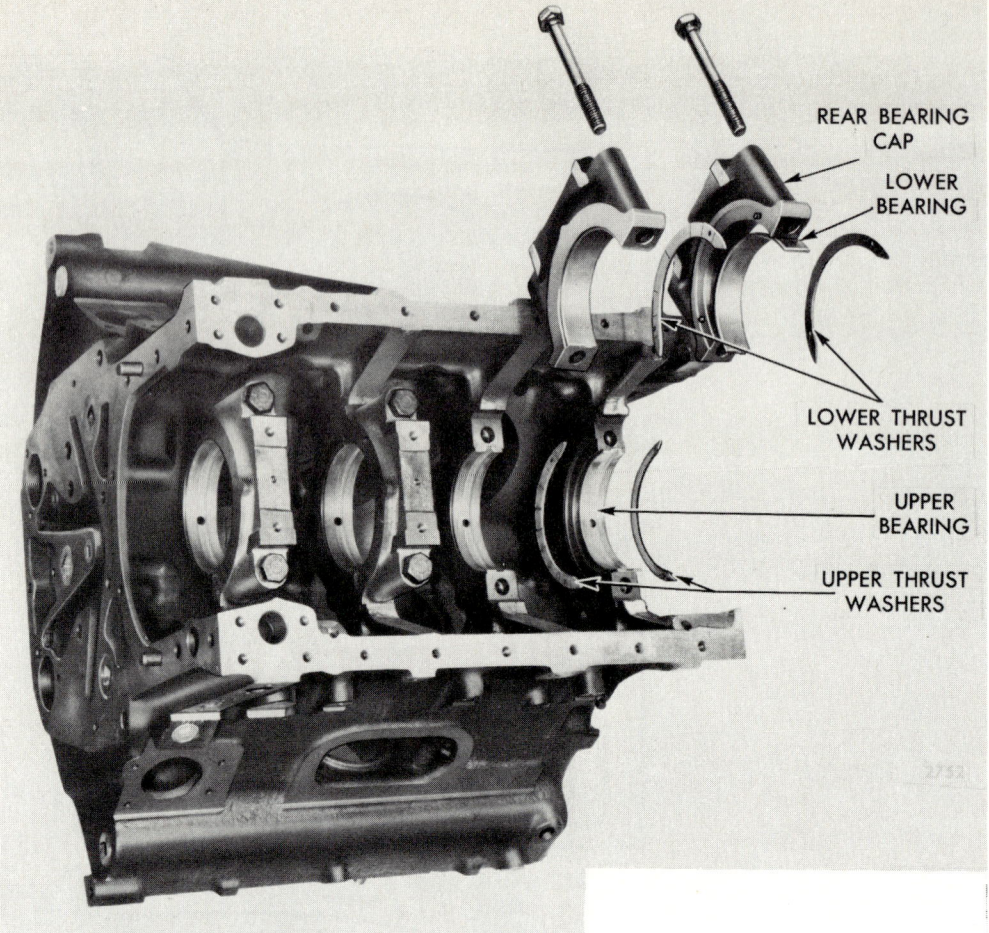

FIGURE 9-4. Main Bearings and Thrust Rings. *(Courtesy of General Motors Corp.)*

4. With the pan removed, observe the general condition of the crank case. Check for bearing particals in the pan.
5. If the oil pump is in the crankcase, remove it and the suction tube. Also remove any parts from the bearing caps.
6. Using ¾-inch drive wrenches, remove one main bearing at a time. A maximum of half of the bearings can be removed, leaving the rest to support the shaft. It is best to remove alternate caps. **[Fig. 9-5]**
7. Check for signs of heat on the bearing caps and block saddle. A severe failure can mean rebuilding the engine, including block repair.
8. If bearing particles have been found, check those bearings first. You may have to replace the crankshaft if a bearing has failed.
9. Keep the bearing halves with the cap in which they were operated.
10. Check bearings for grooves, dirt embedded in the metal, and worn thickness. Copper-lead bearings usually have a thin gray overlay. When this is worn through, the copper will show, and any distress can be seen. Compare the worn thickness with that specified. Use ball micrometers, and be sure of the readings. **[Fig. 9-6]**

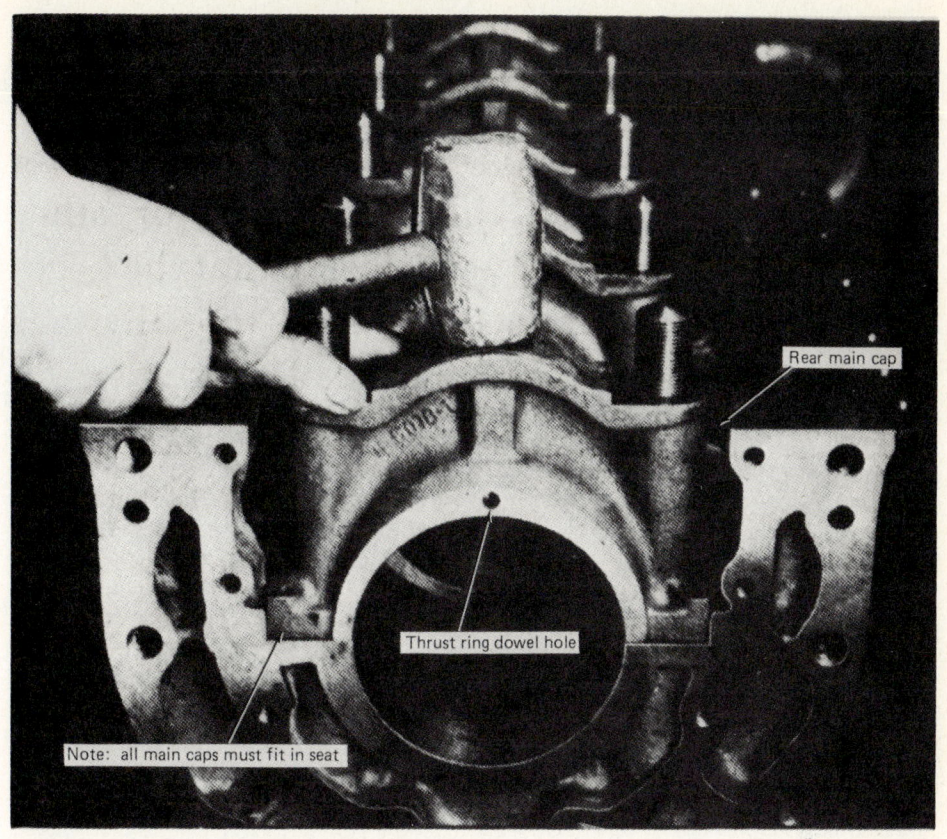

FIGURE 9-5. Main Cap Fit in Block. *(Courtesy of Cummins Engine Co., Inc.)*

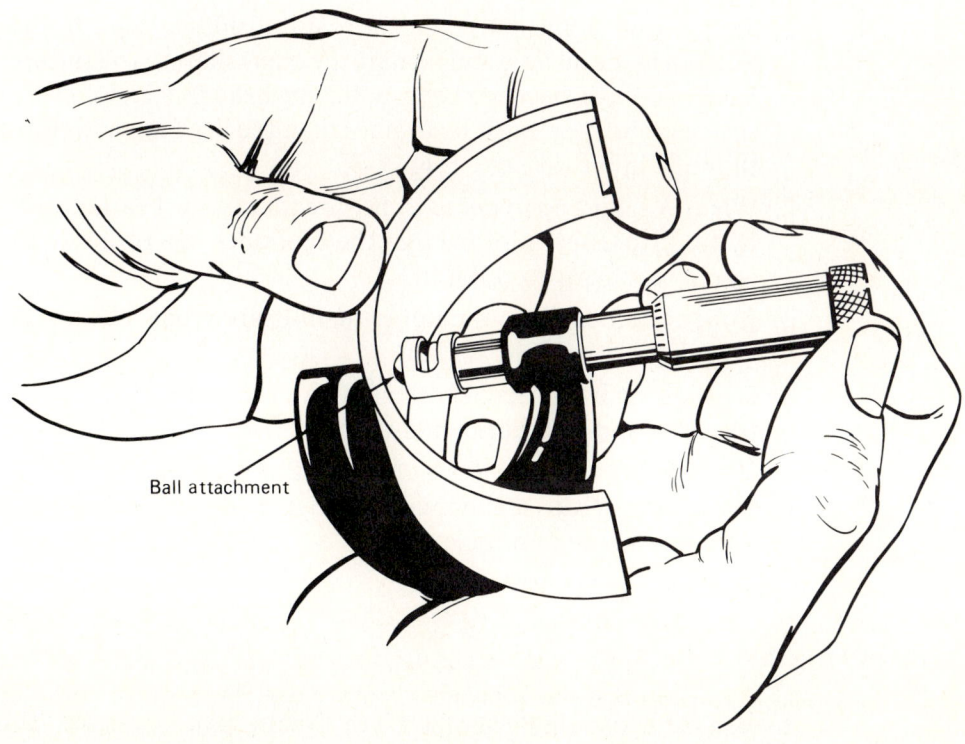

FIGURE 9-6. Measure Bearing Thickness. *(Courtesy of Cummins Engine Co., Inc.)*

170 / Diesel Engine Service

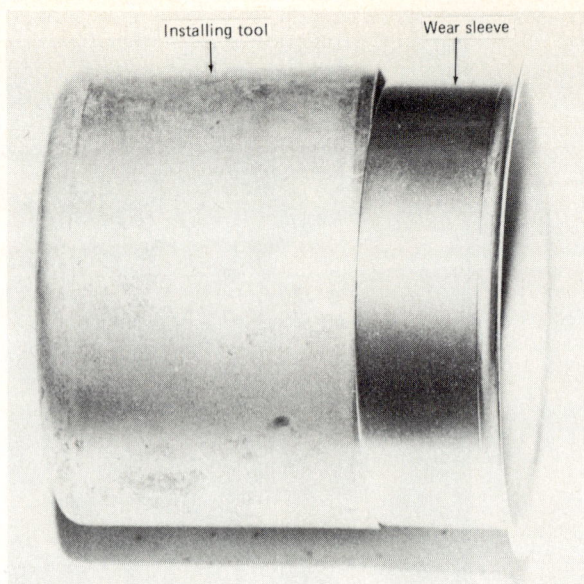

FIGURE 9-7. Sleeve and Installing Tool. *(Courtesy of Cummins Arizona Diesel, Inc.)*

11. If any bearings are distressed, it is well to replace all. Again check the crankshaft size carefully.
12. Any wear sleeve used on the crankshaft to resist seal wear should be replaced while the crankshaft is out of the block. [**Fig. 9-7**]

Main Bearing Assembly

1. Main bearing upper halves can be removed and installed by inserting a special tool in the oil hole and turning the crankshaft to roll the upper half out. Such a tool can be made by grinding the head of a capscrew to make a flat surface. The head must be thinner than the bearing, and the threads must fit freely in the oil hole. [**Fig. 9-8**]
2. When installing an upper half, be sure to start the bearing as far as possible by finger pressure and see that the tool does not go inside the bearing in turning the shaft to roll it in.
3. In nearly all engines, the upper main bearing half has an oil groove, although the lower half is plain.
4. Look at the back of all bearings. Any dark areas indicate poor contact in the bore, and the bearing must be replaced.
5. If bearings are to be reused, they must be reinstalled on the same journal. Main bearing caps are numbered and must be installed as before, with the numbers on the same side.

Bearing Numbering System

In vertical engines the numbers are usually placed on the camshaft side. V-type engines must follow the maker's specification, or have the numbers recorded. Usually the block saddle number guides the installation. [**Fig. 9-9**]

Crankshaft Service / 171

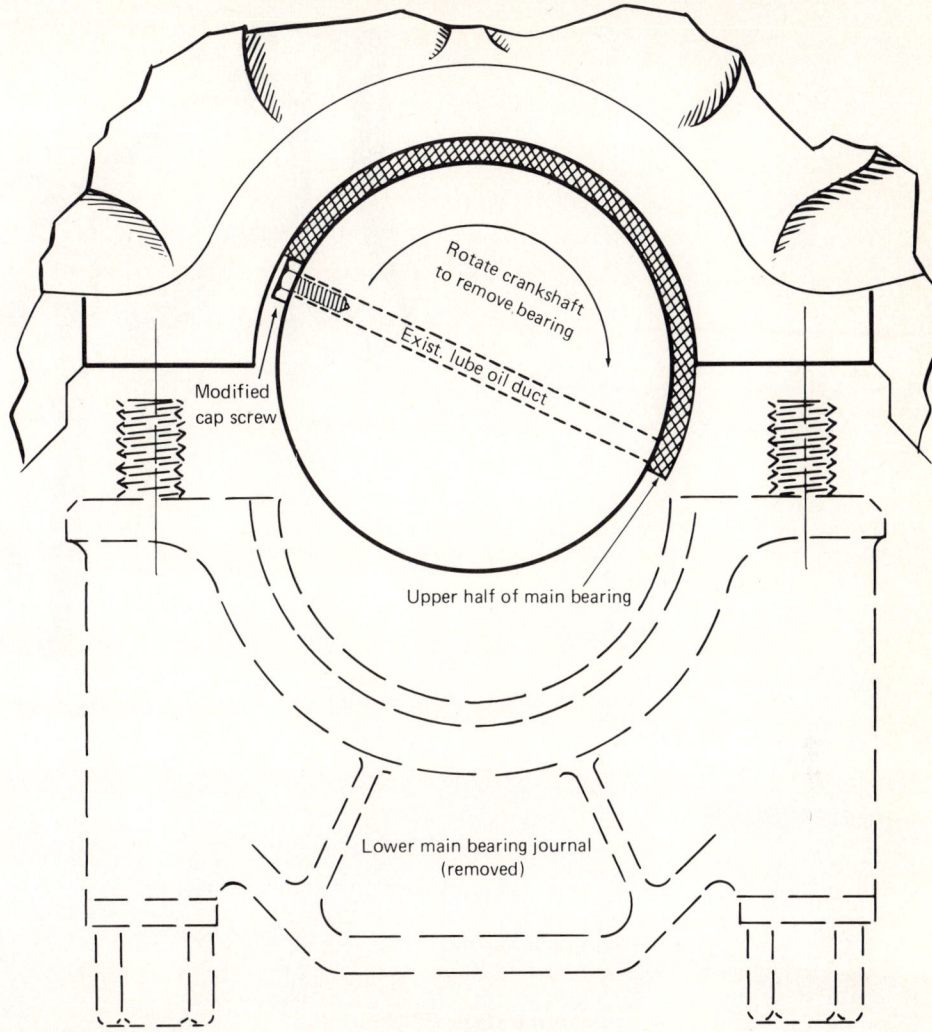

FIGURE 9-8. Upper Main Bearing Removal Tool. *(From a sketch by P.M. Uhl)*

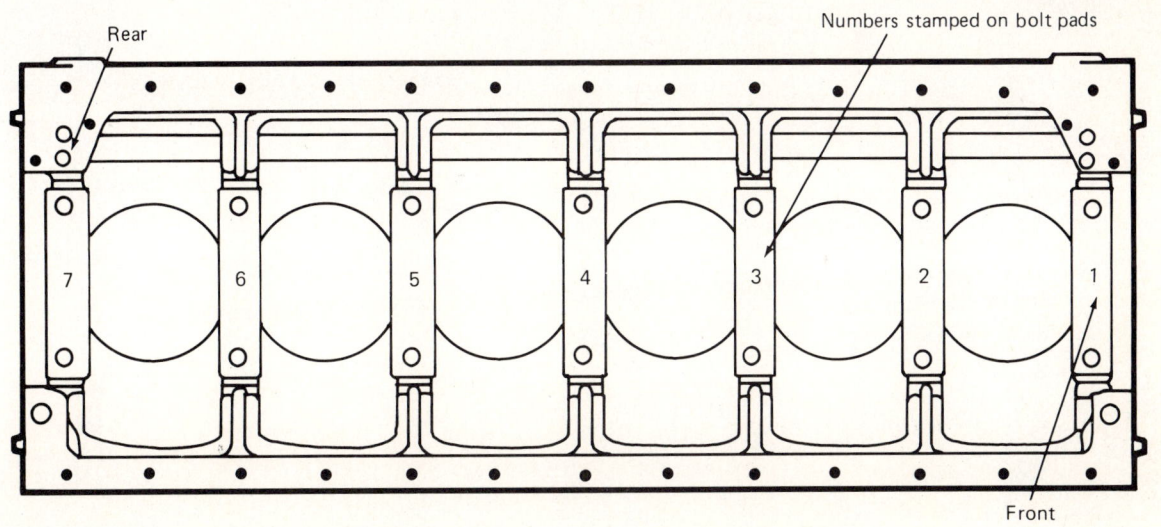

FIGURE 9-9(a). Main Bearing Numbering System. *(Courtesy of General Motors Corp.)*

172 / Diesel Engine Service

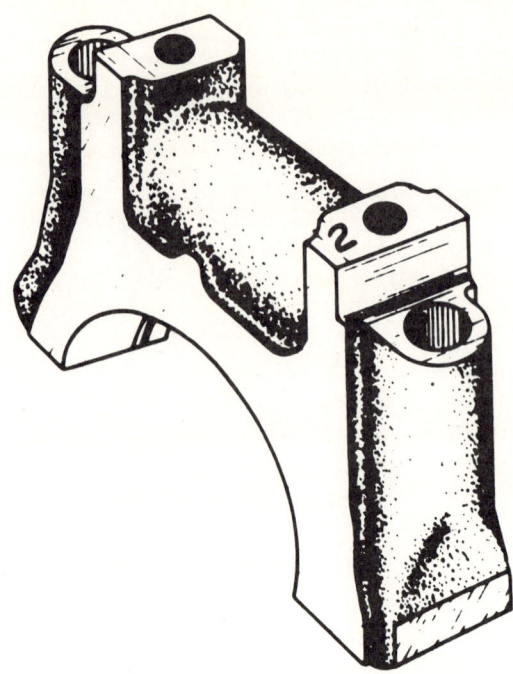

FIGURE 9-9(b). Main Bearing Cap Number. *(Courtesy of General Motors Corp.)*

FIGURE 9-9(c). Main Bearing Number Location. *(Courtesy of Cummins Arizona Diesel, Inc.)*

Crankshaft Service / 173

1. Although some thrust bearings are a part of the main bearing, most are separate. Oversized thrust bearings are available to fit reground crankshafts that have had thrust flanges ground wide.
2. Bearing caps must fit snugly in the saddle. Loose caps indicate distortion and must be replaced with a service cap and rebored. This condition requires rebuilding.
3. Main bearing capscrew threads should be oiled before installation. Just wet the threads; don't fill the hole. Some engine builders use special lubricants for this purpose.
4. Bearings must be lubricated before assembly. A good white grease or clean engine oil can be used. Do *not* oil the back of the bearing.
5. Never attempt to reuse a crankshaft that is badly grooved, discolored from heat, or worn past specified limits. A repeat failure is certain.
6. The capscrews should always be tightened with a torque wrench. Tap the caps with a soft mallet to seat them, then tighten in several passes, alternating from side to side and using the sequence recommended by the manufacturer. If no sequence is given, start at the center main and work both ways. [**Fig. 9–10**] The crankshaft should be hand-free after all caps are secured. Check the end movement.
7. Bend lockplates after all checks are made. [**Fig. 9–11**]

Connecting Rod Service

1. Connecting rod bearings should be checked while the pan is off. A rod bearing failure nearly always follows a main bearing failure.
2. Turn the crankshaft to bring two journals to bottom center, then remove the rod caps from those rods. Use ½-inch drive tools.
3. The same inspection methods and tools are used for rod bearings as for mains, and the same checks apply to the crankshaft rod journals.

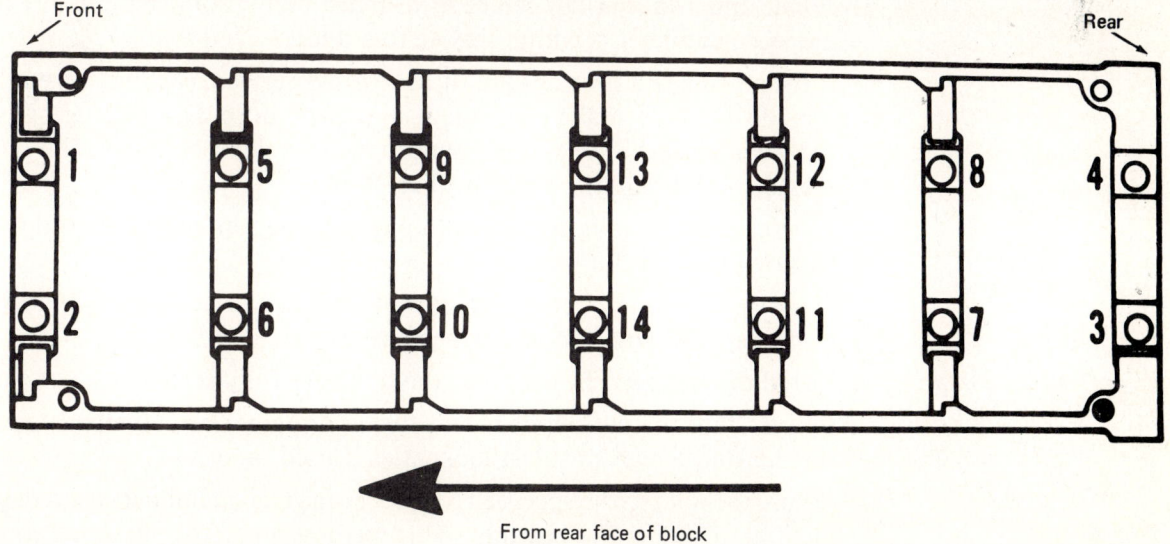

FIGURE 9-10. Main Bearing Tightening Sequence. *(Courtesy of Cummins Engine Co., Inc.)*

174 / Diesel Engine Service

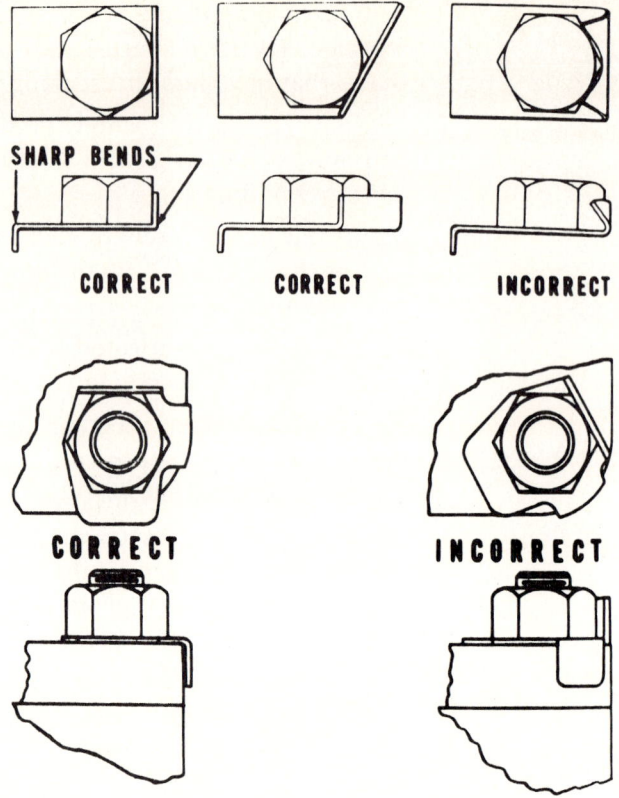

FIGURE 9-11. Lock Plates. *(Courtesy of Caterpillar Tractor Co.)*

4. If any rod bearings are distressed, replace them all with new bearings. No disassembly of the upper part of the engine is needed, unless some fault is found.
5. Connecting rod journals can be measured with micrometers. Be sure that the worn diameter is within service tolerance.
6. If connecting rod bearings are fit for further service, each half should be spread slightly to make it fit snugly in the rod bore. Set the bearing on a flat surface, and use hand pressure to cause a slight bending. Too much pressure will break the bearing. [**Fig. 9-12**]
7. Lubricate the bearing surface with oil. Seat the top half in the rod bore, indexing the locating tang.
8. Bring the connecting rod to the journal, then install the lower half bearing in the cap. Be sure that all surfaces are clean.
9. Install the rod cap. Be sure that the bolts are fully seated.
10. Be sure that the cap number and the rod number are together. Oil the threads and install washers, lockplates if used, and nuts.
11. Secure the nuts to the specified torque. Some engine builders use a degree method of tightening bearings. This method prescribes an initial torque of low value, then turning the fastener through a specified angle to finish tightening. This is an accurate method. [**Fig. 9-13**]

Crankshaft Service / 175

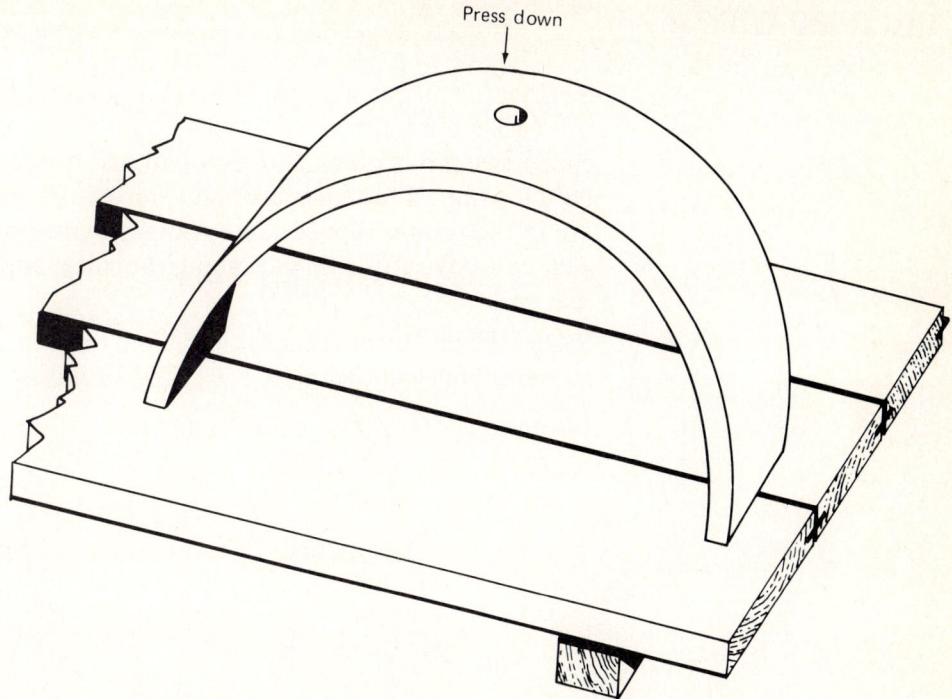

FIGURE 9-12. Spreading a Bearing Shell. *(From a sketch by P.M. Uhl)*

12. Connecting rods should be hand-free on the journal after tightening. Check by moving the rod from side to side.
13. After all bearings have been installed and secured, the oil pump and other parts can be put on.

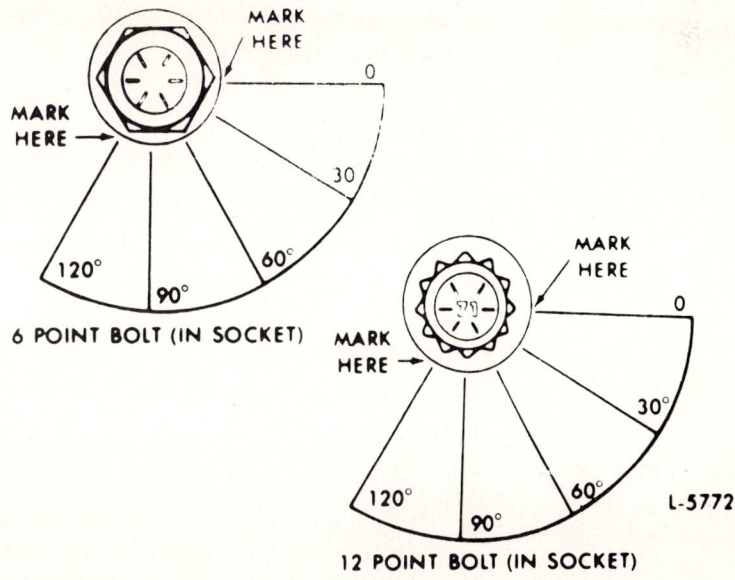

FIGURE 9-13. Degree Method of Tightening. *(Courtesy of General Motors Corp.)*

Installing Oil Pan

1. See that the oil pan is clean. Use a new gasket. It is best not to use sealer on this gasket but to tie it in place with light twine. Remove the twine after locating the pan. To install the oil pan, use a vice grip plier on each side to hold it while all fasteners are hand-started. Be sure that the right fasteners are in the proper tapped holes. Some engines have coarse threads in the gear case cover, fine threads along the sides, and smaller capscrews in the rear seal.
2. Secure the pan by alternately tightening on each side, working from center to ends. This will compress the gasket evenly.

10
CYLINDER, PISTON, AND CONNECTING ROD SERVICE

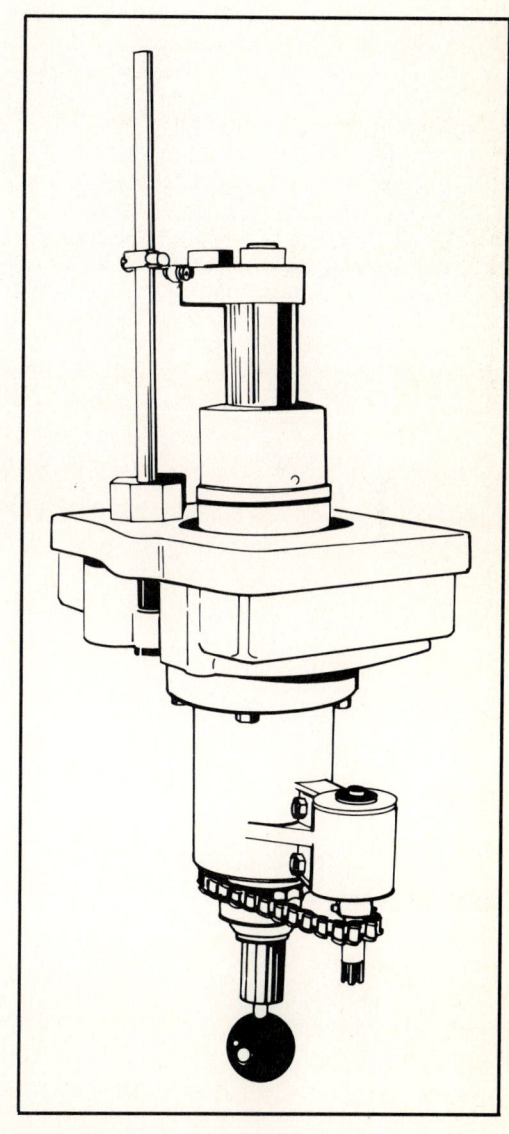

Service to these parts nearly equals an engine overhaul and is usually not done unless an overhaul is required. These parts can be serviced in the machine, without removing the engine.

The extensive use of exchange subassemblies is almost always advantageous. Because special tooling is required, field repairs are usually limited to trouble-shooting and adjustments. When the engine is brought into a shop, complete rebuilding of subassemblies can support the line mechanic as he removes original assemblies and replaces them with rebuilt exchanges.

CYLINDER LINE REPLACEMENT CAPS

The book has described most service operations to engine blocks, crankshafts, and connecting rod bearings. Cylinder liners can be inspected in place, except for the coolant seal rings on the bottom of the liners.

Interior micrometers or dial bore gauges can be used to measure liner wear. Maximum wear occurs at the top of the ring travel, below the top of the liner. In general, wear at this point should never exceed 0.005 inch. Liners worn more than this should be replaced at overhaul. [**Fig. 10-1**]

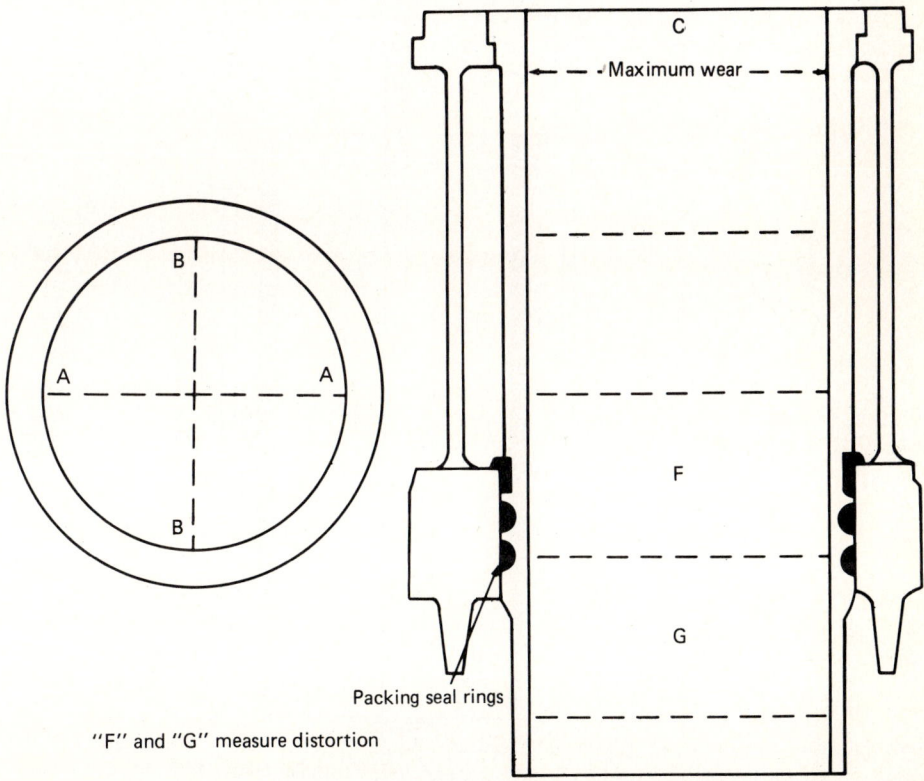

FIGURE 10-1. Cylinder Liner Inspection. *(Courtesy of Cummins Engine Co., Inc.)*

Cylinder Liner Removal

Special liner removal tools are furnished by all engine builders. Such a tool can be made as shown, if the maker's tool is not available.

Cylinder liners are not tight for more than 2 inches. They are usually removed easily, if a good tool is used to start them. [**Fig. 10-2**]

Note: Several small diesel engines are built with cylinders cast as part of the block. This trend is increasing.

Cylinder Liner Honing

Cylinders are finish-honed to establish an angular pattern. The hone marks must make a helical pattern on the wall and must contain oil for lubrication. When repair honing is done, the same pattern should be established by moving the hone up and down. Cylinders are honed not to remove much metal for

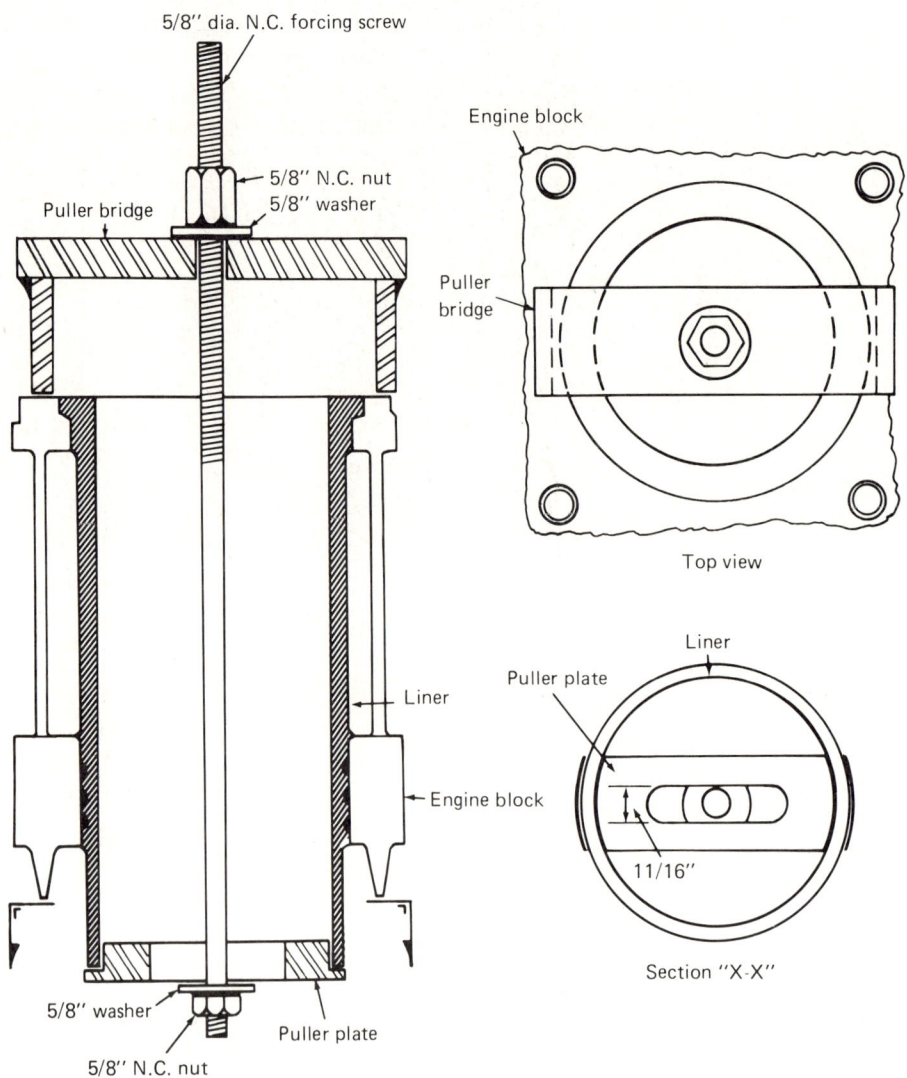

FIGURE 10-2. Liner Puller Tool. *(From a sketch by P. M. Uhl)*

resizing the cylinder but to remove vertical scratches and glaze and to recover the helical pattern.

Cleaning after Honing

Cylinder liners must be thoroughly cleaned after honing. Grit retained in the walls will surely accelerate wear. The mechanic should wash the liners with strong detergent and water, then apply clean oil. A white rag must remain clean after wiping the wall.

The ridge at the top of ring travel must be removed; a ridge of 0.003 inch to 0.005 inch will interfere with the edge of the new ring and may cause damage.

It is necessary only to cut the unworn edge above the worn area. Only enough to clear the ring should be removed.

Other reasons for liner replacement follow [**Fig. 10-3**]:

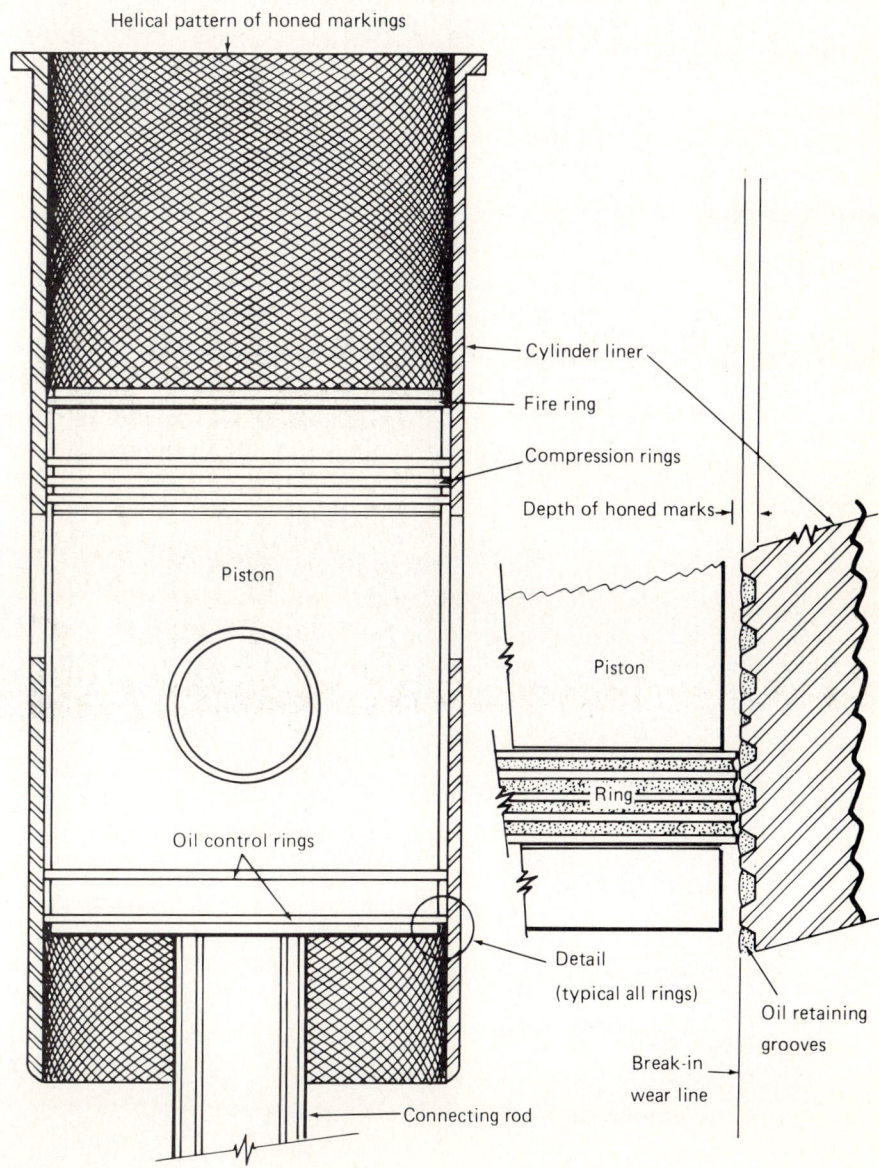

FIGURE 10-3. Honed Cylinder and Rings. *(From a sketch by P. M. Uhl)*

1. Hard and smooth glazed spots on the inner surface.
2. External pitting. This shows up as a line of pits on the side against which the piston bears during the power stroke. These pits may widen at the top seal ring and can go all the way through the liner wall. [**Fig. 10-4**]
3. Scored surfaces. These are almost always caused by overheating. The use of chrome plated rings on a chromed cylinder will result in immediate scoring.
4. Obvious damage from foreign objects, breakage, etc.

We will discuss the basic cause of pitting. As the cylinder fires, the pressure causes the liner to bend slightly. The surrounding coolant is pushed away, then returns to contact as the liner relaxes.

During this action, heat is being conducted through the liner wall, and the outer surface becomes hot enough to make a tiny but violent turbulence as the coolant returns to the hot surface. A tiny flake of metal is torn from the surface. [**Fig. 10-4**]

This action occurs millions of times as the engine runs, and the pits are the result. The liner moves more at the lower end, and the action described can destroy the seals.

Liner Cracks

Cracks can occur just below the upper flange. Liners that are otherwise fit for further use can be inspected magnetically to disclose cracks that would otherwise be invisible. If cracks are found, the liner must be replaced.

If liners are cracked in this area, the block counterbore must be checked. Special tools are available to cut the liner supporting ledge to proper contour, and shims are available to bring the liner to proper projection above the block.

FIGURE 10-4. Pits on Liner. *(Courtesy of Cummins Arizona Diesel)*

Cylinder, Piston, and Connecting Rod Service / 183

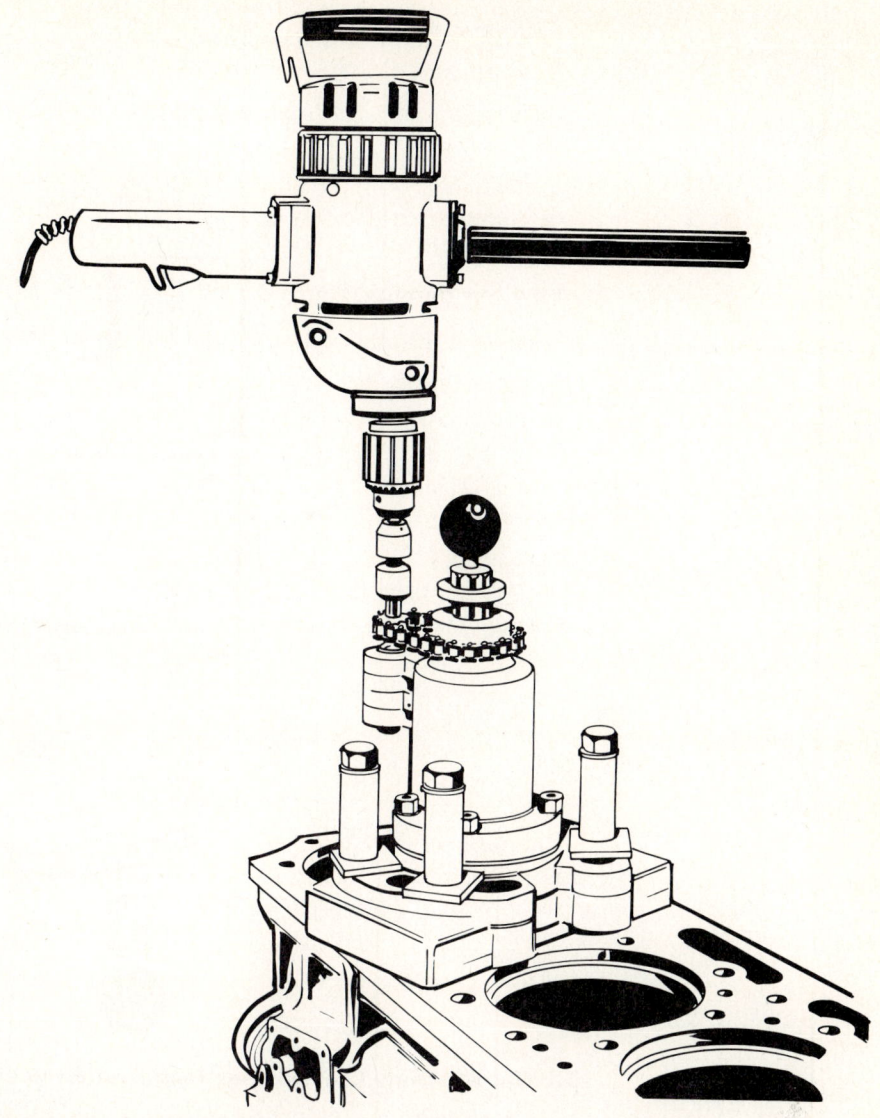

FIGURE 10-5. Counterboring Tool. *(Courtesy of Cummins Engine Co., Inc.)*

[**Fig. 10-5**] If needed, a completely new counterbore can be installed with special equipment.

When the seal ring area at the lower end of the liner is found badly pitted, a new section can be installed with proper equipment. [**Fig. 10-6**] Small pits can be cleaned with sandpaper and filled with an iron cement-like "Smooth-On." When the area dries, it can be sanded smooth.

The crankshaft must be covered during this repair; the crankcase should be washed down with solvent after repairs are completed.

CYLINDER LINER INSTALLATION

After all repairs are made and the block is cleaned, the liners can be installed.

1. Assemble new seal rings on the liner, making sure that they are not twisted. Each engine builder gives details of this assembly.

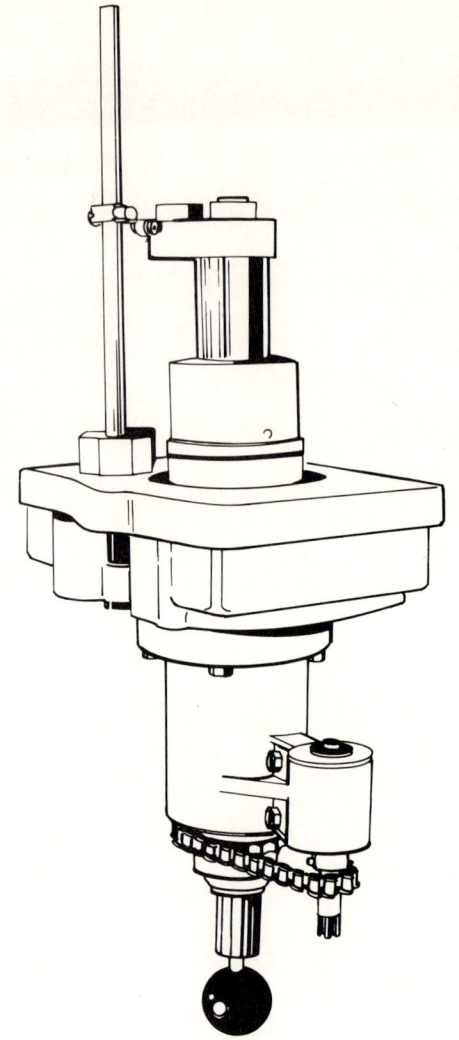

FIGURE 10-6. **Lower Counterboring Tool.** *(Courtesy of Cummins Engine Co., Inc.)*

2. Some engines use a *filler band* under the liner flange. Dip it in clean oil, then assemble it on the liner, and slide it up to the flange. [Fig. 10–7]

3. Wet the lower seal rings with oil, and wet the lower seal seat in the block.

4. Insert the liner, positioning any small pits on the engine center line.

5. Place hands on the top of the liner, fingers in. Push the liner and seals in by giving a sharp hunch with your shoulders. The liner should go in almost all the way. Engine builders offer special drivers or devices to seat the liners. One can use a piece of 4-inch by 4-inch wood and a mallet.

6. Install all liners the same way.

PISTON INSPECTION AND CLEANING CAPS

Pistons and liners are usually replaced during engine overhaul. If pistons are to be reused, they must be cleaned and inspected carefully.

Used pistons can be removed from the connecting rod by first removing

FIGURE 10-7. **Filler Band and Seal.** *(Courtesy of Caterpillar Tractor Co.)*

the retainers from the piston pin (wrist pin), then heating the piston in boiling water to expand it. Aluminum pistons expand about twice as much as iron. The wrist pin should never be driven out, if the piston is to be saved.

Cleaning Used Pistons

Aluminum pistons can be cleaned in a glass-bead cleaner. They must never be put into a caustic cleaning tank with iron or steel parts.

The hard carbon in the ring grooves can be scraped out by grinding a ring end to a sharp edge and manually removing the carbon. One must be careful not to remove any metal from the grooves. The ring used should be from the same sized ring groove.

Piston Inspection

After the pistons are cleaned, they can be inspected to see if further use is possible.

1. Check the piston visually for cracks in either the top or the wrist pin bore. Look at the corners of the valve reliefs. A highly loaded engine often has heat cracks at overhaul periods. If any pistons are damaged, all should be replaced. [**Fig. 10-8**]
2. Ring grooves wear wide. Check them with a new ring, held flush with the piston surface. Insert a feeler gauge beside the ring. More than 0.006-inch clearance all around indicates wear. Reject the pistons. Wear of the grooves accelerates rapidly.

Connecting Rod Inspection

While the pistons are off the rods, the mechanic should check the rods completely.

1. Steel parts can be inspected for cracks by magnetic inspection systems. Rods, rod bolts, wrist pins, and many other parts can be checked with these devices. [**Fig. 10-9**]

186 / Diesel Engine Service

FIGURE 10-8. Piston Inspection Points. *(Courtesy of Cummins Engine Co., Inc.)*

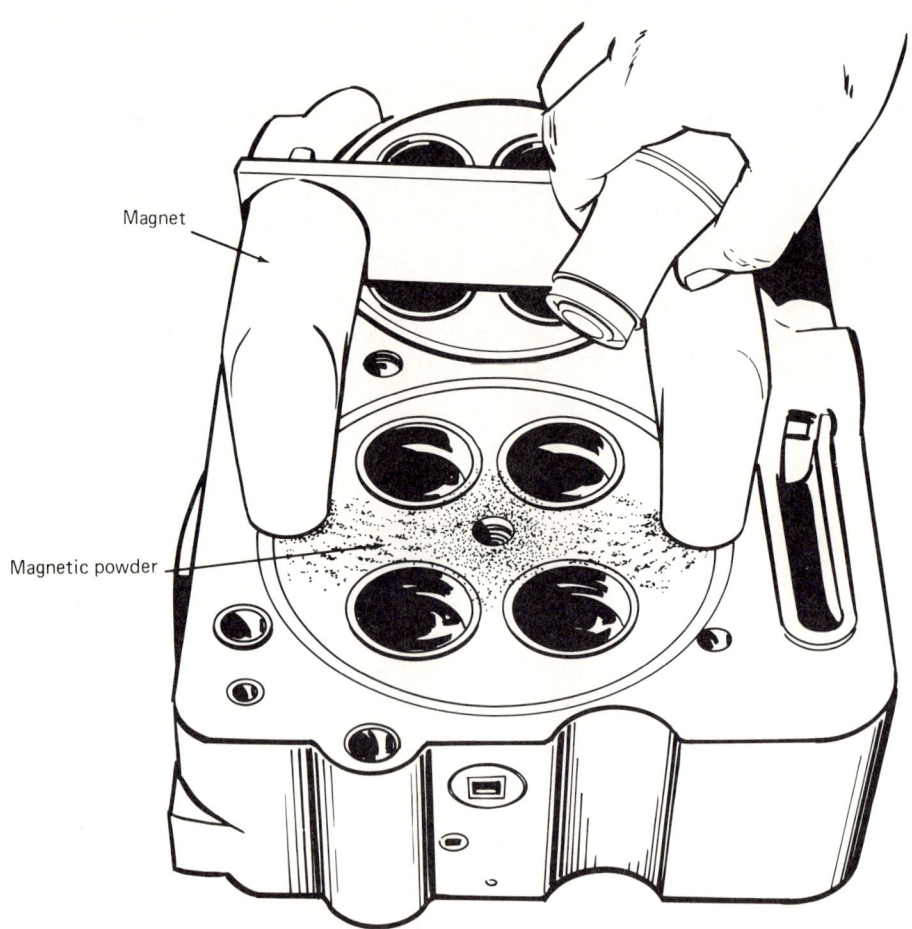

FIGURE 10-9. Magnetic Inspection. *(Courtesy of Cummins Engine Co., Inc.)*

Cylinder, Piston, and Connecting Rod Service / 187

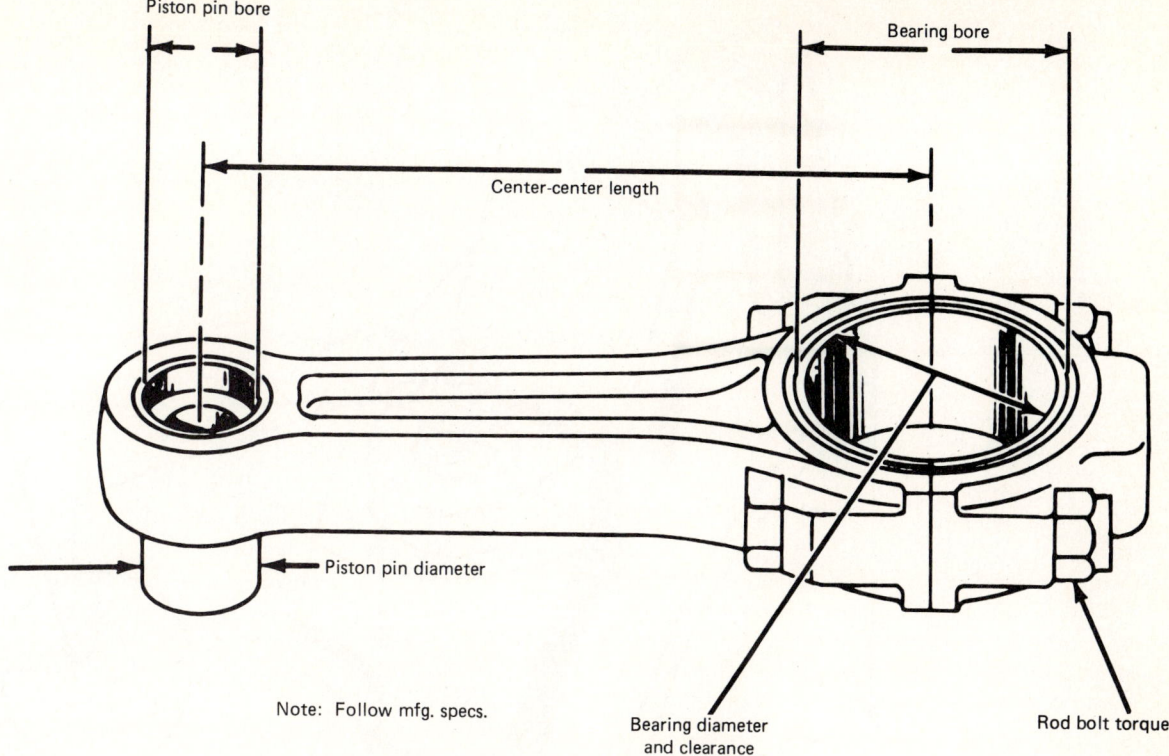

FIGURE 10-10. Connecting Rod Inspection. *(Courtesy of Caterpillar Tractor Co.)*

2. Check for rod twist and bend in checking fixtures, which use dial indicators to show side-to-side differences.

3. Examine the rod bolts closely for stretch. Compare their length to the standard for new bolts. **[Fig. 10–10]**

4. Check the wrist pin bushing diameter, and the bearing bore. Wrist pin bushings can be replaced if worn out-of-round. Some engine builders offer service bushings, which can be bored to correct small rod length faults.

5. Check the rod length in a suitable fixture.

6. When rod bolts are removable, the pad where the bolt head seats should be checked for cracks in the corner. Small cracks can be removed by machining, if care is taken to produce a radius in the finished corner. Avoid sharp edges. **[Fig. 10–11]**

Assemblying Pistons on Rods

After all cleaning and inspections are complete, the pistons can be assembled on the rods. For all aluminum pistons, a heating device must be used to expand the piston for insertion of the wrist pin. This can be a special oven or a bucket of water heated by a fire in a drum of coal or wood. **[Fig. 10–12]**

1. Provide a clean surface on which to work. Use a board, plywood, or a clean bench.

2. Install the pin retainer in one side of the piston bore. Arrange the pistons so that part numbers are all one way.

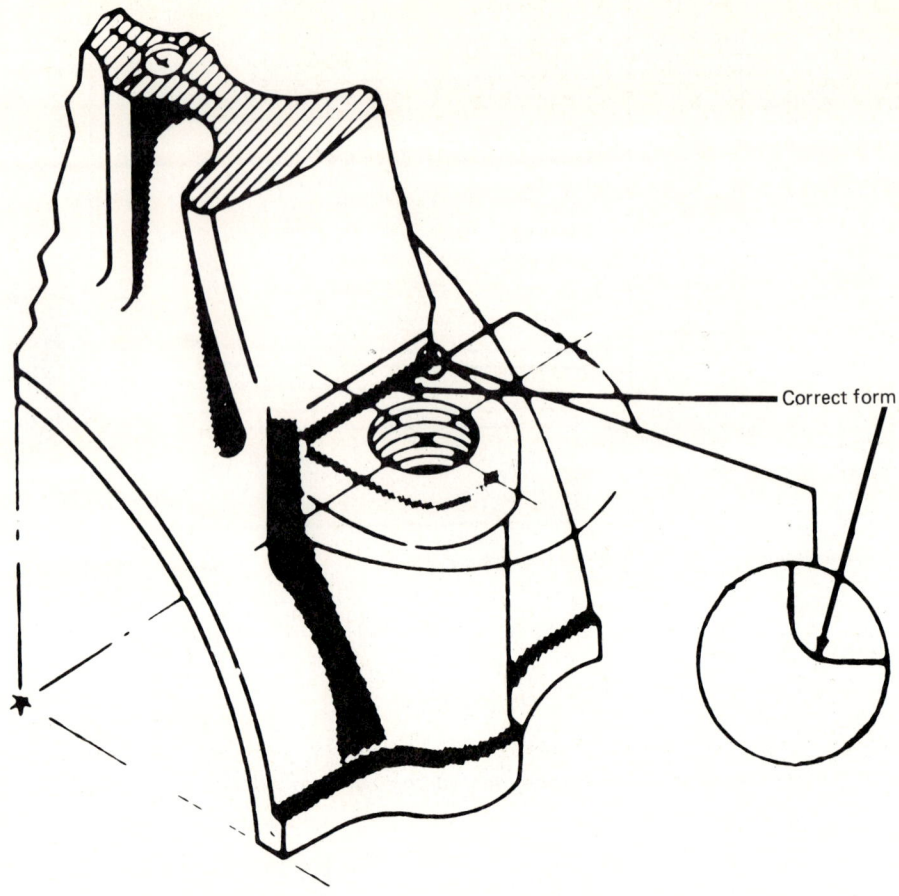

FIGURE 10-11. Bolt Pad Radius. *(Courtesy of Cummins Engine Co., Inc.)*

3. If boiling water is used, use a wire hook to hang the piston in the water.
4. After about five minutes of heating, remove the piston, set it on the board or bench, and quickly insert the wrist pin through the piston bore and rod end.
   Note: Be sure to orient the rod number the same way on all assemblies.
5. Install the pin retainer in the piston.
6. Assemble all pistons to their rods.

Installation of Rings on Pistons

1. Set a piston and rod in a vise, so that the bottom of the piston rests on the vise. Tighten the vise only enough to hold the rod in place.
2. Ring expanders open the ring just enough to slip over the piston. [**Fig. 10-13**] Some rings have a top mark, while others have notches in the ends. The notches go up. If no expander tool is available, rings can be expanded with a shop rag hem. Make two loops in the rag hem, about 5 inches apart. Hook the ring ends in the loops, and pull with your thumb and second finger, extending the first finger along the ring to guide it. Spread the ring just enough to clear the piston. [**Fig. 10-14**]

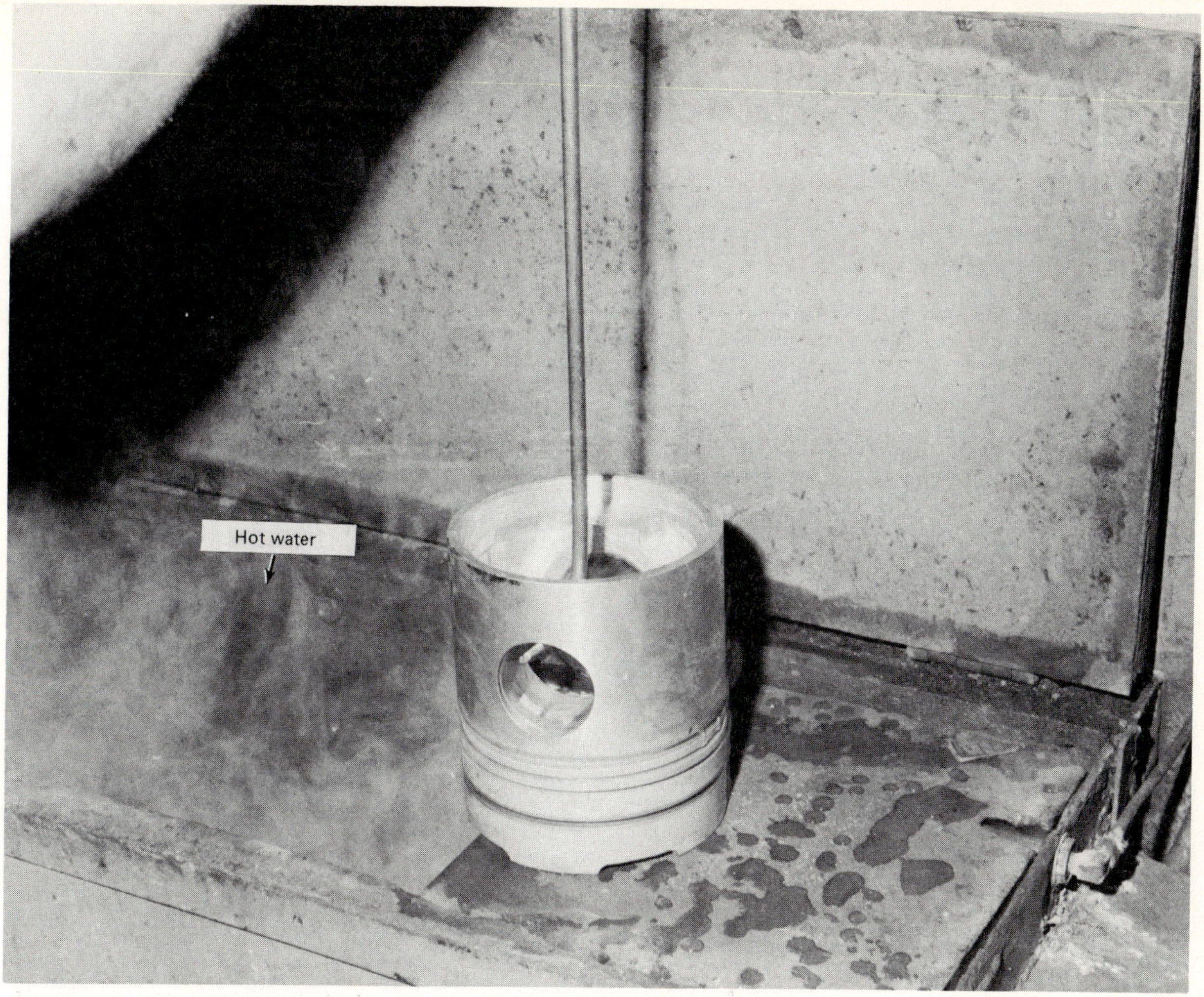

FIGURE 10-12. **Heating Piston.** *(Courtesy of Cummins Arizona Diesel)*

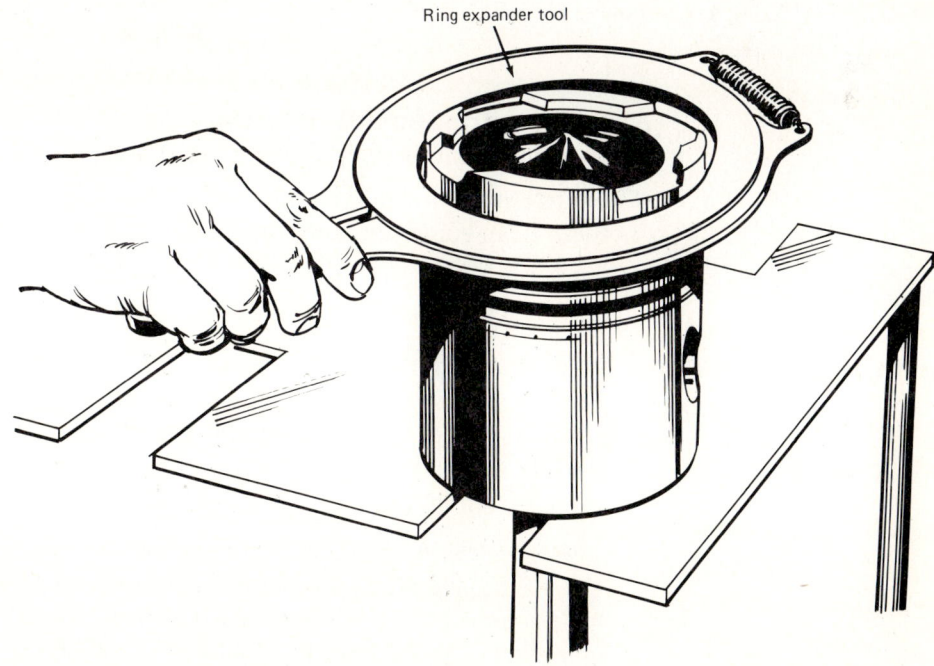

FIGURE 10-13. **Using Ring Expander.** *(Courtesy of Cummins Engine Co., Inc.)*

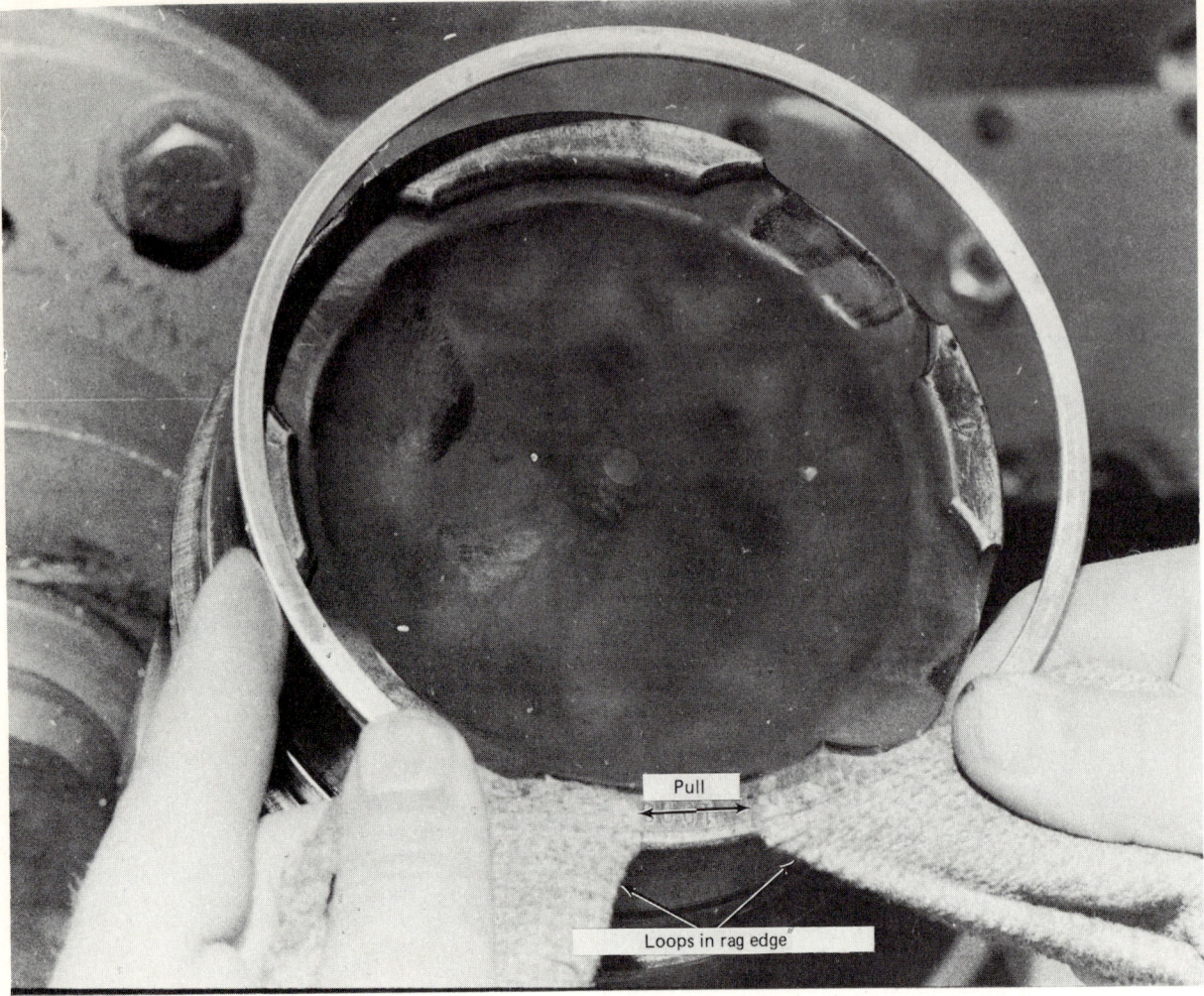

FIGURE 10-14. Using Rag to Expand Rings.

3. Put the oil ring on first. If an expander is used behind the oil ring, its ends must butt, not be overlapped. [Fig. 10-15]
4. Do not store piston, ring, and rod assemblies on their sides. Either hang them from some structure or install them as each is assembled.

Installing Piston and Rod Assemblies

1. Several types of ring compressors are offered. The common band type is flexible as to size but often gives trouble. A ring compressor can be machined from an old liner. [Fig. 10-16]
2. There are two ways to make such a compressor. The one shown in Figure 10-16 uses the lower end of the liner; and it is taper-bored for about 3 inches. When this device is installed on the piston, the connecting rod can be pulled to insert the piston and rings, working from the bottom.

Cylinder, Piston, and Connecting Rod Service / 191

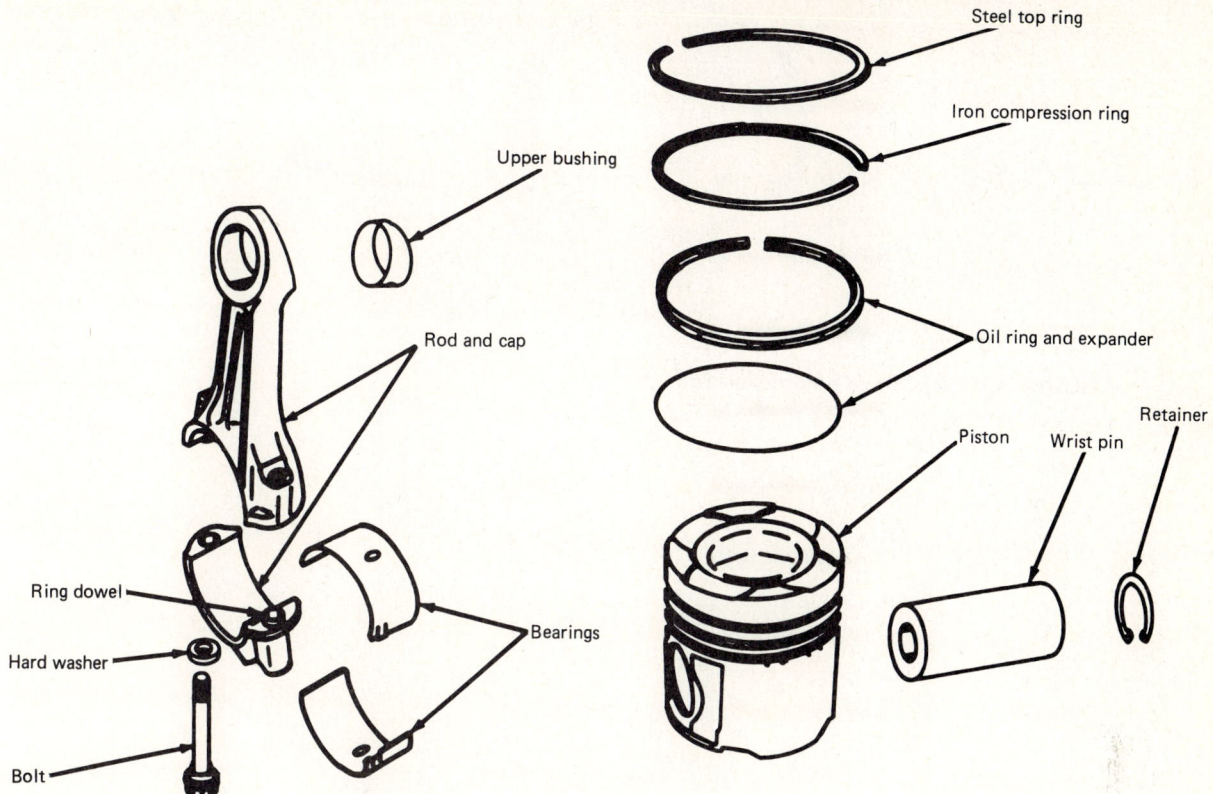

FIGURE 10-15. Ring Set. *(Courtesy of Cummins Engine Co., Inc.)*

3. Another way to use an old liner is to cut it in half and weld a hinge on one side and handles on the other. The lower unworn end is used. With the latter device, the piston is pushed through it into the cylinder.

4. Position the ring gaps so that none are over the wrist pin and no two are in line. The rings will stay as they are installed because there is no force to turn them. **[Fig. 10-17]**

5. The rod bolt threads should be protected by short pieces of hose. They should not contact the crankshaft journal. Some rods have threaded holes for capscrews. A short dowel sized to fit the holes can be inserted to guide the rod end over the journal.

6. Rod bearing upper halves should be assembled just before installing the rod and piston. Wet the surface with oil. Also oil the rings and cylinder bore.

7. Rods and caps have matching numbers and are placed on the camshaft side of vertical engines, and to the outside on V-engines. **[Fig. 10-18]**

8. When the connecting rod contacts the crankshaft journal, insert the lower bearing half into the cap, wet with oil, and apply the cap to the rod. Be sure that the numbers match. Numbers are stamped to match the cylinder number.

   Note: If the liner or piston does not assemble easily as described, remove the unit and correct the interference.

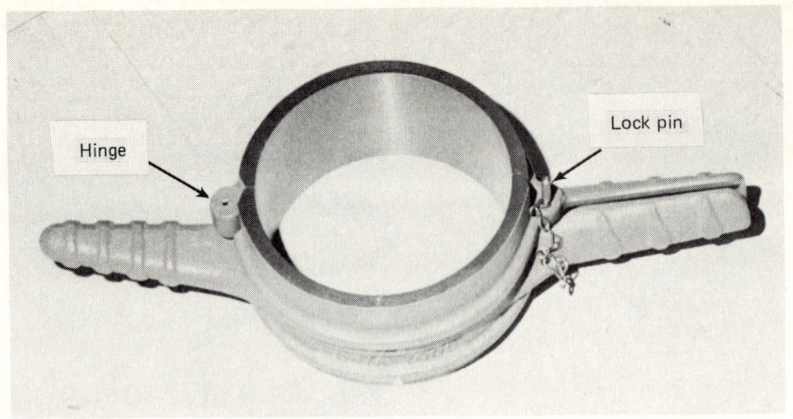

FIGURE 10-16(a). Ring Compression Tool.

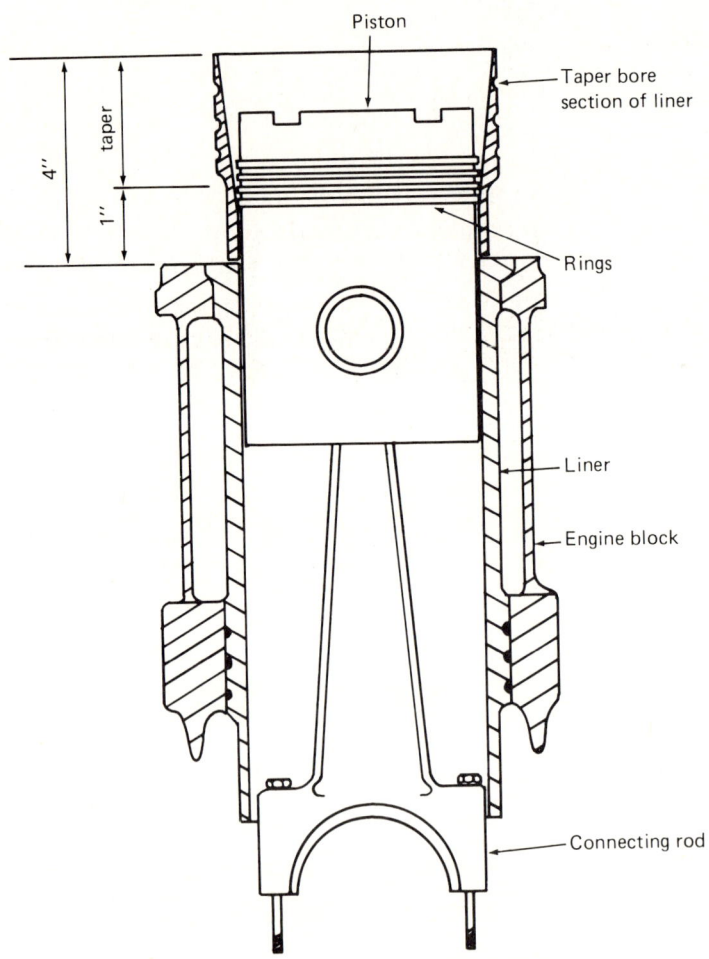

FIGURE 10-16(b). Ring Compression Tool. *(From a sketch by P. M. Uhl)*

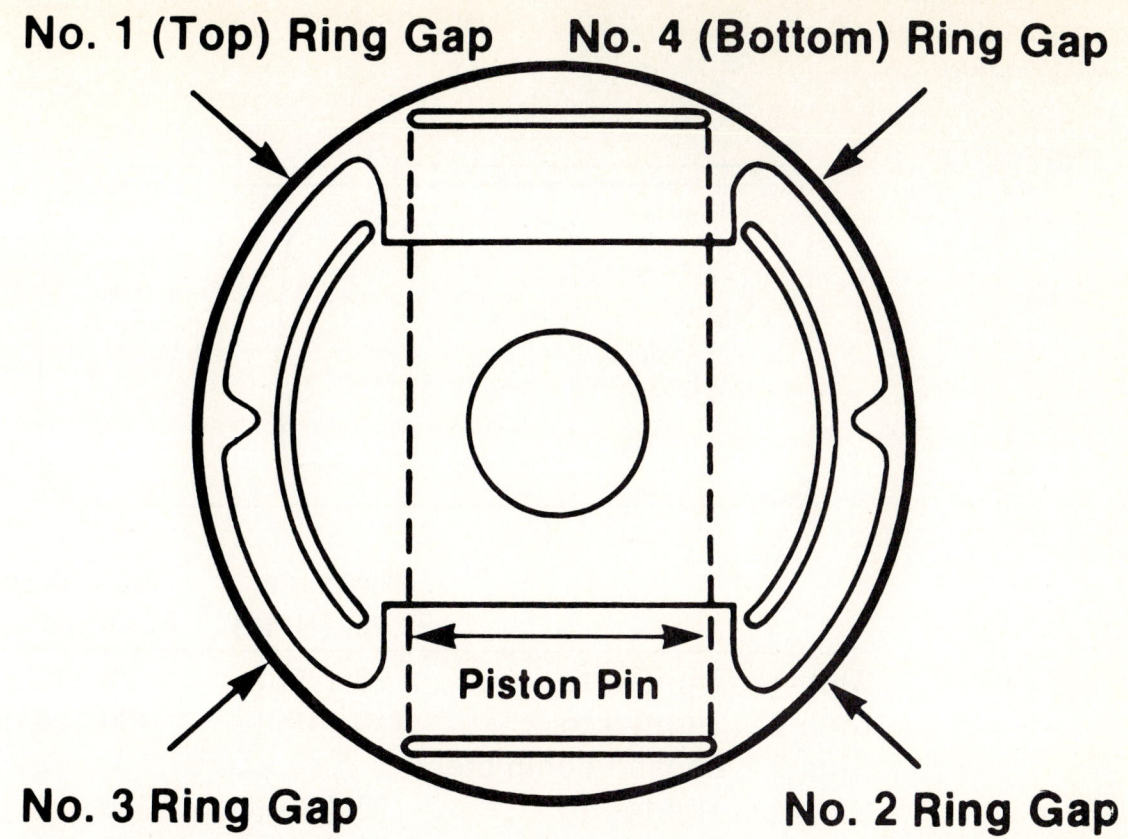

FIGURE 10-17. Ring Gap Placement. *(Courtesy of Cummins Engine Co., Inc.)*

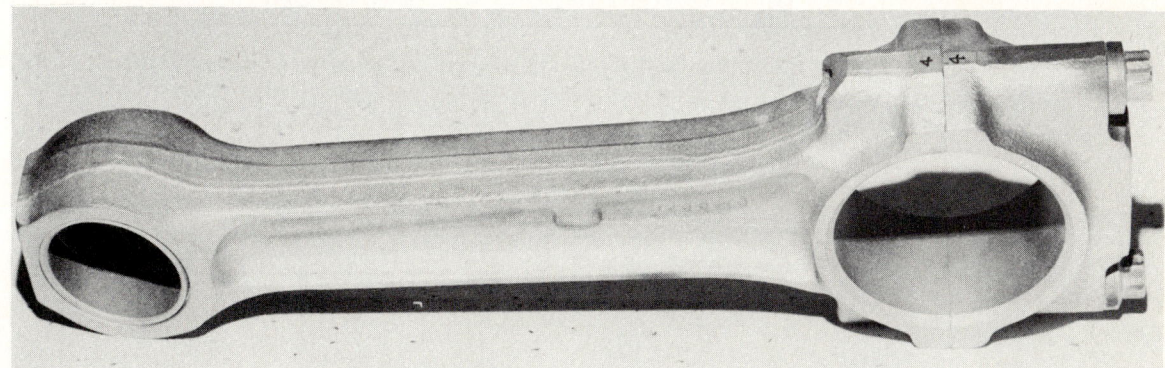

FIGURE 10-18(a). Road and Cap Numbers. *(Courtesy of Cummins Arizona Diesel, Inc.)*

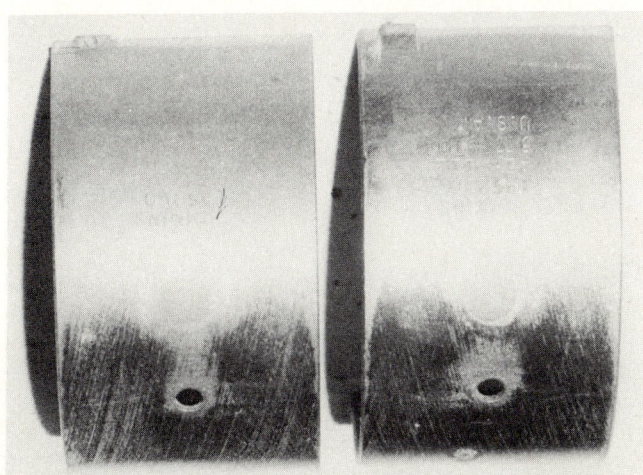

FIGURE 10-18(b). Improperly Installed Rod Bearings. *(Courtesy of Cummins Arizona Diesel)*

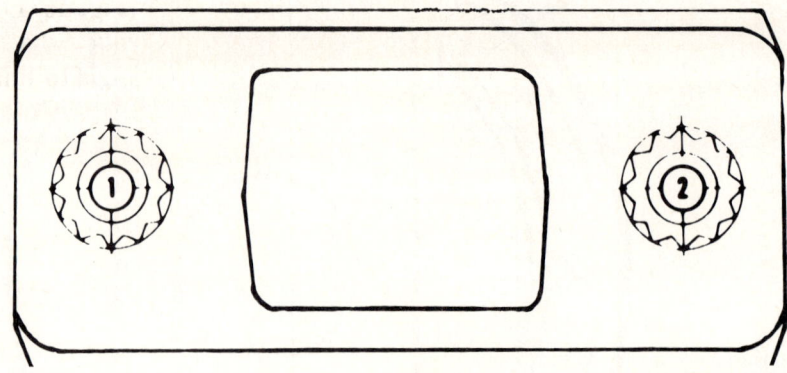

	Minimum Ft.-lb. [N•m]	Maximum Ft.-lb. [N•m]
Step 1. Tighten to	70 [95]	75 [102]
Step 2. Tighten to	140 [190]	150 [203]
Step 3. Loosen completely		
Step 4. Tighten to	25 [34]	30 [41]
Step 5. Tighten to	70 [95]	75 [102]
Step 6. Tighten to	140 [190]	150 [203]

FIGURE 10-19. Rod Bolt Tightening Sequence. *(Courtesy of Cummins Engine Co., Inc.)*

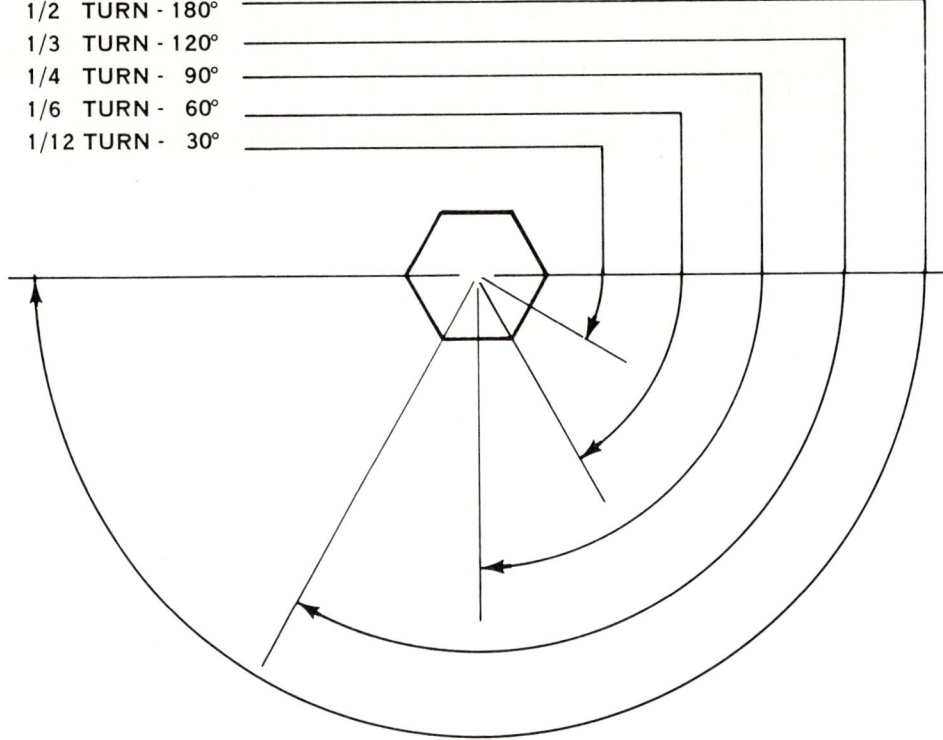

FIGURE 10-20. Degree Tightening. *(Courtesy of Caterpillar Tractor Co.)*

9. Wet the fastener threads with oil, and secure the rod bearing as described for the engine being serviced. [**Fig. 10-19**] Alternate from side to side to draw the assembly together evenly. Some engines require the fasteners to be turned through a specified angle to final tension. [**Fig. 10-20**]
Note: Observe any piston markings that are specified during assembly.

11

TWO-STROKE-CYCLE ENGINES

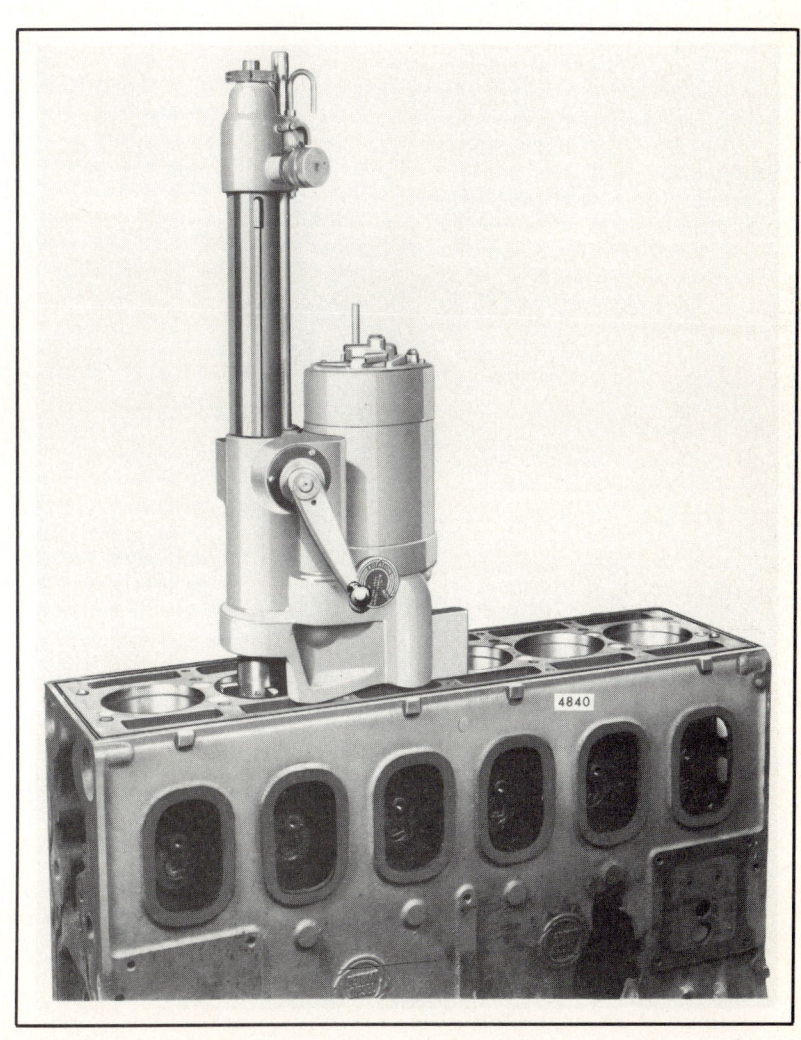

The foregoing instructions apply to four-stroke-cycle engines. Detroit Diesel two-stroke-cycle engines differ in that they have steel liners that fit in block bores. The liners have openings called *ports*, through which air enters the cylinders.

LINER INSPECTION

These cylinders must fit closely in the block bore. Aside from the usual checks for wear, we must oberve the outside surface for dark areas that indicate poor contact. [**Fig. 11-1**]

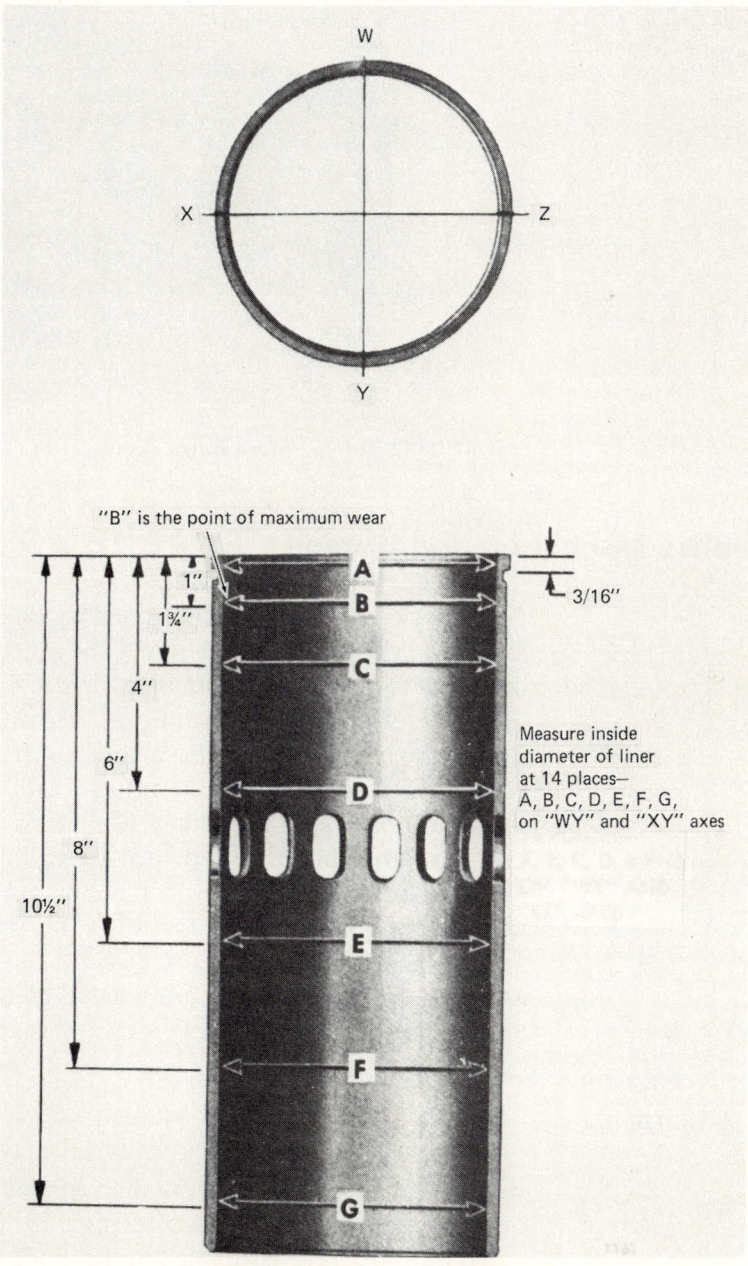

FIGURE 11-1. General Motors Liner. *(Courtesy of General Motors Corp.)*

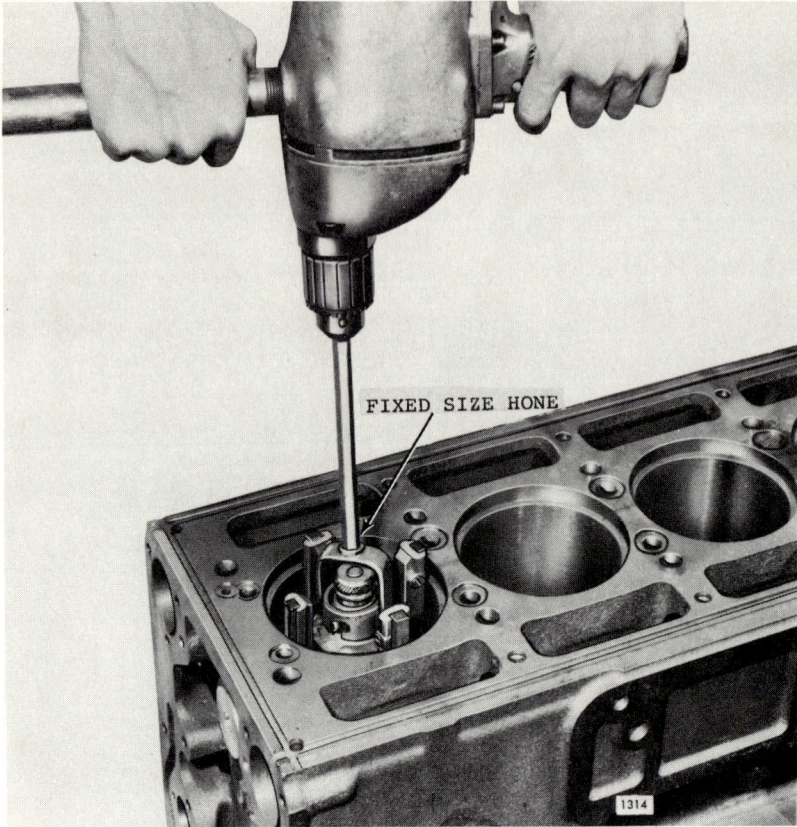

FIGURE 11-2. **Honing Block Bores.** *(Courtesy of General Motors Corp.)*

Block Bore Inspection and Honing

The block bore can be reconditioned with a hone that can be set to a fixed size, rather than a spring-loaded hone. The object is to remove irregularities in the bore, not to establish a particular surface pattern. The hone oil prescribed by the maker should be used, although use no oil on dry hones. [**Fig. 11-2**]

This operation must be guided by feel. That is, a high spot will need greater effort to hold the hone motor. Light cuts should be made to "feel out" the areas needing more honing.

The hone should be kept moving up and down in both the zone above the air ports and that below the ports. Short strokes should be used in tight areas, and the hone will run more easily as the high spot is reduced.

After the bore is straightened, the block should be thoroughly washed, and bore diameters checked with a dial bore gauge. [**Fig. 11-3**] The honed diameter should be compared with the standard for the engine model.

Block Boring

Blocks that need more resizing than can be provided by honing can be bored oversize. Oversize liners are offered to fit.

Note: We have assumed that the block has been stripped and checked for cracks. These tests are detailed in Detroit Diesel manuals. [**Fig. 11-4**]

Two-Stroke-Cycle Engines / 201

FIGURE 11-3(a). Measure Bore Sizes. *(Courtesy of General Motors Corp.)*

FIGURE 11-3(b). Dial Bore Gauge. *(Courtesy of Cummins Engine Co., Inc.)*

202 / Diesel Engine Service

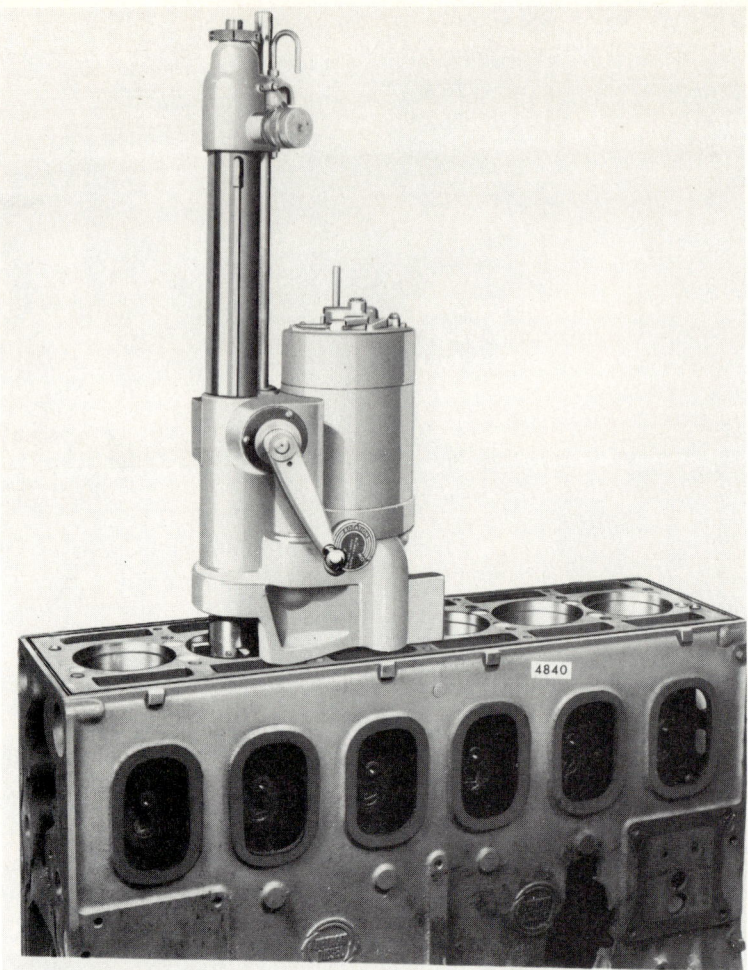

FIGURE 11-4. Block Boring Tool. *(Courtesy of General Motors Corp.)*

Service Main Bearing Caps

Service main bearing caps are offered for use in blocks, when the original bearing cap has been damaged or lost. The caps for all bearing saddles except the rear one have stock so that they can be line-bored. [Fig. 11-5]

Several different caps may be tried to achieve proper alignment. The final bore size and alignment must be within builder's tolerance for that model.

CRANKSHAFT INSTALLATION

The installation of Detroit Diesel crankshafts follows the described practice. All surfaces must be kept clean, with any burrs removed. The tightening standard requires turning the fasteners through a specified angle from an initial torque to final tension. The crankshaft should be hand-free after securing all main bearings.

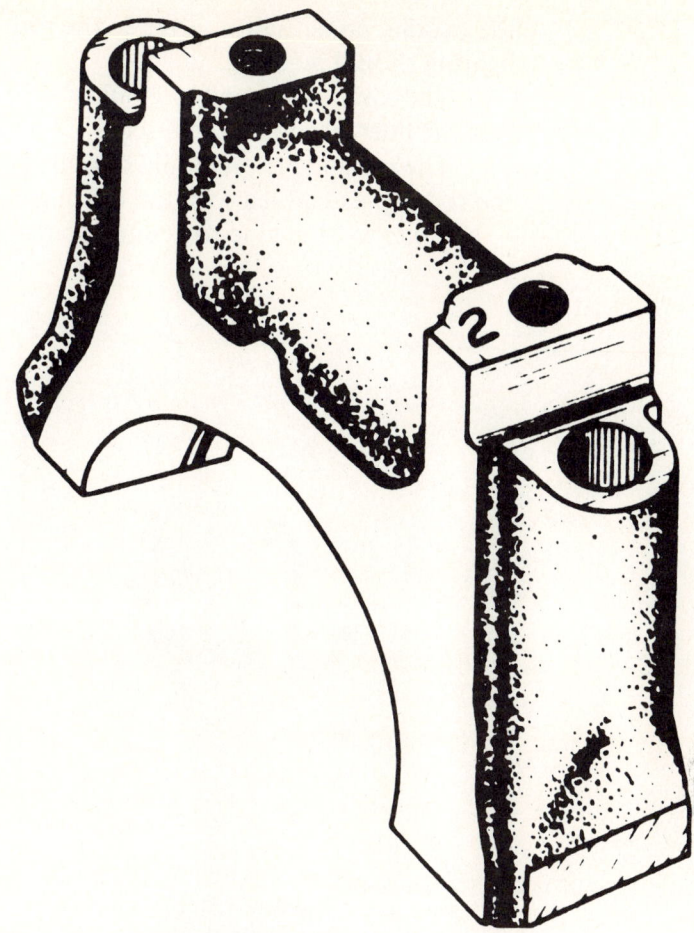

FIGURE 11-5. Main Bearing Cap. *(Courtesy of General Motors Corp.)*

CYLINDER LINER INSTALLATION—DDA ENGINE

Because the big end of the connecting rod will not go through the liner, the piston, ring, and rod assembly must be inserted in the liner from the bottom, and the complete cylinder assembly installed in the block. A special ring compression tool is offered for inserting the rings and piston into the liner. A clamp-type tool can also be used. This has been described. [Fig. 11-6]

Detroit Diesel Series 71 liners are supported by an insert of fixed thickness. Shims are furnished to adjust the height and must be inserted *under* the insert. [Fig. 11-7]

No seal rings are used on the liners of Series 71 engines. The complete assembly should slide smoothly into place on the insert. The installed height of all liners should be from 0.045 inch to 0.050 inch, below the block surface, with no more than 0.002-inch variation in liner height under one head.

The rod bolt threads should be protected with short pieces of tubing and the bearing upper half inserted in the rod, seating the tang in the notch. The bearing should be oiled.

The complete assembly should slide into the bore and be pushed down to seat the liner flange on the insert.

The rod and piston should be pulled into contact with the crankshaft and the rod cap assembled with the bearing over the bolts.

Note: V-71 engine rods must be placed side by side on a journal.

FIGURE 11-6. Cylinder Assembly. *(Courtesy of General Motors Corp.)*

Two-Stroke-Cycle Engines / 205

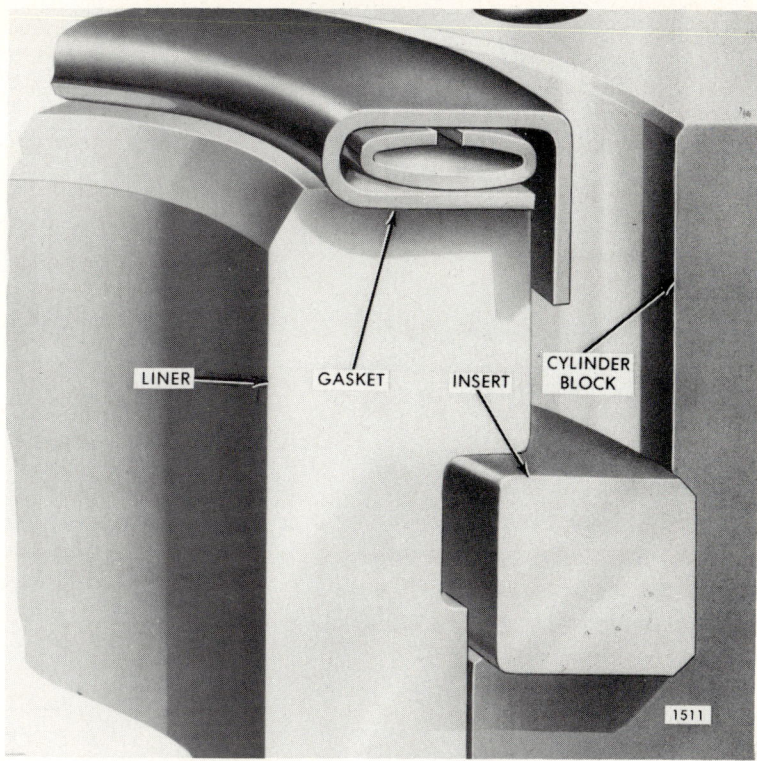

FIGURE 11-7. Liner Insert Detail. *(Courtesy of General Motors Corp.)*

The chamfered side goes to the side, and the flat side faces the other rod. Numbers on both rod and cap face the outside. **[Fig. 11-9]**

Details of this operation can be found in Detroit Diesel manuals. The general principles of service have been described but are not intended to replace the manufacturer's manual.

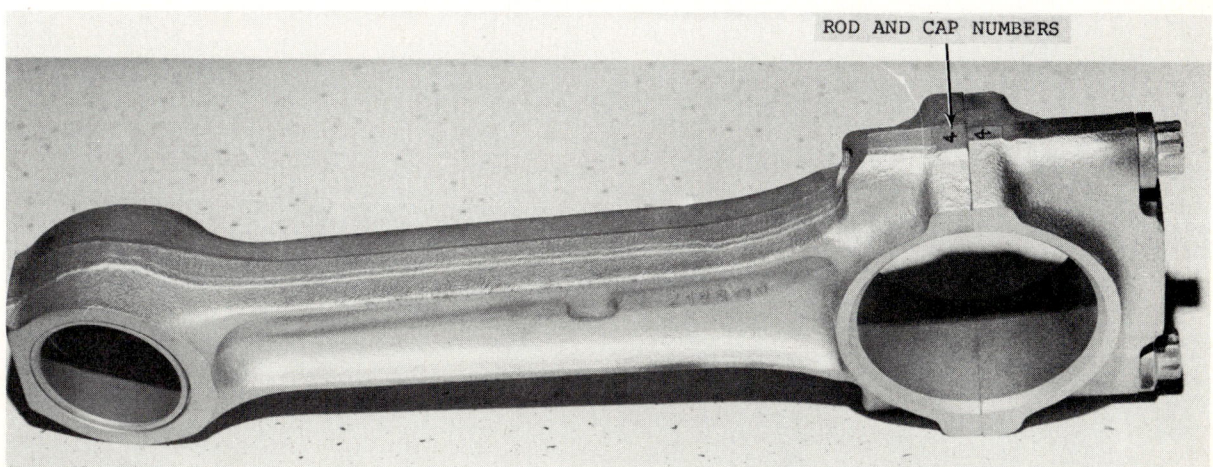

FIGURE 11-8. Rod and Cap Numbers. *(Courtesy of Cummins Engine Co., Inc.)*

12

ROCKER LEVERS: TAPPETS, FOLLOWERS, AND PUSH RODS

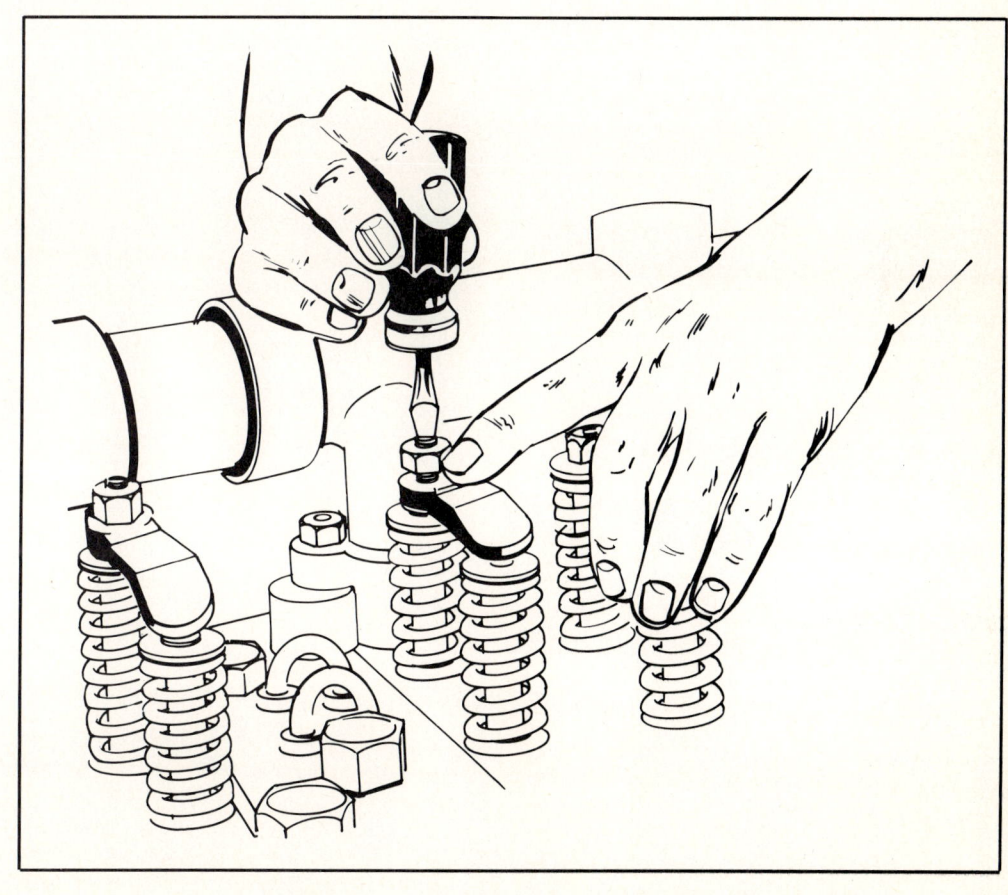

During the repair of an engine, the rocker levers that carry push rod movement to valves and injectors require attention. Wear occurs in their bushings and on the shaft on which they move. [**Fig. 12-1**]

Usually the levers are adjustable for clearance with the valve stems, and the adjusting screws should be checked for distorted threads and worn ends. Overtightening of adjusting screw locknuts can distort the threads and make accurate adjustment impossible.

Such distortion can be detected by removing the adjusting screw, taking off the locknut, and screwing it into the rocker lever bottom side up. Any distortion will increase resistance.

ROCKER LEVER INSPECTION

Rocker levers and shafts can be checked visually. Usually the assemblies are turned in for bench repair, but they should be checked in the field for need of rebuilding.

1. Look at the surface that contacts the valve stem (rocker lever "nose"). Many of these are hard, but any sign of wear should be noticed.
2. Shake the levers on the shaft. They must be free; if they rock sideways, wear is present on both lever bushings and shaft.
3. Look for signs of strain on the underside of the levers. Sticking valves can break levers and bend push rods.
4. Look for any chipping of contact area.
5. Consider the length of service and the general need of the engine for repair.

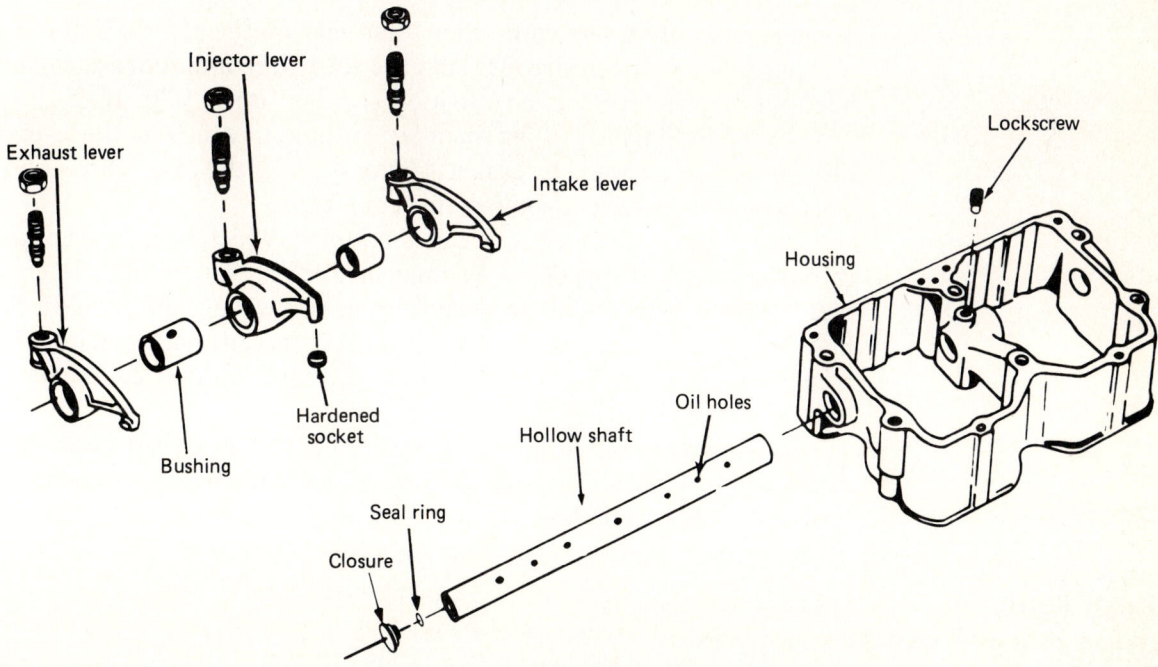

FIGURE 12-1. **Typical Rocker Level Assembly.** *(Courtesy of Cummins Engine Co., Inc.)*

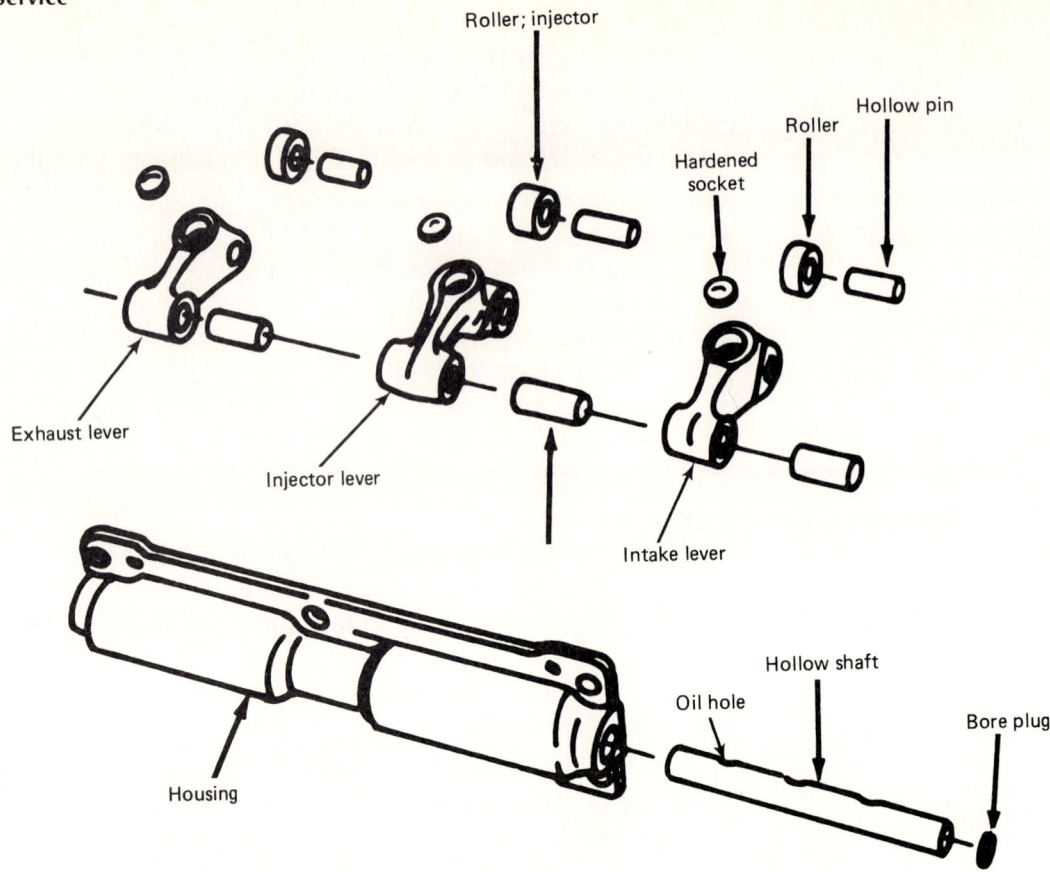

FIGURE 12-2. Cam Follower Assembly. *(Courtesy of Cummins Engine Co., Inc.)*

Cam Follower Levers and Tappets

Some engines use levers with rollers that bear on the camshaft and transmit the cam action to the push rods. They are always reconditioned at engine overhaul but require little service attention at other times. [**Fig. 12-2**]

Cummins engines adjust the injector timing by moving the lever assemblies inward or outward by gasket thickness under the cam follower housing. This process is described later in this chapter.

Many engines use a device called a *tappet,* which bears on the camshaft through a roller. Tappets are distinguished from cam follower levers by the fact that the tappet moves on the centerline of the camshaft, while a cam follower lever is pivoted at a short distance from the camshaft, and its roller contacts the cams inside of the centerline. The cam follower can be moved for timing adjustment; the tappet cannot. [**Fig. 12-3**]

Both types have rollers turning on pins that are oiled under pressure. Aside from roller pin breakage, there are few failures or normal inspection procedures. Magnetic inspection can be used.

Push Rods

Push rods or tubes carry the camshaft lobe force to the rocker levers. They are under compression during operation and must be stiff enough to resist springing out of line. [**Fig. 12-4**]

Rocker Levers: Tappets, Followers, and Push Rods / 211

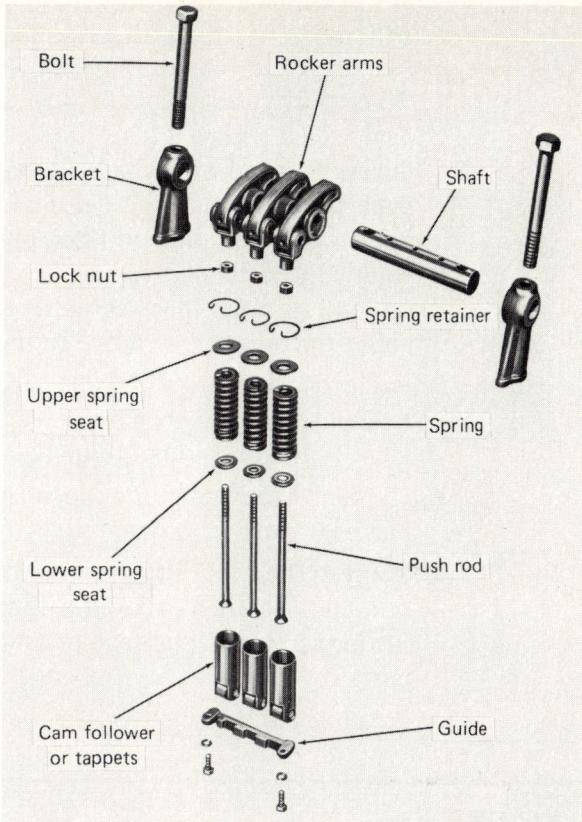

FIGURE 12-3. Valve Operating Train. *(Courtesy of General Motors Corp.)*

Most engines suspend the push rod between the tappet or cam follower at the lower end and the rocker lever at the top. Detroit Diesel push rods screw into the rocker lever clevis, and this thread is used for adjustment.

Others use ball-ended adjusting screws, with a socket on the top of the pushrod. The cam end is fitted with a hardened ball to contact the socket in the tappet or cam follower lever.

In some engines these sockets are hardened pieces that are replaceable. Both ends are pressure-lubricated.

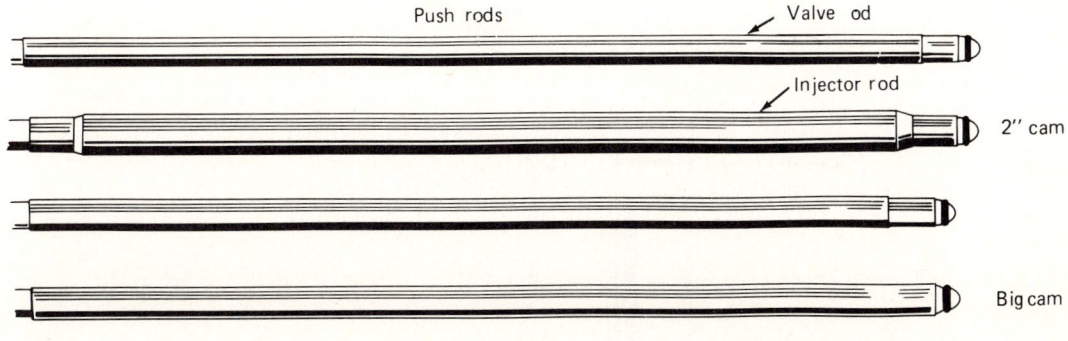

FIGURE 12-4. Push Rods. *(Courtesy of Cummins Engine Co., Inc.)*

Push Rod Inspection

1. When push tubes are made with hardened inserts in the end, they may become filled with oil. This does no harm but increases the weight and reduces the speed with which the valve spring can return the valve to its seat. Such push tubes can be detected by dropping on the floor. Empty push tubes will ring, while oil-filled ones give a dull sound. Push rods may be emptied by drilling a $\frac{1}{16}$-inch hole in the lower end, just above the hard inset, and we must check carefully to be sure that the hardened ends are not about to come out or move enough to ruin the push tube.
2. Ball ends wear and should be checked with a gauge called a *radius gauge*. Ball ends that are worn beyond the engine builder's limits require replacement of the push tube. [**Fig. 12–5**]
3. Hardened sockets in the tappet or cam follower lever can usually be replaced if worn or broken.
4. If any of these mating parts are worn beyond limits or have less than 80 percent contact when checked with mechanic's Blue, replace the socket, adjusting screw and push rod or tube. Avoid using a worn part against a new part.

VALVE CROSSHEADS

An engine that has two exhaust and two intake valves uses a bridge so that both valves can be operated with one rocker lever. This part is called either a *bridge* or a *crosshead*. [**Fig. 12–6**]

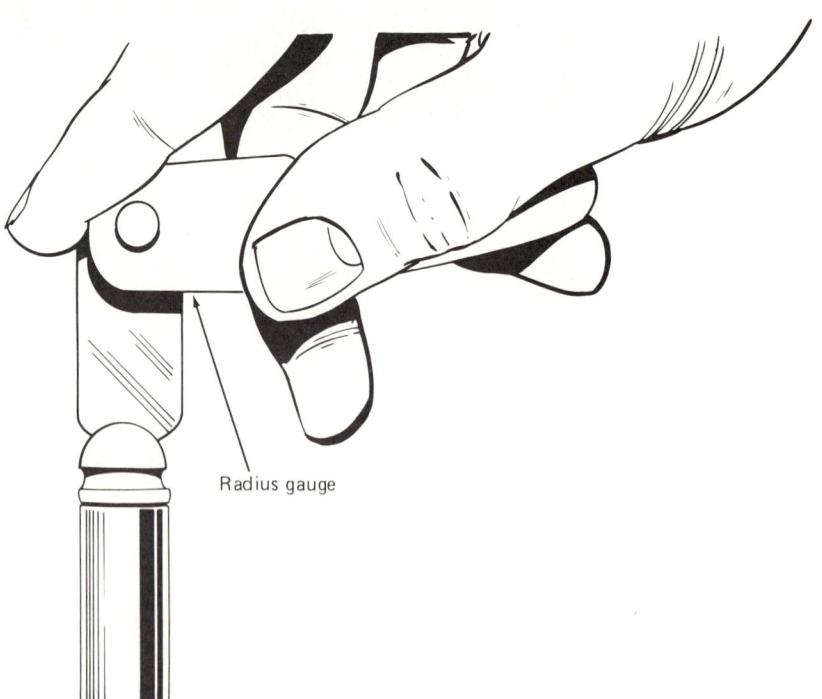

FIGURE 12-5. Check Ball Ends. (*Courtesy of Cummins Engine Co., Inc.*)

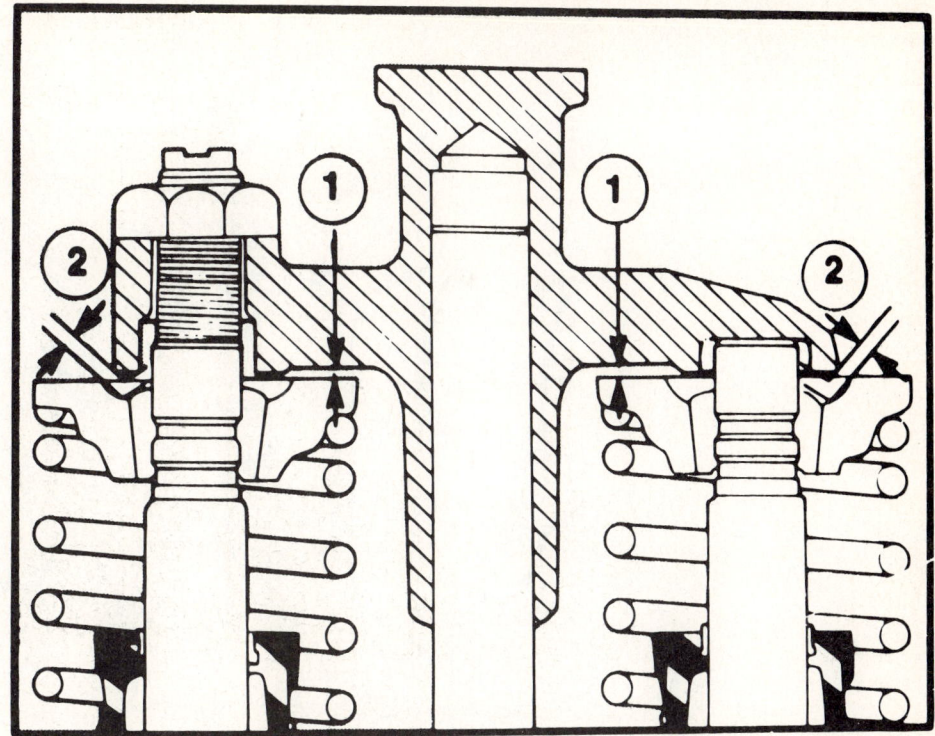

FIGURE 12-6. Valve Bridge or Crosshead. *(Courtesy of Cummins Engine Co., Inc.)*

Detroit Diesel and Cummins both use crossheads on four-valve engines. Older Detroit engines used a light spring on the crosshead. Newer engines have no spring, and Cummins never used one.

Crosshead guides are finely finished pins that are pressed into holes in the head, on the center-line between valves. One side of the crosshead is hardened to bear on the valve stem end.

One end of each crosshead is threaded for an adjusting screw, which is used to set the crosshead level, so that both valves are moved at the same time. The adjusting screw has a hardened end, and a locknut keeps it from turning.

Adjustment of Valve Crossheads

As mentioned, crossheads are adjusted to contact both valve items at the same time. This is done when the crossheads are assembled on their guides, before rocker levers are installed.

They can be adjusted with the rocker levers in place, but it is easier to do it before putting on the rocker levers. The crosshead adjustment has nothing to do with valve clearance, and is simply a leveling operation. [**Fig. 12-7**]

1. Loosen the adjusting screw locknuts.
2. Press down on the center pad of the crosshead.
3. Use a small screwdriver and *fingertip* effort to turn the adjusting screw until you just feel it contact the valve stem. If you are pressing down properly, the adjusting screw will be harder to turn when the valve stem is contacted.

214 / Diesel Engine Service

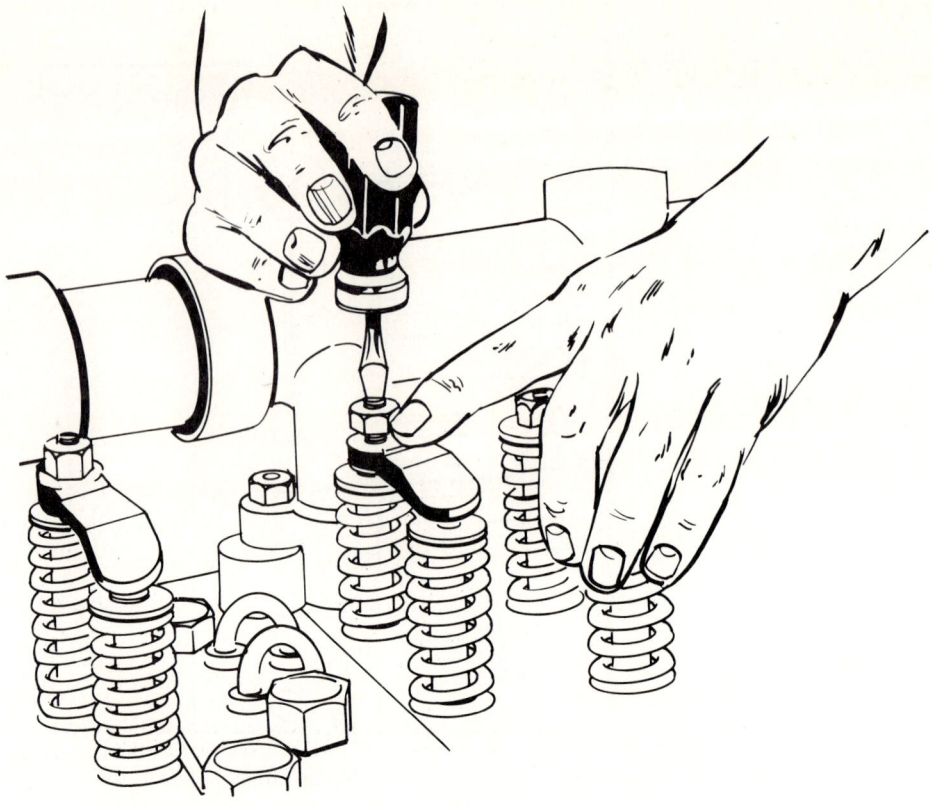

FIGURE 12-7. **Crosshead Adjustment.** *(Courtesy of Cummins Engine Co., Inc.)*

4. Turn the screw down about a one-third turn to straighten the crosshead on its guide. Tighten the locknut.

5. After the crosheads are adjusted, do *not* change them to another guide. Crossheads do not change adjustment unless the valves under them change height. Because newly ground valves will seat slightly and move up into the port, both valve clearance and crosshead adjustment should be checked after a short operating period. Once the valves are stabilized, their clearance can be checked at tuneup periods, with no adjustment of the crossheads unless valve clearance is found disturbed.

ASSEMBLY SUGGESTIONS

With injectors installed and secured (if cam-operated), and crossheads adjusted, the push rods and rocker levers can be installed.

1. Locate each push rod in its socket in the cam follower. Cummins push rods use a collar on the exhaust push rod on 5½-inch bore engines. The same collar is used on the intake push rod on 5⅛-inch bore engines. Some models have push rods of a different length than older engines. Be sure to use the right ones. The injector push rod is heavier than the valve push rods.

2. See that all adjusting screw locknuts are loose.

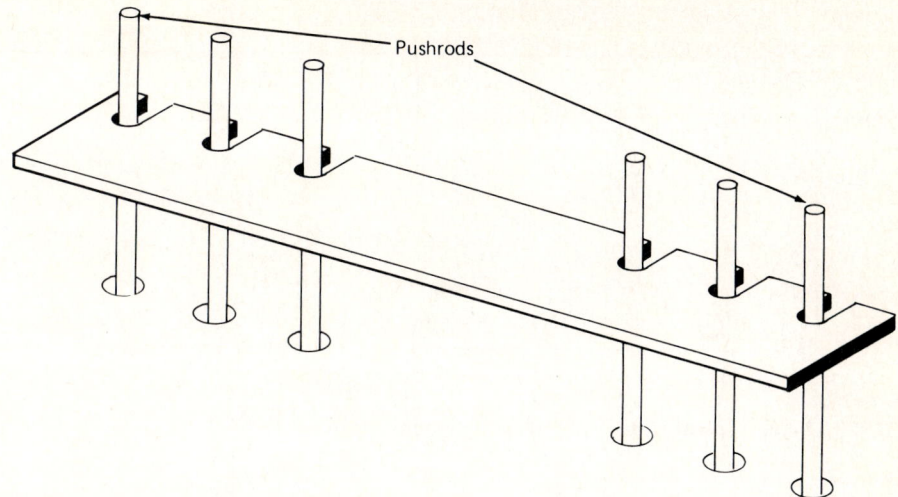

FIGURE 12-8. **Template to Position Push Rods.** *(From a sketch by P. M. Uhl)*

3. Put new gaskets on the head, if used. Use a template **[Fig. 12-8]** to hold the push rods in place. Set the rocker lever assembly on the head.
4. Locate each adjusting screw ball end in its push rod socket, then remove the template.
   Note: This method works well on Cummins engines. Other engine makes vary.
5. Do not tighten the rocker lever fasteners until all push rods are engaged. Follow the tightening sequence in **Figure 12-9,** and be sure to back off any adjusting screws that are on cams, valves opening.

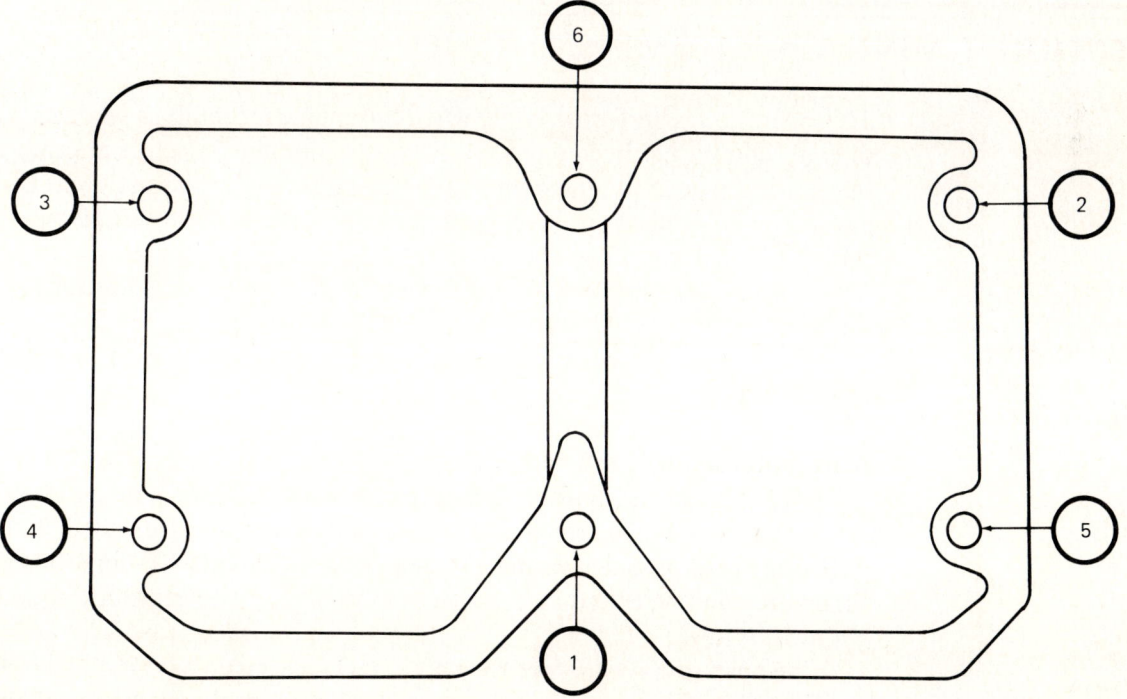

FIGURE 12-9. **Cummins Tightening Sequence.** *(Courtesy of Cummins Engine Co., Inc.)*

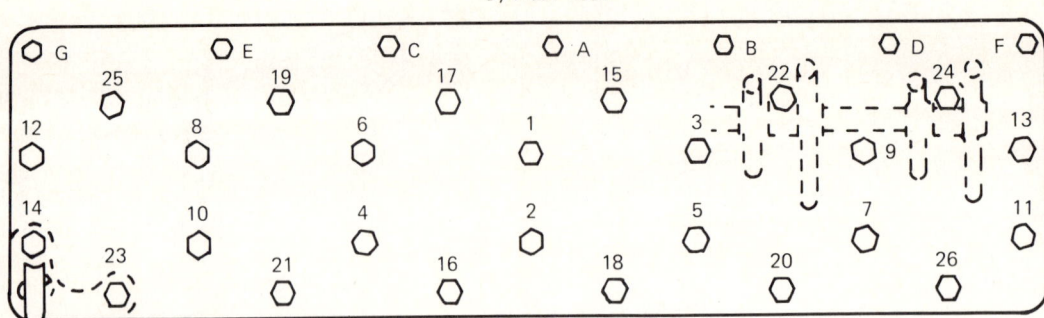

FIGURE 12-10. **Typical Head Tightening Sequence** *(Courtesy of Caterpillar Tractor Co.)*

6. On Detroit Diesel engines, screw the push rods into the rocker lever clevis to increase valve clearance.

7. In general, start to tighten the fasteners in the center of the rocker assembly, and alternate from side to side toward the ends.

8. Tighten rocker lever fasteners to the specified torque.
   Note: Tuneup specifications cover valve and injector adjustment. These are given for Cummins, Caterpillar, and Detroit Diesel engines. [**Fig. 12-10**]

INJECTION TIMING ON CUMMINS ENGINES

This operation is performed only when the engine is rebuilt or when camshaft or cam follower has been replaced. Because Cummins injectors are camshaft-operated, the push rod movement must be timed to the piston travel. H, NH, and NT engines have variable gaskets behind the cam follower levers. Other series engines have offset cam gear keys for timing.

Because piston travel and push rod travel are measured in thousandths, dial indicators are used with the standard holding fixture.

Given: A standard holding fixture, available as a Cummins Service Tool. [**Fig. 12-11**]

Two good dial indicators with 0.250-inch travel. Longer travel is not needed and can be confusing.

This operation is performed on one cylinder of each pair on H-NH engines. It is done before the injectors and rocker levers are installed. V-12 engines are timed on one cylinder on each bank, while other V-engines need be timed only on one cylinder, because they have only one camshaft and adjust with offset keys.

The piston travel *before* top center (BTC) is given as 0.2032 inch for all models. This dimension has been calculated to place the piston at the proper degree BTC. Injector push rod travel for all current engine models is given in the specification lists. The procedure follows.

Rocker Levers: Tappets, Followers, and Push Rods / 217

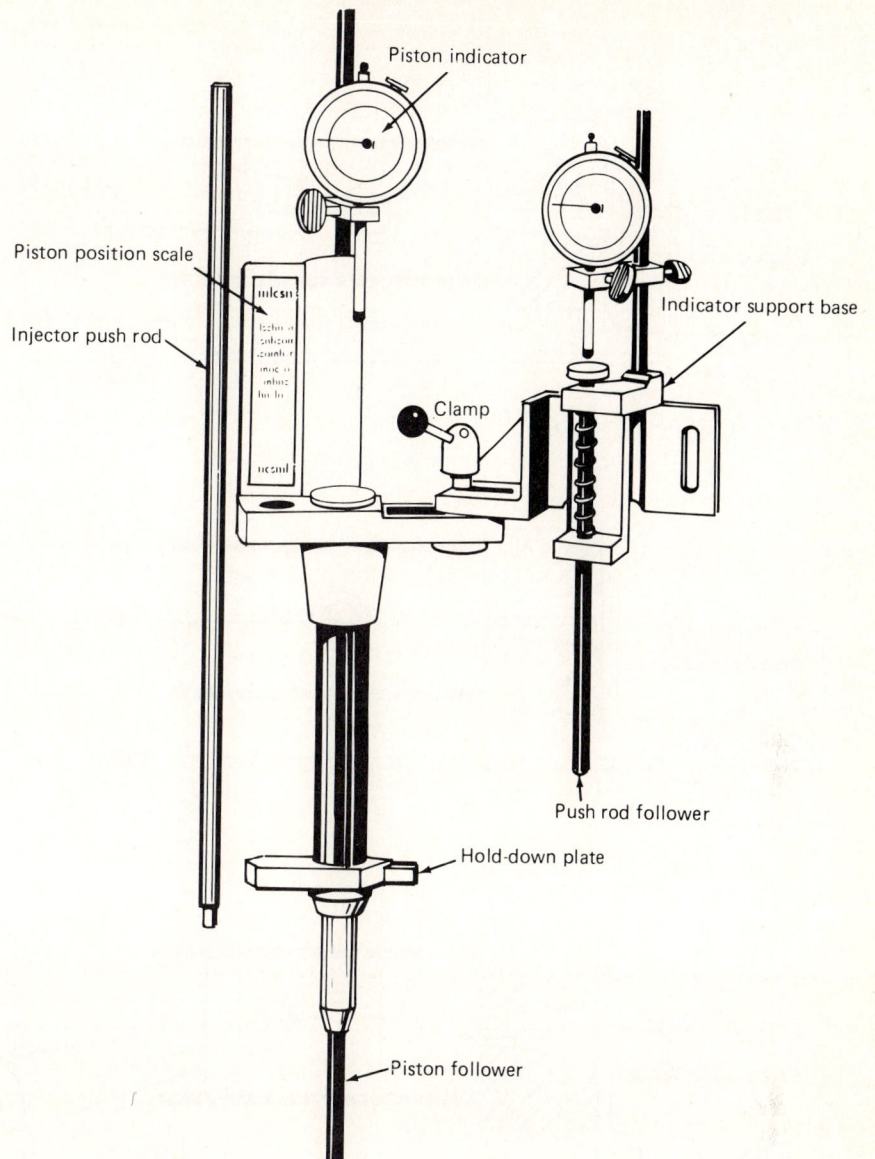

FIGURE 12-11. Injection Timing Tool. *(Courtesy of Cummins Engine Co., Inc.)*

1. Install all cam follower assemblies with one gasket.
2. Position the fixture with an injector push rod in the injector socket on No. 1 cylinder. [Fig. 12-12]
   Note: We will have to turn the engine less if we use the first three cylinders in firing order, Nos. 1-5-3.
3. See that the push rod is in place in the cam follower socket, and engage the indicator rod in its sockets.
4. Bar the engine to absolute top center on that cylinder. Set the indicator so that it is near the top of its travel, and zero the dial. Be sure to rock the engine back and forth until you are sure that you have top center.
5. Because the piston reaches top center twice each cycle, we must be sure that we are on the right stroke. The push rod indicator will move on the compression stroke, while it will not move on the exhaust stroke.

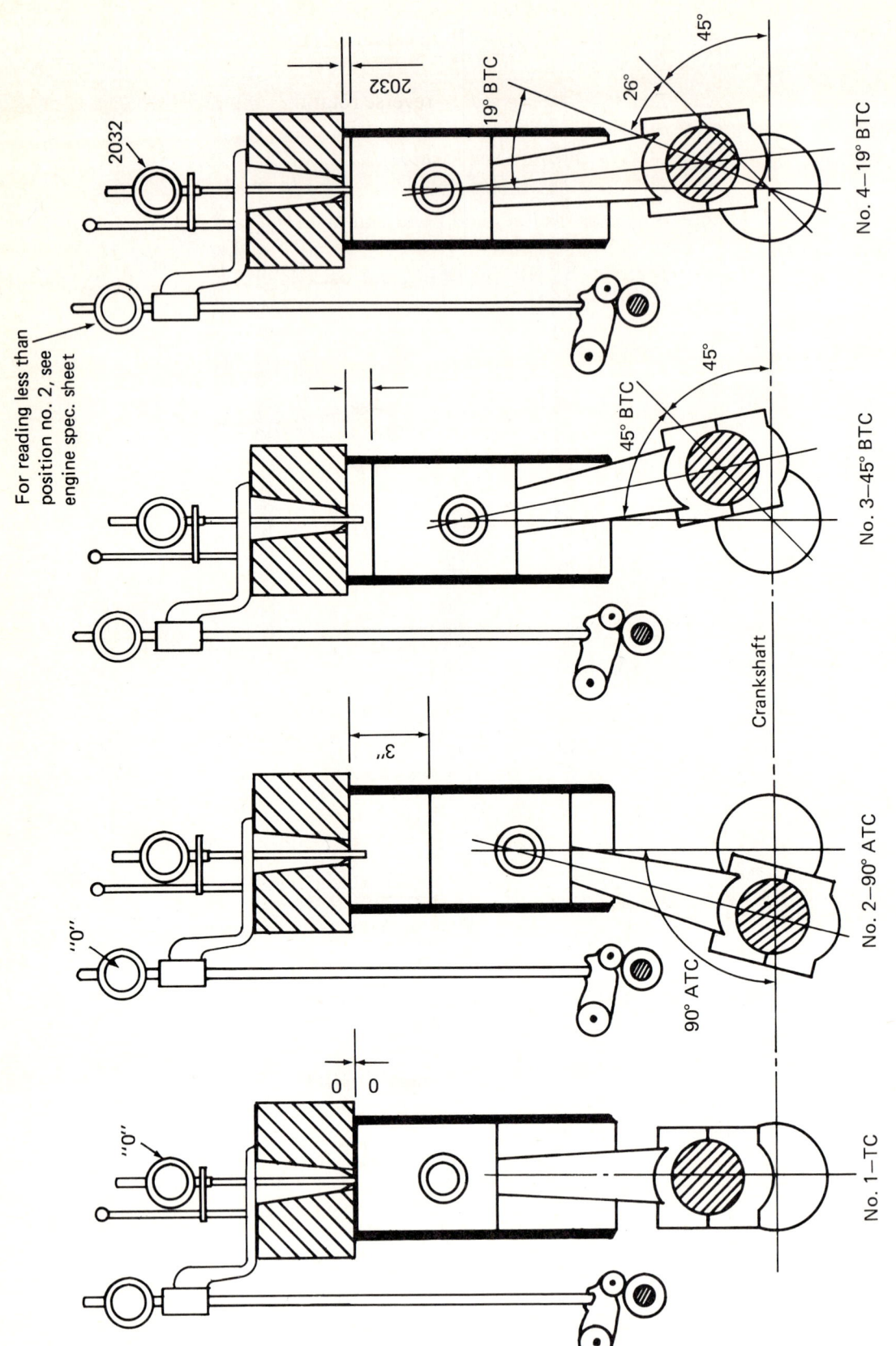

FIGURE 12-12. **Timing Steps (Looking Forward).** *(From a sketch by P. M. Uhl)*

Rocker Levers: Tappets, Followers, and Push Rods / 219

6. Bar the engine to 90° after top center (ATC). Zero the push rod indicator in this position.

7. Bar the engine backward, in reverse rotation, to a point where the piston indicator loses contact with the fixture stem.
Note: With long-travel indicators, bar from top dead center TDC back about 0.300 inch. This allows us to take up gear lash.

8. Bar the engine carefully forward until the piston indicator hand is at 0.2032 inch before zero. You must "bump" the barring tool the last few thousandths. If you overshoot, bar back to take up gear lash, and try again. [Fig. 12-13]

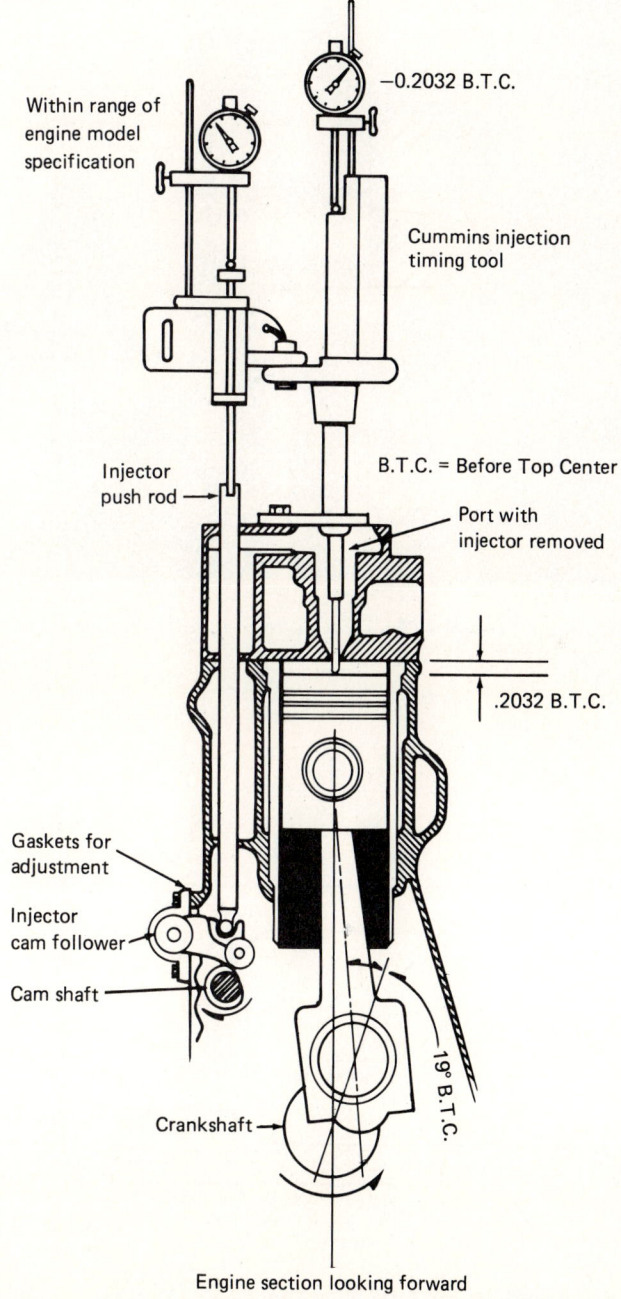

FIGURE 12-13(a). Step 4 Detail. *(Courtesy of Cummins Arizona Diesel, Inc.)*

220 / Diesel Engine Service

FIGURE 12-13(b). Step 4 Detail—Choice. *(Courtesy of ABC Technical and Trade Schools)*

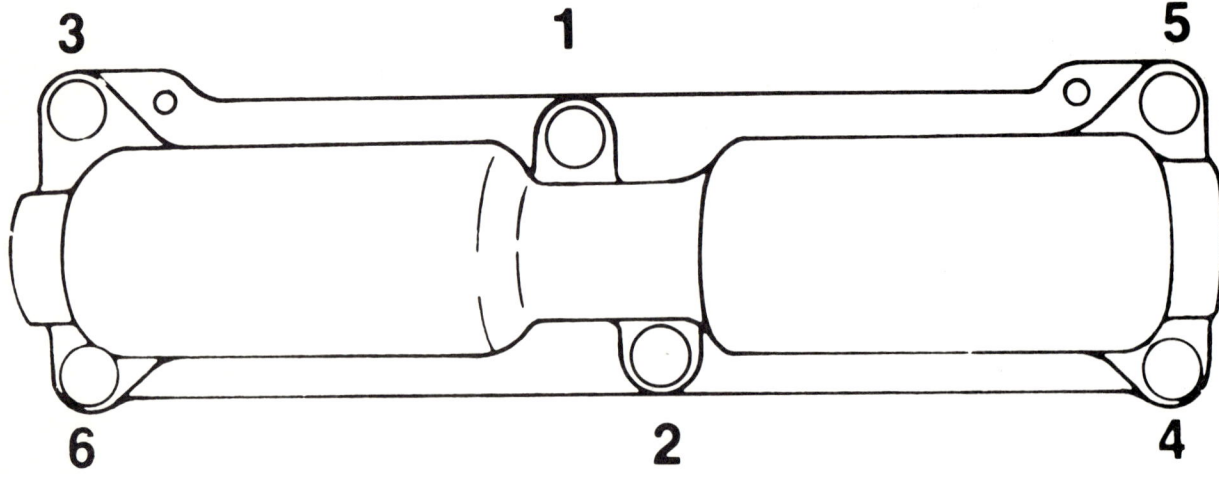

FIGURE 12-13(c). Cam Follower Tightening Sequence. *(Courtesy of Cummins Engine Co., Inc.)*

Rocker Levers: Tappets, Followers, and Push Rods / 221

9. Read the push rod indicator. It should be close to the specifications for that model.
10. To correct, add gaskets behind the cam follower to reduce the push rod travel for earlier timing. Gaskets are furnished in several thicknesses. Work to bring the push rod travel within the tolerance range for that model. Continued trial and correction may be necessary.
11. Be sure to stay within the range. Truck engines may be timed 0.001 inch fast, with that much less push rod travel. Industrial engines should be timed on the specification, or 0.001 inch more, slower.

MECHANICAL VARIABLE TIMING

In the continuing attempt to reduce exhaust emissions, a method of varying injection timing during engine operation has been devised. This system of mechanical variable timing (MVT) uses vehicle service air pressure to move a plunger that is connected to the cam follower lever shafts through a toothed rack to a gear on the shaft. [Fig. 12-14]

All three cam follower shafts are connected by couplings. The complete assembly is installed, as a unit, on a single gasket that is used behind all three housings.

An adjustable spring retainer allows the basic injector timing to be set without changing the gasket thickness. However, the same method of checking timing with the indicators is used. The injector cam follower levers are mounted on eccentric bushings and are moved inward or outward on the camshaft by the plunger movement.

When the plunger is all the way down, the timing is retarded. Air pressure to move the plunger is controlled by an electric solenoid, current to which is controlled by a pressure-sensing switch in the fuel line to the injectors. Thus, injection timing is advanced when the air pressure is above 80 psi (551 kPa) and is moved toward retard when pressure is below that value.

Because exhaust emissions are reduced by retarded injection timing, any load below 25 percent of full load, being reflected in lower fuel pressure, will operate the switch and solenoid to close off air supply to the MVT operating plunger, which will then be spring-driven downward, turning the cam follower shafts to retard injection timing.

The position of the cam follower lever rollers on the camshaft injector lobe determines the travel of the injector plunger in relation to piston travel. On a right-hand rotating engine, movement of the roller away from the camshaft advances timing, while movement toward the camshaft, retards timing.

Basic timing checks are made with the operating plunger down, timing retarded. The engine starts in this mode. As air pressure increases, the plunger is forced up, turning the cam follower shaft and advancing the timing during acceleration.

When the switch is opened by fuel pressure, the solenoid closes the air supply, and the spring pushes the plunger down to retard injection timing. The engine runs most of the time with injection retarded.

The fuel pressure switch is normally closed, opening by fuel pressure. The solenoid circuit is normally closed and is opened to shut off the air flow.

222 / Diesel Engine Service

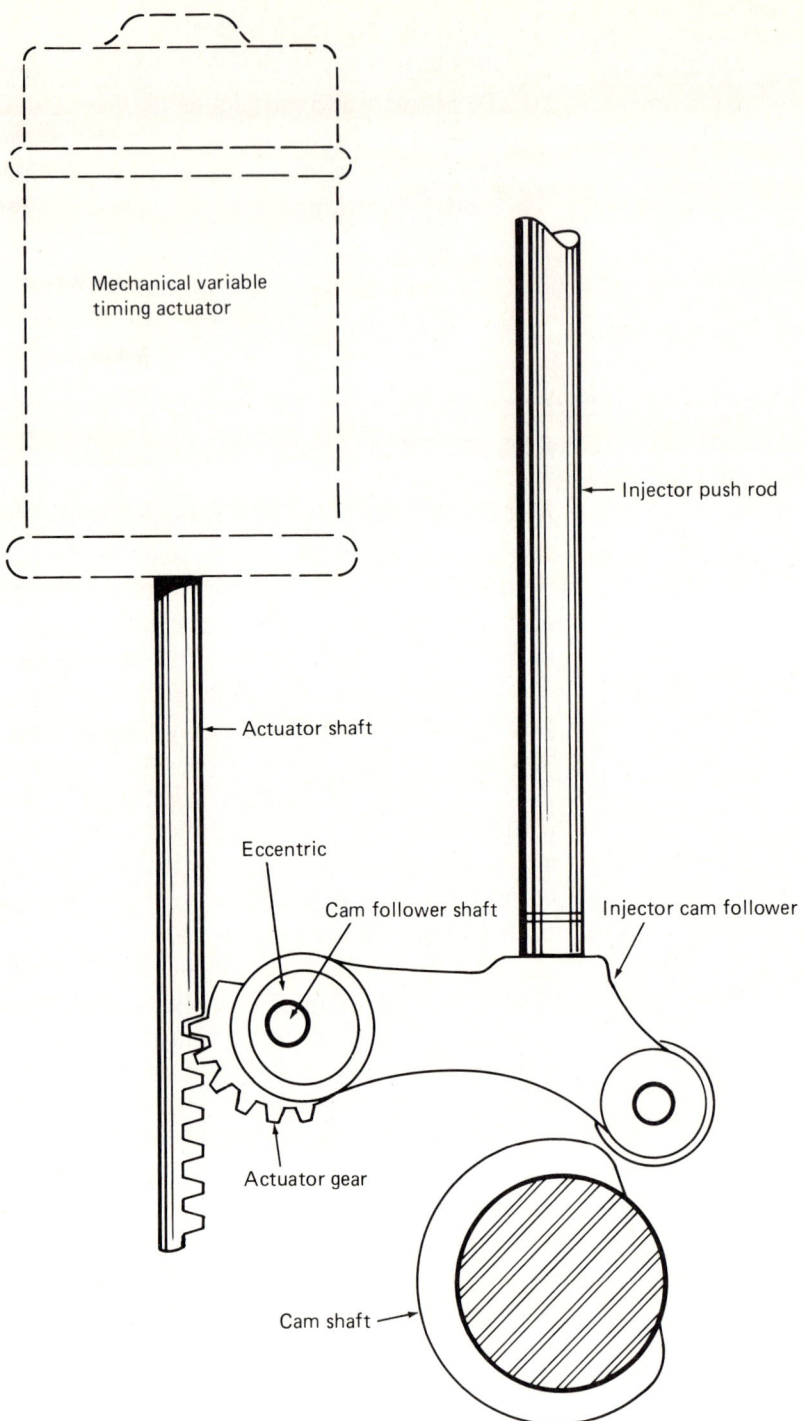

FIGURE 12-14(a). MVT Assembly. *(Courtesy of Cummins Engine Co., Inc.)*

The air connection to the solenoid is through a special series of fittings; screens keep dirt out of the 0.015-inch orifice in the end of the elbow.

A kit of repair parts is available to service the actuator assembly. Although this assembly can be serviced without removing the cam followers, it is advisable to install a rebuilt unit. The Cummins engine manual gives complete repair instructions.

Rocker Levers: Tappets, Followers, and Push Rods / 223

FIGURE 12-14(b). MVT Assembly Demonstration Unit.

Injection timing is checked by the same method as for plain cam-follower assemblies. For either type, on either big cam or small cam engines, limits are specified for minimum and maximum gasket thickness. Should the correct timing not be attainable within the gasket thickness specified, an offset key can be used in the cam drive gear. [**Fig. 12-15**]

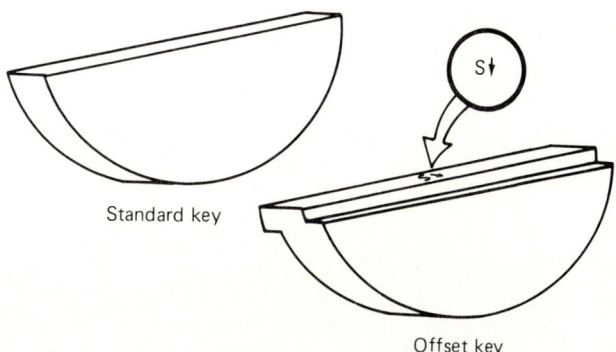

FIGURE 12-15. Offset Key. *(Courtesy of General Motors Corp.)*

Only one gasket is used with the MVT assembly, and a special camshaft is used. Timing adjustment is made by turning the spring retainer with a special tool. The retainer is turned counterclockwise, left, to advance timing. One complete turn will change the push rod travel about 0.004 inch (0.10 millimeters).

Servicing the MVT Timing Unit

Some cautions must be observed in servicing the MVT timing unit.

1. Be sure that the unit is correctly assembled. Follow the manual instructions.
2. Install the solenoid so that the wire terminal is at the top.
3. Install the air inlet elbow so that it points straight down. Use teflon tape on the threads, keeping the tape away from the orifice in the end of the elbow.
4. Follow all torque standards carefully. Avoid overtightening and distortion.
5. Be sure that the actuator gear is positioned correctly and that couplings and seals between cam follower housings are in place.
6. When installing the actuator, be sure that the toothed rack engages the gear. Turn the shaft a little to help engagement.
   Note: The cam follower shaft must be turned to the right, clockwise, as far as possible.
7. Lubricate the seal ring with clean oil.
8. Get help to install this assembly. It is too heavy to handle alone.
9. The actuator plunger must be down all the way before checking timing. Turn the crankshaft to bring the plunger down.
10. Set up the timing fixture on the No. 3 cylinder. Be sure that the indicators are zeroed exactly.

All other commercial engines do not time their spray valves (injectors) but do require adjustment of valve clearance and other service adjustments generally included in the tuneup procedure.

13

GEAR TRAIN AND TIMING MARKS

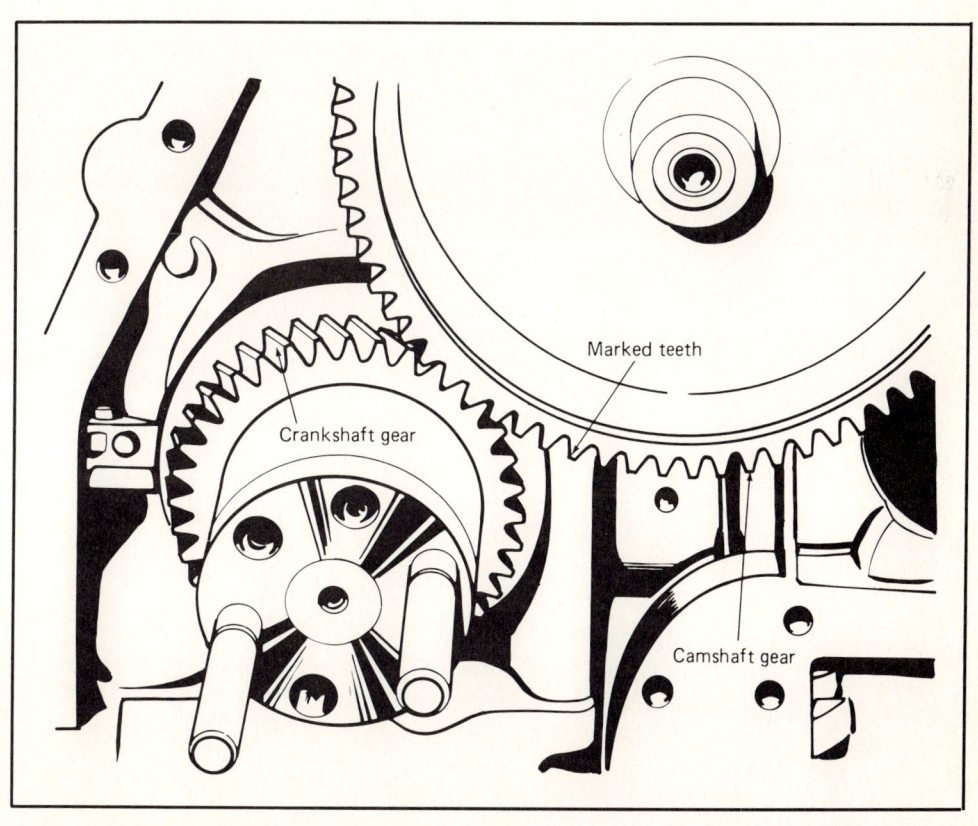

GENERAL DESCRIPTION

Commercial engines drive all accessory subassemblies except water pumps and alternators by a series of gears, called the *gear train,* from the crankshaft gear.

Gears have marked teeth so that the correct relationship can be established between the crankshaft and camshaft. During engine assembly, marked gear teeth are engaged together, and once assembled, do not change.

The gear train is lubricated by oil thrown off as the engine turns. Gear-attached shafts turn in bushings that are pressure-lubricated.

Gear lash, or the clearance between teeth, is established by the position of the bores in which the shafts turn. The only change that can occur in this clearance is due to wear in the bushings or wear on the gear teeth.

Some engines have *idler gears,* which turn on short shafts or pins that are pressed into fitted holes in the block. These gears have bushings in their hub that are pressure-lubricated from oil holes in the shaft. Some of these shafts are attached to the block by a flange and capscrews. [**Figs. 13-1, 13-2, and 13-3**]

The function of an idler gear is to convey driving force from the crankshaft gear to a driven gear positioned at some distance from the crankshaft. The difference in the number of teeth on the driving gear and the driven gear determines the number of times the driven gear turns for each revolution of the drive gear. This is called the *ratio*.

When an idler gear is between the crankshaft gear and the camshaft or fuel pump drive gear, the drive ratio must include it. Some engines have idler gears that do not index the marked teeth on every revolution but come together only once in several revolutions.

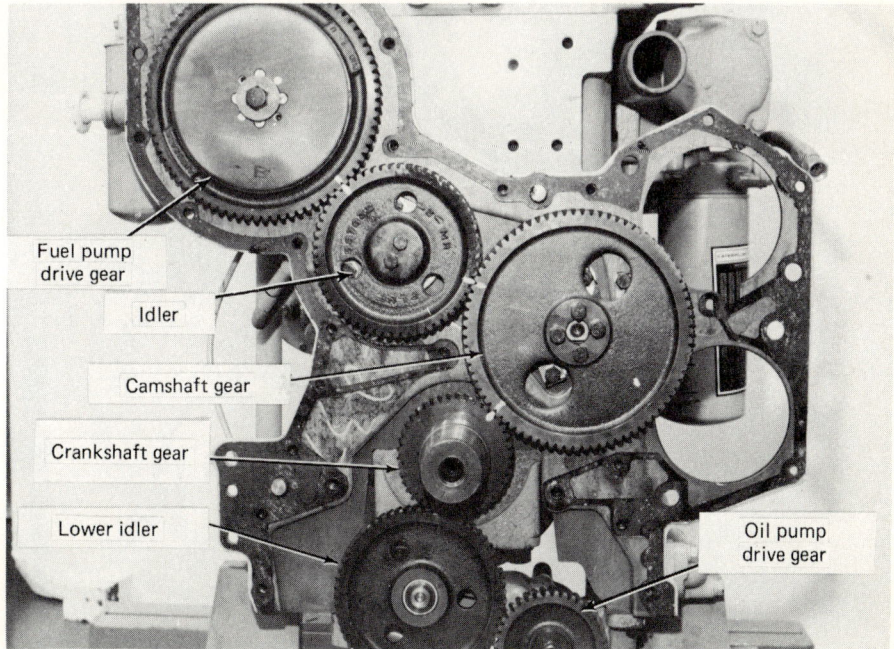

FIGURE 13-1. Typical Gear Train. *(Courtesy of Caterpillar Tractor Co.)*

228 / Diesel Engine Service

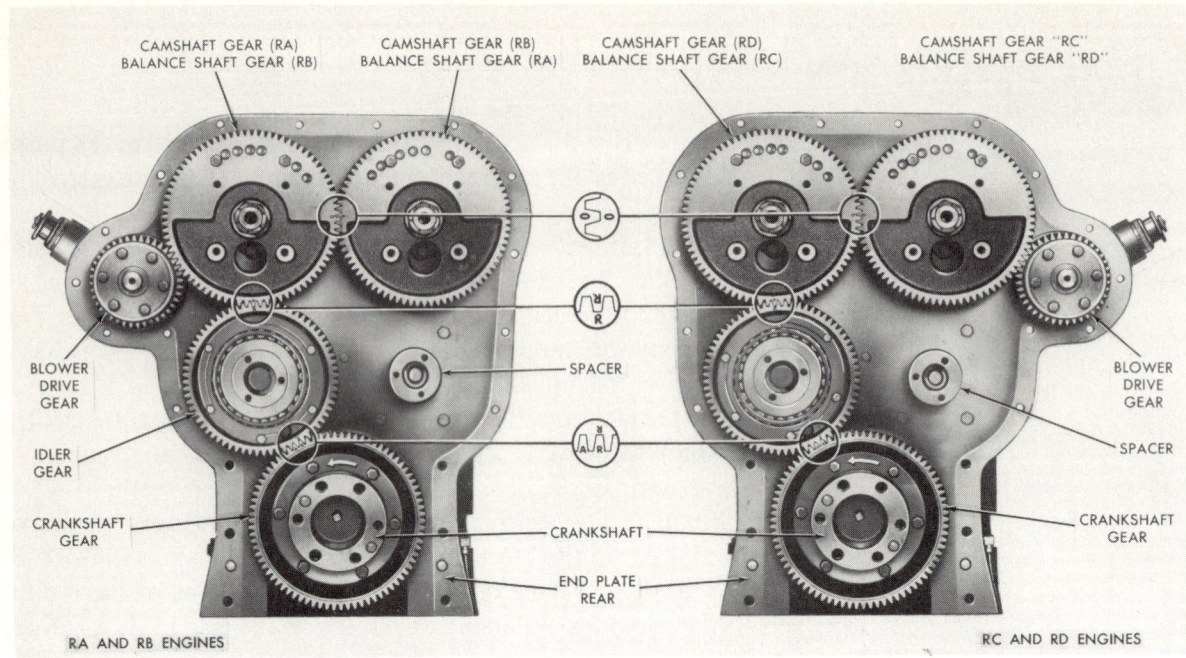

FIGURE 13-2. General Motors Gear Train. *(Courtesy of General Motors Corp.)*

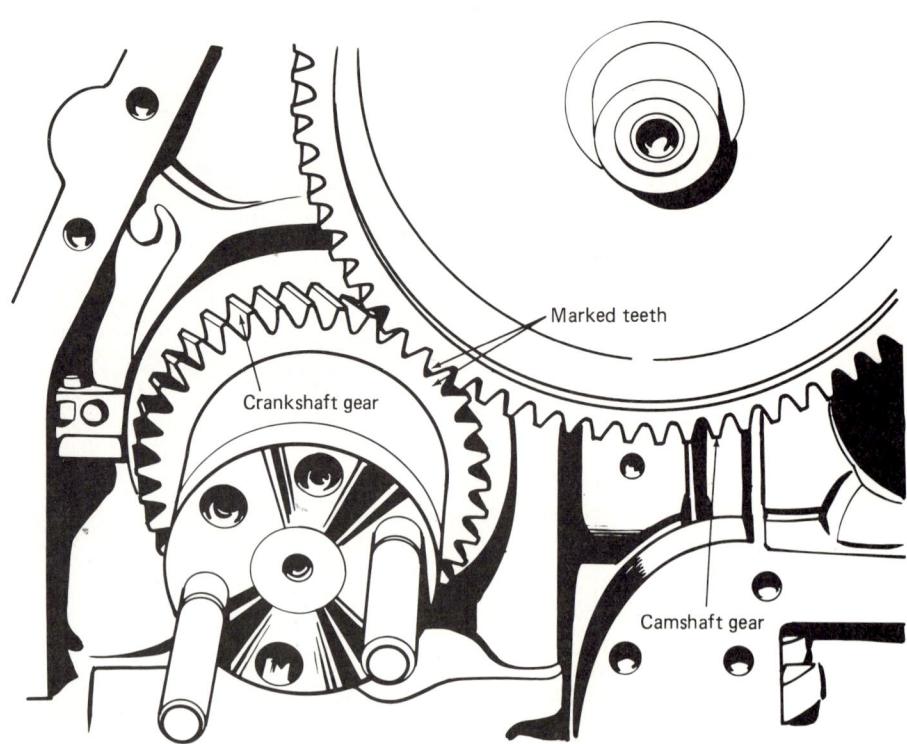

FIGURE 13-3(a). Cummins Gears—1. *(Courtesy of Cummins Engine Co., Inc.)*

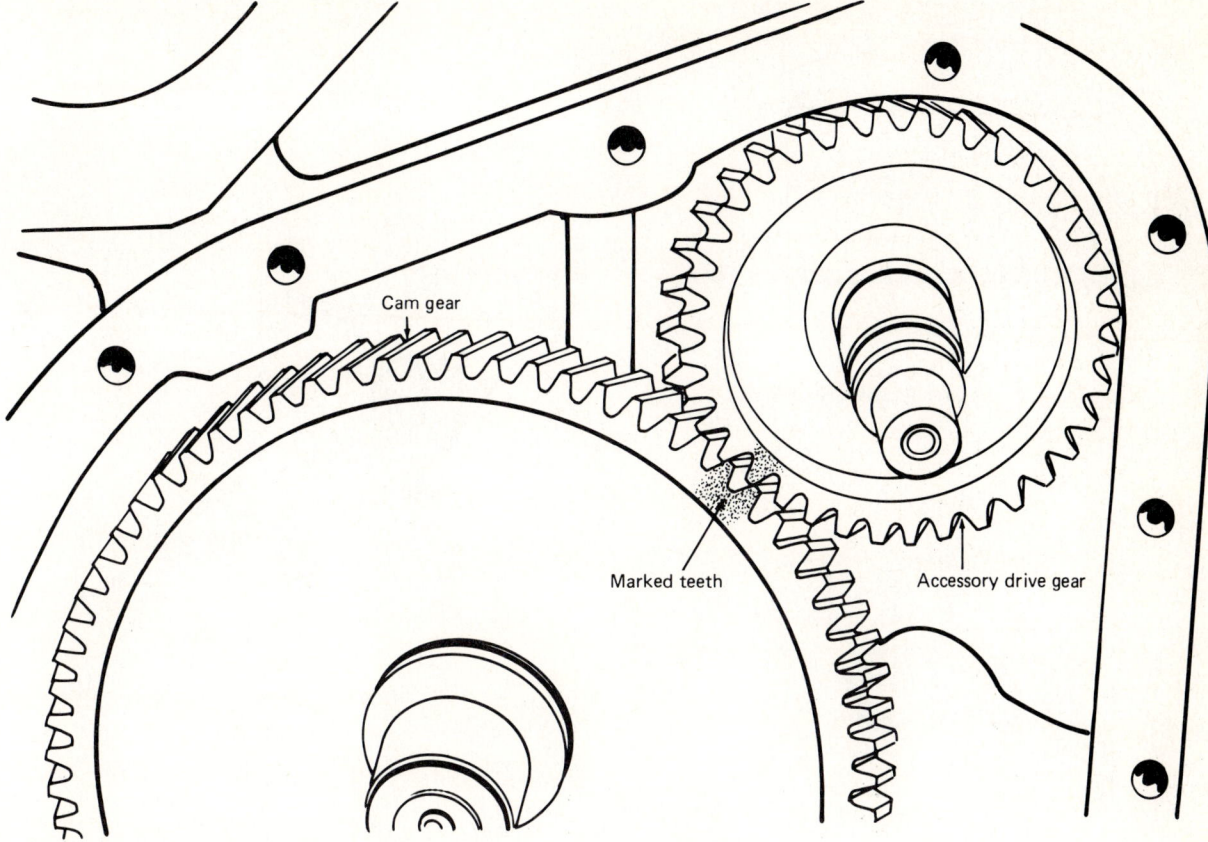

FIGURE 13-3(b). Cummins Gears—2. *(Courtesy of Cummins Engine Co., Inc.)*

Thus it is recommended that gears once assembled correctly not be checked during trouble-shooting.

GEAR TRAIN SERVICE

Gears on which timing is critical are keyed onto their shafts. When replacement is required, a suitable puller must be used to remove the gear. The principle to be observed in this operation is to use the right puller size and engage its jaws to the gear hub, not to the tooth edge. **[Fig. 13-4]**

Puller Variety

Pullers are made in a variety of sizes and shapes. Some use a hydraulic ram, while others use a forcing screw. The latter should contact the shaft end through a dead center, which is a hardened spacer placed between the forcing screw and the shaft. The forcing screw turns on the spacer rather than on the shaft end, preventing damage to the shaft. **[Fig. 13-5]**

230 / Diesel Engine Service

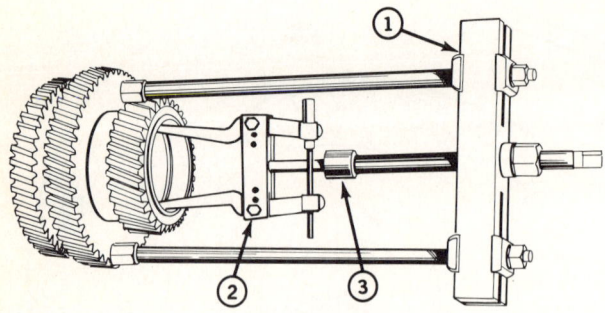

Typical example
1. Push puller. 2. Bearing cup pulling attachment. 3. Reducing adapter.

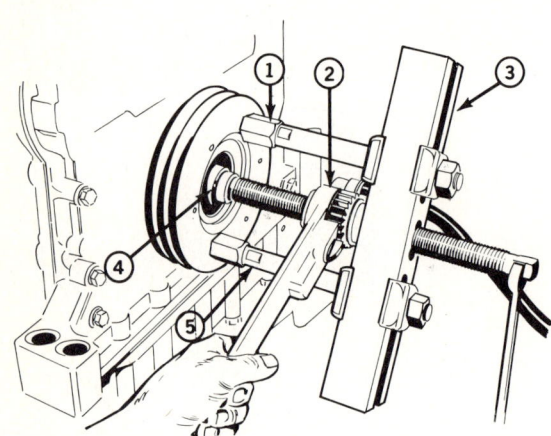

Typical example
*1. Adapters. 2. Ratchet box wrench. 3. Push puller. 4. Step plate. 5. Legs.
*Use as required.

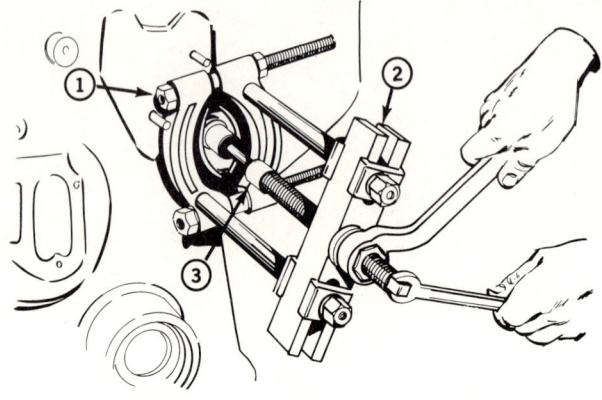

Typical example
1. Bearing pulling attachment. 2. Push puller. 3. Reducing adaptor.

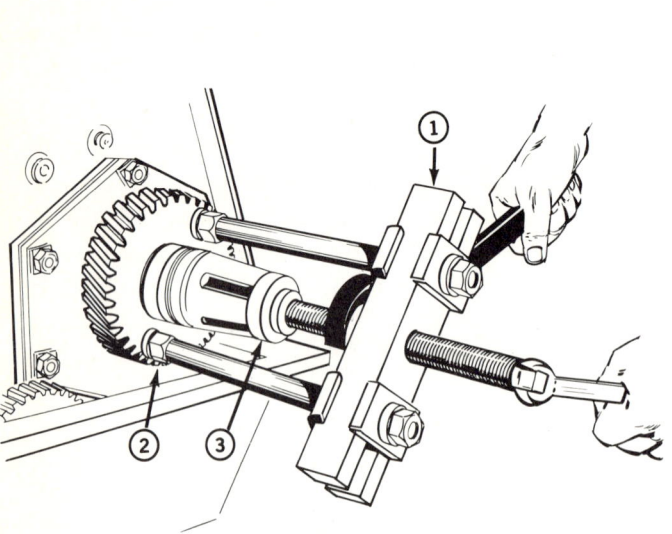

Typical example
1. Push puller. 2. Adaptor. 3. Step plate.

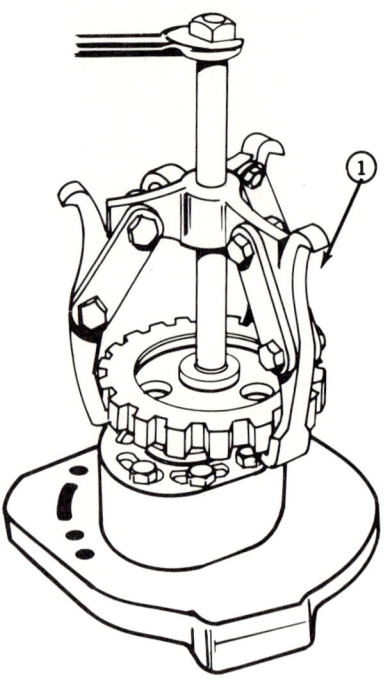

Typical example

Gear Train and Timing Marks / 231

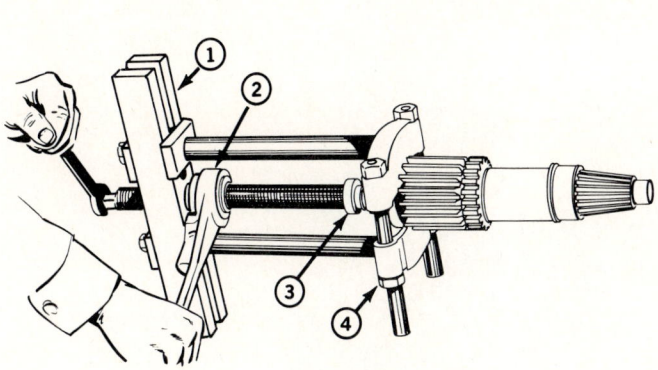

Typical example

1. Push puller. 2. Ratchet box wrench. 3. Step plate. 4. Bearing pulling attachment.

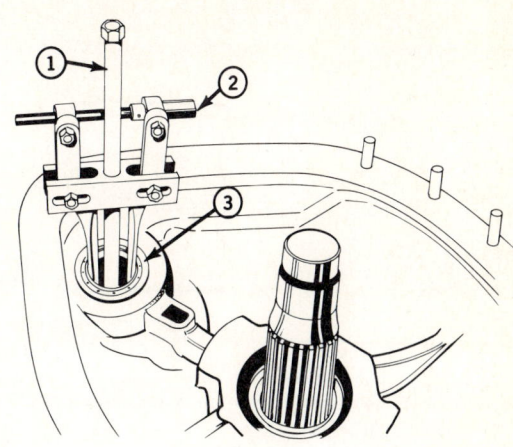

Typical example

1. Screw 2. Bearing cup pulling attachment. 3. Step plate.

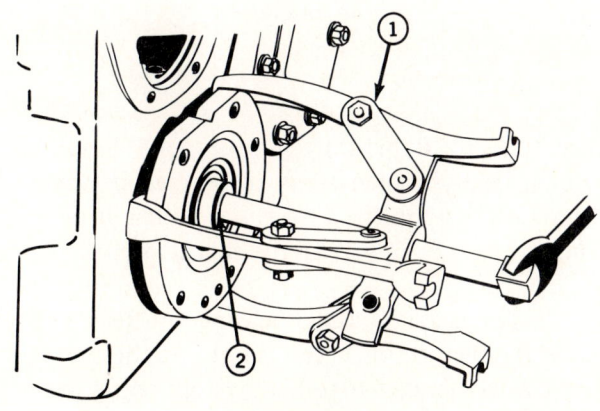

Typical example
1. Puller 2. Step Plate.

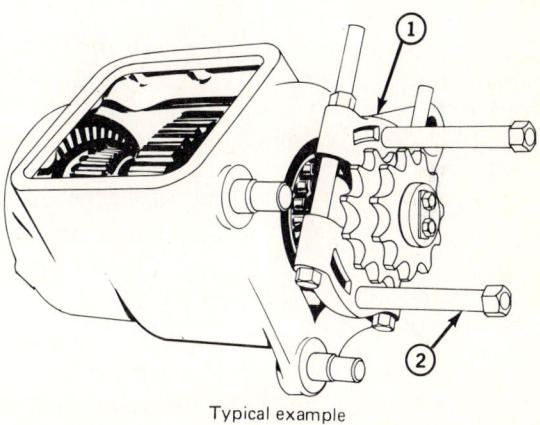

Typical example
1. Bearing pulling attachment. 2. Forcing bolts.

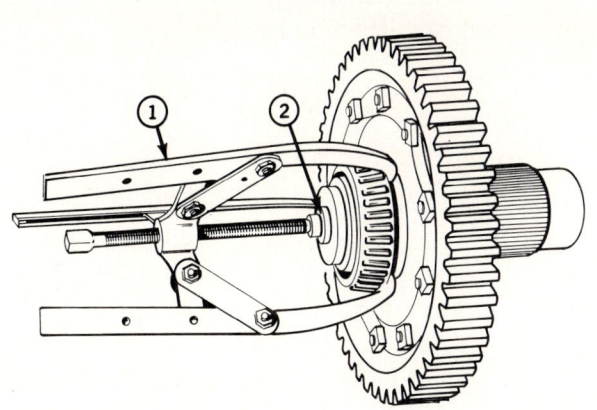

Typical example
1. Puller 2. Step Plate.

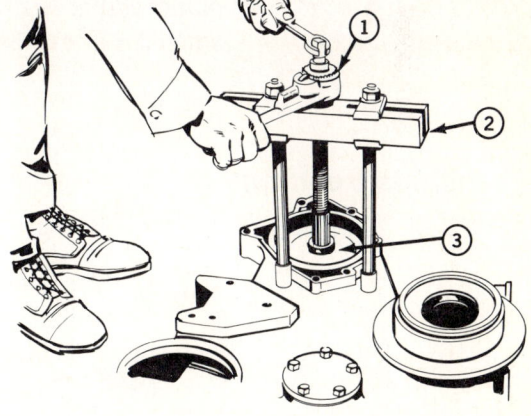

Typical example
1. Ratchet box wrench. 2. Push puller. 3. Reducing adaptor.

Push Pullers

Push Pullers can be used to remove pulleys, gears, shafts, etc. and can be used in a variety of pulling combinations.

FIGURE 13-4. Typical Examples of Pullers. *(Courtesy of Caterpillar Tractor Co.)*

232 / Diesel Engine Service

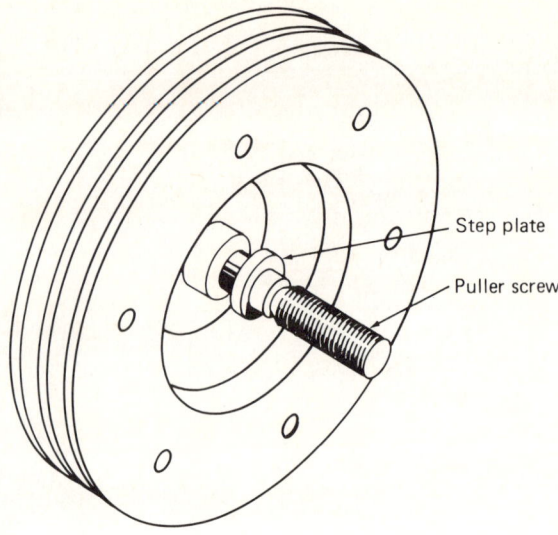

FIGURE 13-5. Puller Dead Center. *(From a sketch by P. M. Uhl)*

If no adapter or dead center is available for the puller, one should use a piece of steel ¼ inch thick or a heavy washer. Any piece used between the puller screw end and the shaft end should be lubricated with a little white grease on the screw side. The screw should turn on the spacer, not the spacer turn on the shaft end.

When pulling a ball or roller bearing, the "bearing splitter" type of puller jaws should be used and the bearing race that is tight on the shaft or in the bore engaged. The jaws can be forced together by tightening the nuts evenly, then forcing the jaws behind the bearing race.

Sleeve-type bearings, such as cam bushings, are usually replaced using a mandrel that fits the inside diameter of the bushing, although its outside diameter is a few thousandths of an inch smaller than the bore.

Camshaft Bushings

Cummins engines use such a mandrel; the bushings are forced into place by striking the mandrel while holding it solidly against the bushing. [**Fig. 13-6**]

Caterpillar camshaft bushings are removed and installed with a special puller tool. [**Fig. 13-7**]

Detroit camshaft bushings are separate units, and the bushings at each end are flanged for attachment with capscrews. The intermediate bearings are two-piece and are held together with snap rings. They are held in the block by lock screws. [**Fig. 13-8**]

Camshaft bushings are oiled from drilled holes, which lead pressure oil from the main oil gallery. Oil holes in the bushings must index with the oilfeed holes, and their position must be attended to during installation. Each engine builder gives details of cam bushing installation.

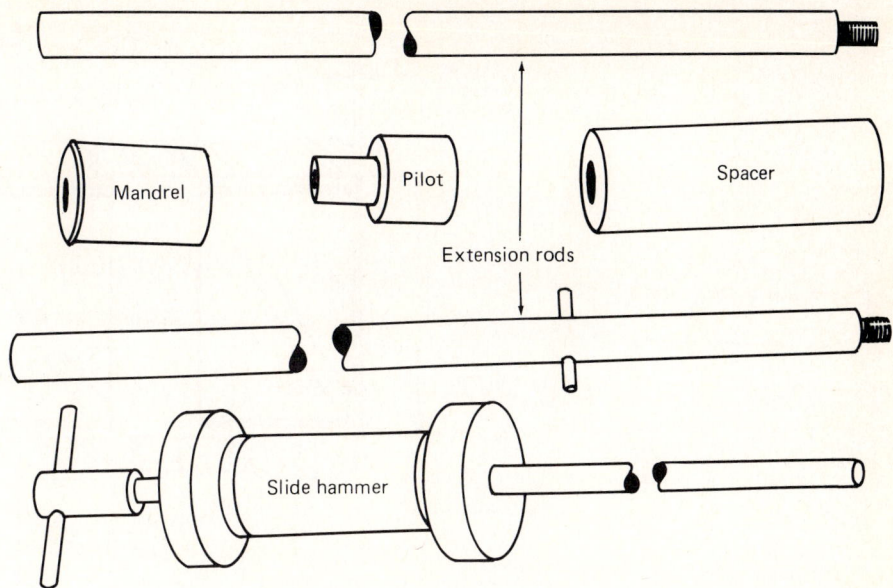

FIGURE 13-6. Camshaft Bushing Installing Tool. *(Courtesy of Cummins Engine Co., Inc.)*

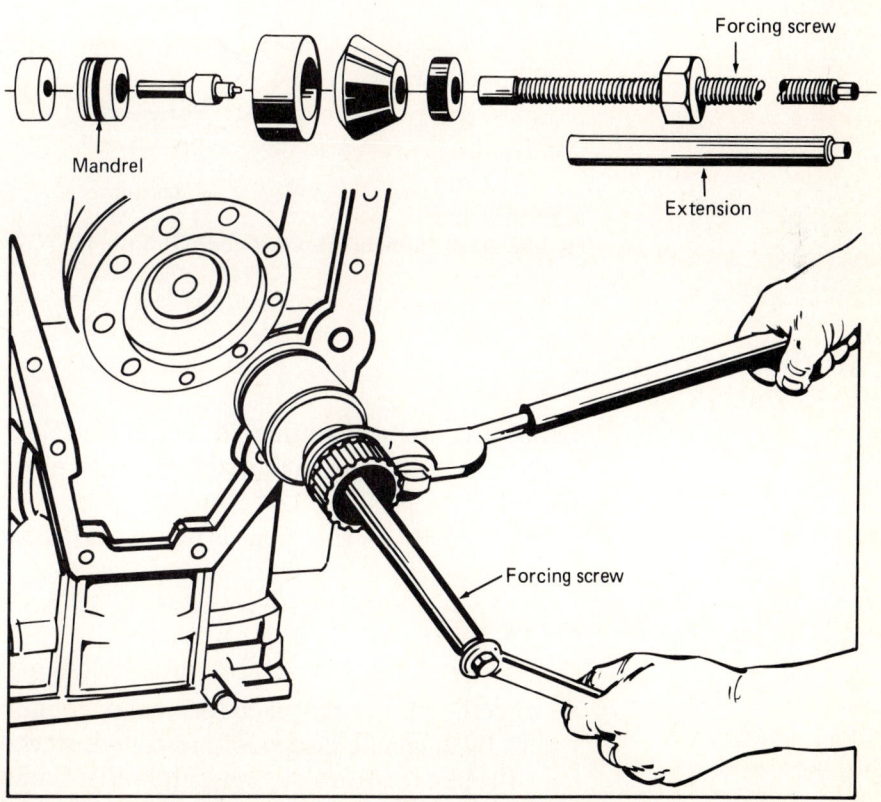

FIGURE 13-7. Caterpillar Bushing Tool. *(Courtesy of Caterpillar Tractor Co.)*

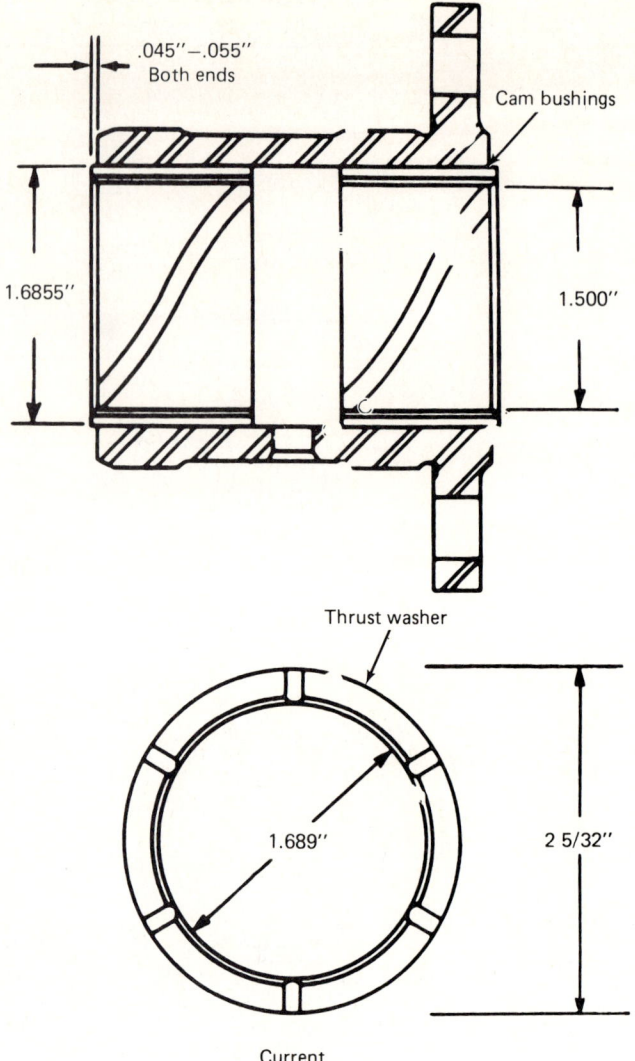

FIGURE 13-8. General Motors Camshaft Bushings. *(Courtesy of General Motors Corp.)*

Camshaft Inspection

Camshafts are not considered repairable. Visual inspection for lobe failure must be made during engine service, and a new camshaft installed if failure is found.

Camshaft journal wear is not usually a cause for replacement.

Gear Inspection

Gear teeth should be checked for wear or roughness. Gears can be checked for cracks by magnetic inspection, as can all steel forged parts.

Broken teeth are obvious, and other damage is not common. When internal bearings or bushings are used in gears, they should be checked for wear and replaced if required.

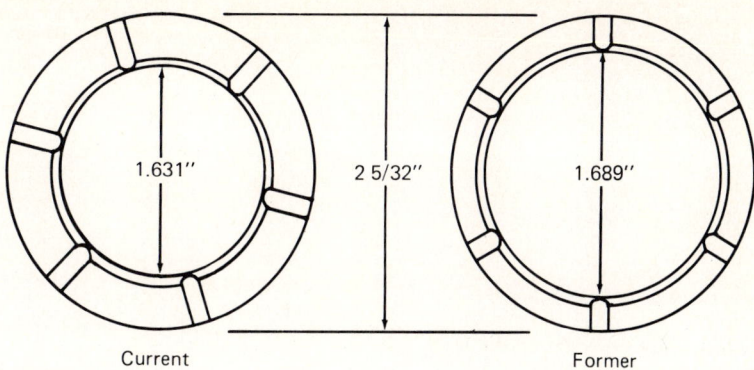

FIGURE 13-9. Thrust Rings. *(Courtesy of General Motors Corp.)*

Thrust Washers

Bronze or brass washers are used to take any endwise force on shafts, or gears. Gears that are cut at angle, called *helical gears,* generate a force that tries to move the gear endwise.

Thrust washers resist this force. They are often made with oil grooves on one surface, which is always installed to face the moving part, usually the side of the gear. This placement assures that the internal chamfer or bevel on the washer will be toward the shoulder of the shaft or against the face of the gear. [Fig. 13-9]

Gear Fasteners

There is almost always a heavy washer on the capscrew or behind the nut that holds a gear on its shaft. These washers get lost or damaged and can be renewed during assembly.

Fastener Torque

The tension to which such fasteners are tightened is specified and should not be exceeded. Different grades of capscrew and bolt strength are normally identified by markings on the head. (See the Appendix.)

The tension of all fasteners depends on both the application and the strength of material. The objective is to tighten enough to prevent loosening in service but not enough to exceed the stretch limit of the fastener or threads.

Torque values are given in pound-inches (lb.-in.), pound-feet (lb.-ft.), or Newton-meters (N-m), the latter for use in areas using the metric system.

A conversion chart is included in the Appendix.

14

VIBRATION DAMPERS

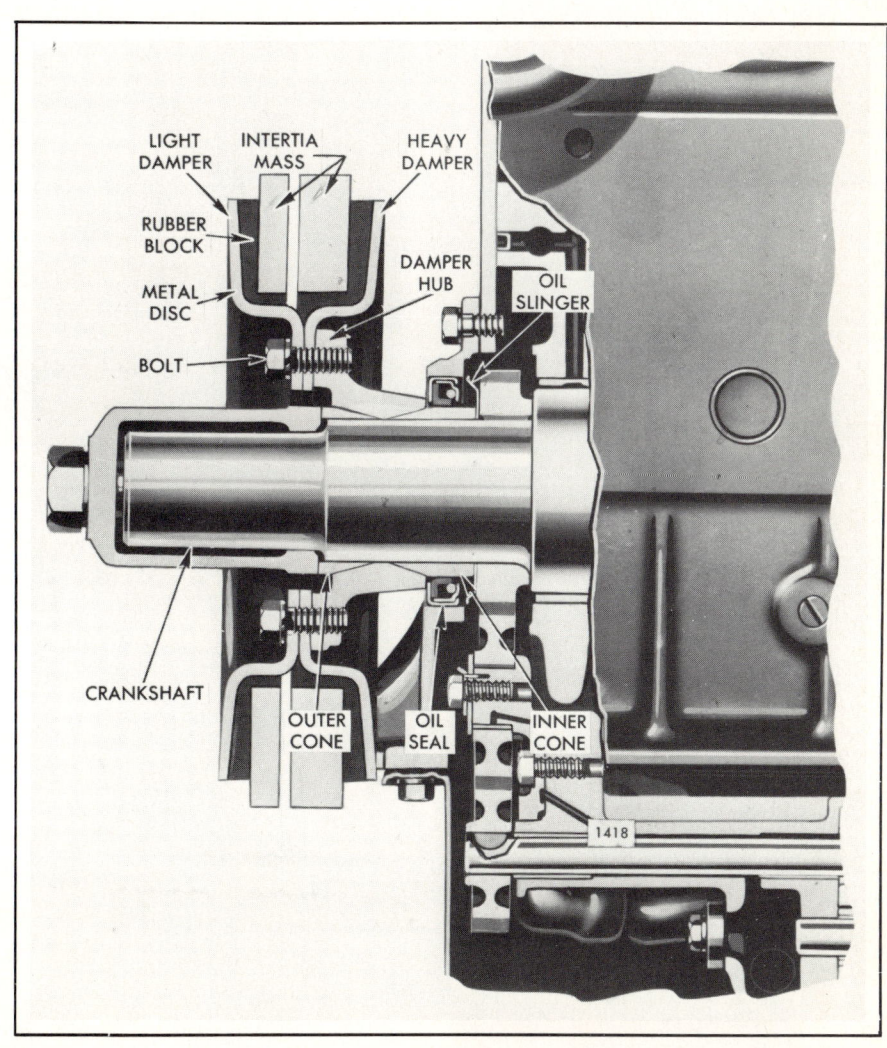

FUNCTIONAL DESCRIPTION

The torsional (circular) vibration of the crankshaft has been described. The vibration damper is designed to reduce the extent, or *amplitude,* of these vibrations, thus preventing shaft breakage.

Commercial engines use one of three types of vibration damper, attached to the front of the crankshaft.

Double Dampers

This unit consists of two elements, one light and one heavier. The elements are iron rings and are supported in rubber blocks. The light damper is active at lower speeds, while both elements act at high speed and load. **[Fig. 14-1]**

The damper turns with the crankshaft, but the elements turn slightly in the rubber blocks in order to resist the acceleration and deceleration of the crankshaft by their own inertia.

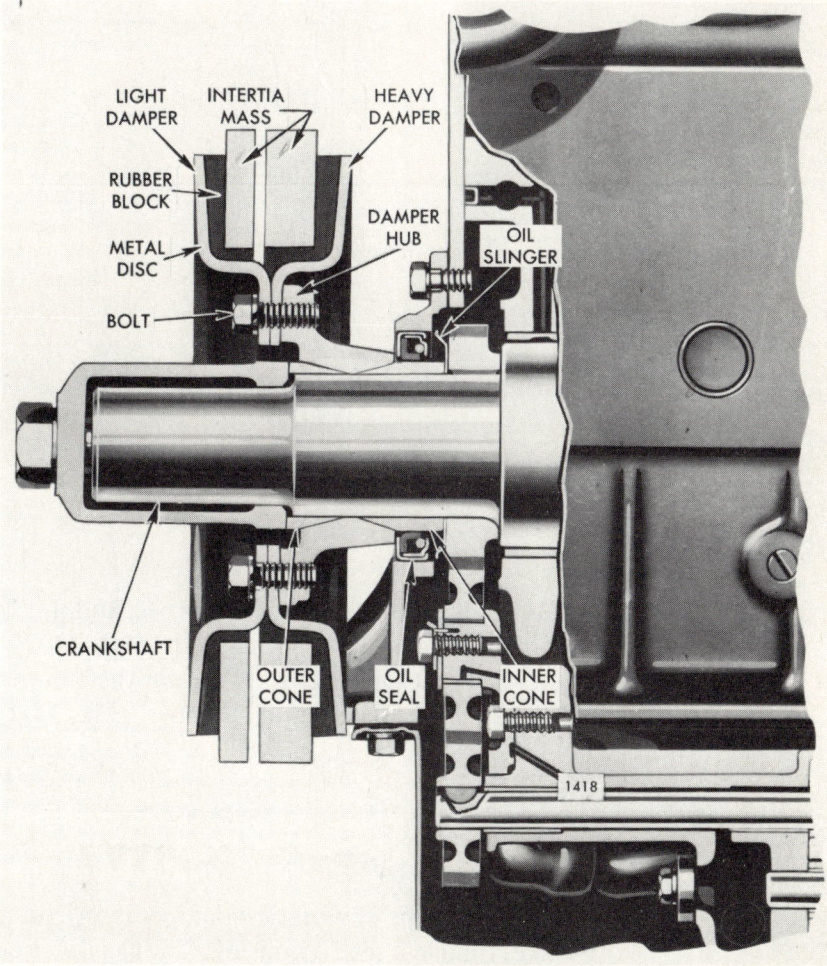

FIGURE 14-1. Double Damper. *(Courtesy of General Motors Corp.)*

239

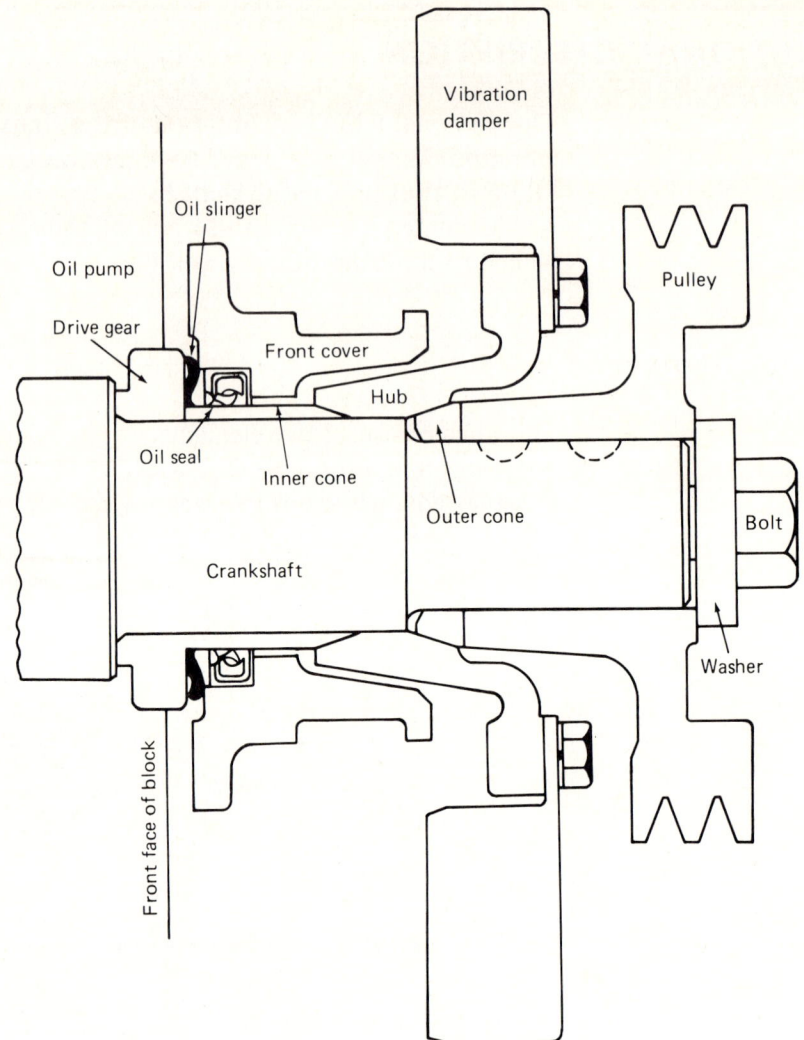

FIGURE 14-2. **Single Damper.** *(Courtesy of General Motors Corp.)*

Single Dampers

Some engines use only one damper, and it is bolted to the crankshaft flange, which is keyed to the crankshaft. [**Fig. 14-2**]

Detroit Diesel engines mount the damper on the crankshaft nose by attaching it between two steel cones. [**Fig. 14-1**]

Viscous Dampers

The third type of vibration damper is an iron ring suspended by viscous (jellylike) fluid in a sealed housing. The clearance between the ring and the housing is only about 0.010 inch. These dampers are effective but cannot stand a dent or any distortion. [**Fig. 14-3**]

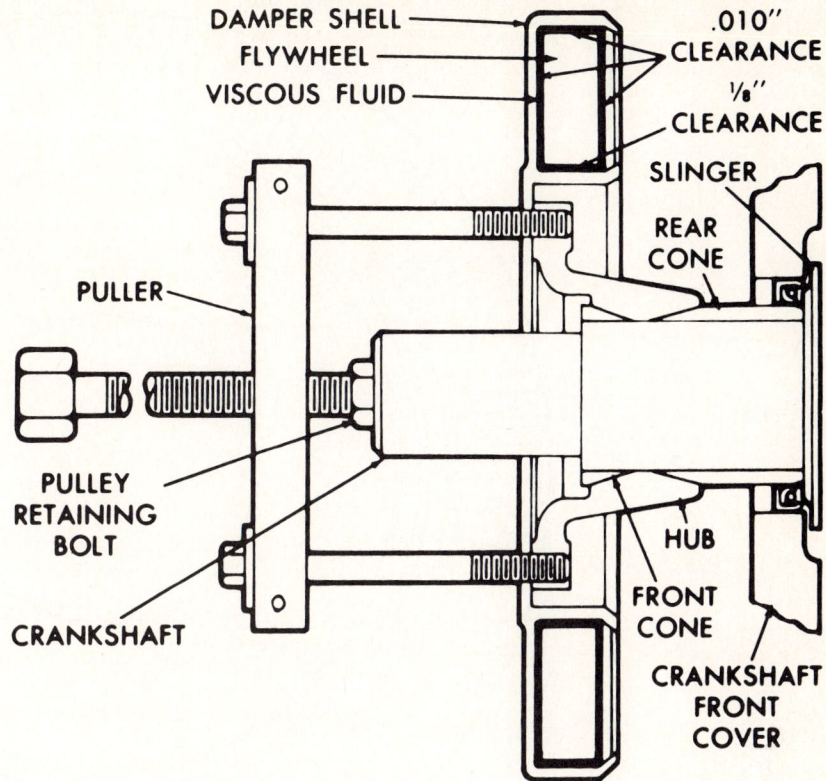

FIGURE 14-3. Viscous Damper. *(Courtesy of General Motors Corp.)*

VIBRATION DAMPER SERVICE

Dampers are removed whenever the crankshaft is serviced or when access is needed to the gears or front seal. They should be cleaned and inspected when removed. They cannot be repaired but can be replaced with a new damper if damaged.

The damper should be cleaned with fuel and a brush. Fuel should not remain on it longer than the cleaning process; residue should be washed off with detergent and water. The surface must be dry and free from any fuel. Rubber is sensitive to petroleum products.

Alignment Marks

Some rubber dampers have alignment marks on the hub and iron ring. If they are out-of-line more than $\frac{1}{16}$ inch (1.59 millimeter), the damper should be replaced. **[Fig. 14-4]**

In removal of any vibration damper, it should not be struck with a hammer or pried loose. The correct puller should be used; heat must never be applied. **[Fig. 14-5]**

Viscous dampers should be examined for dents, alignment with the hub, and distortion. Any leakage of fluid is reason for renewal. **[Fig. 14-5]**

242 / Diesel Engine Service

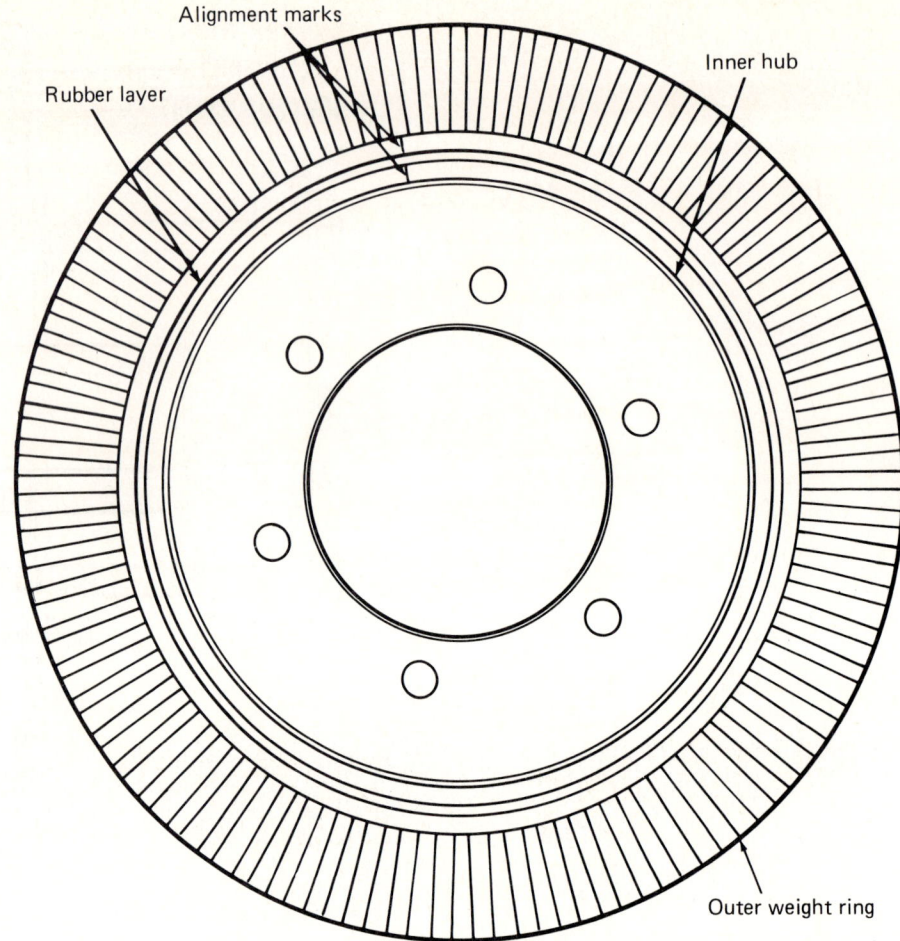

FIGURE 14-4. Damper Alignment Marks. *(Courtesy of Cummins Engine Co., Inc.)*

Front Oil Seal Installation

Before installing the vibration damper, the front crankshaft seal should be checked for leakage. Oil streaks, indicating oil leakage, indicate the need for a new seal.

On some engines the front flange and a pulley are in one piece. After loosening the retaining screw, a puller should be used to remove the adapter.

On other engines the damper is fastened to the flat-front of the crankshaft with capscrews and lockplates. The assembly can be lifted from its mounting.

Seal Removal

If only the front seal is to be replaced, the old seal can be removed in the following way.

1. Using a drill motor, drill two small holes opposite each other in the outer edge of the seal. [**Fig. 14-6**]

Vibration Dampers / 243

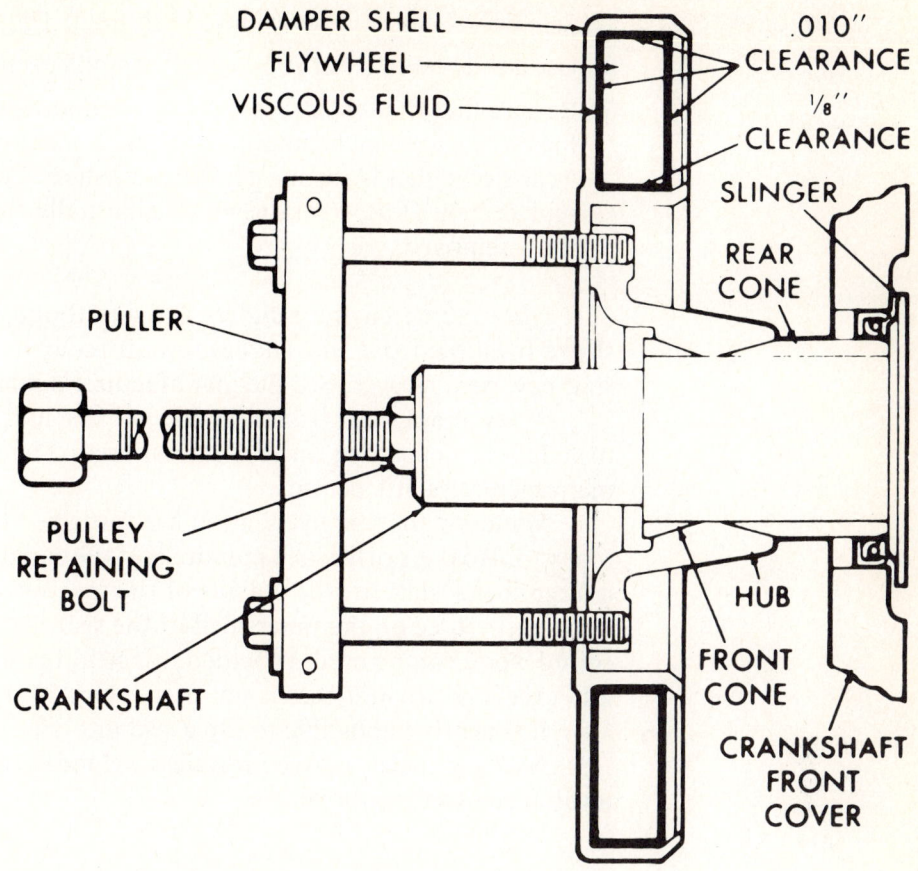

FIGURE 14-5. Remove Viscous Damper. *(Courtesy of General Motors Corp.)*

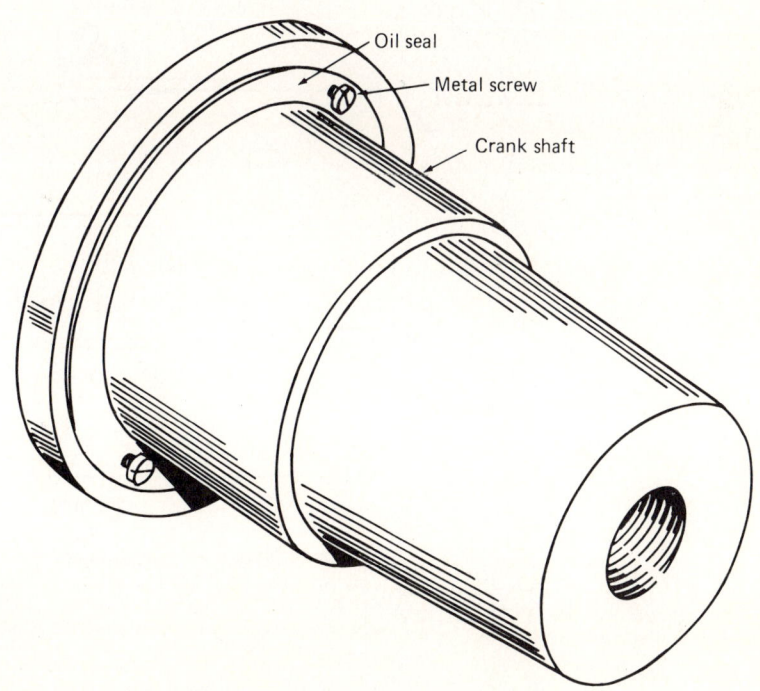

FIGURE 14-6. Seal Removal. *(From a sketch by P. M. Uhl)*

2. Screw a self-tapping screw into each hole. They must fit tightly.
3. Use a slide-hammer or roll-head bar to pull each side, and remove the seal.
4. Be sure that the seat is cleaned of any residue, and check the journal surface where the seal bears for grooving. Such grooves are often found on a wear sleeve that is shrunk on the crankshaft. Severe grooves will need the replacement of the wear sleeve, which usually must be done with the crankshaft removed.

Note: Some engine builders provide tooling that can remove a wear sleeve from the front without crankshaft removal. Tools are available to install new wear sleeves. See the manufacturer's manual.

A new crankshaft front seal may require lubrication or may require dry installation, depending on the seal material. The mechanic must follow the manufacturer's instructions.

Mandrels for seal installation are furnished by all engine builders, and many tool makers offer such mandrels. If the correct mandrel is not available, a large socket that fits the outside of the seal may be used. Any driving tool *must* exert force on the outer shell of the seal. As simple a tool as a piece of round wood can be used, provided it is held firmly and squarely to the seal, and excessive driving force is not used. [**Fig. 14–7**]

It is nearly impossible to tap a seal into place with a hammer or punch.

Note: A spacer may be installed behind a seal to move the sealing lip away from a wear groove.

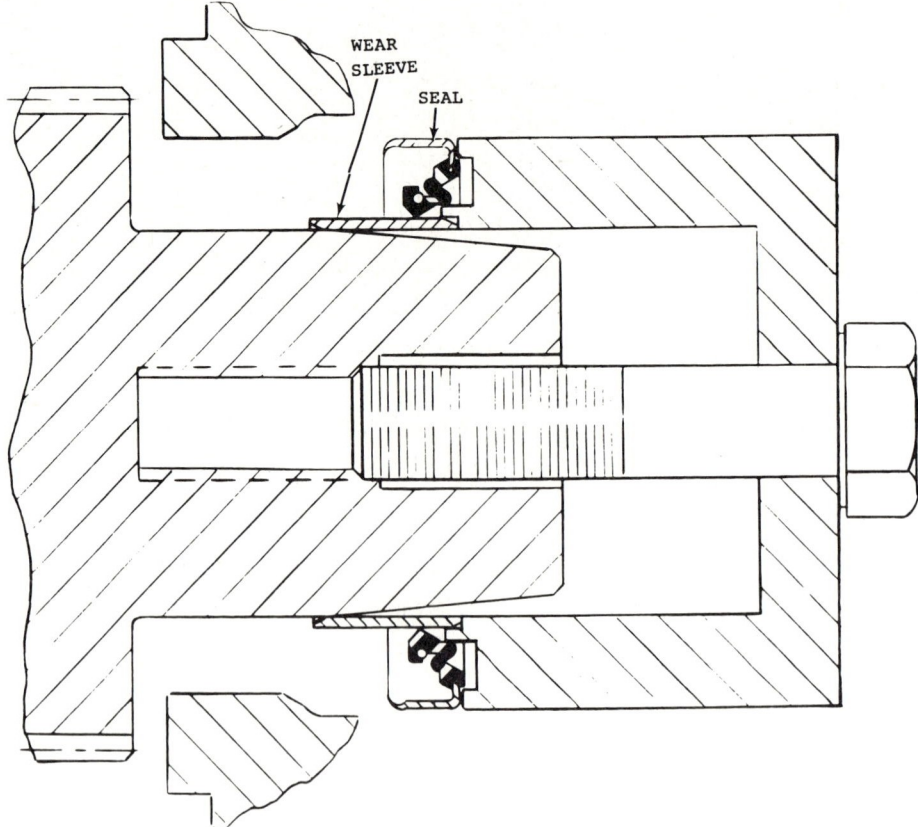

FIGURE 14-7. Install Seal and Wear Sleeve. *(Courtesy of Caterpillar Tractor Co.)*

Installation of the Damper Hub or Flange

1. See that no burrs or dirt is on the mounting surface.
2. Check tapered crankshaft ends for roughness or evidence of a loose flange. If the surface is not too bad, the flange can be lapped to a good contact.
   a. Use Grade A (280 GRIT) lapping compound, and apply a thin layer to the tapered bore in the flange hub.
   b. Put the flange on the crankshaft nose, and turn it about a half turn in both directions. Continue until the tapered surfaces match.
   c. Clean the surfaces, then check the fit with mechanics blue. There must be 80 percent full contact.
3. Be sure to clean both surfaces to the metal.
4. Wet the crankshaft flange surface with SAE 30 oil for steel flanges only. Use *no* lubricant for cast-iron flanges.

Installation of the Engine Front Support

5. If a solid (undivided) front support is used, put it on the trunnion first. This support usually wears, and you should check it before installation.

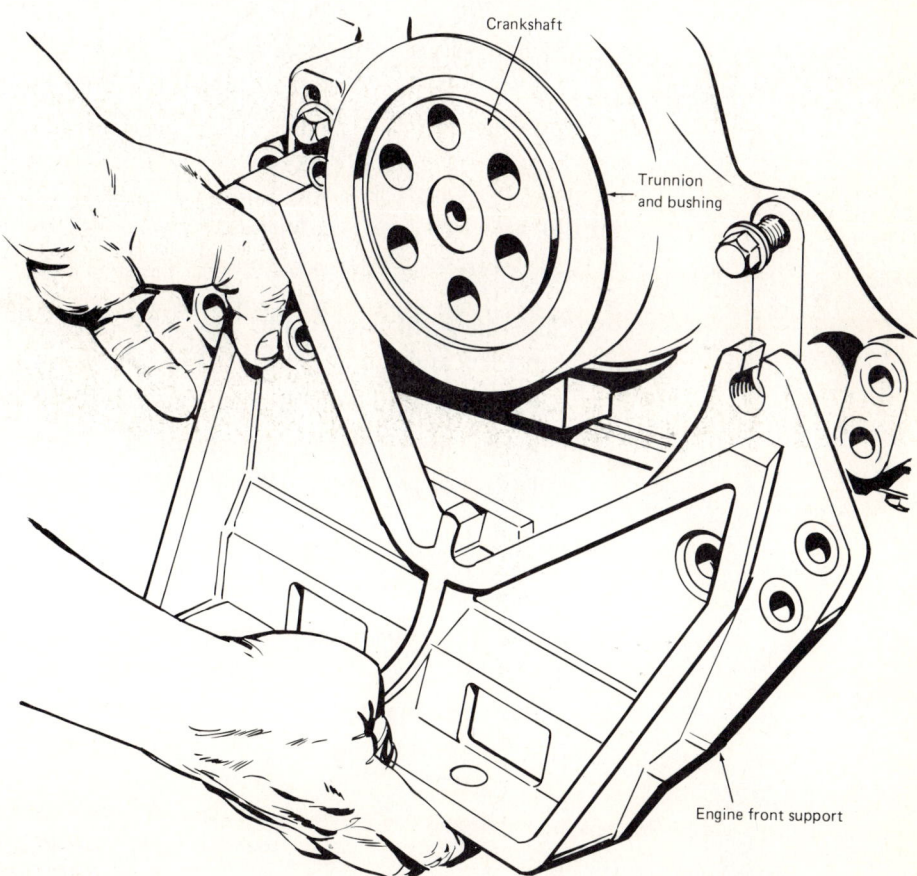

FIGURE 14-8. Front Engine Support. *(Courtesy of Cummins Engine Co., Inc.)*

246 / Diesel Engine Service

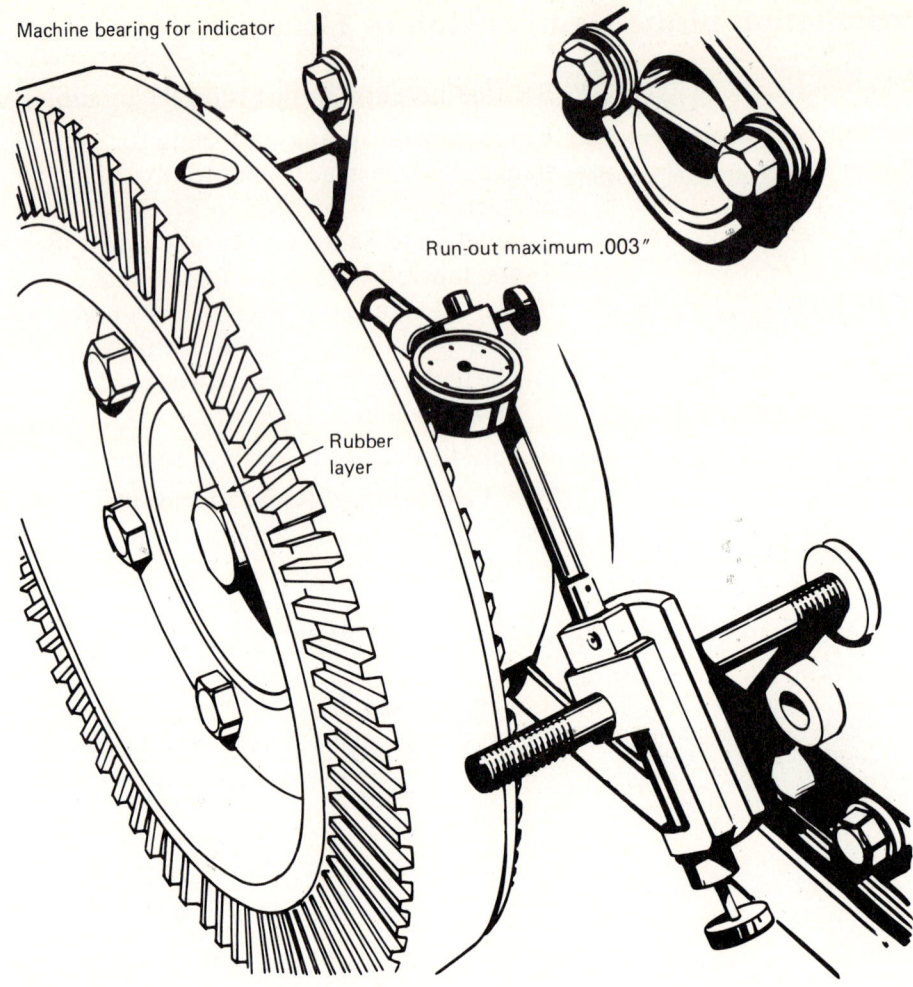

FIGURE 14-9. Check Damper Run-out. *(Courtesy of Cummins Engine Co., Inc.)*

6. Slide the front flange onto the crankshaft nose, and engage the retaining capscrew and heavy washer. [**Fig. 14-8**]

7. Tighten the capscrew to manufacturer's specified tension.

8. Position the vibration damper on the flange, and install the capscrews with new lockplates. Tighten them to 55 to 60 pound-foot (74.5 to 81 N-m).

9. Total run-out of the damper edge must not exceed 0.003 inch (0.08 millimeter). [**Fig. 14-9**]

10. Total face run-out must not exceed 0.0025 inch (0.064 millimeter) per inch of distance from the center to the point of measurement.

11. After checking run-out, bend lockplates to prevent capscrews loosening.

15

COOLING SYSTEM SERVICE

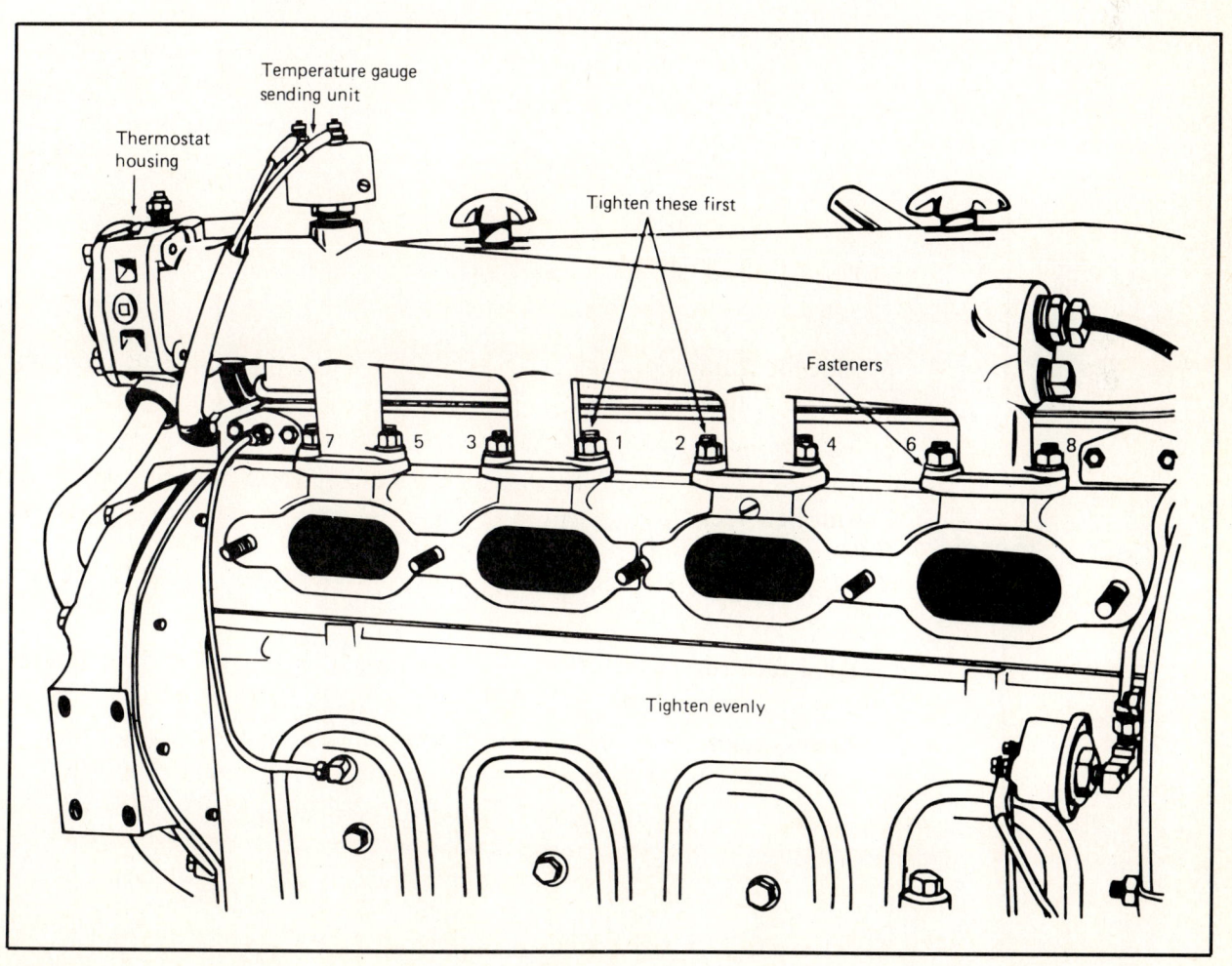

WATER PUMP ASSEMBLY

We have described general service to water pumps. This subassembly can be exchanged for a rebuilt unit.

As described, the water pump may be gear or belt driven. The initial points to observe in servicing it are the mounting fasteners and the piping connections.

1. Remove water pump. Drain coolant. Be sure that the block is empty.
2. Remove the inlet connection. This may require the removal of the hose from the radiator or the removal of a flanged connection from the water pump body.
3. Remove any bypass connection from the thermostat housing to the water pump. Discard old hoses and gaskets. [**Fig. 15-1**]
4. Remove the mounting capscrews and withdraw the water pump. Be careful

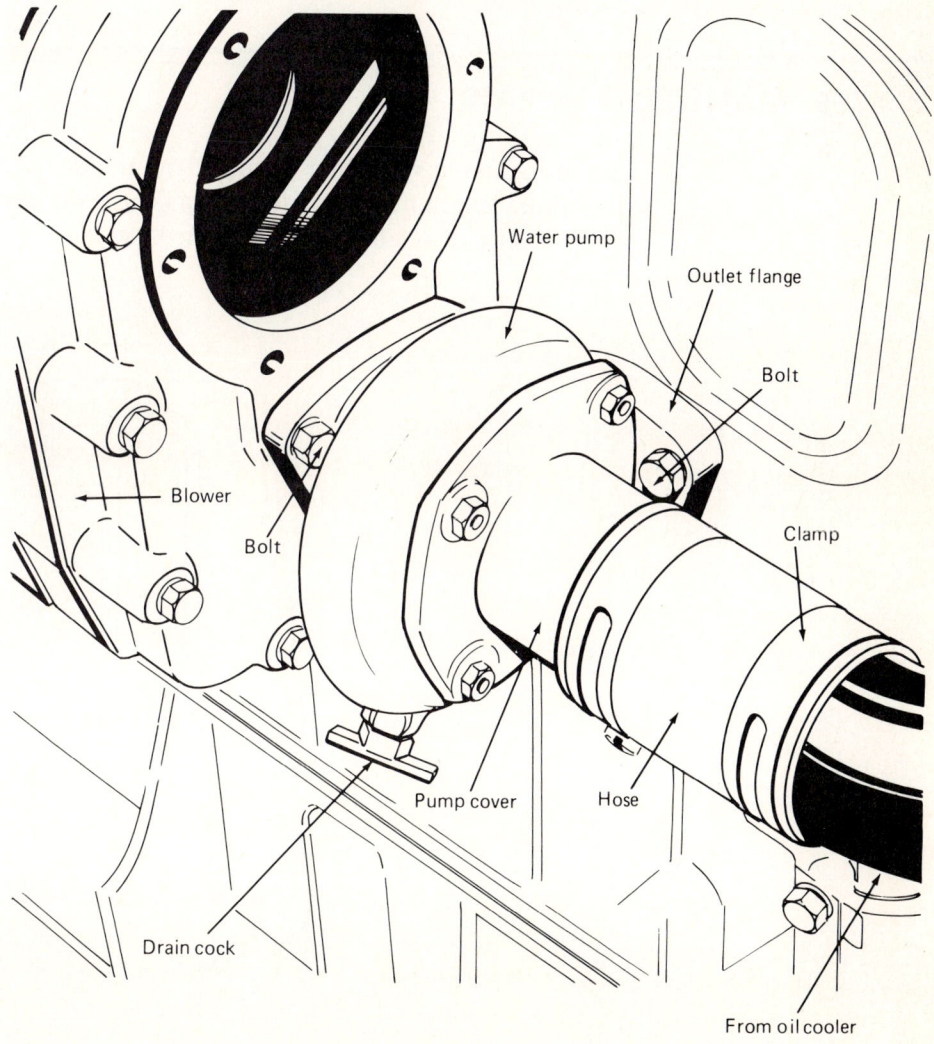

FIGURE 15-1. Typical Wear Pump. *(Courtesy of General Motors Corp.)*

249

250 / Diesel Engine Service

not to bump the impeller and lay the pump on a clean bench or board, impeller up. Water pump seals are usually ceramic and can be broken by rough handling.

INSTALL WATER PUMP

1. Position the water pump on its mounting, and start capscrews by hand. Be sure that the right capscrews are used in correct positions. Some are longer than others and must not "bottom" in the threaded hole.
2. Attach any hoses or flanges. Leave all fasteners loose until all connections are made.
3. Tighten fasteners evenly, alternating from corner to corner. [Fig. 15-2]
4. After the water pump is secured in position, tighten flanges, hose clamps, and any other connections.

WATER MANIFOLD REPAIR

Some engines have a water manifold mounted at each cylinder head, running the whole length of the engine. Others route the coolant back to the thermostat housing through a passage in the head.

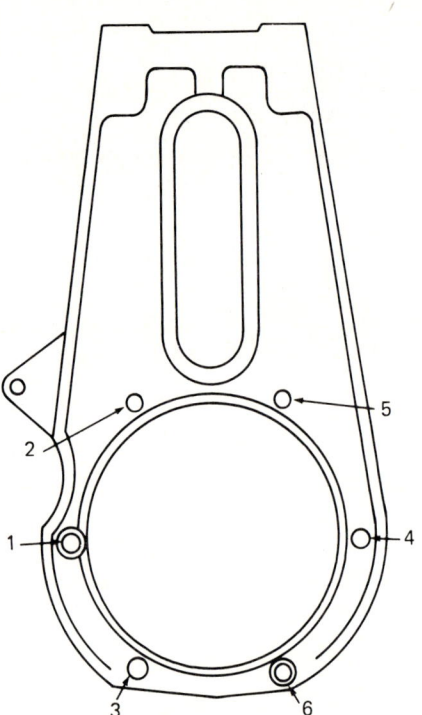

FIGURE 15-2. Tightening Sequence. (Courtesy of Cummins Engine Co., Inc.)

Installation of Coolant Filter

8. When a coolant filter is used, install a new element. This unit provides chemical treatment of the coolant, and retards liner pitting and other corrosion. [**Fig. 15-7**]

16

EXHAUST MANIFOLD AND TURBOCHARGER

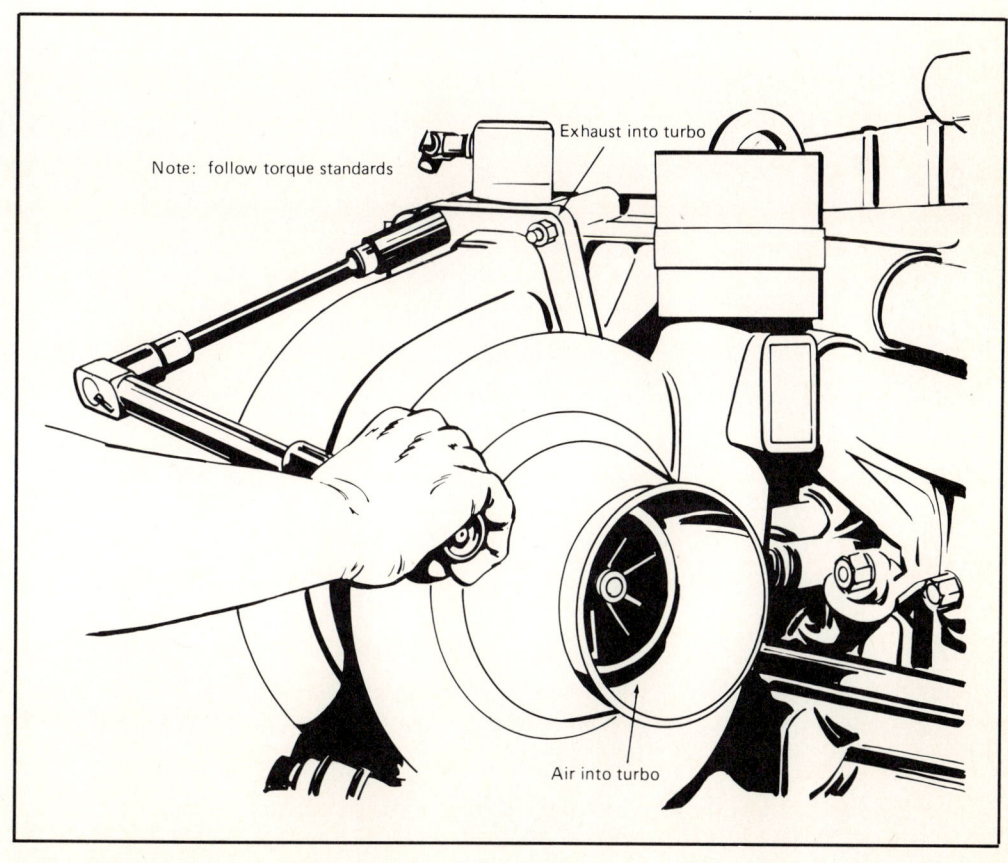

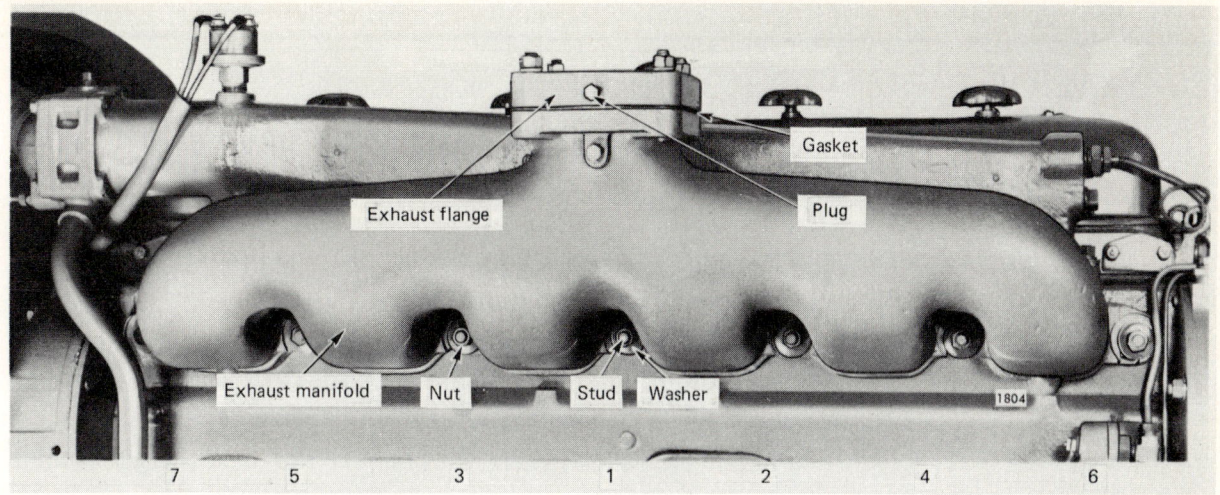

FIGURE 16-1. Typical Exhaust Manifold. *(Courtesy of General Motors Corp.)*

The exhaust manifold is attached to each cylinder head exhaust outlet by studs or capscrews. The manifold must be positioned so that no restriction is imposed on the outlets.

An engine manifold without a turbocharger has a flanged outlet to the exhaust piping. [**Fig. 16-1**] When a turbocharger is used, it is mounted on a pad in the center of the manifold, and the exhaust piping attaches to the turbine outlet. [**Fig. 16-2**]

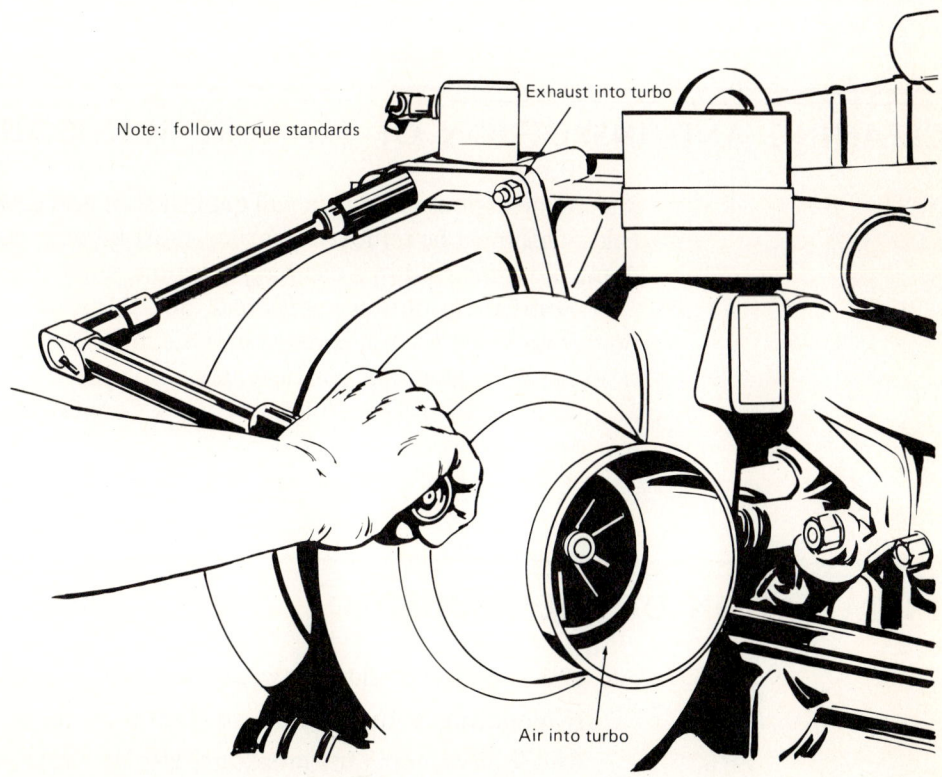

FIGURE 16-2. Turbocharger Mounting. *(Courtesy of Cummins Engine Co., Inc.)*

REMOVAL OF PARTS

Removal of the Turbocharger

1. Remove the air connection to the engine inlet manifold.
2. Remove the exhaust pipe from the turbine end.
3. Remove oil feed and drain tubing.
4. Attach a rope sling to the turbocharger, and connect a hoist to it. Some turbochargers are heavy enough to need a hoist to handle.
5. Remove mounting pad fasteners, and remove the turbocharger as an assembly.
   Note: Some engines are built with two turbochargers in series. They are both removed as described.

Removal of Exhaust Manifold

1. Remove the exhaust piping, if attached to the manifold.
2. Remove all fasteners, and take the exhaust manifold off. Store it in a safe place.
3. Remove old gaskets and discard.

CLEANING AND INSPECTION OF EXHAUST MANIFOLD

An engine in poor condition will deposit soot and carbon in the exhaust manifold. This must be removed. Carbon must be scraped from the parts, and the manifold immersed in a cleaning solution, such as used for other iron parts.

Some manifolds are sectioned, and the joints must be clean enough to permit movement from expansion. One must check for cracks, flange distortions, and broken pads for fasteners.

Exhaust manifolds run hot. They seldom fail but may bend enough to allow exhaust leaks.

INSTALLATION OF THE MANIFOLD

1. Use a scraper to clean any residue from the mounting pads.
2. If mounting studs are used, see that they are clean and straight.
3. Check any clamps for breakage. Use new lockplates.
4. Run a die or thread restorer onto the stud threads. Apply a small amount of Never-Seize compound onto the threads.

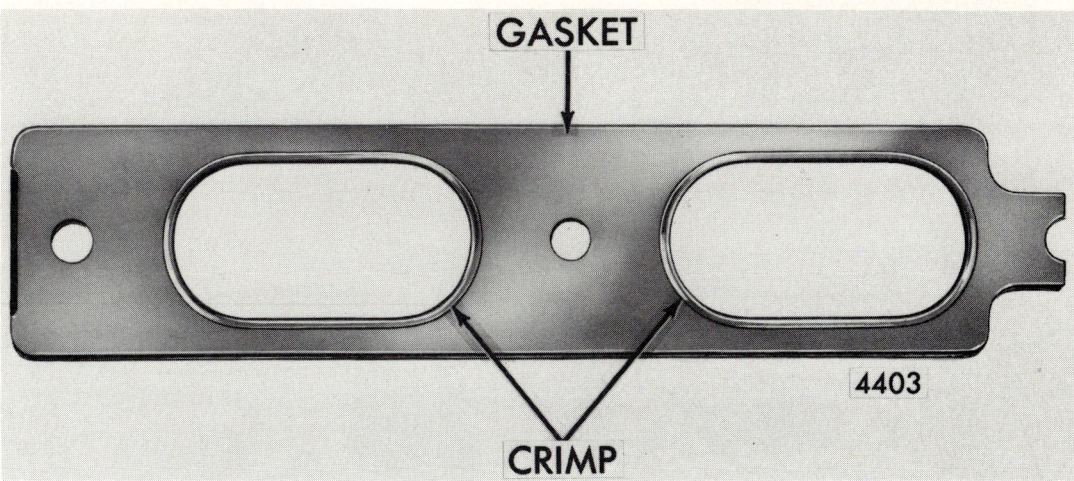

FIGURE 16-3. Exhaust Gasket Marks. *(Courtesy of General Motors Corp.)*

5. Apply new gaskets over the studs. These gaskets may be marked "out", which is to be installed toward the manifold. Some have a crimp in the metal, which goes to the head surface. **[Fig. 16-3]**
6. Position the cleaned manifold onto the studs, and assemble one nut and washer in the center to hold it.
7. Start all the nuts or capscrews, seeing that the right washer, lockplate, or clamp is in place.
8. If a heat shield is used, install it with the manifold.
9. Check any bar clamps used, and be sure that they are straight.

Tension Specifications

10. Tighten the fasteners to the specified tension. Tension for various engines is found in the manufacturer's manual, but this text has given standards for most fasteners. **[Fig. 16-4]**

Water-Cooled Exhaust Manifold

All marine applications require a water-cooled exhaust manifold to reduce the fire hazard. Such manifolds are cast iron, with a cored water jacket around the exhaust passages.

Coolant is routed through the jackets from the cylinder heads. Thus the cooled exhaust manifold replaces the water manifold, on most engines. If coolant flow is routed through a bypass tube to the exhaust manifold jackets, the water manifold is retained.

These exhaust manifolds are heavy. A hoist or adequate help is needed to both remove and install them.

262 / Diesel Engine Service

FIGURE 16-4. Tightening Sequence *(Courtesy of General Motors Corp.)*

Clean Water-Cooled Exhaust Manifolds

Any scale or carbon must be removed from both the exhaust side and the water side of the manifold. The manifold should be immersed in the cleaning solution used on other iron parts and may require a block-descaler compound to remove scale from the water jacket.

Some good cleaning compounds are branded Oakite, Turko, and Wyandotte G.

Caution: Solutions of these compounds should not be splashed onto skin. The mechanic must wear protective clothing and be careful.

Commercial pickling solutions are made by several companies. They are acidic in composition, and parts so cleaned must be immersed in a neutralizing solution afterward. A final rinse in plain hot water is effective.

Water cooled exhaust manifolds are installed in the same manner as dry manifolds. New gaskets should be used; a mechanic needs help or a hoist to position them.

INSTALLATION OF THE TURBOCHARGER

The important points to be observed when mounting an exchange turbocharger are to keep it clean and to use new fasteners and gaskets. Any tape

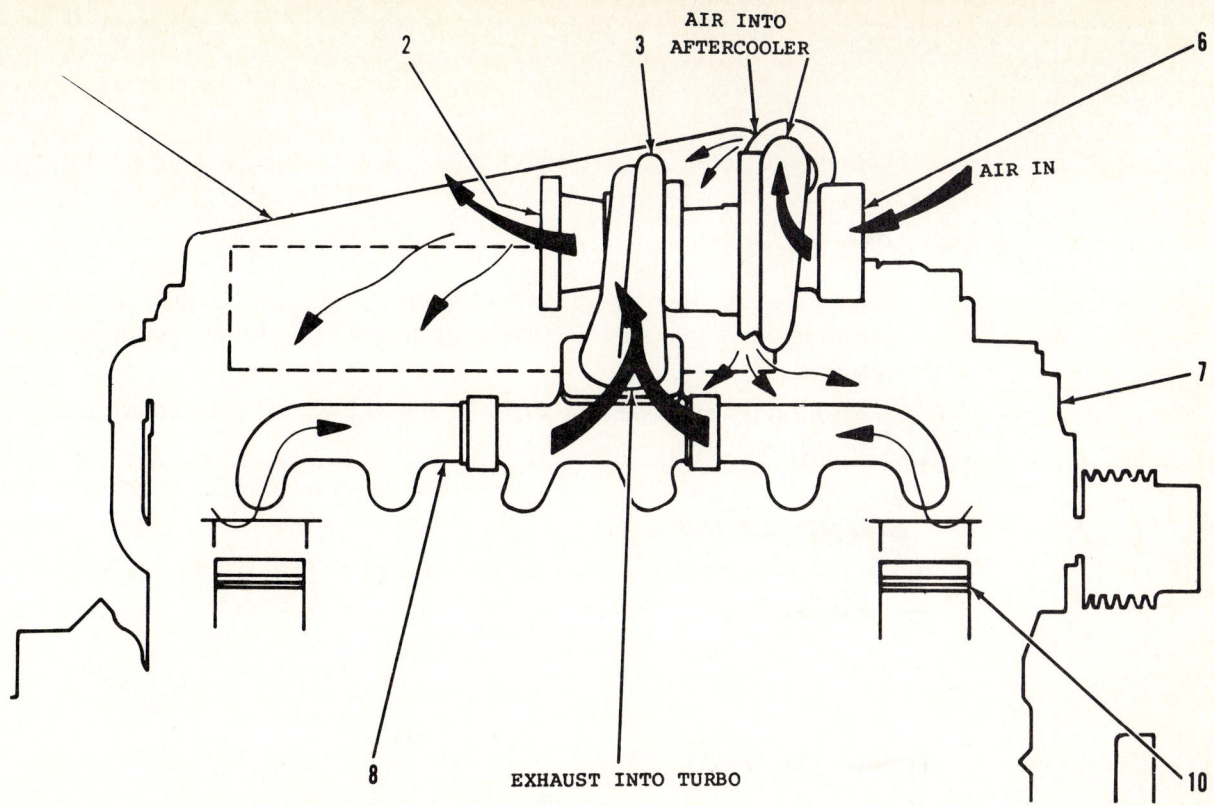

FIGURE 16-5(a). Turbocharger Air Flow. *(Courtesy of Caterpillar Tractor Co.)*

FIGURE 16-5(b). Turbocharger Installed. *(Courtesy of Cummins Engine Co., Inc.)*

263

264 / Diesel Engine Service

covers should not be removed from openings until ready to connect those openings.

1. See that the mounting pad on the exhaust manifold is clean. Either use new bolts or studs or clean the threads and inspect for distorted threads on the old studs. New fasteners are specified by some engine manufacturers.
2. Position a new gasket on the pad. Watch for marks. Place any crimp to the turbocharger.
3. Use a hoist and sling to lift the turbocharger into position. Start the fasteners to hold it.
4. Start all fasteners by hand. Do not tighten until air crossover and exhaust connections are aligned.

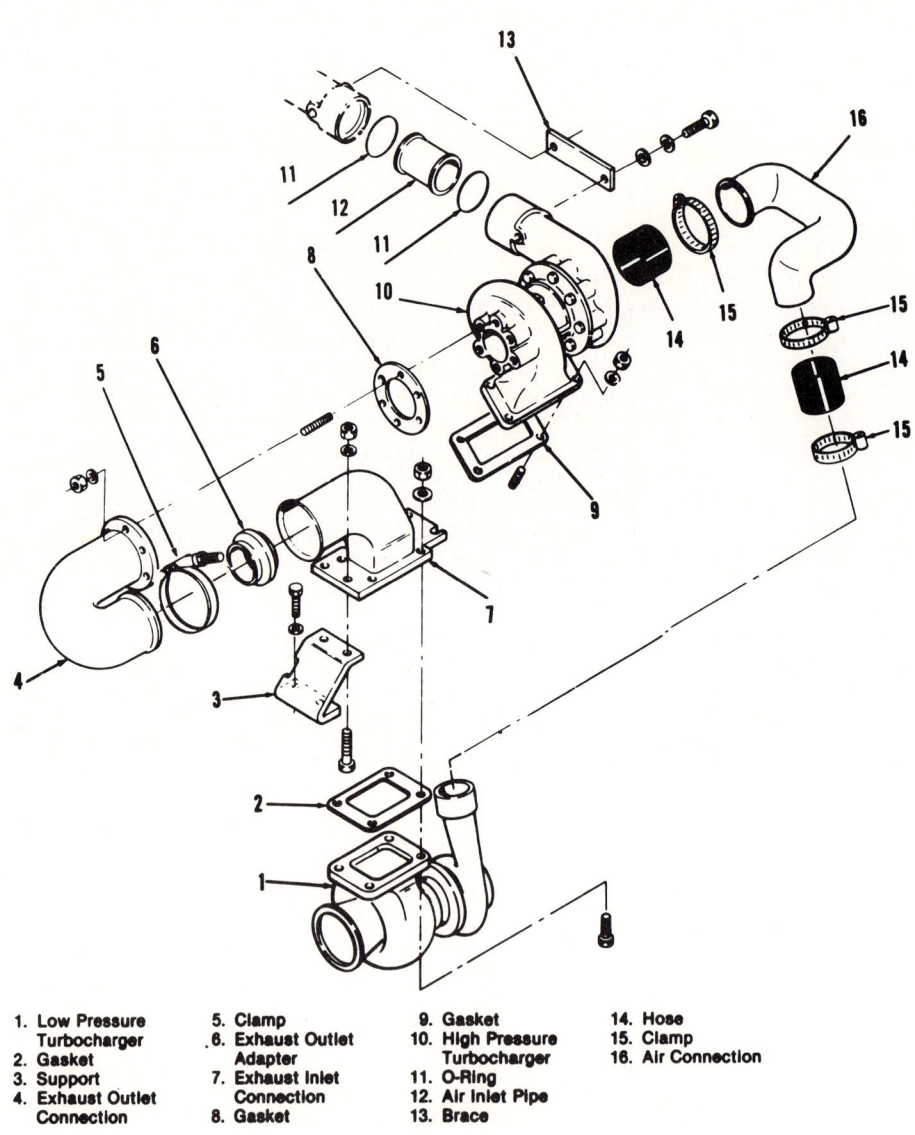

1. Low Pressure Turbocharger
2. Gasket
3. Support
4. Exhaust Outlet Connection
5. Clamp
6. Exhaust Outlet Adapter
7. Exhaust Inlet Connection
8. Gasket
9. Gasket
10. High Pressure Turbocharger
11. O-Ring
12. Air Inlet Pipe
13. Brace
14. Hose
15. Clamp
16. Air Connection

FIGURE 16-6(a). Series Turbochargers. *(Courtesy of Cummins Engine Co., Inc.)*

5. Align the air crossover. Connect the hose, and use a new gasket on the inlet manifold or aftercooler. Start all fasteners by hand.
6. Tighten the turbocharger mounting fasteners evenly to the specified tension.
7. Tighten the air crossover-to-air manifold fasteners evenly to specified tension.
8. Secure the exhaust pipe connection. The exhaust pipe must not exert strain on the turbocharger. A flexible section should be applied close to the turbocharger. The exhaust pipe should be supported by a clamp, so that it does not hang from the turbocharger housing.
9. Connect the oil feed and drain tubes. [Fig. 16-5]
   Note: If the engine is to be started at once, use a squirt can to fill the bearing housing with oil. With the air inlet open, turn the rotor by hand to distribute oil over the bearings. Then install the oil feed tube.
10. Connect the air tubing from the air cleaner. Check to be sure that the air cleaner element is clean or new.
11. Start the engine, and run at idle for 10 minutes.

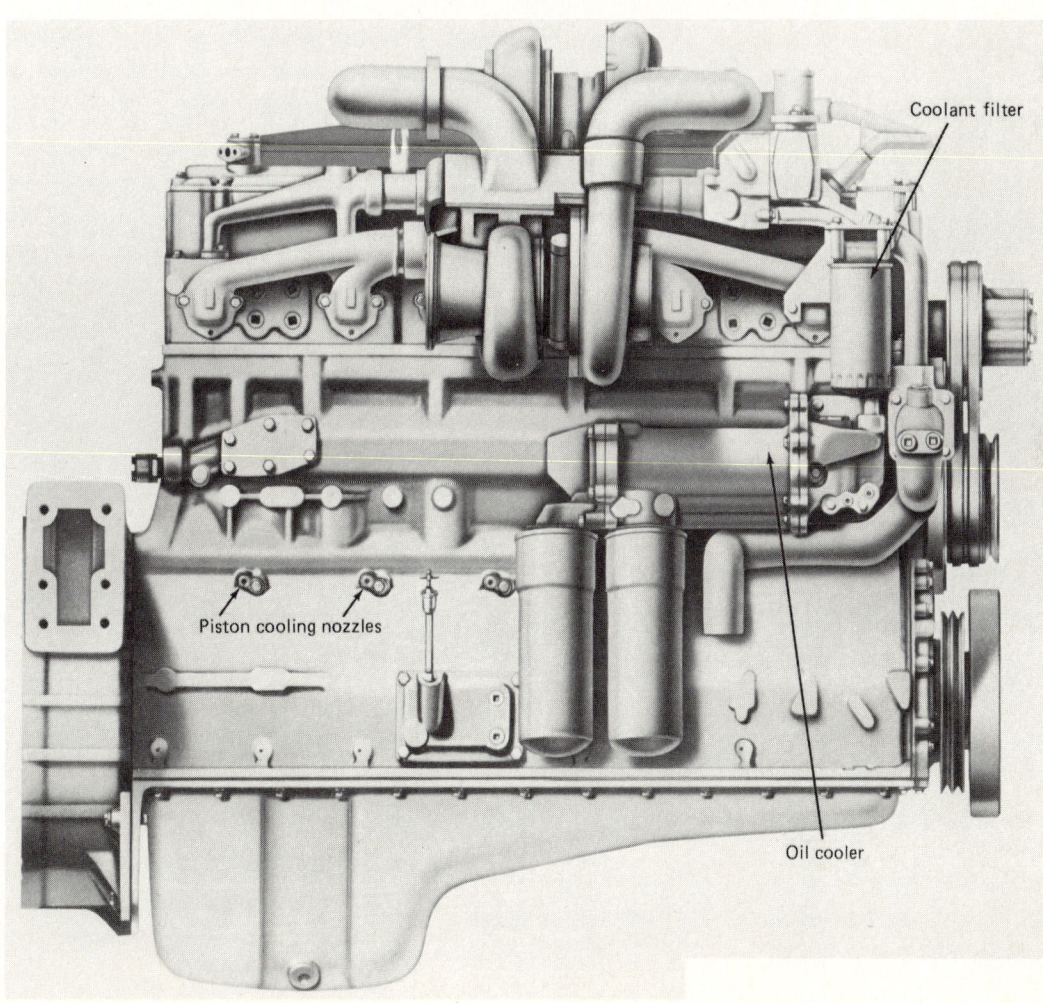

FIGURE 16-6(b). Series Turbochargers Mounted. *(Courtesy of Cummins Engine Co., Inc.)*

Caution: Never snap-accelerate a turbocharged engine after changing turbochargers. Be sure that oil is flowing before raising engine speed.

Installation of Series Turbochargers

Turbochargers in series are installed the same way, and all fasteners are left loose enough to permit final alignment of all supports. The exhaust gas is directed through the high-pressure turbine, then through an elbow to the low-pressure turbine, then to the exhaust pipe. [**Fig. 16–6**]

The air flows into the low-pressure impeller, then through a connection to the high-pressure impeller, then through the connecting pipe to the aftercooler and engine. Thus, the intake air is pressurized once in the low-pressure turbocharger, then again in the high-pressure unit. Air density is increased considerably; greater engine efficiency and power result. Compression-created heat is reduced by passage through the aftercooler.

17

AFTERCOOLER SERVICE

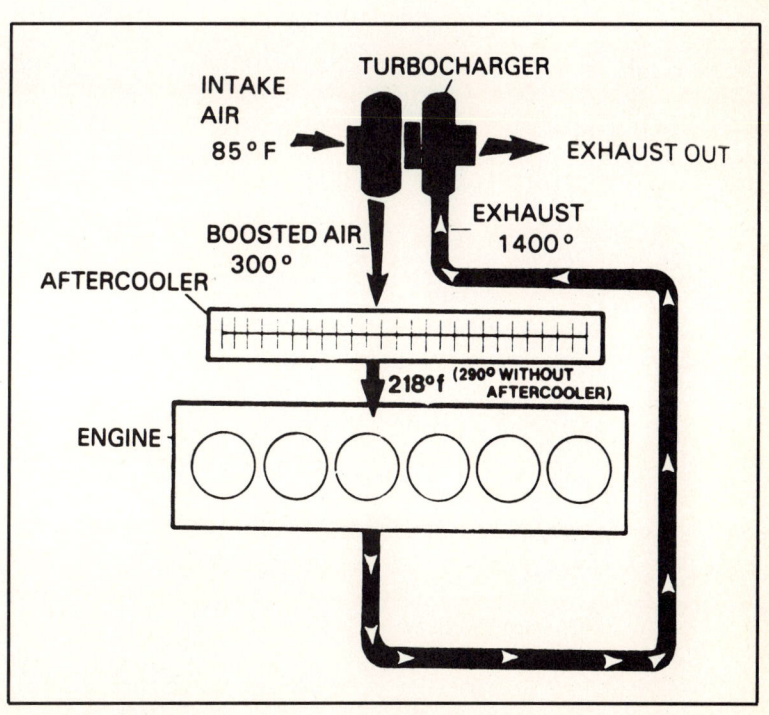

REMOVAL, CLEANING, AND INSPECTION

The text has described the aftercooler and its function. Should it develop faults, it can be removed, cleaned and repaired or replaced.

1. Drain the coolant from the engine.
2. Clean upper exterior with steam.
3. Remove coolant connections. Discard old hoses, gaskets, and seal rings.
4. Remove air crossover connections.
5. Remove air compressor connection.
6. The aftercooler housing and intake manifold are one piece. Remove the fasteners holding the assembly to the cylinder head. Because considerable weight is involved, either attach a sling and hoist, or get help to remove the aftercooler from the engine. [**Fig. 17–1**]

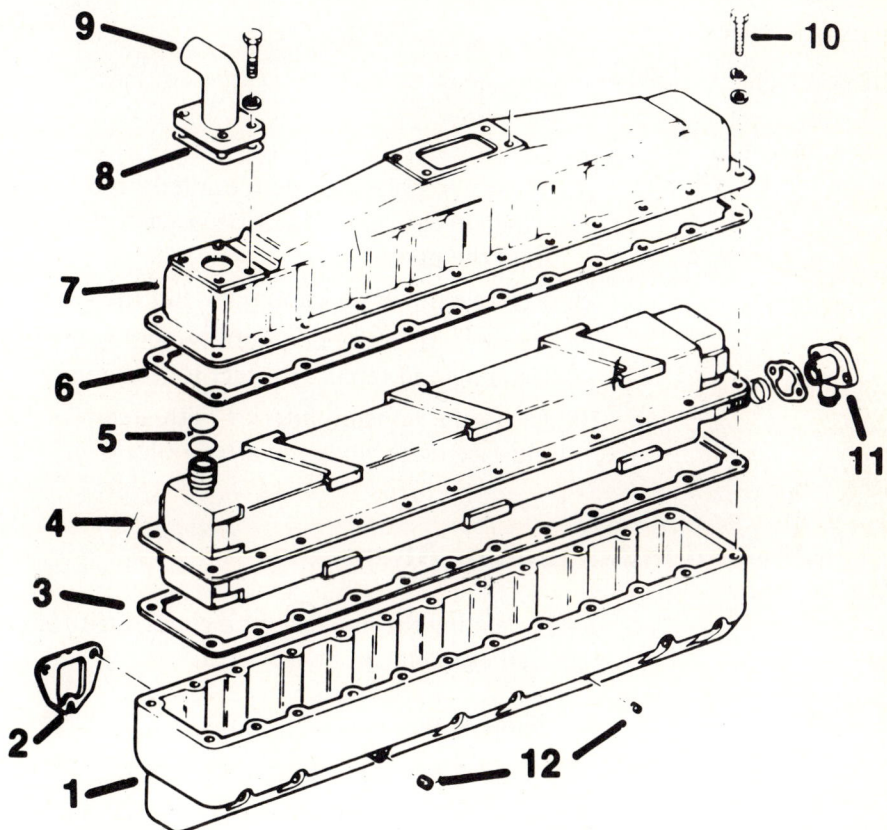

1. Housing
2. Intake Manifold Gasket
3. Gasket
4. Element
5. O-rings
6. Cover Gasket
7. Cover
8. Gasket
9. Water Outlet Connection
10. Capscrews
11. Water Inlet Connection
12. Plugs

FIGURE 17-1(a). Exploded View of Aftercooler. *(Courtesy of Cummins Engine Co., Inc.)*

270 / Diesel Engine Service

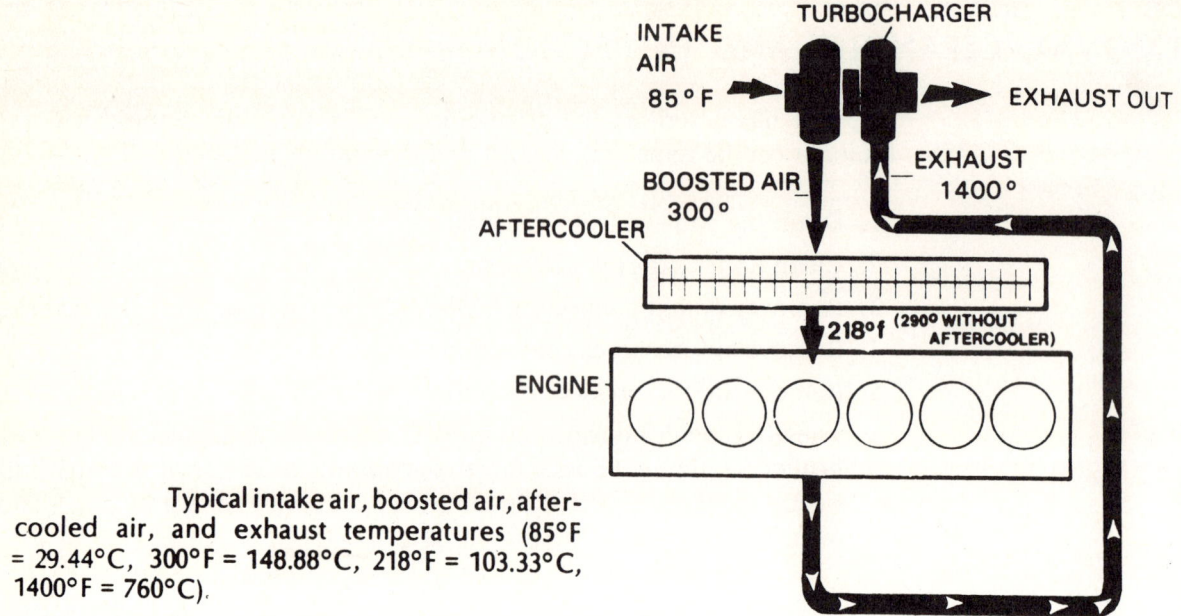

Typical intake air, boosted air, aftercooled air, and exhaust temperatures (85°F = 29.44°C, 300°F = 148.88°C, 218°F = 103.33°C, 1400°F = 760°C).

FIGURE 17-1(b). Aftercooler Function. *(Courtesy of Cummins Engine Co., Inc.)*

7. The aftercooler can be mounted on a board or other suitable support, which in turn can be grasped in a bench vise. Never use a vise directly to hold the unit.
8. Remove the cover capscrews and the cover.
9. Remove the element carefully, and route it to a good radiator shop for cleaning and repair. Protect it after removal to prevent damage.
10. Clean the housing interior with steam or a mineral spirit-air gun. This must be done while the aftercooler is off the engine.

Assembly and Installation of Aftercooler

After cleaning the housing, and checking it for cracks or damage, the new or repaired element can be installed.

1. Support the housing as described.
2. Apply a new gasket, making sure that all holes line up. [Fig. 17-2]
3. Wet new seal rings with oil, and put them on the inlet and outlet water connections.
4. Fit the assembled element into the housing carefully. Check to see that the seal rings are not disturbed.
5. Put a new gasket on the element flange, and assemble the inlet fitting with a new gasket. Install the capscrews finger tight.
6. Position the cover, and finger-start all the capscrews.
7. Install the outlet fitting with a new gasket. Finger-start the capscrews.
8. Snug-tighten the cover capscrews.

FIGURE 17-2(a). Install Aftercooler Element. *(Courtesy of Cummins Arizona Diesel, Inc.)*

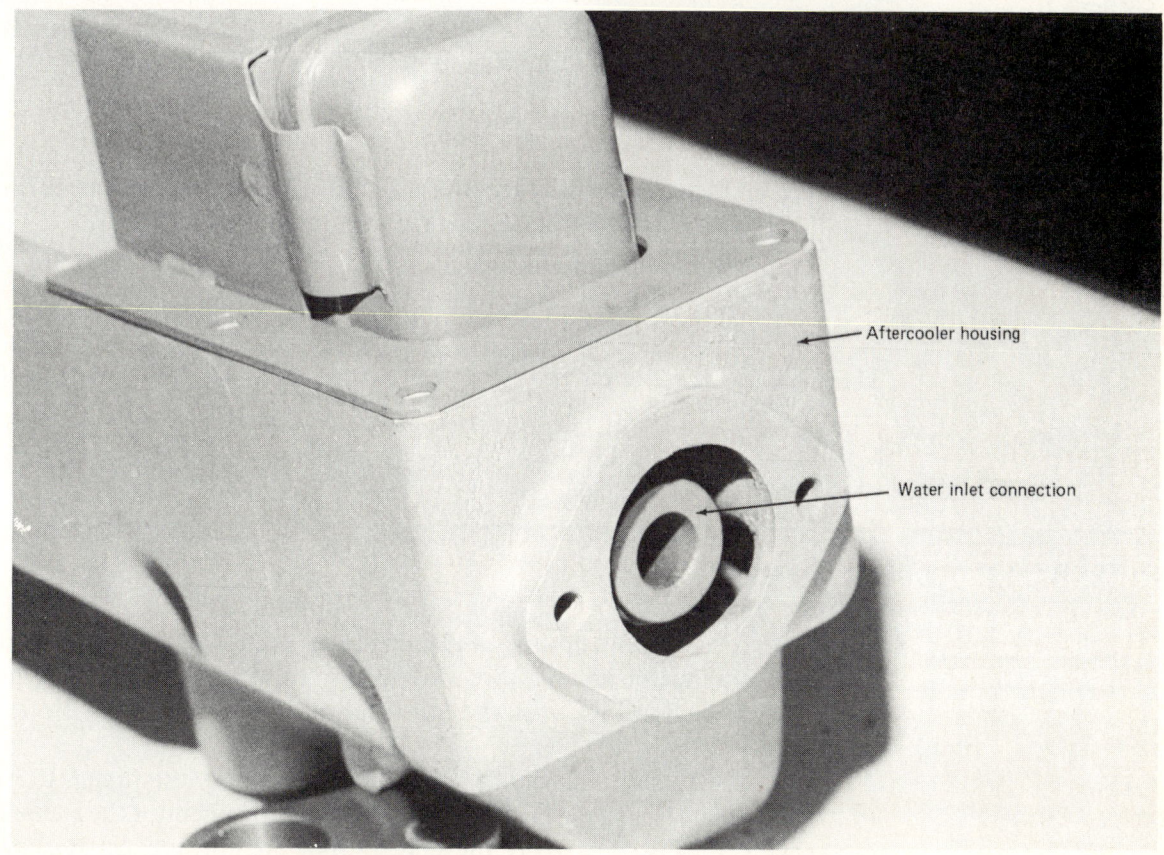

FIGURE 17-2(b). Aftercooler Water Inlet. *(Courtesy of Cummins Engine Co., Inc.)*

271

272 / Diesel Engine Service

FIGURE 17-2(c). Cross Bolt Aftercooler Removed. *(Courtesy of Cummins Arizona Diesel, Inc.)*

9. Tighten the inlet connection capscrews to 27 to 32 pound-foot (37 to 43 N-m) tension with a torque wrench.
10. Tighten the cover capscrews to 25 pound-foot (34 N-m), alternating from side to side from the center toward each end. Make at least three passes to reach full tension.
11. Tighten the water outlet connection capscrews to 15 to 20 pound-foot (21 to 27 N-m) evenly.

CROSSBOLT AFTERCOOLER

Some aftercoolers elements are supported in the housing by crossbolts that are engaged from the outside into threaded holes in solid members in the element. Assembly is as described, with the following exceptions: [Fig. 17-3]:

Note: The element fits closely in the housing, with no more than 0.013-inch (0.07 millimeter) clearance. This requires careful movement of the element. A cord sling can be used to lower the element into the housing. Such a cord or nylon rope can be withdrawn after placing the elements.

Aftercooler Service / 273

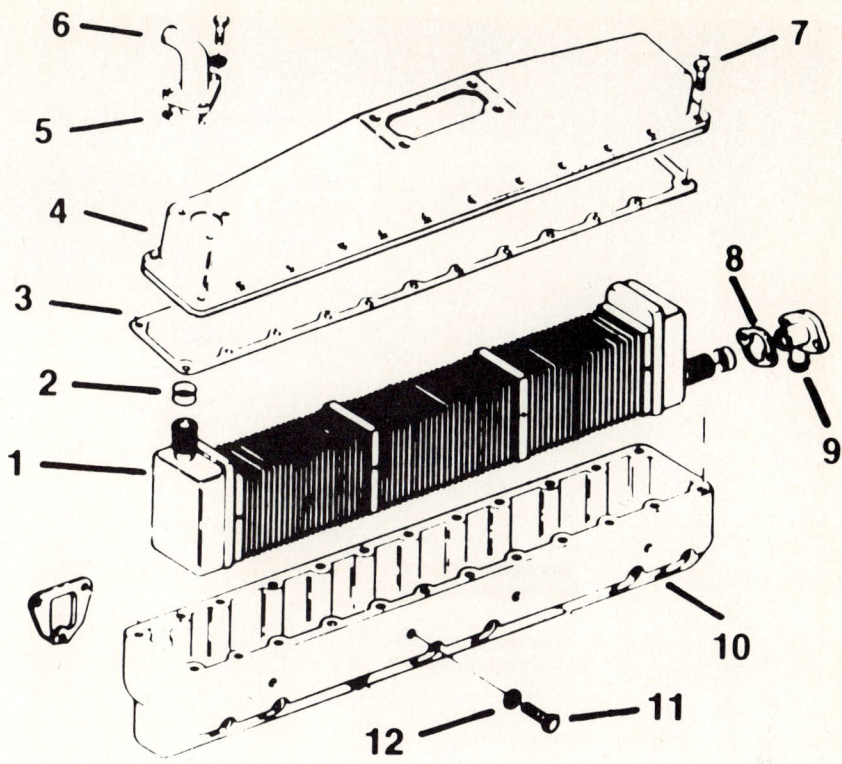

1. Core
2. O-ring
3. Cover Gasket
4. Cover
5. Gasket
6. Water Outlet Connection
7. Capscrew
8. Inlet Connection Gasket
9. Inlet Connection
10. Housing
11. Cross-Bolts
12. Hardened Washer

FIGURE 17-3. Exploded View of Cross-Bolt Aftercooler. *(Courtesy of Cummins Engine Co., Inc.)*

Note: Only one cover gasket is used. The rest of the assembly routine is the same as the one described. Torque specifications are the same.

Assembly of Crossbolt Aftercooler

1. Carefully install the element into the housing. Check the side clearance at the crossbolt pads. It must be from 0.003 inch to 0.013 inch (0.007 to 0.033 N-m). **[Fig. 17-4]**
2. Align the element and install the crossbolts and hard washers.
3. Tighten the crossbolts to 15 pound-foot (21 N-m), starting at the two center bolts and alternating to the ends.
4. Install the water inlet with a new gasket. Finger-start the capscrews.

FIGURE 17-4. Cross-Bolt Aftercooler Clearance. *(Courtesy of Cummins Engine Co., Inc.)*

5. Place a new gasket on the housing and install the cover. Finger-start the capscrews.
6. Install the water outlet with a new gasket and copper washers on the capscrews. Start them with fingers.
7. Tighten the water inlet to 27 to 32 pound-foot (37 to 43 N-m).
8. Tighten the cover capscrews to 25 pound-foot (34 N-m).
9. Tighten the outlet connection to 15 to 20 pound-foot (21 to 27 N-m).

DETROIT DIESEL AFTERCOOLER

Because these are two-stroke-cycle engines, which use a mechanical air blower as well as a turbocharger, the aftercooler is located in the blower discharge opening in the engine block.

Aftercooler Service / 275

Vertical Series 6-71 Aftercoolers

In vertical Series 6-71 engines, the water inlet for the aftercooler is in an adaptor in the block water jacket below the cylinder ports. A connection on the aftercooler seats in this adaptor. [**Fig. 17–5**]

Water leaves the aftercooler through a tube connected from the aftercooler rear to the water manifold.

Removal of the Aftercooler

1. Steam clean the engine, at least on the blower side.
2. Disconnect the turbocharger-to-blower inlet hose. The blower air inlet adaptor can be removed also.
3. Remove the blower and attached parts as an assembly.

FIGURE 17-5. General Motors 6-71 Aftercooler. (Courtesy of General Motors Corp.)

4. Drain the coolant from the engine.
5. Remove water outlet tube from the rear hand hole cover. Then remove the cover.
6. Remove the outlet connection from the aftercooler.
7. Remove the capscrews that hold the aftercooler in place.
8. Carefully take the aftercooler out through the block opening. Be sure that the fins do not catch and get bent.
9. The capscrews that have the nylon lock are to be discarded. Use new capscrews for assembly.
10. Clean the block hole, and remove any gasket residue.
11. Remove and discard all inlet and outlet seal rings.
    Note: The aftercooler can be routed to a good radiator shop for cleaning and repair.

Installation of the Aftercooler in Vertical Engines

1. Apply new seal rings on both inlet and outlet connections. Wet the installed rings with oil just before assembly.
2. Carefully install the aftercooler in the block. Protect the seal rings and fins.
3. Secure the aftercooler in position with six new $\%_{16}$-inch-18 by $\%_{16}$-inch nylon-patched capscrews. Tighten evenly to 18 to 20 pound-foot (25 to 30 N-m).
4. Install the outlet tube and adapter through the rear hand-hole. Install the hand-hole cover.
5. Position the outlet elbow on a new gasket and seal ring onto the aftercooler outlet connection. Secure it evenly with two nylon-patched capscrews, and tighten them evenly.
6. Connect the water outlet to the cylinder head water connection through the tube and hose. Tighten the hose clamp.
7. Install the blower assembly on a new gasket. See that all blower-driven units are properly aligned, and secure the blower.
8. If the air inlet is removed, attach it to the blower with a new gasket $4\%_{16}$-inch capscrews and lock washers. Tighten to 46 to 50 pound-feet (62 to 68 N-m).

Note: It is usually not necessary to remove the turbocharger when repairing the aftercooler. Aftercooler repairs are always made during engine overhaul and when water is found in the air box. This is not a common occurrence.

Detroit Diesel Vee Engine Aftercoolers

These units are mounted in the vee of the block, under the blower. Repairs can be made when water leakage is found, which shows up as water in the air box or white steam in the exhaust, with frequent coolant addition necessary.

Water flow is from rear to front, through an adaptor in the block,

FIGURE 17-6. General Motors 8-V Aftercooler. *(Courtesy of General Motors Corp.)*

through the aftercooler, then to a tube to the thermostat housing. Other than that, there are only a few differences in the repair procedure. [**Fig. 17-6**]

Removal of Aftercooler from Vee

1. Steam clean the top of the engine.
2. Remove the turbocharger, blower, and attached parts.
3. *Do not remove* the four capscrews in the body of the aftercooler. [**Fig. 17-7**]

Installation of the Aftercooler in Vee Engines

This unit is installed in the same way as just described. The mechanic must be warned against rough handling of the element.

In this case the element water outlet must be installed end first, with the

278 / Diesel Engine Service

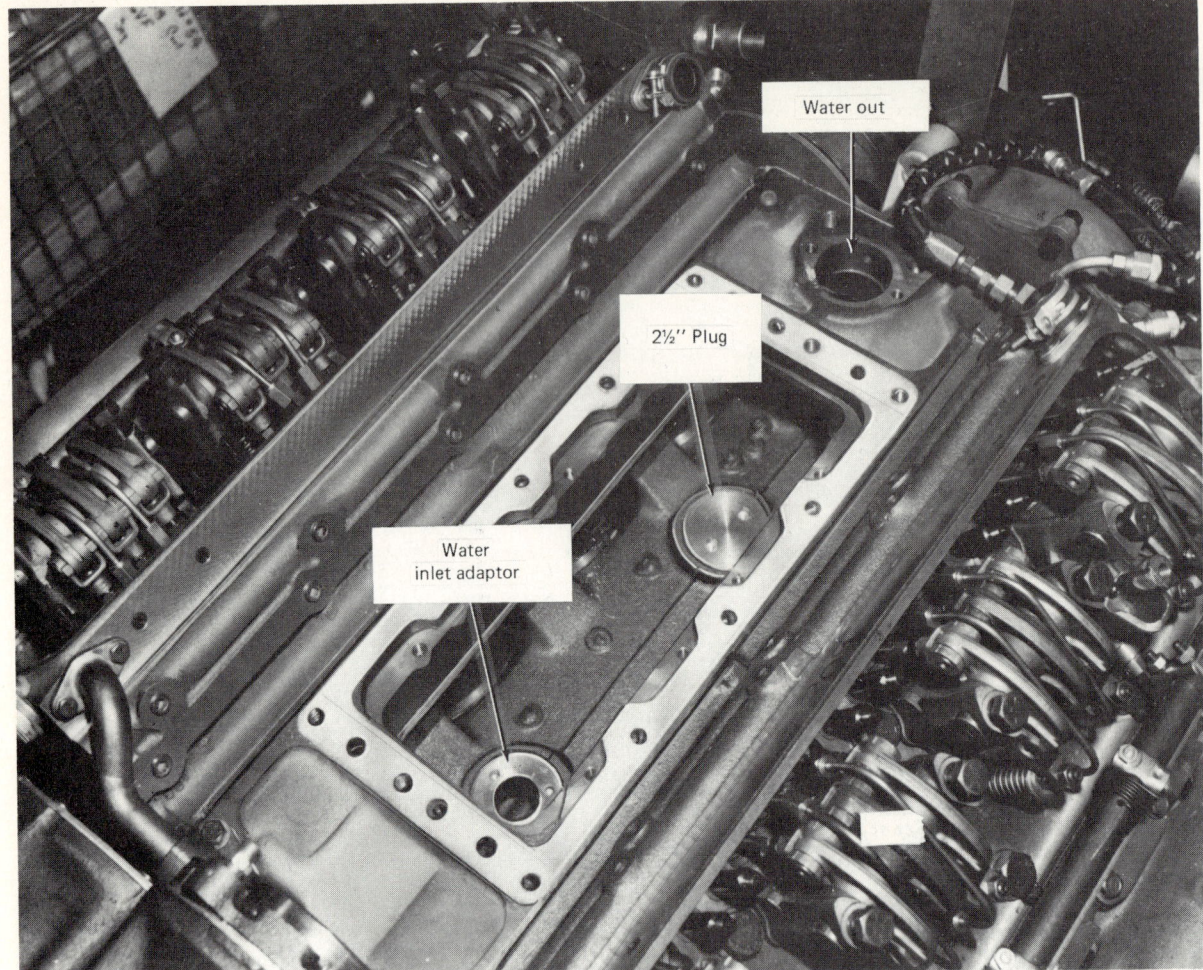

FIGURE 17-7. 8-V Aftercooler Connections. *(Courtesy of General Motors Corp.)*

fasteners loose until the water outlet tube has been positioned. Then the two attaching capscrews are tightened.

New nylon-patched capscrews should be used at all points where they were used before. The blower and attached parts must be attached with new gaskets.

Eight $7/16$-inch -14 x $1\frac{1}{2}$-inch capscrews are used on the air inlet adapter of the 6-V engine, and 10 on the 8-V engine. These and the capscrews for the turbocharger mounting are to be tightened evenly to 46 to 50 pound-feet (62 to 68 N-m).

18
GEAR COVER ASSEMBLY

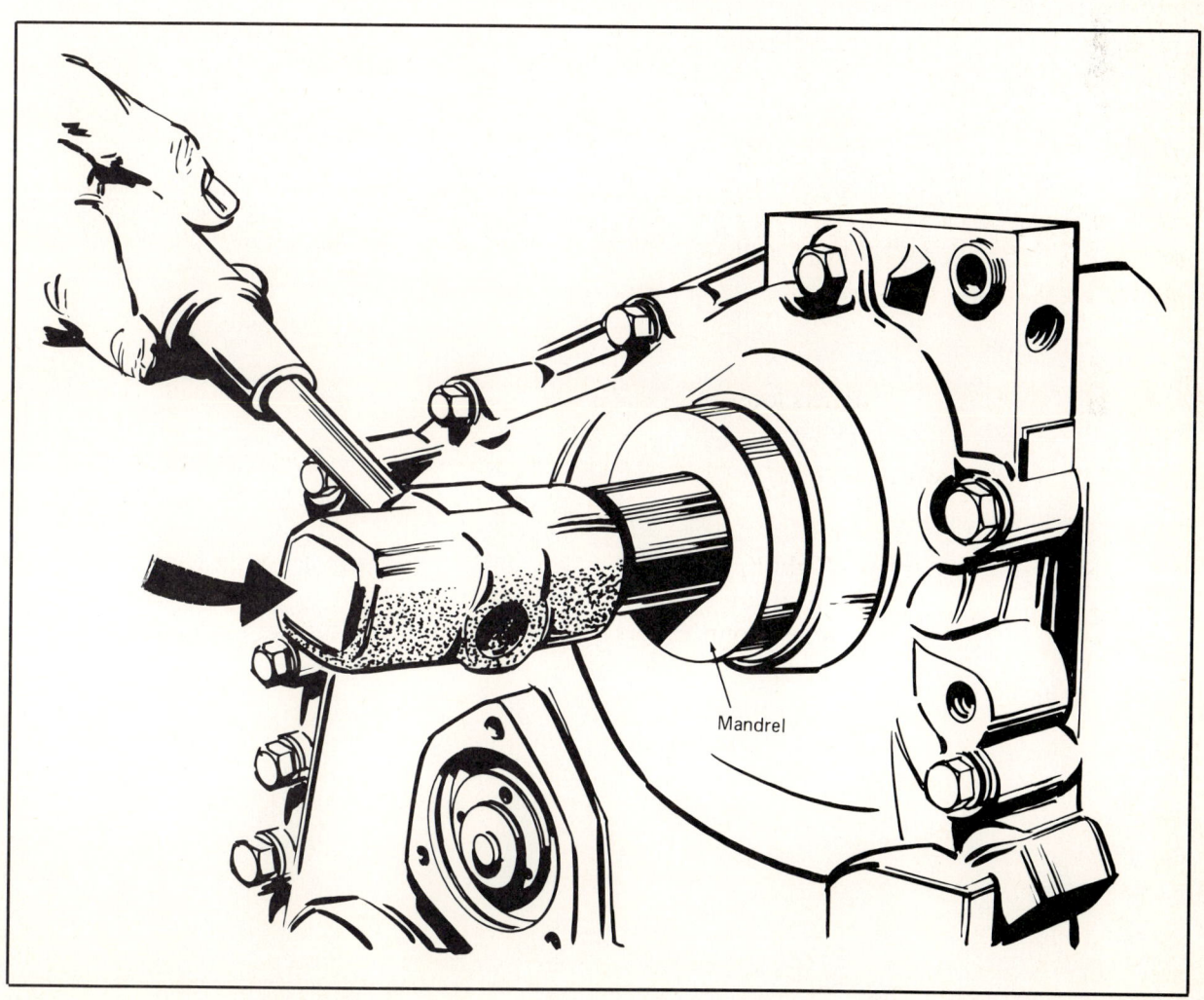

GENERAL DESCRIPTION

Some engines have the gear train at the front, some at the back of the block. The gear cover may be a simple plate with mounting pads for accessories, or it may contain bearings and bushings for the support of shafts. The chapter treats the general principles of gear cover service, with references to the various subassemblies such covers may carry. [**Fig. 18-1**]

Whatever form they take, front covers are usually dowel-located to the block. The only time that redoweling is necessary is when a new cover is to be installed. [**Fig. 18-2**]

BUSHINGS, BEARINGS, AND SEALS

All such support or sealing parts are renewable. Seals can be replaced without removing the cover. Most bearings and bushings require cover removal for service. [**Fig. 18-3**]

The common metal-cased oil seal is used where shafts or pulleys must go through a hole in the cover. Removal of such seals has been described under "Vibration Damper" in Chapter 14. [**Fig. 14-6**]

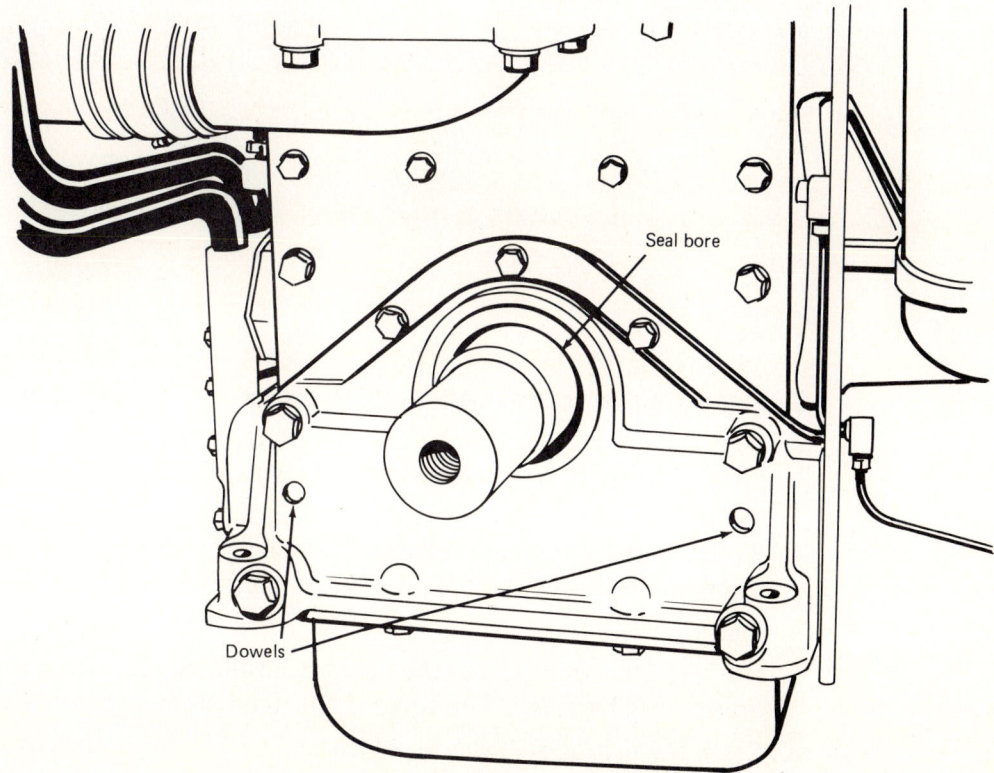

FIGURE 18-1. Front Cover Plate. *(Courtesy of General Motors Corp.)*

282 / Diesel Engine Service

FIGURE 18-2. Front Cover Dowels. *(Courtesy of Cummins Engine Co., Inc.)*

Removal of Front Cover

1. Remove any accessory parts from the cover.
2. Remove pulleys, crankshaft flange, alternator, and any other interfering parts. The parts to be removed vary with different engine makes and models.
3. Be sure that any oil pan capscrews that engage the bottom of the cover are taken out. Some engines require the oil pan to be removed.
4. Remove all cover attaching capscrews.
5. Use a soft hammer to tap the cover from the dowels. Do not use screwdrivers, pry bars, etc. The cover is not tight enough to need wedging.
6. Clean the mating surfaces of all gasket residue. Store the cover in a protected place.

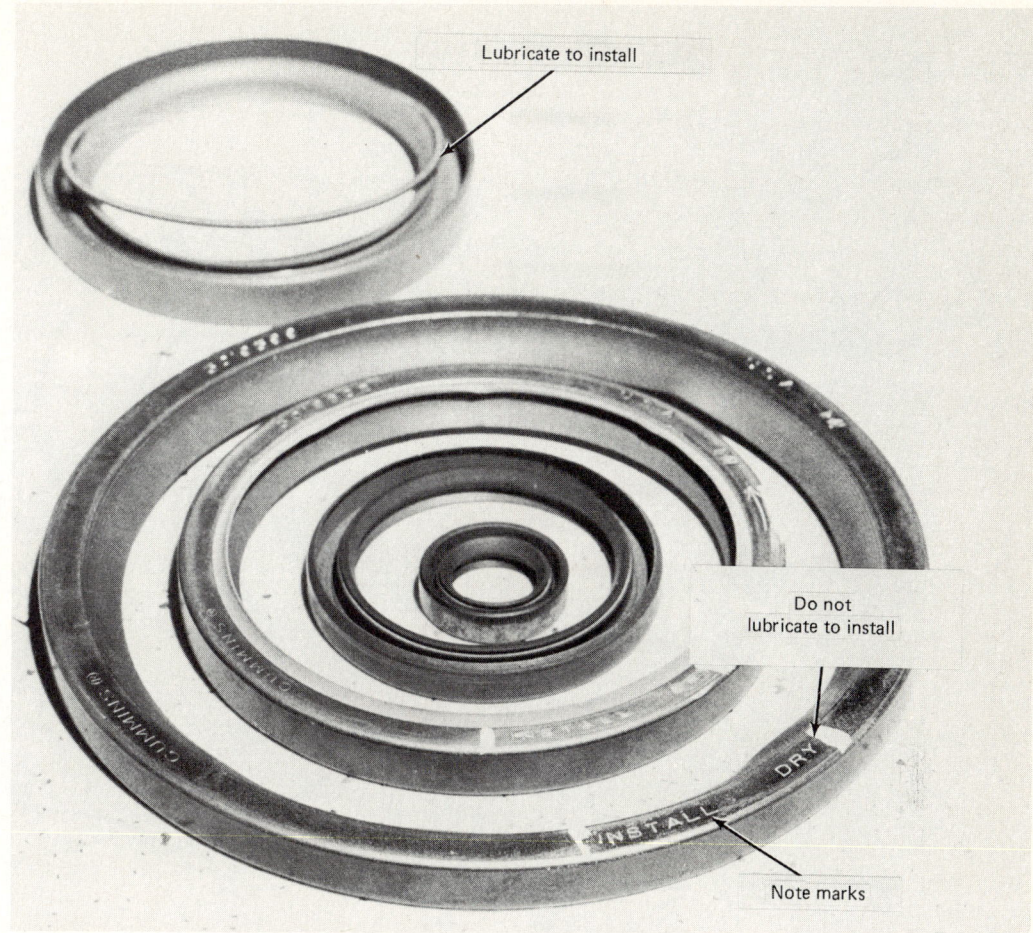

FIGURE 18-3. Seal Details. *(Courtesy of Cummins Engine Co., Inc.)*

Inspection of Front Cover

Front covers seldom crack, and any wear is confined to the engine support trunnion, if used.

The trunnion is a circular projection from the cover, and the front engine support is located on it. Any wear is usually absorbed by either a wear ring or a separate front engine support bracket, which is bolted to the cover.

Front Cover Bushings

There is an increasing design trend toward relieving the front cover of any bushings by using separate housings bolted to the cover. When bushings are pressed into cover bores, they must be checked for wear, using small hole gauges and micrometers. [Fig. 18-4]

Such inspections should be performed with the cover on a clean bench. Worn bushings should be renewed with a press and the correct mandrel. The must be straight and seated to the right depth. [Fig. 18-5]

These bushings receive oil under pressure. They do not wear rapidly.

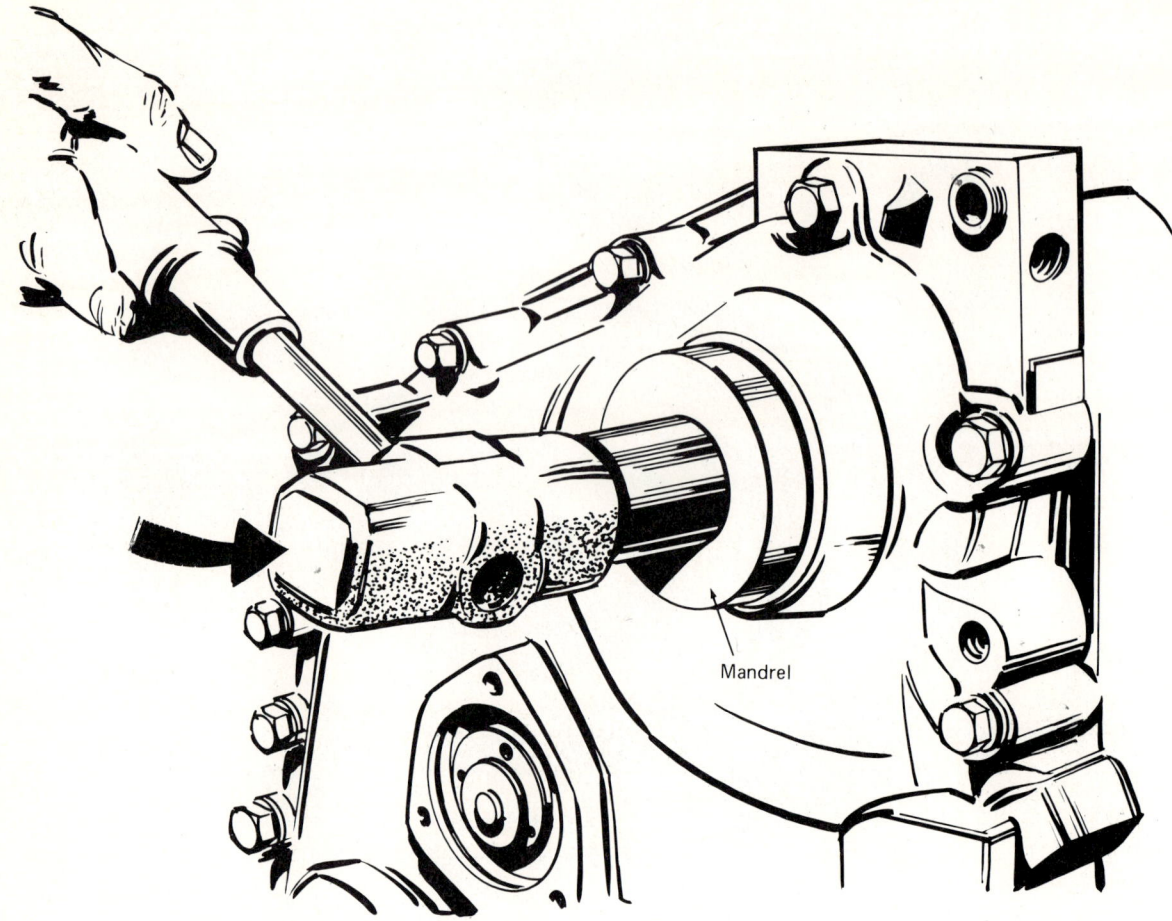

FIGURE 18-4. Install Seal with Mandrel. *(Courtesy of Cummins Engine Co., Inc.)*

Installation of Front Cover

The front cover or gear cover may be installed after all gears have been secured and timed and after it has been cleaned and checked. A new cover would require alignment and redowelling. The original cover will be in correct alignment on the dowels.

1. Place a new cover gasket on the dowels.
2. Position the cover over the dowels, and handstart all fasteners. Be sure that the correct capscrews are used in each hole.
   Note: If the oil pan is not removed, be careful when you install the cover to keep from damaging the pan gasket end. A thin coating of grease on the pan gasket may help. It is best to loosen the oil pan from the block to lower it.
3. Draw the cover fasteners up evenly to a snug tension. Do not final tighten them at this time.
4. Cut any protruding ends of the cover gasket flush with the metal surface. Check to see that the cover is even with the block edge.

Gear Cover Assembly / 285

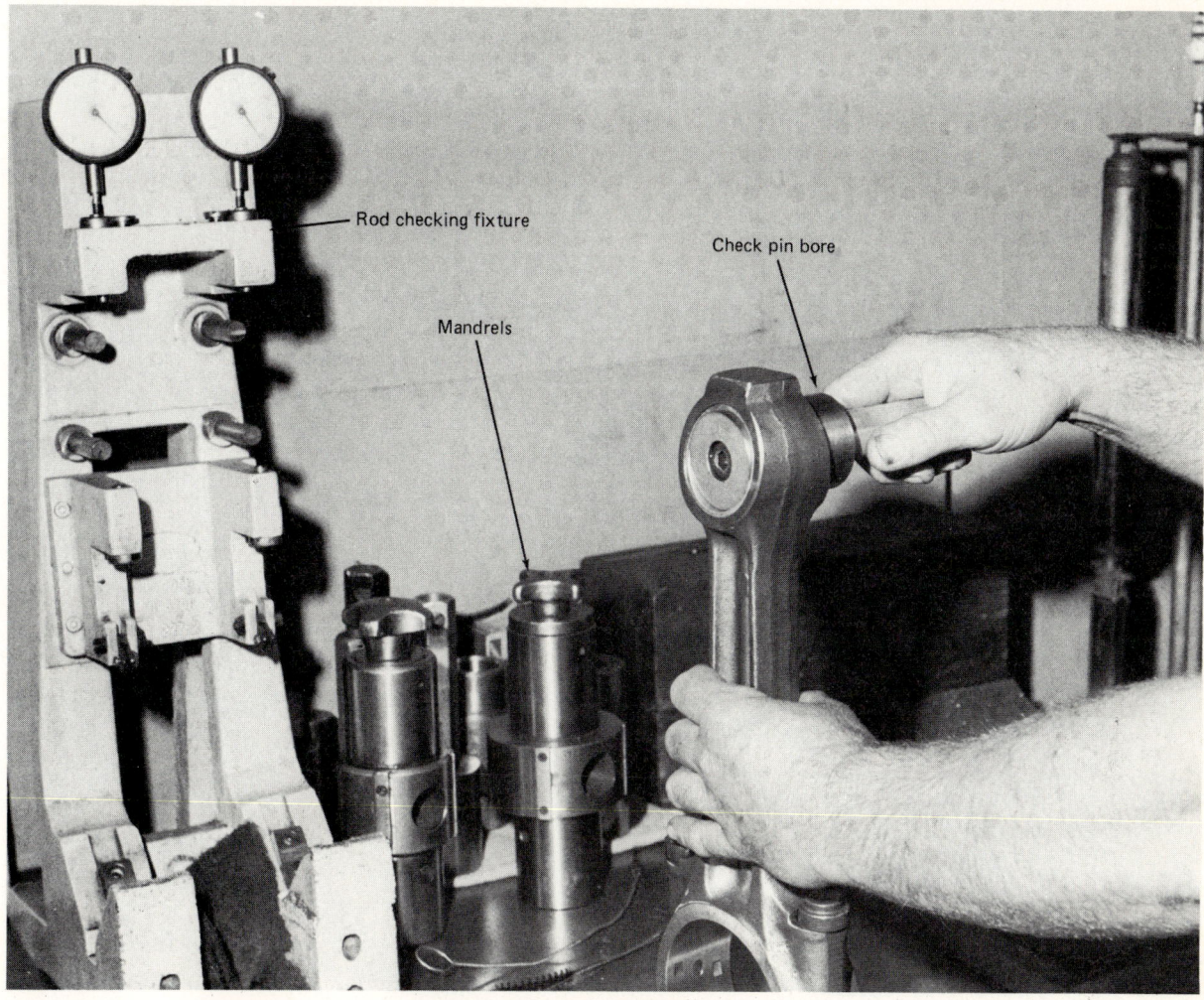

FIGURE 18-5. Check Bushing for Wear. *(Courtesy of Cummins Engine Co., Inc.)*

5. Alternate from side to side in tightening the cover fasteners. Follow torque settings recommended by the engine builder.

6. Capscrews from oil pan to cover are often of different thread pitch from those from oil pan to block. Use the right capscrews in the various holes.

7. If the oil pan is removed, install it with a new gasket. When the block has a flange, hold the pan in position with two Vise Grip pliers, one on each side. If there is no flange, engage a capscrew on each side with your fingers. Tighten as described when all are started.

8. Some oil pans are attached to the rear crankshaft seal housing with capscrews, which often are not in view from the side of the engine. Do not forget them.

Camshaft Thrust Plate or Support Bearing

This flanged unit fits in a bore in the front cover either against the end of the camshaft or over it. Its clearance with the camshaft is established with spacer shims.

FIGURE 18-6. Camshaft Thrust Plate. *(Courtesy of Cummins Engine Co., Inc.)*

If it has a grooved flat surface to bear on the camshaft, it is a thrust plate. If it has a bushing to fit over the end of the camshaft, it is a support bearing. [Fig. 18-6]

This part is found on Cummins six-cylinder engines. Other engines have different systems for controlling camshaft endwise movement.

It is important to determine accurately the spacer shim thickness required.

1. Remove the seal ring and shims from the thrust plate or support bearing.
2. Position the part into the cover bore, and push it in until it is against the camshaft. Hold it squarely, and use a thickness gauge to find the distance of the flange from the gear cover pad.
3. Add to this measurement 0.008 inch to 0.013 inch (0.20 to 0.33 millimeter) for a support bearing 0.001 inch to 0.005 inch (0.013 millimeter to 0.03 millimeter) for a thrust plate. The total is the thickness of shims to install in each case. [Fig. 18-7]
4. Assemble the shim pack that you just found needed onto the thrust plate.

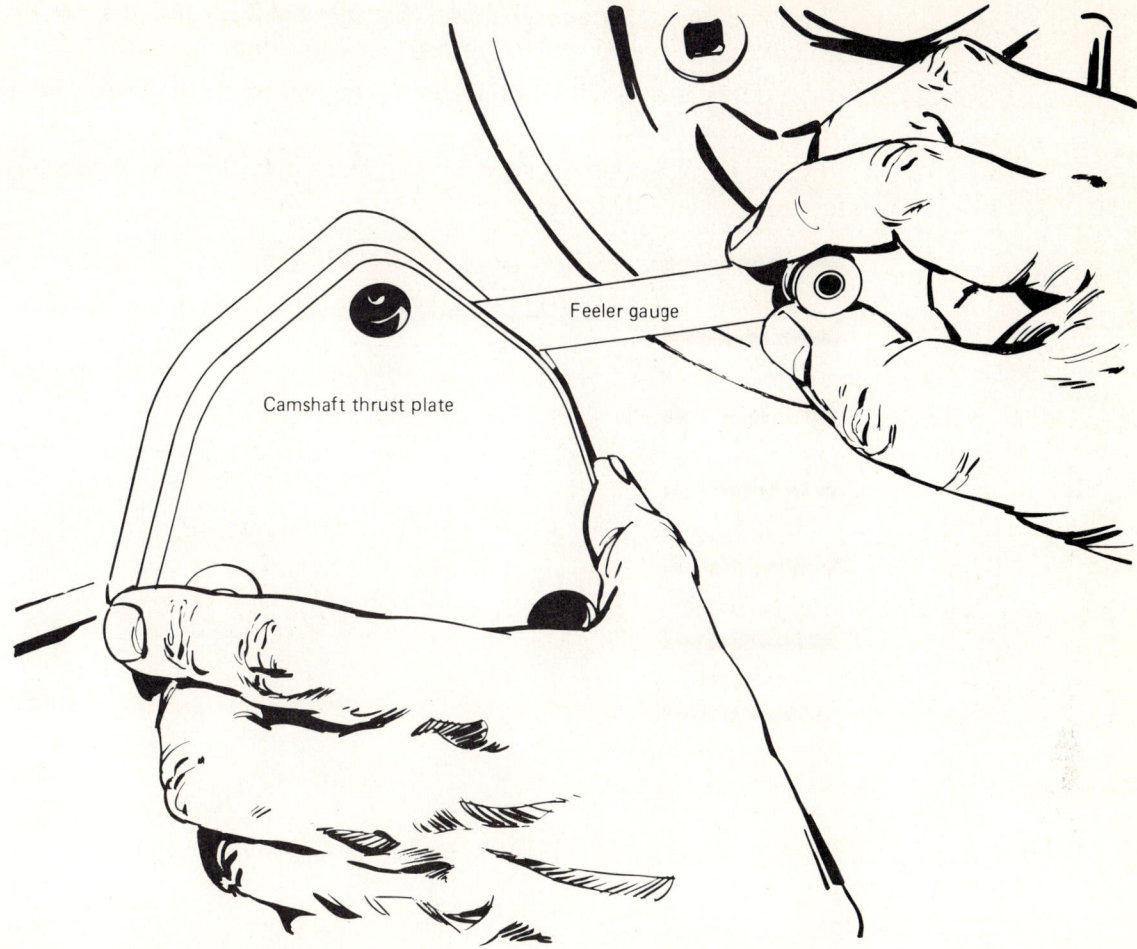

FIGURE 18-7. Check Shim Thickness. *(Courtesy of Cummins Engine Co., Inc.)*

5. Put a new seal ring into the groove, and wet it with oil.
6. Position the part into the gear cover, and install the three capscrews and lockwashers.
7. Tighten the capscrews evenly to 15 to 20 pound-foot (20 to 27 N-m).

Installation of New Gear Case Cover

As mentioned, this will not be required often. Certain things must be done as needed.

1. The new cover must be aligned with both the crankshaft and the lower edge of the block.
2. The old dowels must be removed, and new oversize dowels installed.
3. With the new cover in place, held by a few capscrews, set up an indicator on the crankshaft end, with the probe set in the seal bore of the cover.
4. Turn the crankshaft, observing the indicator. It may take two men to do this. Run-out should not exceed 0.005 inch.

5. Check the edges of the cover, to see that they are flush with the block edge. An oil leak will surely follow disregard of this item.
6. If the edges and the bore are aligned, you can resize the holes and insert new oversize dowels.
7. Position a new gasket over the dowels, and secure the fasteners evenly, alternating from side to side.

19

REAR CRANKSHAFT SEAL, FLYWHEEL HOUSING, AND FLYWHEEL

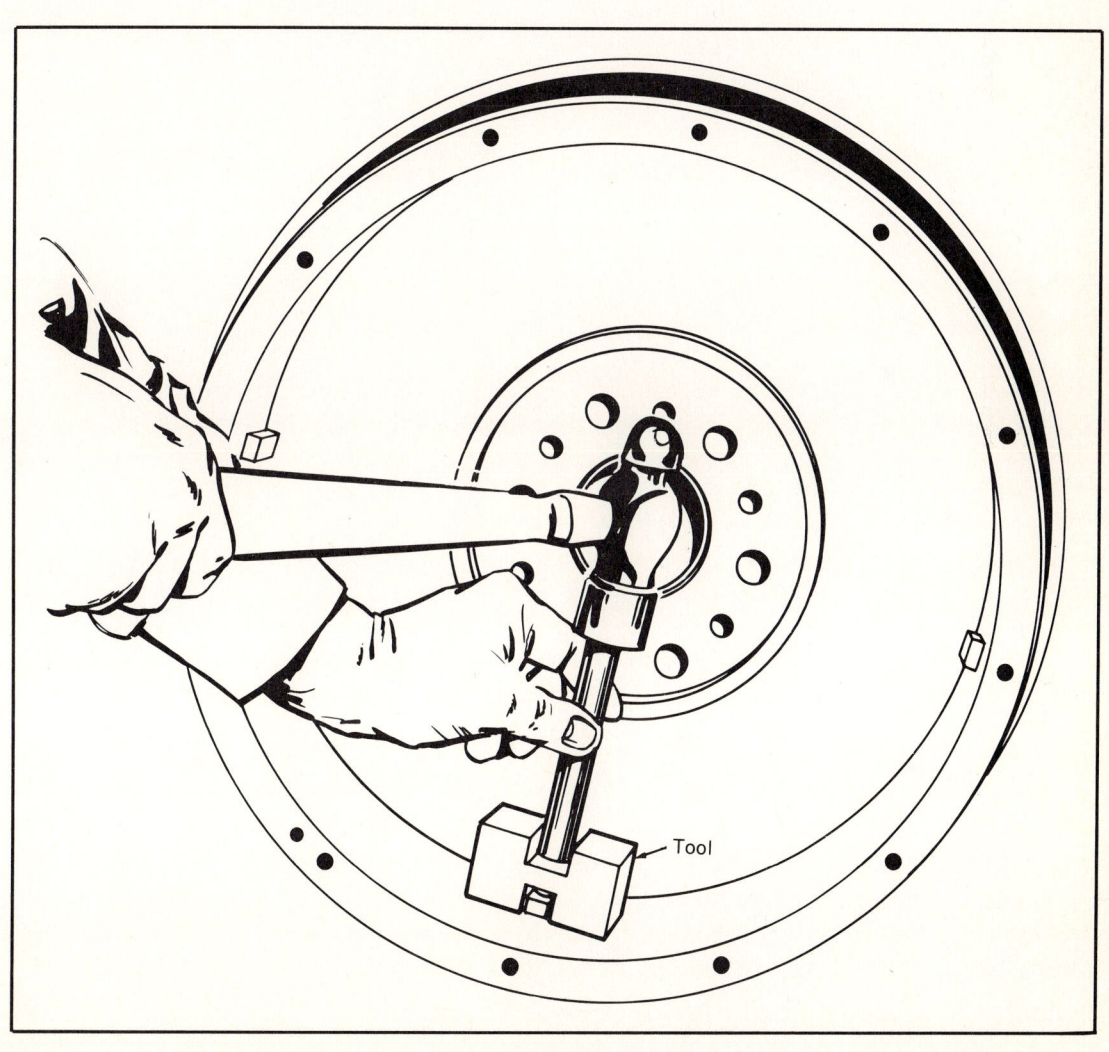

The rear or flywheel end of the engine is always repaired during rebuild. A leaking rear crankshaft seal may need service at other times.

Clutches, throw-out bearings, and crankshaft thrust bearings can fail due to rough driving techniques. All of these parts except the thrust bearings and flywheel housing must be removed to replace the rear seal.

REPLACEMENT OF REAR SEAL

One starts the job by removing the transmission:

1. Remove all air tubing, shifting rods, shift tower, and driveshaft from the transmission.
2. A simple transmission support can be made to hold the transmission during engine repair. **[Fig. 19-1]**

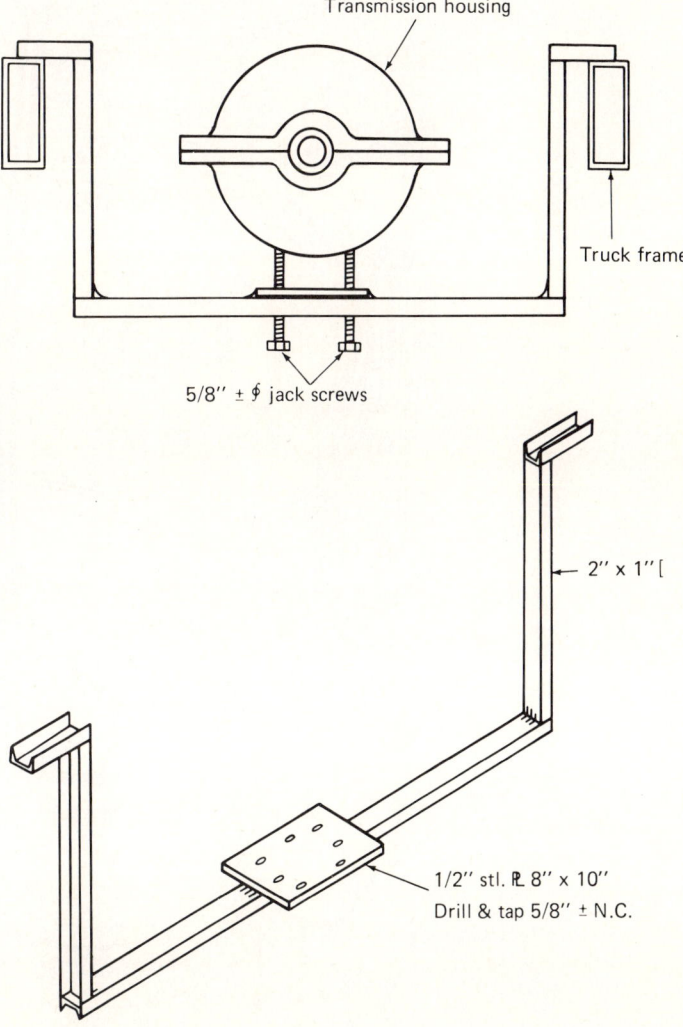

FIGURE 19-1. Transmission Support. *(From a sketch by P. M. Uhl)*

292 / Diesel Engine Service

3. After the transmission support is in place, remove the rear support to the cross member.
4. Disconnect the clutch operating linkage.
5. Remove all capscrews from the clutch housing to the flywheel housing. See that the transmission is supported by the hanger as described.
6. Use a floor jack to move the transmission back enough to clear the clutch. Then take the floor jack out, returning the heavy transmission to the hanger.
7. Remove the clutch pressure plate, being careful to loosen the capscrews evenly to prevent warping the plate.

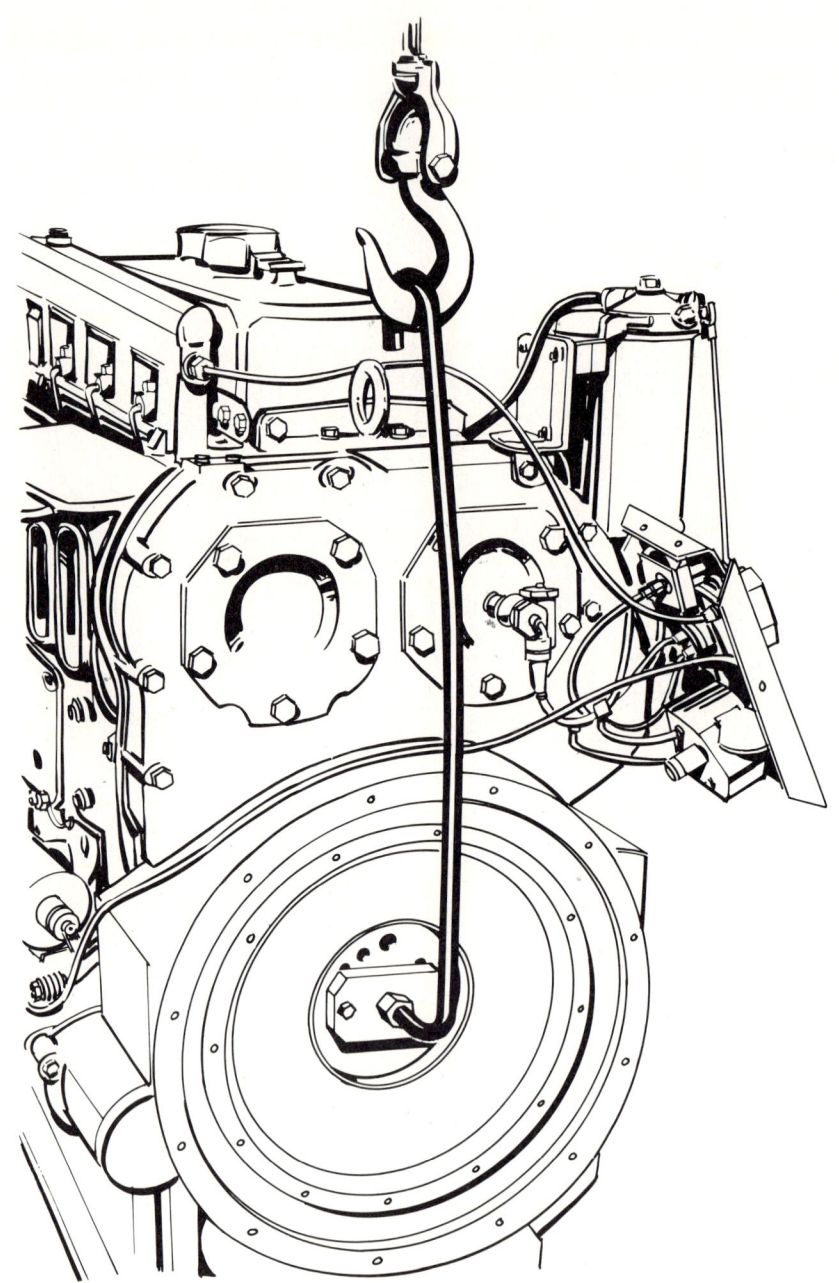

FIGURE 19-2. Hook on Flywheel. *(Courtesy of General Motors Corp.)*

8. Cut the lockwires in the flywheel capscrews, or bend the lockplates. Remove the capscrews.
9. The flywheel will be retained by the dowels, or one capscrew can be used to keep it in place.
10. Flywheels are heavy. A hoisting hook can be made from scrap steel and used with a hoist to handle the flywheel. [**Fig. 19-2**]
11. With the flywheel removed, the rear crankshaft seal is in view. Some of these ride on a wear sleeve on the crankshaft.
12. Tools are available to "crease" the wear sleeve, expanding it for removal. It can be renewed if worn badly. You may put a spacer behind the new seal to move it away from the worn groove.
13. The use of metal screws to remove seals has been described. Clean the seal bore, and check the wear sleeve as described.

INSTALLATION OF NEW REAR SEAL

Wear sleeves and seals are driven in with special mandrels. The manufacturer's instructions must be followed. Such mandrels can be machined locally, but the cost may be high. However the new seal is installed, it and the wear sleeve must be properly seated in their respective locations. [**Fig. 19-3**]

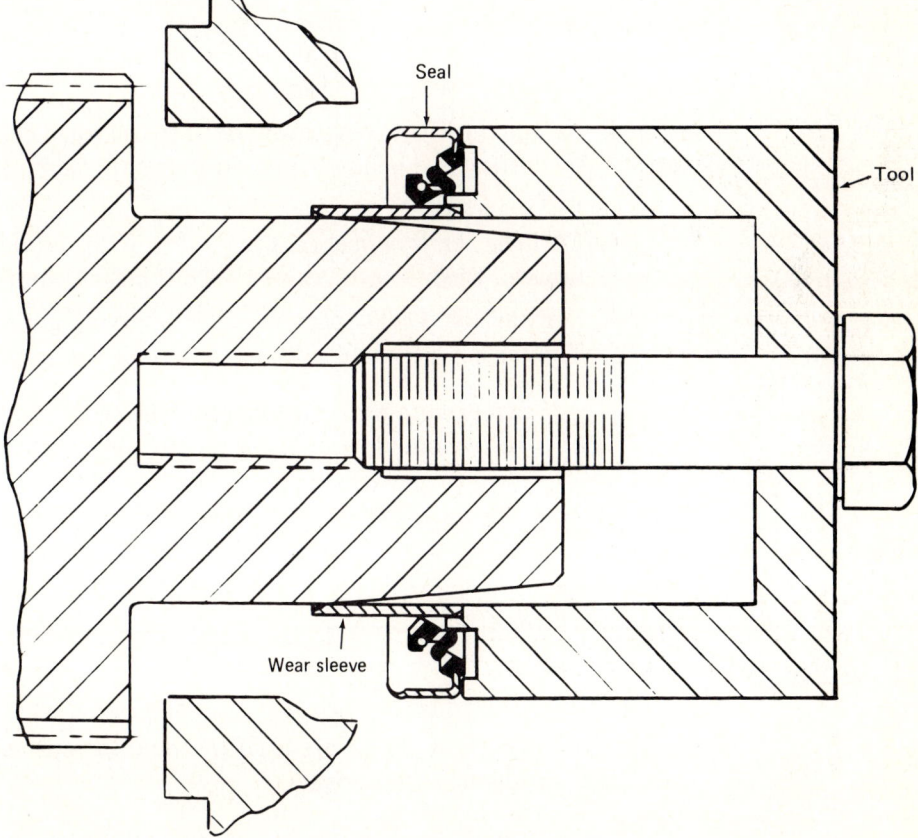

FIGURE 19-3. Install Seal and Wear Sleeve. *(Courtesy of Caterpillar Tractor Co.)*

Some seals are lubricated with oil, while some are installed dry. Check the seal package for instructions. Do not assume that all seals must be oiled. The assembly sequence follows.

Inspection of Flywheel

1. Check or replace the pilot bearing in the flywheel.
2. Clean contact surfaces, and remove any burrs. If the flywheel is severely heat-cracked, it may need to be reground or replaced.
3. Look at the ring gear teeth. If they are damaged, the ring gear can be replaced.
   a. Lay the flywheel on a wood block, ring gear down.
   b. Use a drift and mallet to drive the ring gear from the flywheel. Work around the gear, striking it from all sides.
   Note: Notice any chamfer on the ring gear teeth. Be sure that the new gear is the same as the original.
   c. Turn the flywheel over, ring gear seat up.
   d. Support the new gear on a nonflammable surface, and heat it by moving an acetylene torch around it. Keep the torch moving, and do not overheat the gear. Temperature-indicating crayons can be used to guide the heating.
   e. As the gear is heated, it expands. When it is at 350° to 400° F, grasp it with tongs and place it on the flywheel, chamfer *up*. This must be done quickly.
   f. Tap the ring gear all around to seat it firmly. If the gear cannot be seated, remove it and reheat.
4. Attach the lifting device and hoist to the flywheel. Guide studs can be made by cutting the heads from capscrews the same size and thread as the flywheel fasteners.
5. Position the flywheel on the guides, with the dowel holes in line with the dowels. Push it into contact with the crankshaft.
6. Start all flywheel capscrews with the locking plates or other part in place. Tighten on alternate sides to manufacturer's specifications.

When driving pins must be removed from the flywheel to permit the surface to be machined, they must be installed perfectly square with the center line to allow proper clutch engagement and release. A tool for driving these pins can be made in any good machine shop. [**Fig. 19-4**]

CLUTCH DESCRIPTION

Several types of clutch are used by the different vehicle makers. Any clutch has two distinct functions:

1. To provide a cushioned or yielding connection between the moving crankshaft and the stopped transmission so that the load can be started smoothly.

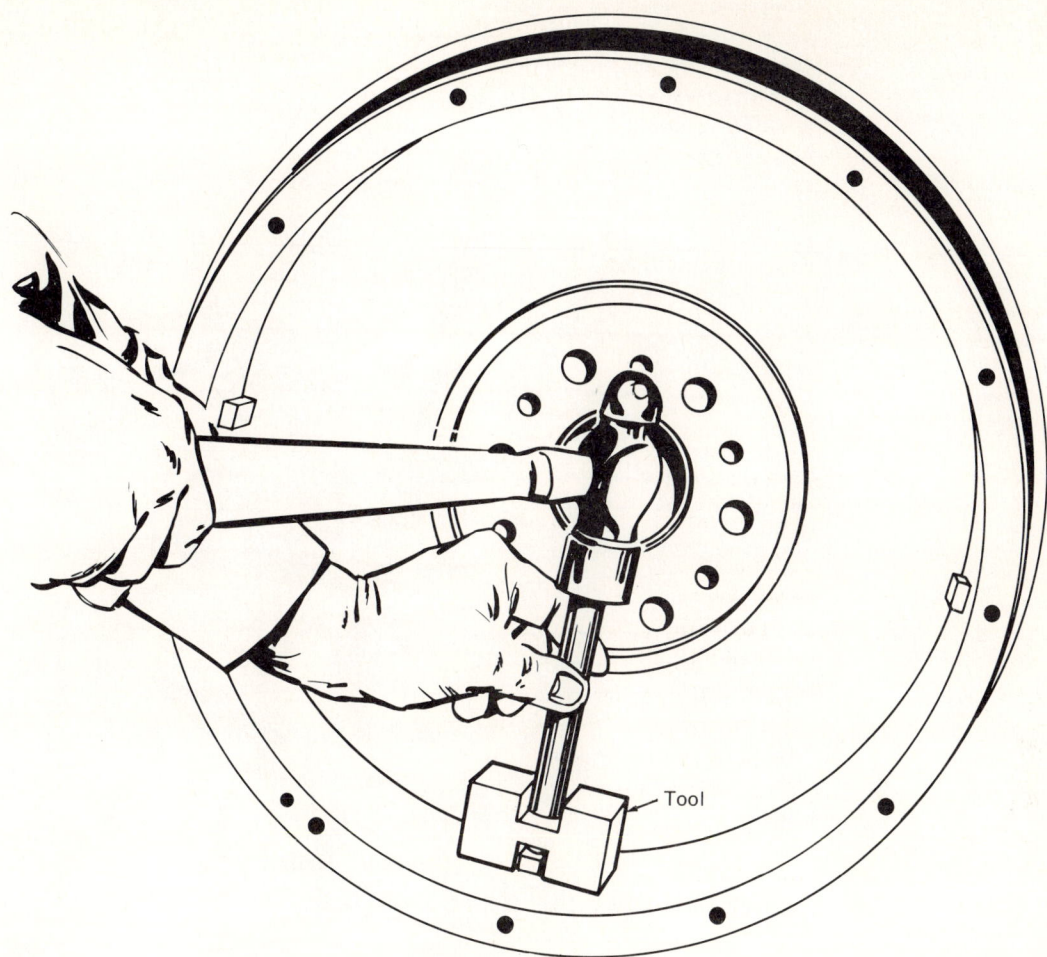

FIGURE 19-4. Tool for Aligning Drive Pins. *(Courtesy of Lipe-Rollway Corp.)*

    2. To break the torque flow to the gears, so that they can be shifted while in motion.

The clutch consists of a driven disc, which is covered with a friction lining. The hub of this disc is splined to fit the input shaft of the transmission. In some clutches, springs are contained in the driven disc plate, which yields slightly to encourage smooth engagement. Some installations have two-plate clutches, in which two driven discs engage the input shaft, separated by a cast-iron intermediate disc.

The intermediate disc fits into the flywheel bore, and turns with the flywheel. It is driven by four or six lugs that drive-fit into holes in the flywheel, engaging notches in the edge of the disc. **[Fig. 19-6]**

The clutch pressure plate consists of a plate that bolts to the flywheel and a set of heavy springs that are suspended between the bolted plate and the movable friction plate. Levers attached to the movable plate turn about pivots on the spring cover and are pressed on by the throw-out bearing to compress the springs. This action draws the movable plate away from the driven disc against the springs. This releases the clutch from its clamped position on the flywheel face. **[Fig. 19-5]**

296 / Diesel Engine Service

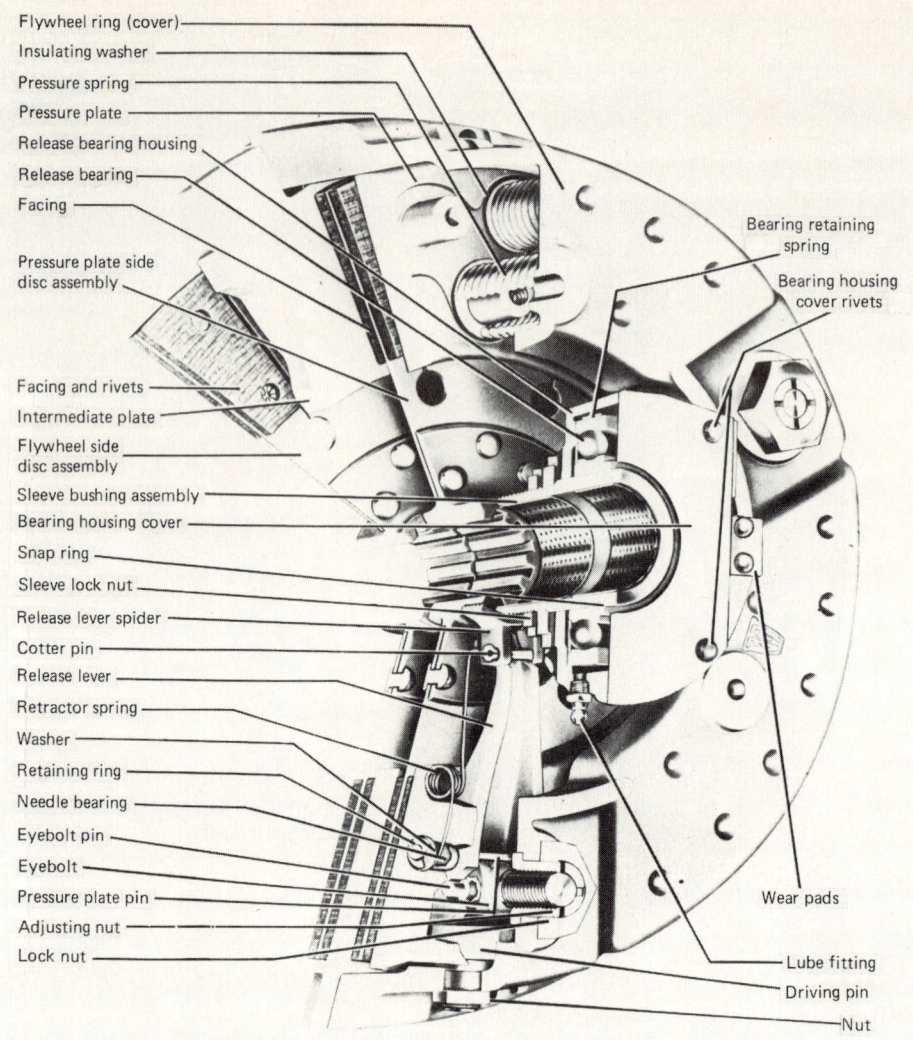

FIGURE 19-5. Typical Clutch Assembly. *(Courtesy of Lipe-Rollway Corp.)*

The pressure plate is an assembly that is removed and installed as a unit.

The clutch is installed in the flywheel by positioning the driven disc, the intermediate plate, if used, the second driven disc, and the pressure plate assembly.

Installation of Clutch

Before tightening the fasteners to the flywheel, it is necessary to move the driven disc to center on the pilot bearing. A special alignment tool is required to position the driven disc. It consists of a center bar on the end of which adapters can be used to fit a range of pilot bearing inside diameters. Over this bar are placed cones or adaptors to hold the driven disc splines. [Fig. 19-6]

A clutch disc alignment tool can be the input shaft of the transmission. Use of this shaft ensures that the end fits the pilot bearing and that the splined shaft fits the clutch discs.

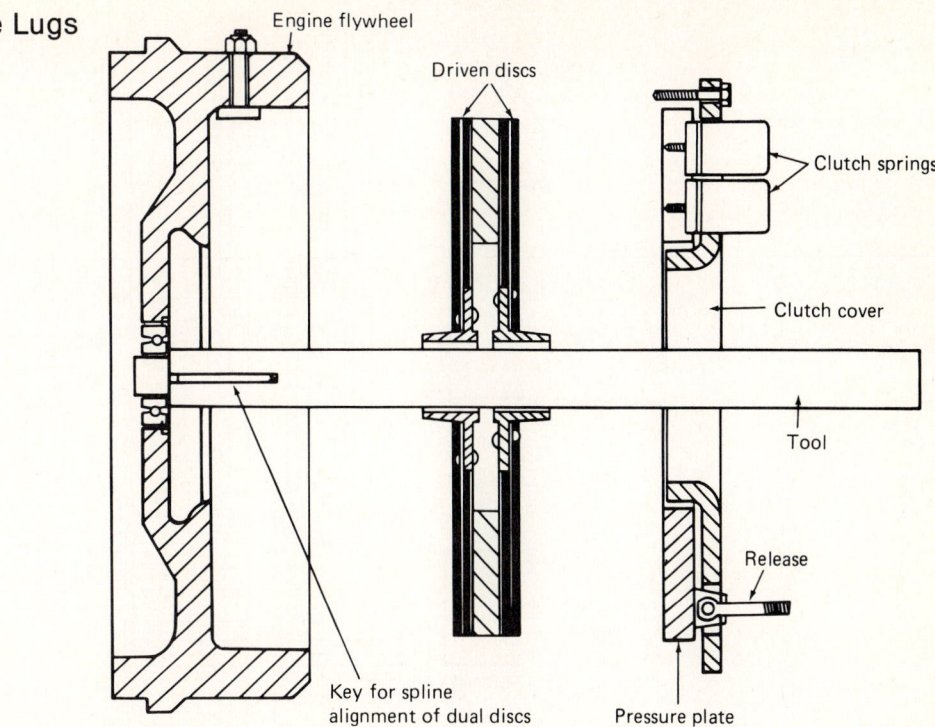

FIGURE 19-6. Clutch Alignment Tool. *(From a sketch by P. M. Uhl)*

Alignment of Clutch

1. With the clutch positioned in the flywheel and the pressure plate loosely attached, insert the alignment tool or input shaft through the driven disc splines.
2. Insert the pilot bearing adaptor into the bearing.
3. Slide the cone adaptor or input shaft into the splines in order to force the driven disc to the center.
4. When the driven disc is centered, start to tighten the capscrews holding the pressure plate to the flywheel.
5. Tighten these slowly, working side to side in order to bring the pressure plate into square contact.
6. After all capscrews are tight, withdraw the alignment tool.

Note: Two plate clutch discs must be installed with the long side of the hubs away from each other. Thus the inner disc has the long side in, and the outer disc has the long side out. [Fig. 19-7]

Wet Clutches

Modern heavy-duty trucks sometimes use clutches designed to operate in oil. The result is cooler clutch operation and less wear while transmitting high forces.

Most of these clutches use a form of air assist to make release easier; they

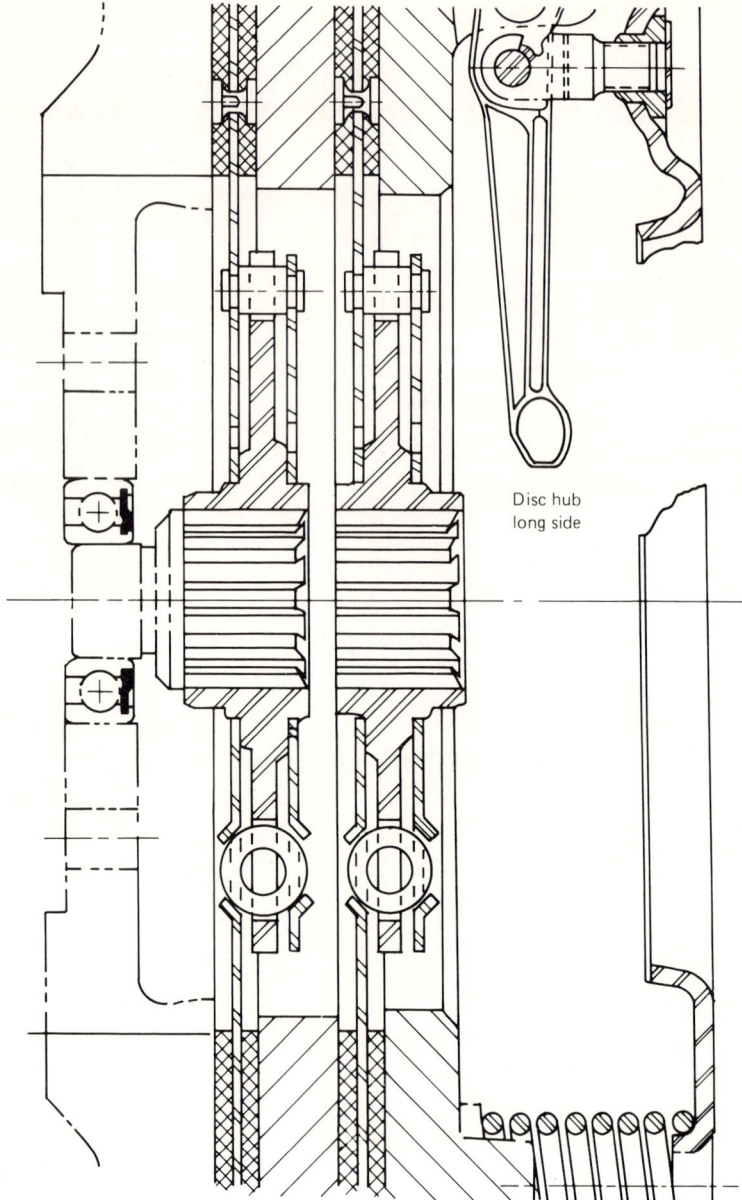

FIGURE 19-7. **Clutch Plate Position.** *(Courtesy of Lipe-Rollway Corp.)*

make clutch adjustments less frequently required. In fact, external adjustment of pedal free play is all that is usually needed.

Automatic transmission fluid is used in wet clutches; the level can be checked by removing a plug on the housing. Only enough fluid is used to permit the collector ring to be covered. A typical wet clutch takes 7 to 10 quarts, depending on the position of the reservoir. Overfilling the clutch reservoir must be avoided.

Clutch Brakes

When clutches are disengaged while operating, inertial forces cause them to turn, thus making an upshift difficult by delay.

Clutch brakes are used to stop the input shaft and reduce shifting time.

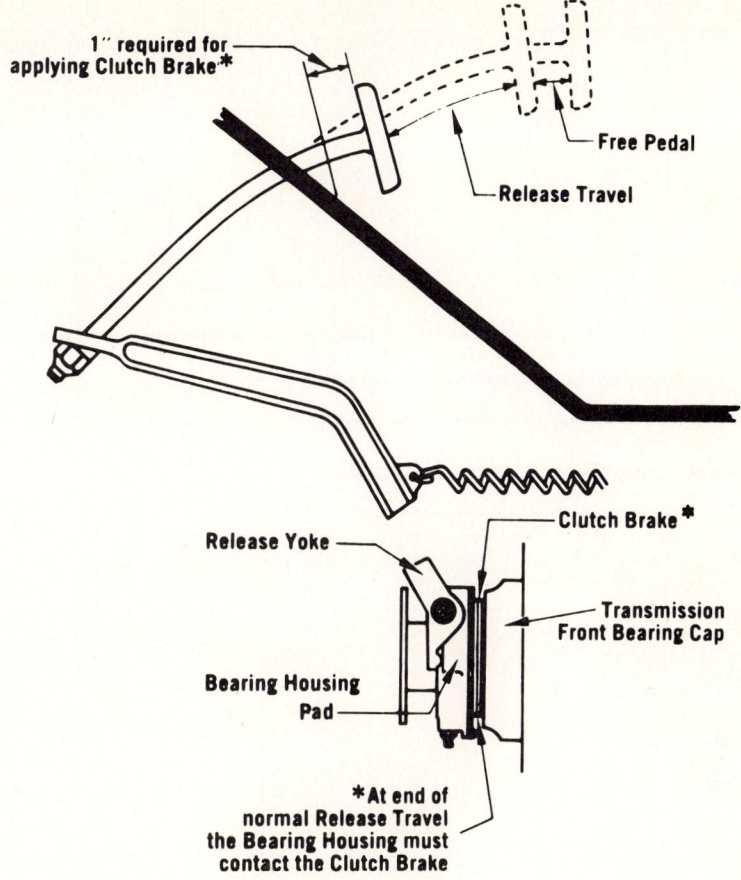

FIGURE 19-8. Allowance for Clutch Brake Action. *(Courtesy of Lipe-Rollway Corp.)*

These brakes consist of alternate layers of fiber friction material and steel washers, which are provided with internal tangs that fit in slots in the input shaft. [**Fig. 19-8**]

The clutch brake is engaged by moving the clutch pedal to the bottom of its stroke, which retracts the release bearing to bring its rear face against the clutch brake. The resulting friction stops the input shaft and eases gear shifting.

Clutch Adjustment

The following is a description of a typical clutch adjustment on the Lipe-Rollway clutch, a popular make. [**Fig. 19-9**]

1. Working from the bottom, remove the cover over the clutch access hole.
2. The release levers are set at the factory and should never be adjusted in the field.
3. The release sleeve is adjusted for pedal free travel by loosening the notched locknut and turning the release sleeve nut in order to obtain correct clearance of the release bearing housing face and the front clutch brake disc. This clearance is ⅜ inch for 14-inch 1 clutches, ½ inch for 14-inch 2, and 15½-inch 2 clutches.

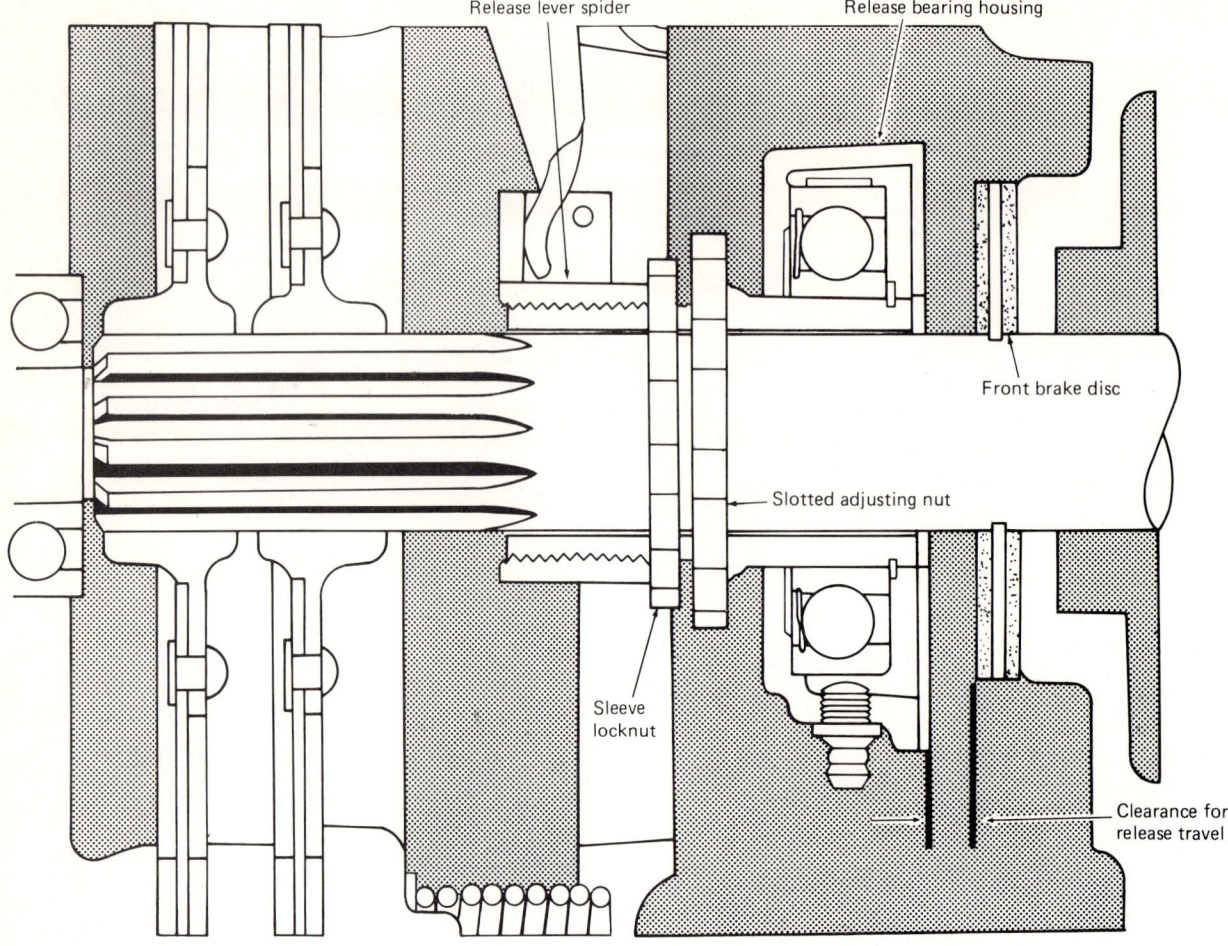

FIGURE 19-9. Clutch Adjustment. *(Courtesy of Lipe-Rollway Corp.)*

4. Leave the locknut loose and adjust pedal travel on the pedal operating lever in order to obtain 1½ inches of free travel.

5. Block the pedal in the released position and secure the release sleeve locknut. Always disengage the clutch to tighten the locknut but engage it to adjust the release sleeve.

6. Once properly adjusted, the clutch will need only periodic checks of pedal travel and adjustment of the release sleeve to maintain correct free travel. *Note:* A spanner wrench is available to turn the release bearing adjustments. If it is not available, a brass drift and hammer can be used.

Clutch Brake Adjustment

This adjustment is made on the external operating linkage. It should be obvious that any adjustment to shorten the linkage will bring the clutch brake into contact with the transmission face sooner in the pedal stroke.

Some adjustments are made on the linkage, some on the operating lever. One must know what type is being serviced.

Adjustments to clutches are simple. Where the free pedal travel cannot be increased beyond ½ inch by adjustment, the clutch is worn out and must be replaced.

The clutch must never be left unadjusted until slipping occurs. Failure will surely follow; one must never try to adjust beyond 1½ inches of free pedal travel. Excessive clutch pressure can overload the release system and cause damage to the pressure plate. [**Fig. 19-10**]

All clutches must be adjusted to take up wear. There are several types of adjusting systems, depending on the clutch make and type. Worn clutch discs should be replaced when the pedal free play can no longer be brought to 1½ inches by adjustment.

One must know the type of adjustment used. Any repair involving transmission removal should include inspection of the clutch, throw-out bearing, and other wearing parts.

INSTALLATION OF TRANSMISSION

1. Check the throw-out bearing on the input shaft. Renew it if it is worn.
2. Lift the transmission on a floor jack to move it forward. Move the hanger with it.
3. Guide the input shaft through the throw-out bearing and splines of the driven disc. You may have to turn the shaft slightly to allow the splines to engage.
4. Install the capscrews in the top of the clutch housing to the flywheel housing. Start as many as possible.
5. Observe the gap between the clutch housing and flywheel housing. Adjust the height of the transmission to make the gap even at all sides. The transmission should slide into full engagement with the flywheel housing without excessive effort. It should not be forced.
6. By manipulating the transmission support and keeping the top fasteners in contact, you will be able to start all capscrews.
7. As you tighten the capscrews, try to keep the gap even, and tighten the point where the gap is widest first.
8. Finally, draw all capscrews evenly to the specified torque.
9. Assemble and tighten the transmission rear support. Then loosen and remove the auxiliary support.
10. Install the driveshaft. Make sure that the universal joints are in the same plane.
11. Install the shifting tower and all air lines, shifting rods, and other attached parts to the transmission.
    Note: Be sure to check crankshaft endwise movement after you have installed the transmission. It should move freely at least 0.006 inch.

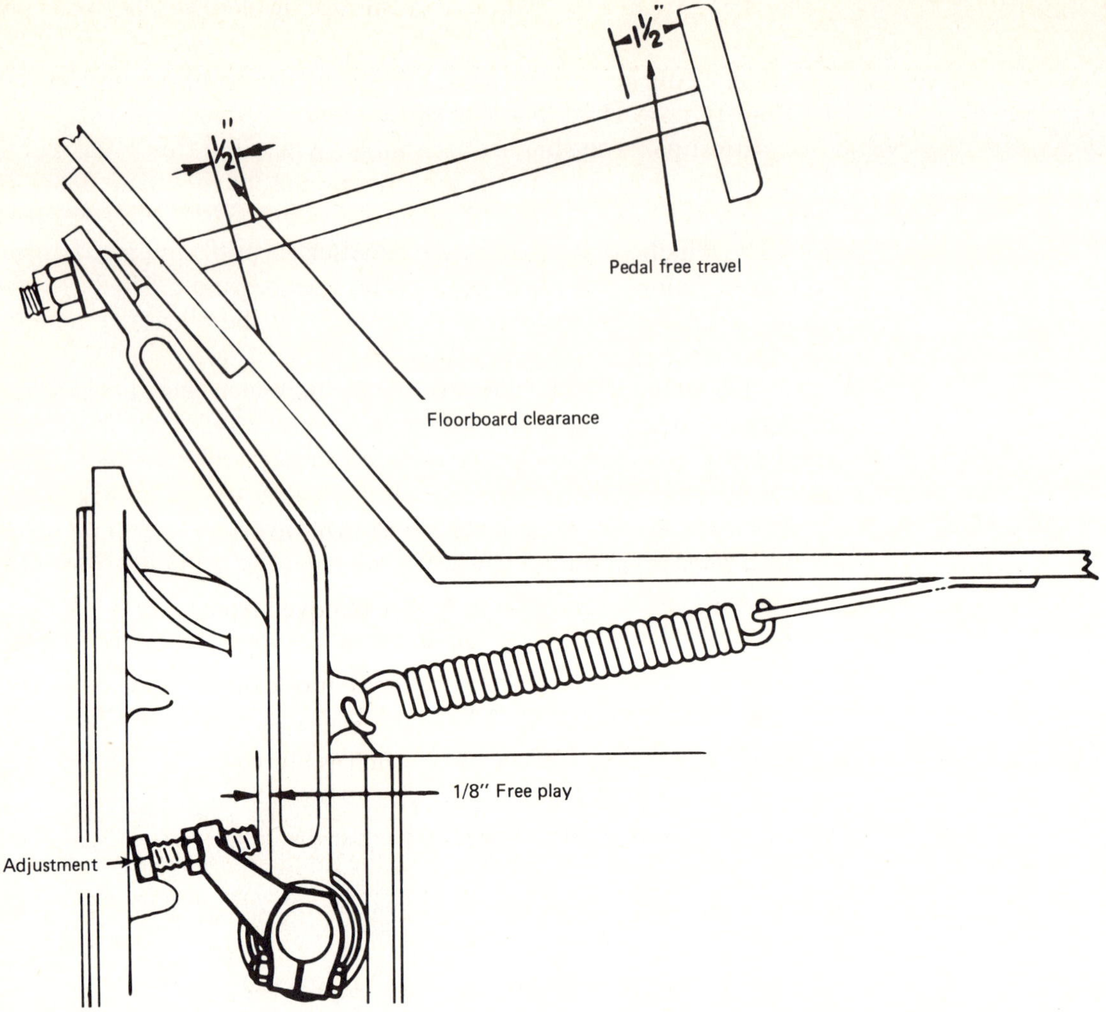

FIGURE 19-10(a). Clutch Pedal Adjustment. *(Courtesy of Lipe-Rollway Corp.)*

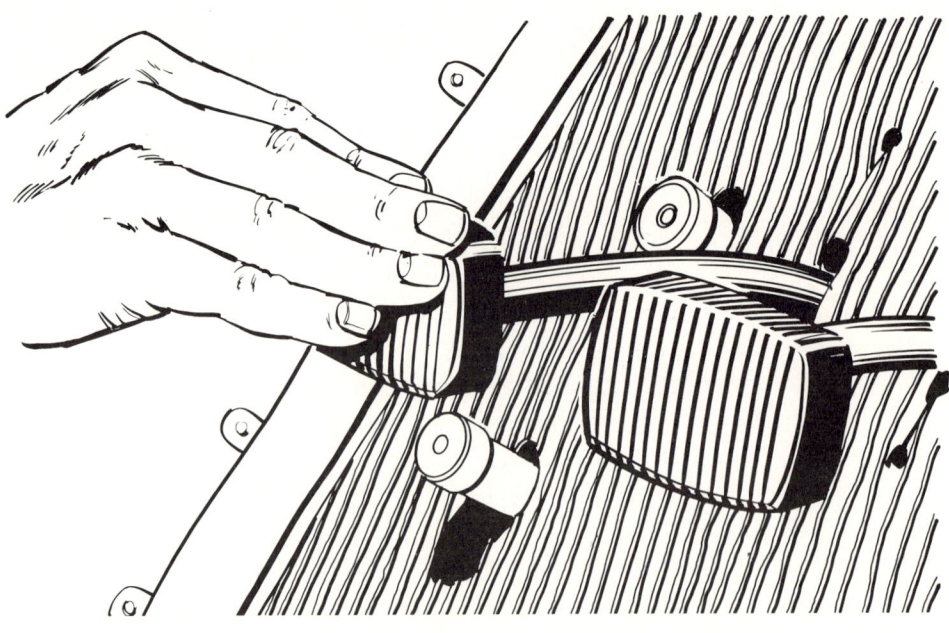

FIGURE 19-10(b). Pedal Free Play. *(Courtesy of Lipe-Rollway Corp.)*

ALIGNMENT OF FLYWHEEL HOUSING

On those rare occasions when a flywheel housing must be replaced it must be aligned with the crankshaft. Replacement of the flywheel housing is not generally necessary during rear seal replacement.

The flywheel housing is doweled in place and can be reinstalled without alignment unless it is damaged.

When a flywheel housing must be changed, alignment with the crankshaft is necessary in order to prevent side loading and wear on the transmission input shaft.

This alignment requires a technique that is now described.

1. Remove the old dowels. New oversize dowels are required. Most old dowels can be removed by twisting them with heavy pliers or vise grips.
2. Be sure that all dirt and residue are removed from the mounting surfaces.
3. Position the new flywheel housing on the mounting surface without any gasket. Hand-start all capscrews except one on each side near the crankshaft center, and at the top. Snug these enough to hold the housing.
4. Mount on a good dial indicator, such as Starrett 196, on a holding fixture attached to the crankshaft with two capscrews and washers.
5. Use a piece of sandpaper to clean the bore of the flywheel housing of all burrs, scale, or dirt.
6. Set the indicator to bear on the bore, taking up about half its travel. Mark the housing at 3, 6, 9, and 12 o'clock with chalk. **[Fig. 19-11]**
7. If you are working alone, mark the front cover at the same points that you marked on the housing.
8. Turn the crankshaft to position the indicator on the horizontal center line. It makes no difference which side. Set the dial to zero.

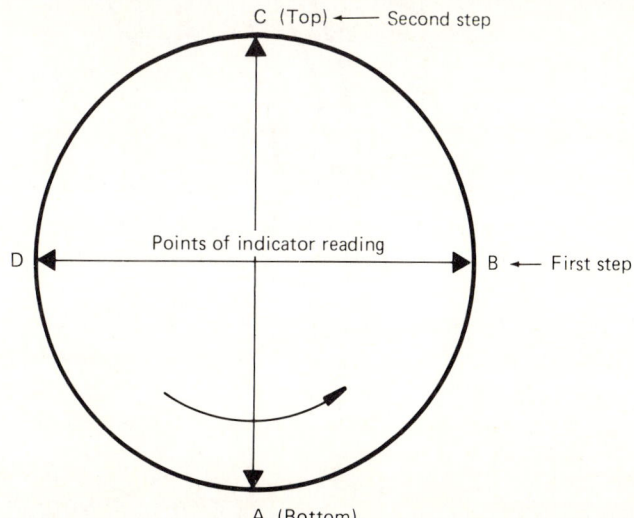

FIGURE 19-11. Alignment Check Points on Flywheel Housing. *(Courtesy of Caterpillar Tractor Co.)*

304 / Diesel Engine Service

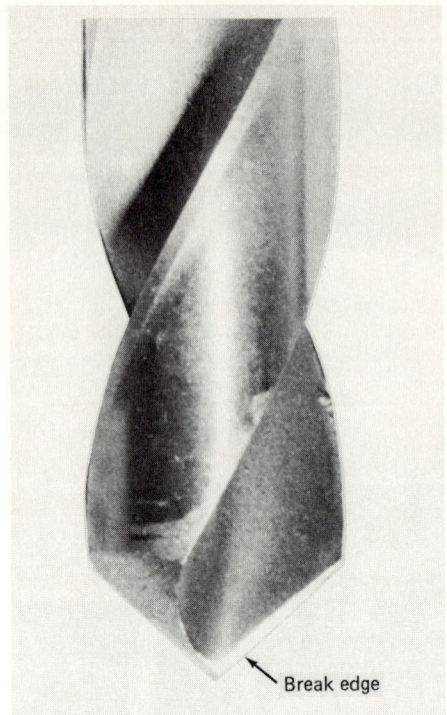

FIGURE 19-12. Modified Drill.

9. Turn the crankshaft 180° in order to move the indicator to the other side. Read the dial.
10. Let's assume that you find the housing 0.020 inch off center. Use a bar in a capscrew hole to move it sideways 0.010 inch.

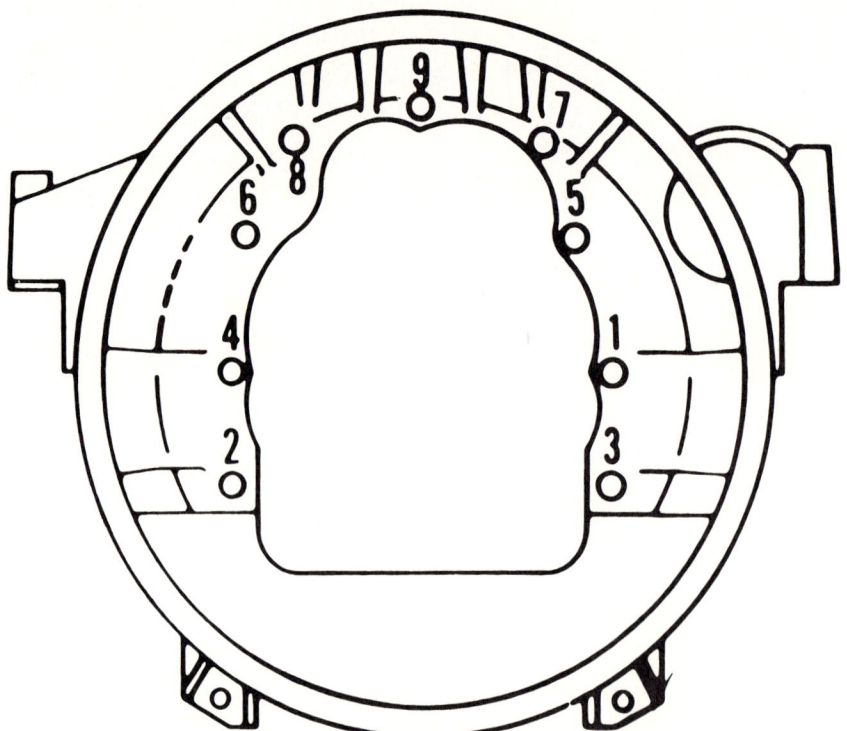

FIGURE 19-13. Tightening Sequence. *(Courtesy of Cummins Engine Co., Inc.)*

Rear Crankshaft Seal, Flywheel Housing, and Flywheel / 305

11. Bring the indicator to either top or bottom center, and zero the dial.
12. Turn the crankshaft one half turn (180°) and read the dial. You read 0.030 inch low.
13. Use the bar to move the housing up 0.015 inch. Zero the indicator.
14. Turn the crankshaft to all four points and read the indicator at each. You should be within 0.005 inch of center.
15. Tighten the fasteners, and remove the indicator.
16. Use a modified drill of one size bigger than the old dowels to enlarge the ½-inch pilot hole in the housing, and make a new oversize dowel hole. Reamers are available for this resizing. A properly modified drill acts as a reamer. **[Fig. 19-12]**
17. New dowels must be a drive fit in the holes. They must not protrude from the housing.
18. After redoweling, remove the housing and apply new gaskets, if used.
19. Secure the housing to the prescribed torque value in the sequence shown. **[Fig. 19-13]**

20

ENGINE TUNEUP AND ADJUSTMENTS

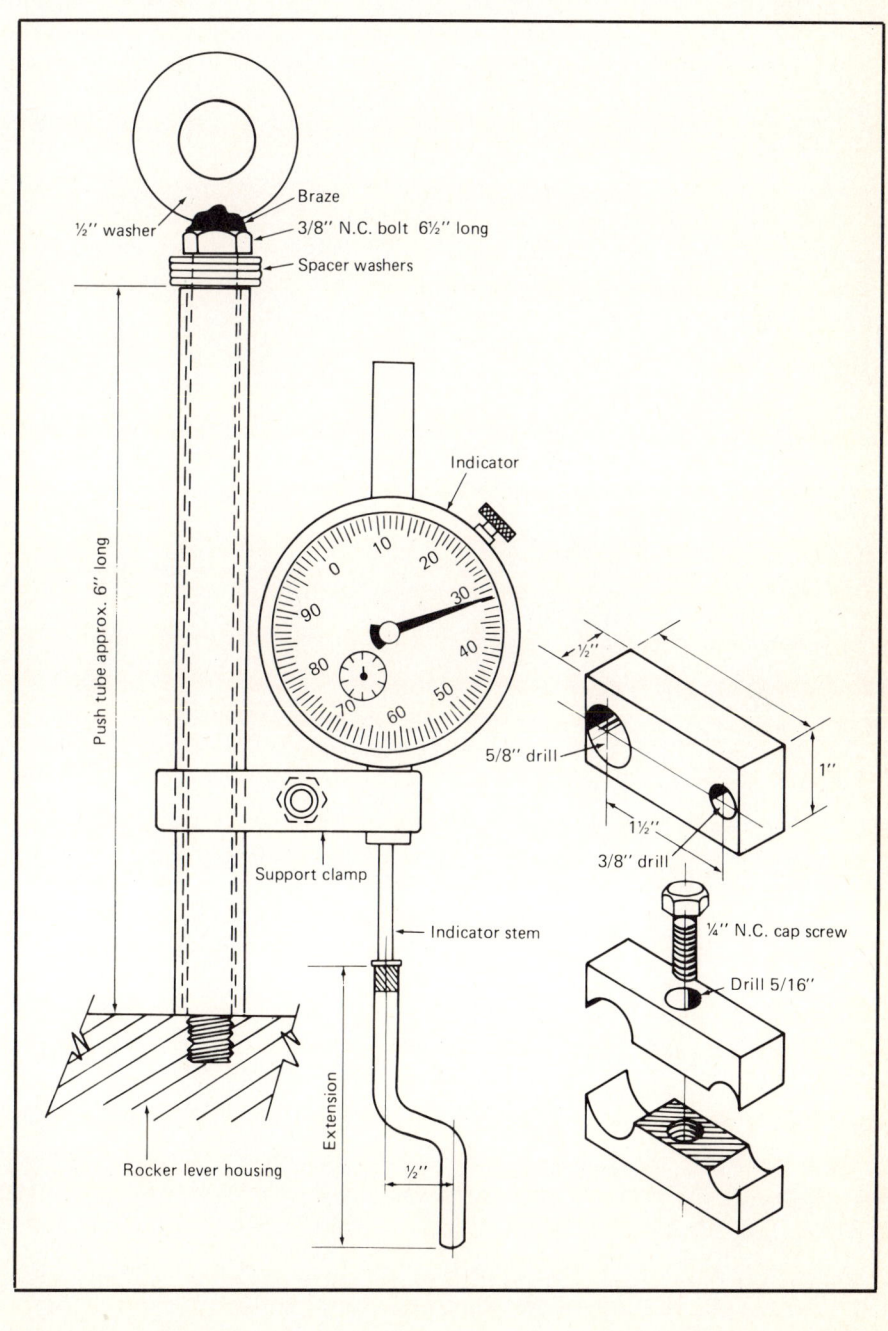

MEANING OF TUNEUP

An engine after overhaul or rebuilding is adjusted during assembly. That is, valve clearance, injector timing adjustments, if applicable, and fuel pump timing should have been done during assembly. (Chapter 3 covered the procedures on Detroit Diesel engines.)

The term *Tuneup* includes service to all parts of the four systems that require periodic attention but does not include major repairs or parts replacement. A tuneup must be done before dynamometer testing and should be the first step in analyzing complaints.

Tuneup on a diesel engine is not the same as on a gasoline engine, because there is no ignition system or carburetor to adjust. There are, however, several important service points; the diesel engine will not run properly unless these points are serviced and correctly adjusted. These points follow.

1. Steam-clean engine.
2. Change oil and filters.
3. Service fuel, air, and coolant filters.
4. Adjust belts.
5. Correct any leaks.
6. Check for loose fasteners, clips, etc.
7. Clean any screens or water traps in the fuel system.
8. Check coolant level.
9. Check throttle linkage for full travel.
10. Adjust valves and injectors.
11. Start engine, warm up, and check governed speed, fuel pressure, and idle speed.
12. Make dynamometer test of power. These items are listed in the approximate order of performance. Injectors need not be adjusted unless they are camoperated, such as Cummins or Detroit. None of these tasks requires much disassembly.

CATERPILLAR TUNEUP

The items listed apply to Caterpillar Model 3306 engines, except for injector adjustment. This covers service to the components that can be serviced by line and field mechanics.

The factory specifications for adjustment are listed in Table 20-1 and Figure 20-1:

1. Oil pressure with No. 10 oil, minimum of 20 psi, above 1,500 rpm.
2. Fuel filters, maximum restriction of 12 inches of water (305 millimeters).
3. Air filter restriction of 25 inches of water (635 millimeters).

309

4. Valve clearance with intake of 0.015 inches (0.38 millimeter), exhaust of 0.025 inch (0.64 millimeter).
5. Idle speed and maximum governed speed. See engine data plate. [**Fig. 20-2**]
6. Fuel pressure in the pump, 25 to 35 psi (170 to 240 kPa).

The description of the Caterpillar fuel system detailed field adjustments that may be made to governed speeds if proper tooling is used by trained mechanics.

Valve clearance must be adjusted last. The engine does not stay at operating temperature for the time required to do this adjustment. One should never try to adjust valve clearance on an idling engine.

These specifications are for the Caterpillar 3306 truck engine. The manufacturer's manual should be used for other engine models.

TABLE 20-1. Engine Firing Order

Right Hand: **1-5-3-6-2-4**	Left Hand: **1-4-2-6-3-5**

Adjustment Limits Using Dial Indicator Method — Inch [mm]

Oil Temp.	Injector Plunger Travel	Valve Clearance Intake	Exhaust
Aluminum Rocker Housing			
Cold	0.170 ± 0.001	0.011	0.023
	[4.32 ± 0.03]	[0.28]	[0.58]
Hot	0.170 ± 0.001	0.011	0.023
	[4.32 ± 0.03]	[0.28]	[0.58]
Cast Iron Rocker Housing			
Cold	0.175 ± 0.001	0.013	0.025
	[4.45 ± 0.03]	[0.33]	[0.64]
Hot	0.170 ± 0.001	0.011	0.023
	[4.32 ± 0.03]	[0.28]	[0.58]
NTE-855 (European Big Cam Only)			
	0.225	0.011	0.023
	[5.72]	[0.28]	[0.58]
NT-855 (Australian Big Cam Only)			
	0.228	0.011	0.023
	[5.79]	[0.28]	[0.58]

Note: Always check the engine dataplate for the injector and valve adjustment values.

Definition of "Cold"

The engine must be at any stabilized water temperature of 140°F [60°C] or below.

Engine Tuneup and Adjustments / 311

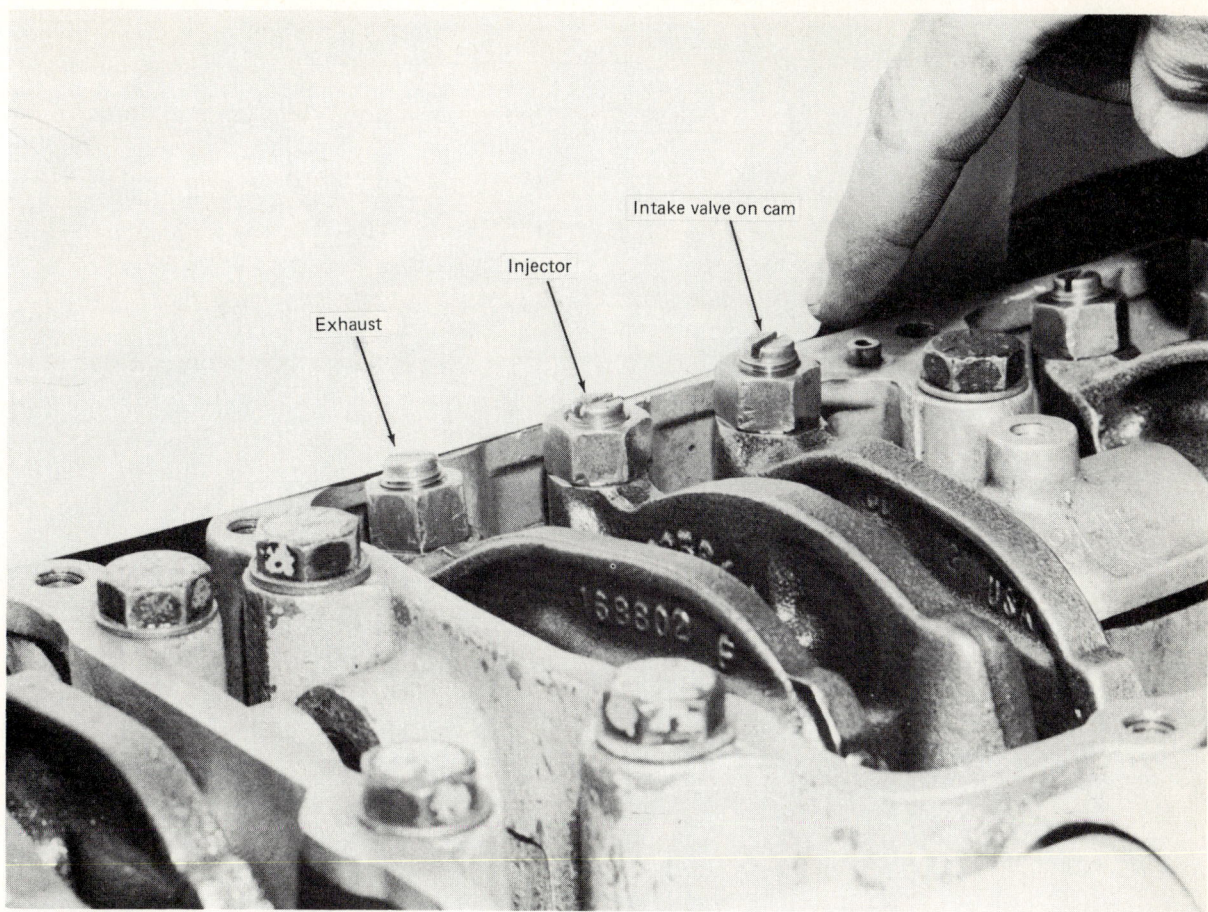

FIGURE 20-1. **Point to Point Intake Valve.** *(Courtesy of ABC Technical and Trade Schools)*

CUMMINS INJECTOR AND VALVE ADJUSTMENT

A tuneup is not an overhaul. The items described take little time but do require observation and logic.

An Indicator Holder for Injector Adjustment

This is the indicator method.

1. *Given:* An old push tube that is not bent and cut 6 inches from its center:
   a. ⅜-inch NC capscrew at least 6½ inches long
   b. a steel block ⅜ inch by 1 inch by 2½ inches
   c. a dial indicator with 1.000-inch travel; a long extension
   d. one ¼-inch NC capscrew.
2. Scribe a center line on the block and lay out 2 holes 1½ inches between centers.
3. Drill 1 hole ⅝-inch diameter, the other ⅜ inch.
4. Split the block along the center line lengthwise through both holes.

312 / Diesel Engine Service

FIGURE 20-2. **Pressure Connection on Fuel Pump.** *(Courtesy of Cummins Arizona Diesel, Inc.)*

5. As shown in the sketch, drill one-half of the block for ¼-inch NC tap.
6. Drill the other half ⁵⁄₁₆ inch to clear a ¼-inch capscrew.

Assembly of Tool

1. Fit the block halves together with a ¼-inch by 1-inch capscrew. [Fig. 20-3]
2. Place the push tube section in the ⅝-inch hole and the indicator into the ⅜-inch hole. Tighten the capscrew to hold them in place.
3. Heat and bend the indicator extension to form a ½-inch offset.

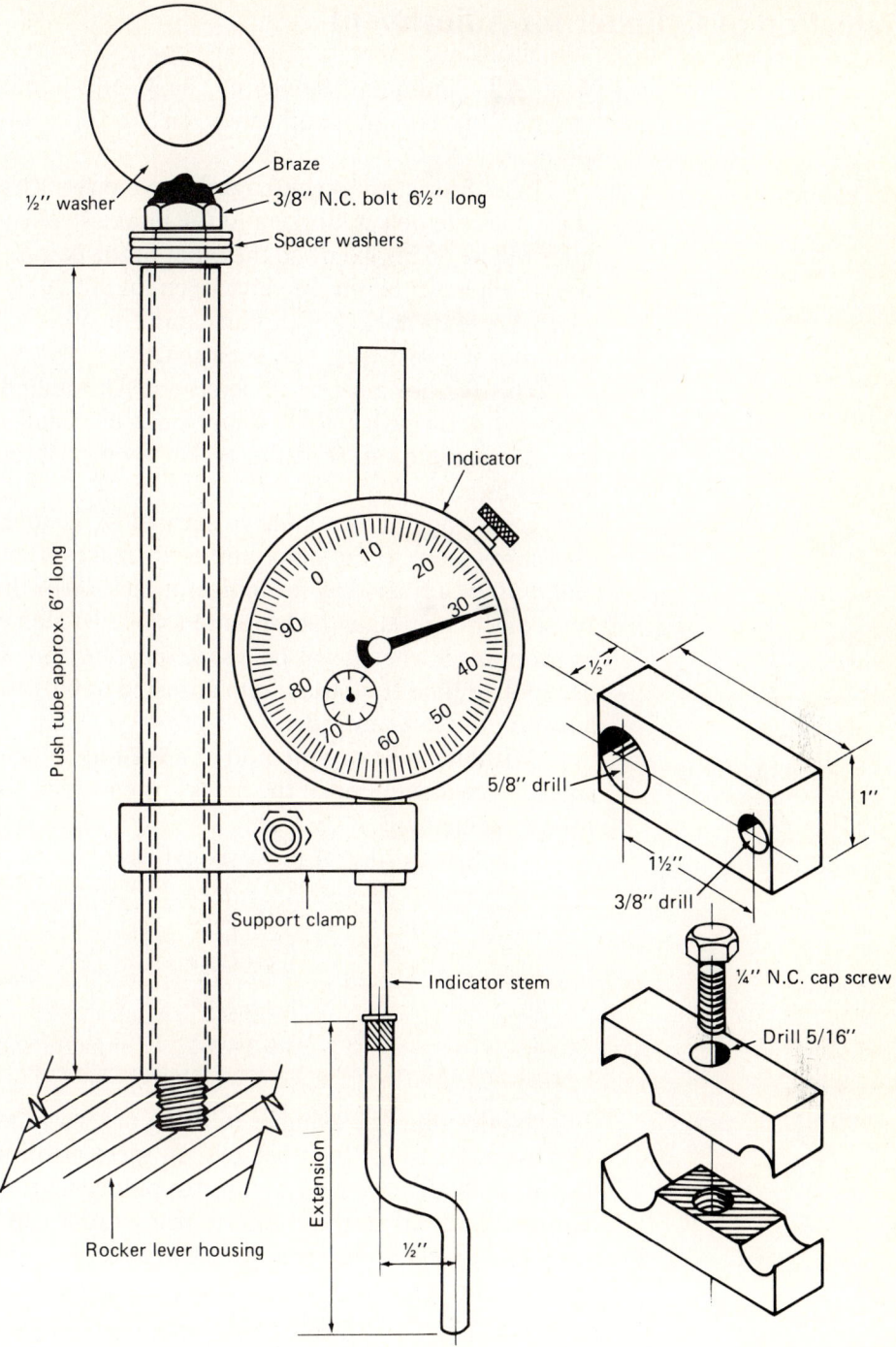

FIGURE 20-3. Injector Adjustment Fixture. *(From a sketch by P. M. Uhl)*

4. Install the extension in place of the contact end of the indicator. It may be necessary to use a small washer to secure it.
5. Braze a flat washer onto the ⅜-inch capscrew head. Use flat washers as needed.
   Note: This fixture will fit any Cummins model by making a suitable support block.

Selection of Cylinder for Adjustment

Note: All Cummins Engines of the H-NH and V-12 series are turned or "spotted" to 90° after top center on the firing stroke. V-8 and V-6 engines spot at 60° ATC.

In all four-stroke cycle engines, two pistons travel in the same plane. Because both the power stroke and the intake stroke are down-strokes, one cylinder will be on intake when the other in the pair is on power. The cylinder on intake will have the intake valve open; at 90° ATC it will be wide open, with the rocker lever noticeably up on cam. The other cylinder in the pair will have both rocker levers free, valves closed.

This feature must be understood. Although marks are provided on the accessory drive pulley, this assembly can be installed out of time with the cam gear. Unless one can read the rocker lever positions, accurate adjustment is impossible.

Later engines have pulleys marked A, B, and C. Although these markings correspond to the old 1 and 6 VS marks, the new marks signal engines that *must* be adjusted by the indicator method or that have top-stop injectors. With older engines, an inch-pound torque wrench was used to preload the injector plunger when it was held seated by the cam. This was done on the cylinder spotted, and the valves were adjusted to 0.014-inch intake, 0.027-inch exhaust on the same cylinder.

With the indicator method of adjusting injector plunger travel, one first spots the engine to one of the marks, then selects the cylinder to adjust by the following method:

The firing order of most six-cylinder, four-stroke cycle engines is

A B C A B C	
1-5-3-6-2-4	
5-3-6-2-4-1	valve set
3-6-2-4-1-5	injector set

Notice that one cylinder in firing order was moved to set the valves and two cylinders to set the injectors. In these positions, the camshaft lobes are down, and valves are closed, injector plungers off cam. It makes no difference which cylinder is started because firing order can be followed after the first one. It is good practice to loosen all rocker lever locknuts before starting the adjustment. Thus as the adjustment of each rocker lever is completed, the locknut can be tightened, so that the mechanic knows which ones have been adjusted.

For these engines it is best to make up thickness gauges pairs by selecting the correct thickness and bending the ends at right angles. Then the mechanic fastens the two together with a small screw or rivet. This allows the gauge to be placed under the rocker lever nose squarely. The correct lash for H-NH engines is 0.011-inch intake, 0.023-inch exhaust. [Fig. 20-4]

Steps for injector adjustment follow:

1. Bar the engine to spot one cylinder. Look at the rocker levers to determine which is on power.

Engine Tuneup and Adjustments / 315

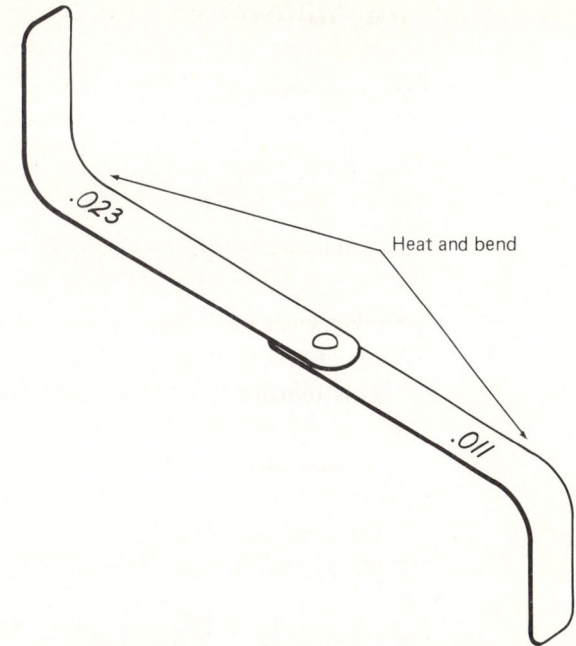

FIGURE 20-4(a). Feeler Gauge Pairs. *(From a sketch by P. M. Uhl)*

FIGURE 20-4(b). Feeler Gauge in Use. *(Courtesy of Cummins Engine Co., Inc.)*

316 / Diesel Engine Service

2. If we are spotted on No. 1, set the indicator up on No. 3. Start the ⅜-inch capscrew in the cover hole nearest that injector.
3. Pick up the indicator stem and place it on the injector plunger top. See that the stem does not touch the rocker lever nose. Now tighten the capscrew.
4. Move the clamp block vertically on the support to position the indicator about one-third into its travel.
5. Turn the adjusting screw down to seat the injector plunger. The specified tension is 40 inch-pounds. Be sure that it is seated. Now set the indicator to zero.
6. Turn the adjusting screw back until the indicator reads exactly 0.170 inch. Tighten the locknut, and check the adjustment by barring the rocker lever to seat the plunger. Recheck the zero setting. This injector is now adjusted.

Note: The injector travel given is valid for older engines with 2″ camshafts. Later engines having 2½″ camshafts use 0.228″ travel, and this value is marked on the engine data plate.

Adjustment of Valve Lash

The indicator is still spotted on No. 3. The valve lash on No. 5 will now be set.

1. Insert the 0.011-inch thickness gauge under No. 5 intake rocker lever. Release the gauge. Turn the adjusting screw down until you feel contact with the gauge. This is a "finger-tip" feel. Lock the nut, and the valve is adjusted.
2. Follow the same method on the exhaust valve, using 0.023 inch.

Leave the indicator on the adjusted cylinder until ready to adjust another. The engine should be barred to the next cylinder in firing order, then the indicator moved to that one. No. 3 has been adjusted and now the mechanic is ready for No. 6. Firing order should be followed for the rest. The valves on the cylinder on which the injector was set, will be adjusted. (No. 3)

Rocker Lever Barring Tool

The chapter has discussed barring the rocker lever to seat the injector plunger after adjusting its travel. While it is possible to bar it using a box wrench on the nut, a simple tool that will not slip off can be made.

1. Obtain a steel rod, ⅜ inch by 36 inches.
2. Heat and bend the rod as shown in the sketch. It may be necessary to position a short bar across the hook end to shorten the bar and prevent its interfering with the indicator. [Fig. 20–5]

Note: Later Cummins engines have injectors that have the plunger travel set during calibration. These are called *top-stop injectors*. In adjusting, the

Engine Tuneup and Adjustments / 317

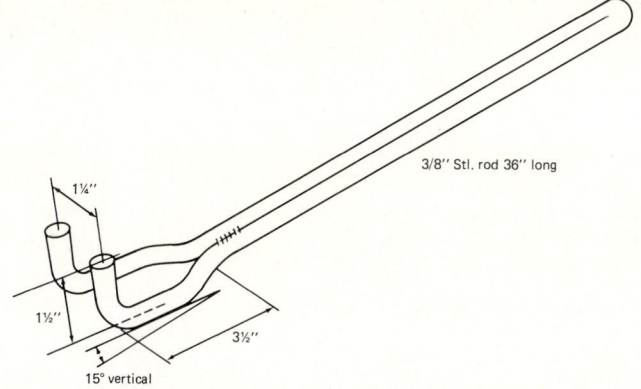

FIGURE 20-5. Barring Tool For Rocker Levers. *(From a sketch by P. M. Uhl)*

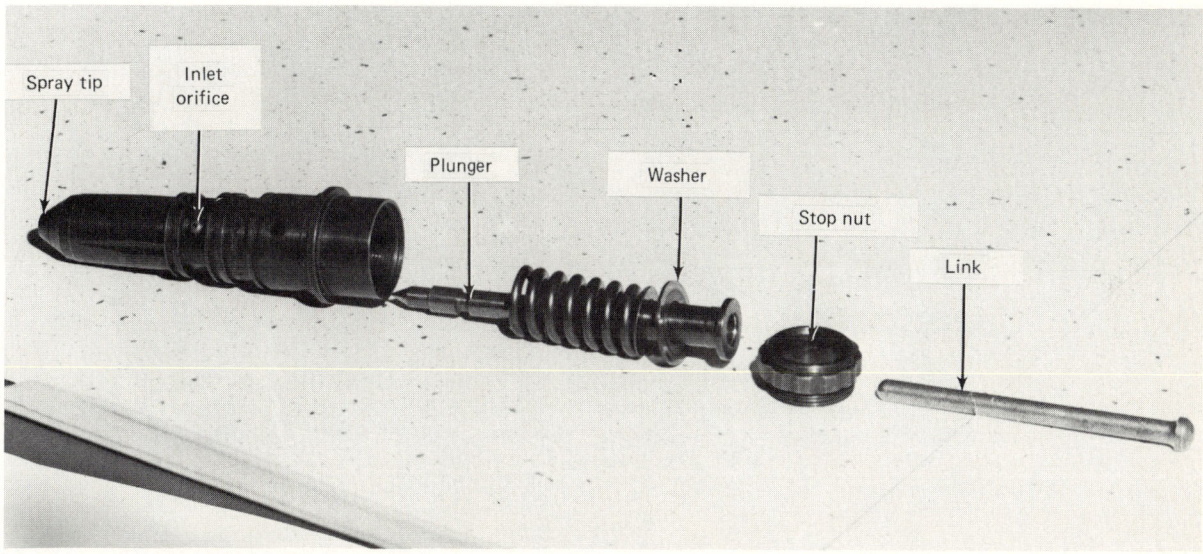

FIGURE 20-6. "Top Stop" Injector—Exploded View. *(Courtesy of Cummins Arizona Diesel, Inc.)*

adjusting screw is turned to just bind the plunger link. Five inch-pounds is specified. [Fig. 20-6]

Note: Occasionally, adjusting screws are hard to turn, leading to false adjustment. Such screws must be cleaned with a top and die or replaced. They are too hard to clean with a thread file.

The same method of cylinder selection should be used for both top-stop and conventional injectors.

An Engine Barring Tool

Nearly all Cummins engines have a gear-driven accessory drive pulley. These pulleys are tapped with puller holes in the web of the pulley, on 3½-inch centers. Because the holes are tapped to ⅜-inch NC, they can be used to make a simple barring tool, which allows one to turn the engine without overtightening the pulley nut.

318 / Diesel Engine Service

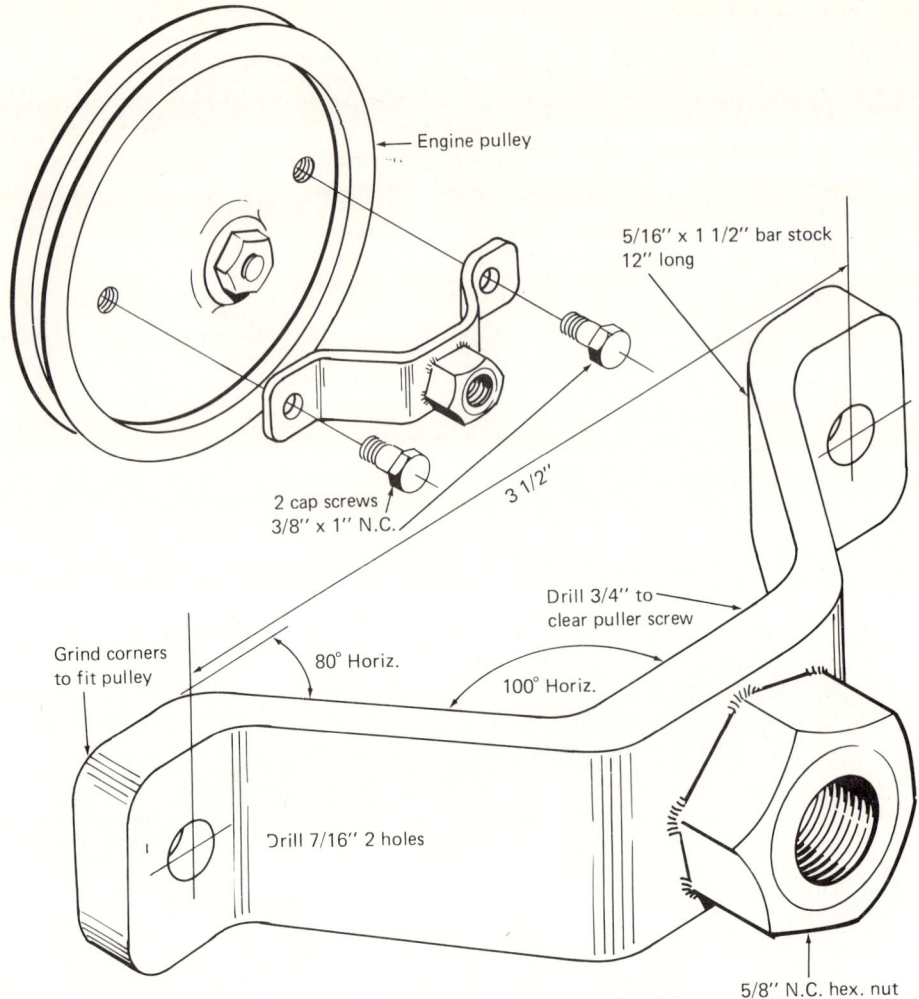

FIGURE 20-7. **Engine Barring Tool.** *(From a sketch by P. M. Uhl)*

1. *Given:* A piece of flat steel stock, ¼ inch or 5/16 inch thick by 1½ inch wide by 10 to 12 inches long.
   a. In the sketch.
   b. One nut 1-inch to 1½-inch hex diameter.
   c. Two ⅜-inch x 1-inch capscrews.
2. Heat and bend the ends of the flat stock to about 80° as shown. Drill each end 7/16 inch, to clear ⅜-inch capscrews. Bend the stock to form a bridge as shown. [**Fig. 20-7**]
3. Drill the center of the stock to a size that will clear the selected nut thread. Weld the nut in place over the hole. A forcing screw can be used in the nut to act as a puller.
4. Grind the ends of the stock to clear the inside of the pulley. Mount the tool on the pulley with the ⅜-inch capscrews. Clean the threads before mounting.

CLEANING OF AIR CLEANER ELEMENTS

Most modern engines are equipped with the pleated-paper type of air filter element. Although they can be replaced, the expense of new elements can become a burden. Several makers have prescribed cleaning methods, most of which work well. The method of cleaning must depend on the texture and adhesion of the dirt.

Where dirt is trapped on the outside of the element, a simple cleaning system is possible. The following method is used to remove dry dust.

1. Provide a pressure vessel of at least 30-gallon capacity. This can be any steel tank that will hold 60 psi. [**Fig. 20-8**]

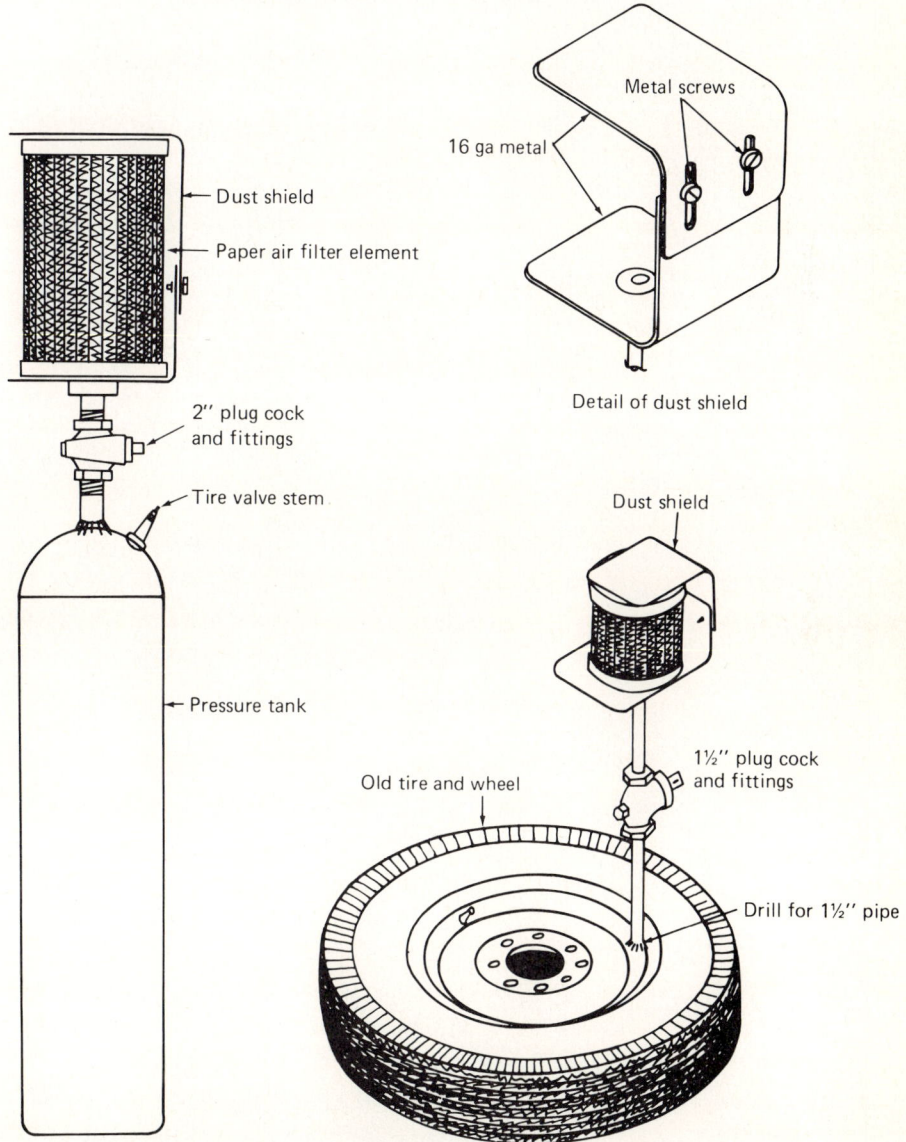

FIGURE 20-8. Air Filter Cleaning System. *(From a sketch by P. M. Uhl)*

2. Install a short nipple in the largest opening in the tank. Assemble a flat support plate and a 90° cock on the nipple.
3. Take a piece of pipe of a size to fit the nipple, and drill ¼-inch holes in it in a random pattern. Install it in the top of the nipple, and cap it with a pipe cap.
4. Provide a tire valve or other means of charging the tank to about 60 psi.
5. A clamping device can be applied to hold the element.

In use the element is placed on the plate over the short pipe. The tank is charged with air, and the cock is snapped open to shake the dust loose and blow the element clean. One must turn the face away to prevent injury. Once set up, elements can be cleaned rapidly.

Note: Instructions on tuneup of Detroit and Caterpillar engines have been covered in chapters applicable to those engines.

21
ENGINE TESTING

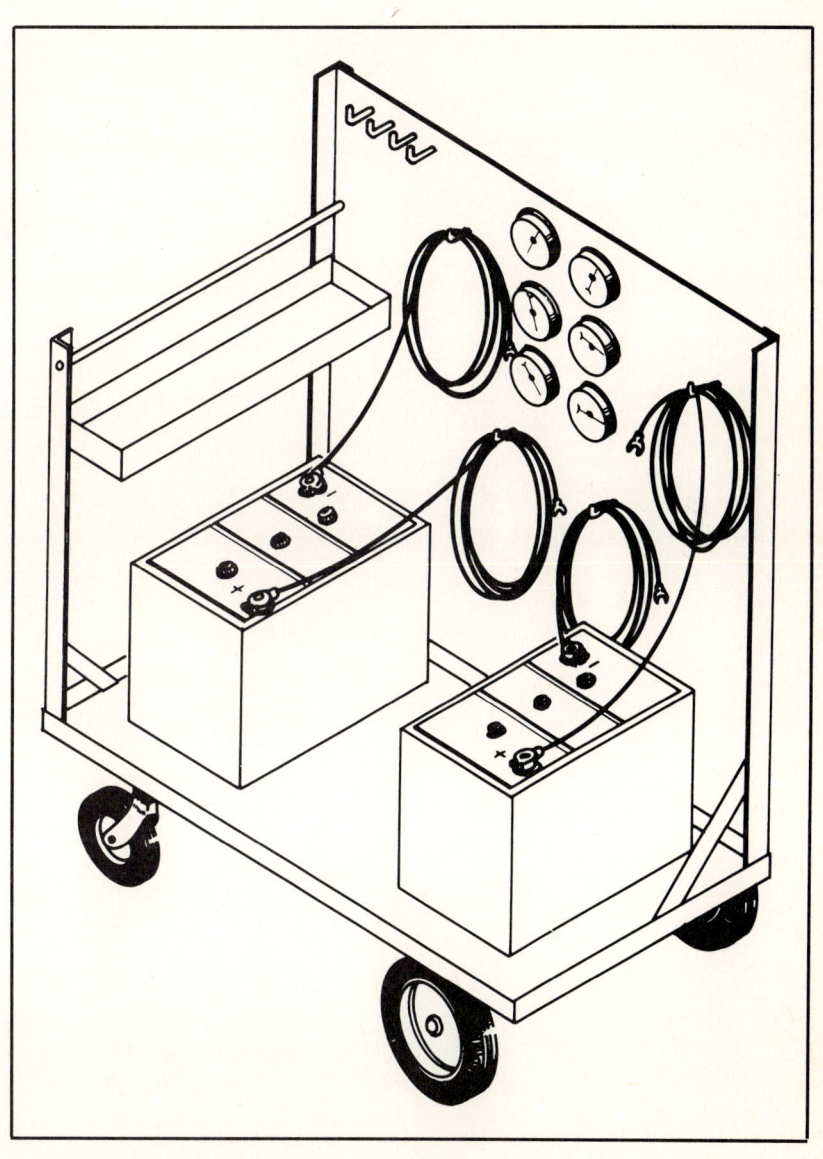

TEST PRINCIPLES

An engine can be proven to produce only the power claimed for it by applying a controlled load. Several methods of doing this are available; such tools are called *dynamometers*. Essentially they are designed as controlled brakes to resist the turning effort of the engine and to measure that effort, either as scale readings of torque, force in rotation, or horsepower directly on meters.

DYNAMOMETER TYPES

The common types of dynamometer are

1. Water brakes, which use water to absorb the force of rotation.
2. Electric dynamometers, which are calibrated generators and furnish current that is read with electrical instruments. Because 1 horsepower is 0.746 Kw, the engine's power is easily calculated.

The water brake machines can be divided into two categories:

1. Vehicle or chassis dynamometers, which carry the drive wheels on rollers, whose rotation is resisted by a water brake. [Fig. 21-1(a)]

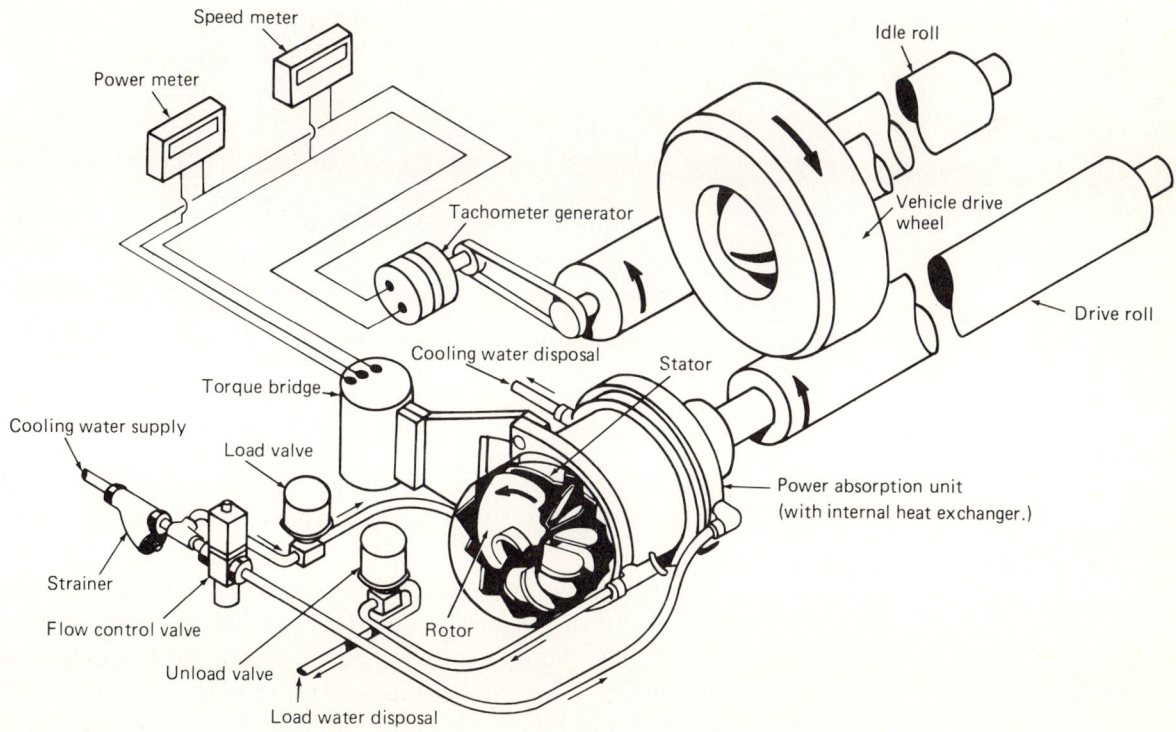

FIGURE 21-1(a). Chassis Dynamometer. *(Courtesy of Clayton Manufacturing Co.)*

324 / Diesel Engine Service

FIGURE 21-1(b). Engine Dynamometer. *(Courtesy of Taylor Dynamometer and Machine Co.)*

2. Engine dynamometers, which provide mountings for the engine only and connect the crankshaft directly to the water brake. [**Fig. 21-1(b)**]

The well-known Clayton dynamometers are used by many engine dealers. Several versions of this design are made, including portable systems for testing passenger cars, permanently installed systems for truck diagnosis, and engine dynamometers for engine testing after rebuild. [**Fig. 21-2**]

The system shown in Figure 21-1(a) is typical. All of these dynamometers are water brakes, with appropriate adaptations.

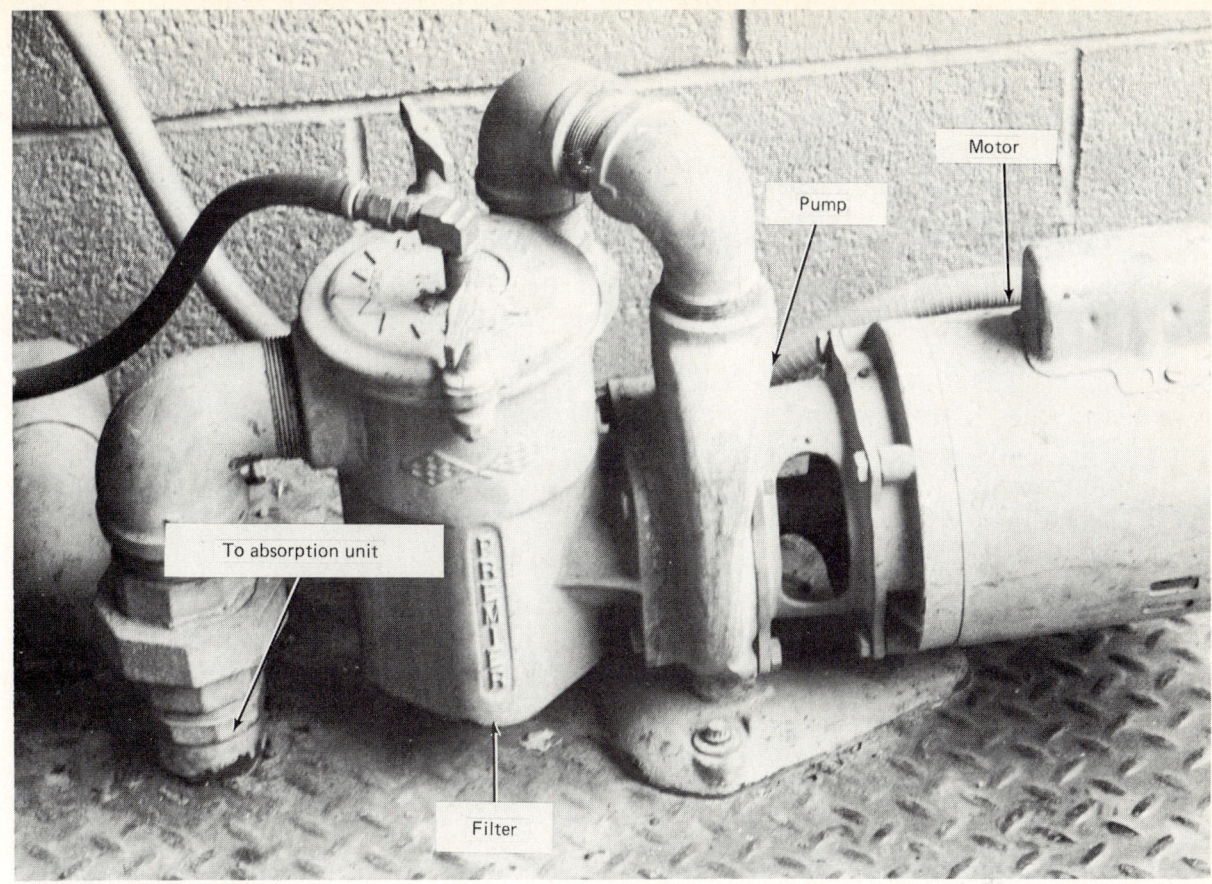

FIGURE 21-1(c). Dynamometer Charging Pump.

FIGURE 21-2(a). Taylor Dynamometer. *(Courtesy of Taylor Dynamometer and Machine Co.)*

325

FIGURE 21-2(b). Engine Dynamometer in Use. *(Courtesy of Taylor Dynamometer and Machine Co.)*

The need to reduce tire wear on chassis dynamometers and to provide accurate and practical testing of all sizes of engine led to the development of the flexible Go-Power dynamometer line. These units can be bolted directly to the engine flywheel housing, mounted on frame rails to fasten to the transmission output shaft, or used as conventional chassis dynamometers with large rolls to reduce tire heating. **[Fig. 21-3]** Again, these are water brake systems, adapted for various models.

These machines are all expensive, and many shops have not changed machines bought several years ago, when 200 to 250 horsepower was common, for modern machines capable of handling up to 500 horsepower.

Any load-absorbing device must be able to load the engine to its rated capacity. Otherwise readings will not reflect the true horsepower or torque output.

Dynamometer Charging Pump

Water dynamometers control load by varying the volume of water in the brake housing. Water pressure to the brake must be furnished by a supply pump to ensure that a steady volume is available.

FIGURE 21-3(a). Go-Power Dynamometer. *(Courtesy of Go-Power Corp.)*

The water used to absorb energy is heated as it resists the engine's torque and is usually drawn from a tank of sufficient size to fill the system, plus a reserve for cooling the return water.

THE TAYLOR DYNAMOMETER

This continues to be a popular system. The power absorption unit has some unique features and is widely used for testing bare engines as well as truck chassis accommodation.

The unit is an assembly of several sections, each receiving water through its own valve. Thus a wide range of capacity is provided. It is possible to regulate the water flow into the unit to provide exact load control. [Fig. 21-4(a)]

Another interesting feature is the resilient coupling, which dampens torsional vibrations from the engine. This feature is useful with engines of two, three, four, and some eight cylinders.

328 / Diesel Engine Service

FIGURE 21-3(b). **Variable Resistance Bank.** *(ABC Technical and Trade School)*

This company also makes chassis dynamometers that are also popular for truck use.

In use the inlet valves should be adjusted to allow all sections to be used. There is no question that many operators use only one or two valves for lower horsepower engines. This use of only one or two valves keeps water from flowing through the other sections and may result in frequent service.

A minimum load can be accommodated by opening all valves a small amount. The load can be regulated closely and all absorbing sections receive some water.

The Taylor system is widely used and has provided accurate engine testing for many years.

ELECTRIC DYNAMOMETER RESISTANCE BANK

Electric dynamometers feed the current generated through banks of resistors that are usually mounted outside in the open air. Roof mounting is common.

FIGURE 21-4. Taylor Chassis Dynamometer. *(Courtesy of Taylor Dynamometer and Machine Co.)*

Diesel engine-driven generator sets, such as locomotives, are often tested using a large tank full of strong salt water, a brine tank. Pieces of scrap iron are hung in the brine from rods, and the current flow is between two such assemblies through the surrounding brine.

These units are crude but effective in absorbing engine power. Load control is the locomotive throttle, although control switches can be wired into the circuit. This system is used to check full power; intermediate settings are not considered.

Meter readings on electric dynamometers are accurate. Resistance in the air-cooled resistance units will increase as they heat up and will stabilize at the temperature created by a steady load. Thus several minutes of operating time must be used at each load change to stabilize the resistors.

ENGINE COOLING

Engine Cooling on the Test Bed

Although a radiator can be used to cool the engine during run-in or test, the most common method is by a vertical coolant tank called a *cooling tower*. [Fig. 21-5] This cooling tower is sized to suit the engine being tested. Flow

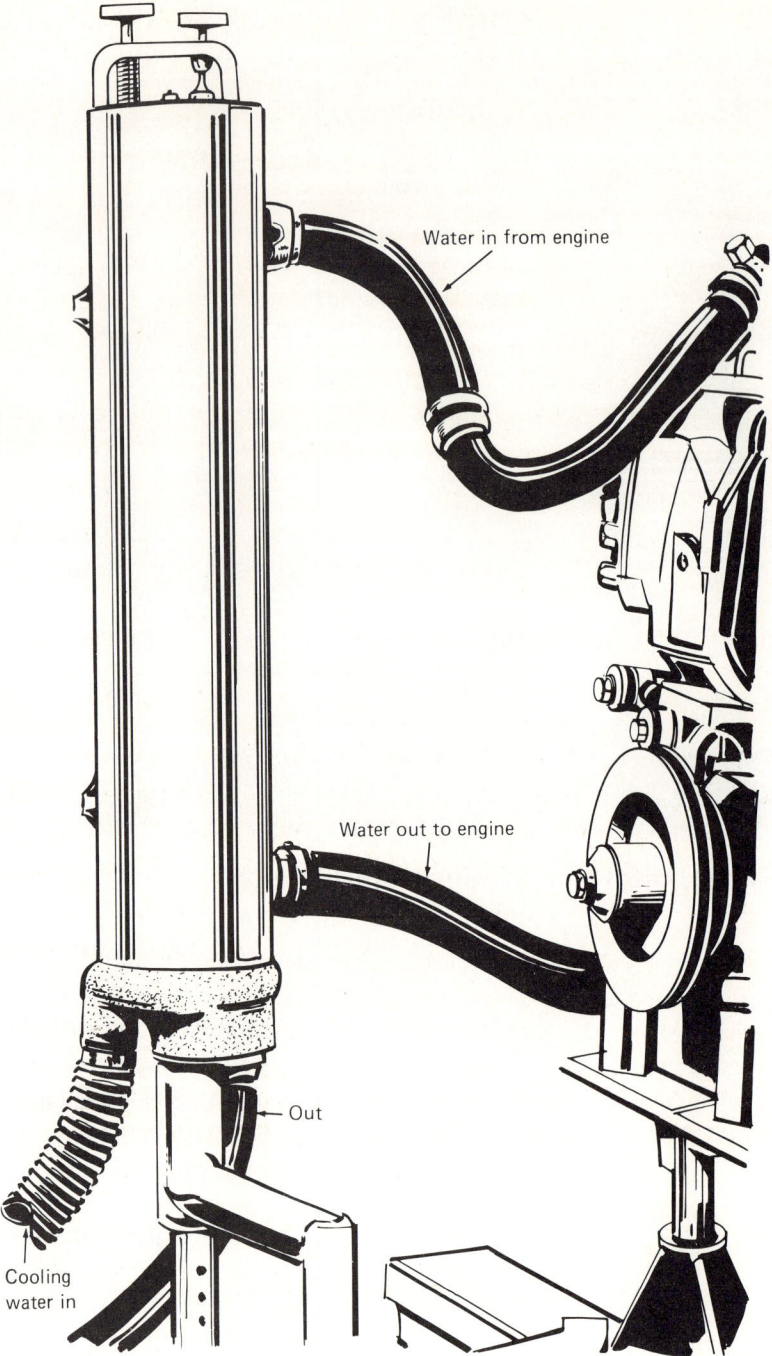

FIGURE 21-5. Cooling Tower and Plumbing. *(Courtesy of Clayton Manufacturing Co.)*

through the cooling tower is the same as through a radiator. That is, coolant to the engine water pump is taken from the bottom of the tower, and coolant from the engine is returned to the top. Some systems have an auxiliary reservoir at the top, fitted with a pressure cap and overflow.

Coolant temperature rise through the engine is about 10°F. The volume of the cooling tower must be high enough to fill the system, plus enough to

Engine Testing / 331

provide heat absorption. Usually two to three times the engine block capacity is used.

Engine Cooling on the Chassis Dynamometer

We do not usually try to make a complete run-in schedule of two or four hours on the chassis dynamometer. This test system is used more for trouble-shooting and complaint solving than engine run-in.

PREPARING FOR CHASSIS TEST

Chassis dynamometers can be dangerous. We must put the vehicle on the rolls with tie-downs and other safety measures.

1. Drive the vehicle onto the dynamometer *straight*. See that the drive wheels are seated on the rolls squarely.
2. Attach tie-down chains or cables to the truck frame and to a secure hook on the floor.
3. Use a small screwdriver or equivalent to remove *all* gravel, stones, or solid articles from the tire treads.
4. If special tires are used for this test, mount them before putting the truck onto the rolls.
5. See that air pressure is adequate and equal in all tires. Check the pressure with the tires cold.
6. Put wheel blocks in front of the front wheels. Ideally the tires should rest on extensions of the blocks. [**Fig. 21-6**]
7. During the test, shift gears smoothly and avoid sudden changes in load or speed. Add load gradually.

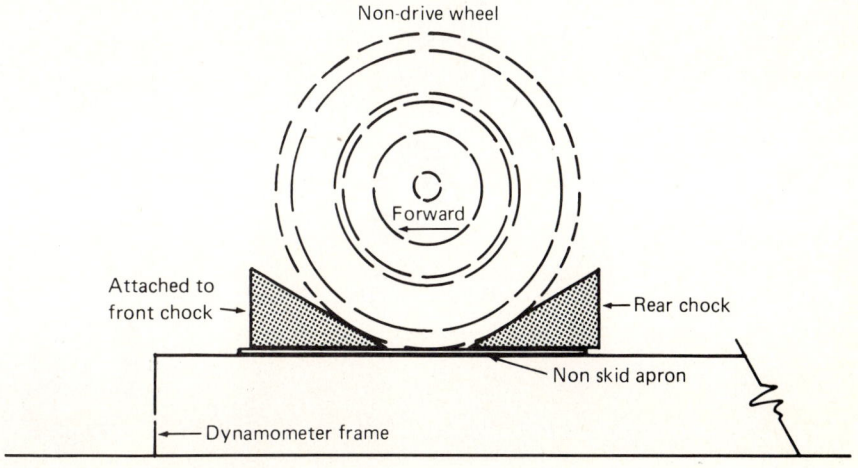

FIGURE 21-6. Wheel Chocks. *(From a sketch by P. M. Uhl)*

332 / Diesel Engine Service

8. Cool the engine from a load run by allowing it to idle at least five minutes, no load. This rule applies to all dynamometer testing.

DYNAMOMETER INSTRUMENTATION

A useful and complete set of instruments is necessary to give the information needed during engine testing on a dynamometer. [**Fig. 21-7**] The necessary instruments follow:

1. Oil pressure gauge, connected to the engine's main oil gallery.
2. Oil temperature gauge, which can be immersed in the oil sump through the dipstick hole.

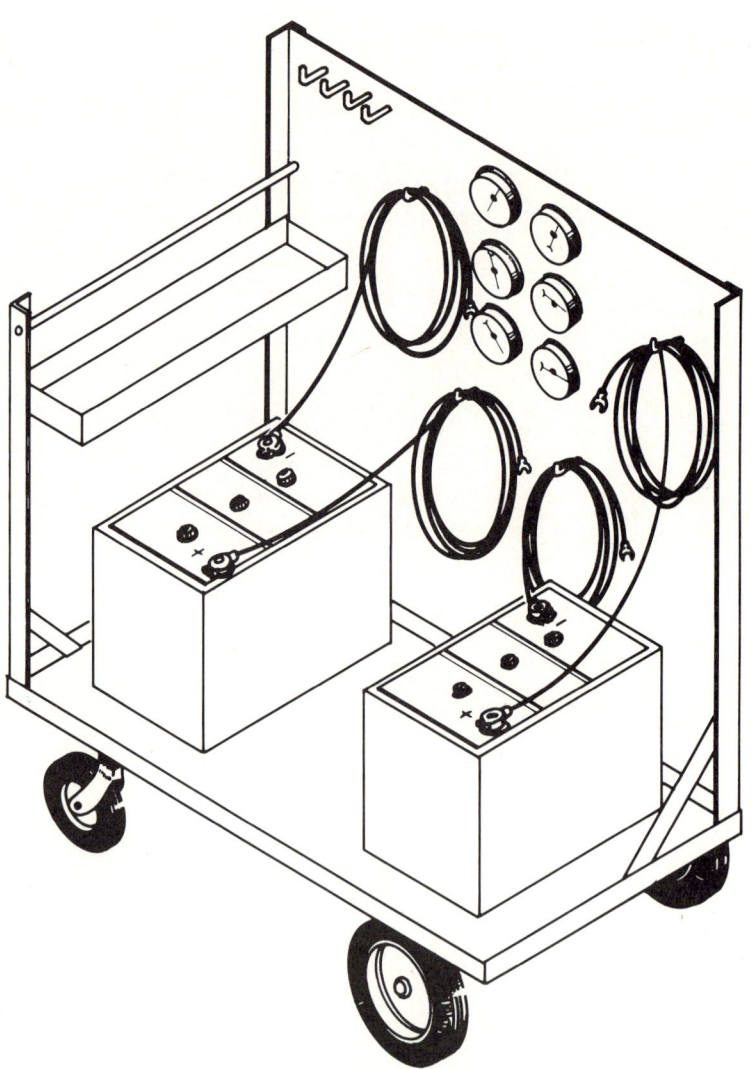

FIGURE 21-7(a). Instrument and Starting Cart. *(From a sketch by P. M. Uhl)*

Engine Testing / 333

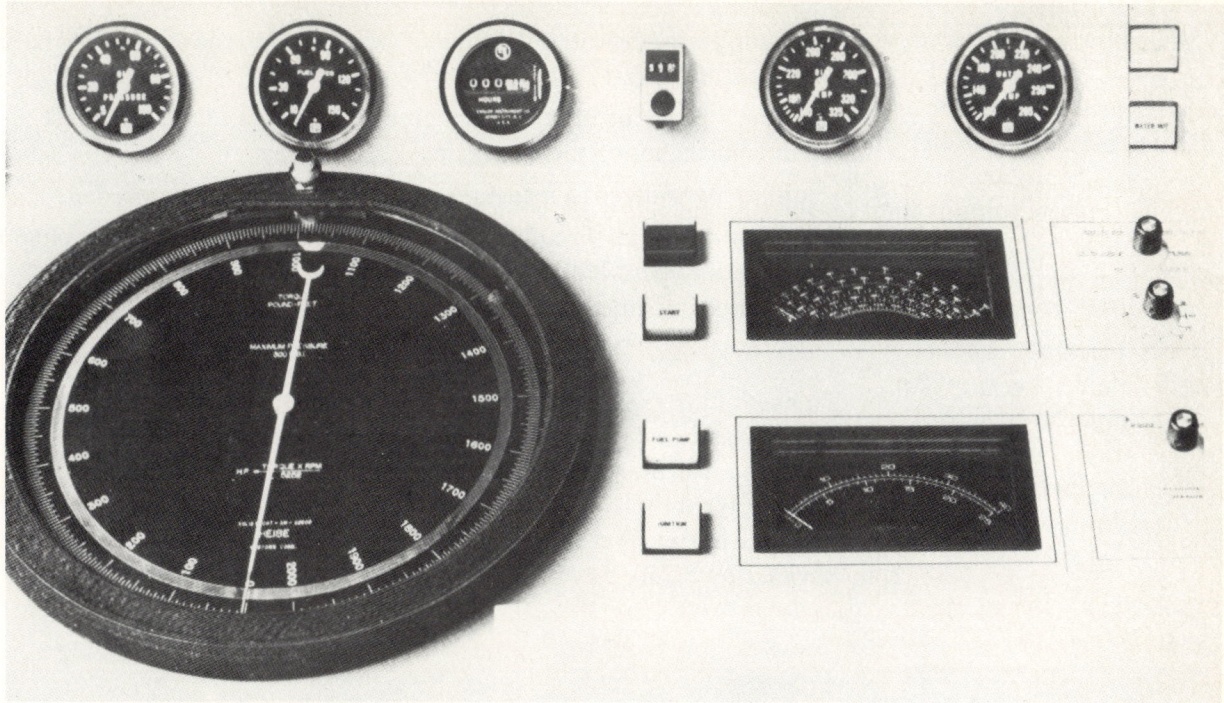
FIGURE 21-7(b). Go-Power Instrument Panel. *(Courtesy of Go-Power Corp.)*

3. A low-reading pressure gauge or a mercury manometer, connected to the air manifold.
4. A coolant temperature gauge, connected to the thermostat housing on the engine side of the thermostat.
5. A coolant pressure gauge, connected to the water manifold or thermostat housing.
6. A low-reading pressure gauge or water manometer connected to the engine crankcase.

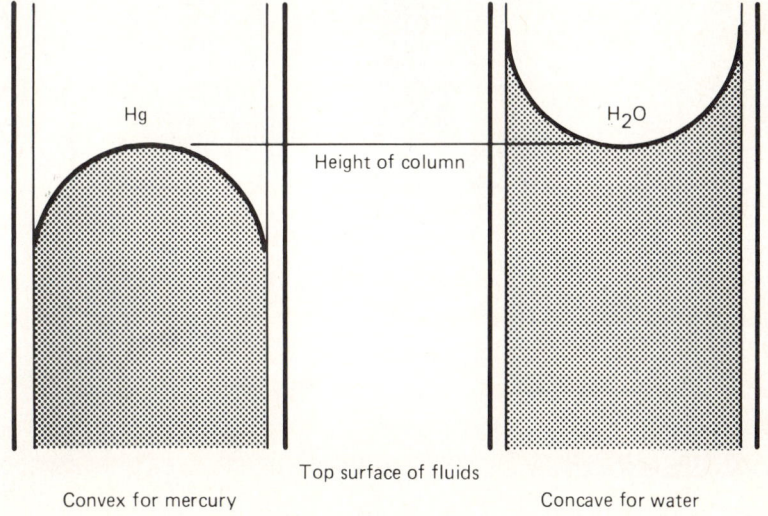
FIGURE 21-7(c). Reading a Manometer. *(Courtesy of General Motors Corp.)*

334 / Diesel Engine Service

7. A vacuum gauge or water manometer, connected to the air inlet to the engine or turbocharger.
8. A pressure gauge or mercury manometer connected to the exhaust pipe within four inches of the flange.
9. A contact pyrometer to measure exhaust temperatures as needed.
10. A 300-pound gauge should be connected to the shutdown valve plug on Cummins Engines. This gauge will read in fuel pressure, by which these engines are controlled. [Fig. 21-8]

This set of instruments can be made up on a portable stand, so that both engine and chassis dynamometers can be served. Flexible tubing can be used to connect the instruments to the engine. [Fig. 21-8]

Chassis dynamometers usually have horsepower and rpm instruments included. Brake horsepower (bhp) is the product of torque in pound-foot, times rpm, divided by 5,250, a constant for all dynamometers.

FIGURE 21-8. Fuel Filters for Test. *(Courtesy of Cummins Arizona Diesel, Inc.)*

TEST PROCEDURE

$$bhp = \frac{(t \times rpm)}{5{,}250}$$

Where it is required to express manometer readings in pounds per square inch (psi), the following conversion factors are useful:

27.7 inches of water = 1 psi
2.04 inches of mercury = 1 psi

When good parts and correct assembly have preceded engine testing, a lengthy run-in is not required. The following schedule is suggested:

1. Start engine with no load, and run at idle speed for at least 10 minutes. Add some load, about 10 hp, or 5 pound-foot torque, after the engine starts. Observe oil pressure. Watch to be sure that both engine and dynamometer are operating properly. After 8 to 10 minutes, observe the coolant temperature. Check for leaks.

2. After the engine temperatures are up to normal, stop the engine, and feel it over for hot areas. This is the only time that valves and injectors can be adjusted using "hot" values. Perform the adjustments soon after shutdown in order to prevent heat loss and inaccuracy.

3. Increase speed to about half the rated maximum. Adjust the load to give the horsepower shown in the schedule.

4. Run at this setting for 10 minutes. Check engine for "hot spots."

5. Increase to full rated speed. Set the load as shown. Run for 30 minutes.

6. Raise the load to the next value, and run for 30 minutes.

7. Continue to raise the load in scheduled stages until full power is reached. Do not raise the speed above maximum.

8. A "green" engine should produce about 95 percent of its rated load at rated speed after the run-in is finished.

The engine must be observed during the run-in, so that any faults can be corrected before complete failure occurs. An alert test mechanic will notice rising temperatures, loss of oil pressure, unusual vibration, etc.

With some engines, particularly Detroit Diesel models, governor adjustments of maximum speed and idle speed may be required after run-in. Run-in reduces friction, and the engine may gain speed.

USE OF CHASSIS DYNAMOMETER

The chassis dynamometer is useful for trouble-shooting much more than for engine run-in. The chapter has described the safety and instrumentation re-

336 / Diesel Engine Service

quirements. Usually a run on the chassis dynamometer is aimed at checking power output, revealing faults in operation, and checking causes for complaints.

Although some faults will not be revealed on the dynamometer, this test will eliminate all but the rare incidents that must be analyzed by logical thought.

Fuel Supply for Testing

It is of considerable advantage to be able to read fuel consumption of engines under test. There are accurate flowmeters that can be connected in the fuel suction system. An inexpensive measuring system can be made, using an ordinary platform scale. The scale is set up on the level area close to the engine, with a clean tank on the platform, and hose connections to the fuel pump inlet and return tubing.

These hoses should be bracketed to the tank and kept with the fuel supply system. In use the tank is filled with fuel, and the weight recorded.

Fuel Consumption Formula

Engine fuel consumption is expressed in pounds per brake horsepower hour (lbs./bhp/hr). Thus, where the weight of fuel consumed is known, the "specific" fuel consumption can be calculated easily:

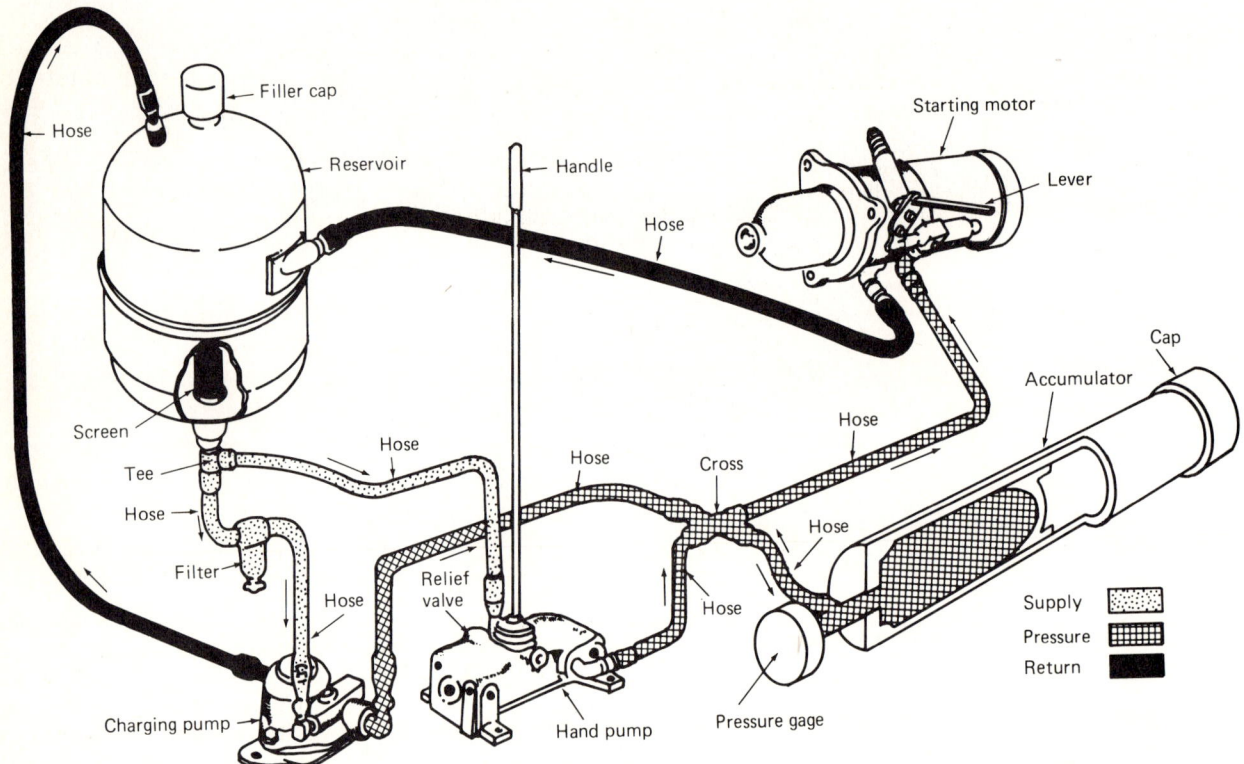

FIGURE 21-9. Hydraulic Cranking System. *(Courtesy of General Motors Corp.)*

fuel consumed in 1 hour - 75 lbs.

Bhp for 1 hour   - 200

divide 75 pounds by 200 bhp = 0.375 pounds per bhp/hr.

This is good consumption for a turbocharged engine under this load for 1 hour. One half hour of test would use 37.5 pounds of fuel for the same specific consumption.

Starting System for Testing

It is desirable to use the cranking system that is on the engine. For engine testing after rebuilding, a battery cart can be made, with the proper switches and wiring on the cart. [**Fig. 21-7(a)**] The following rules should guide the construction of such a cart.

1. Use steel angle stock to make a holder for one heavy-duty 12-volt battery. Most engines use 12-volt cranking motors.
2. Provide a mounting for the necessary switch and any relays used.
3. Make up and install cables *permanently* to the battery and switch. Never try to use a free cable as a switch. The arc will burn any terminal so used and may cause injury.
4. Provide suitable length cables to the cranking motor. These can be made from scrap welding cable, available in any metal scrap year. [**Fig. 21-10**]
5. Use caster wheels to give the cart mobility.
6. When making connection to the cranking motor, be sure to secure the terminals and route the cables away from the exhaust pipe.

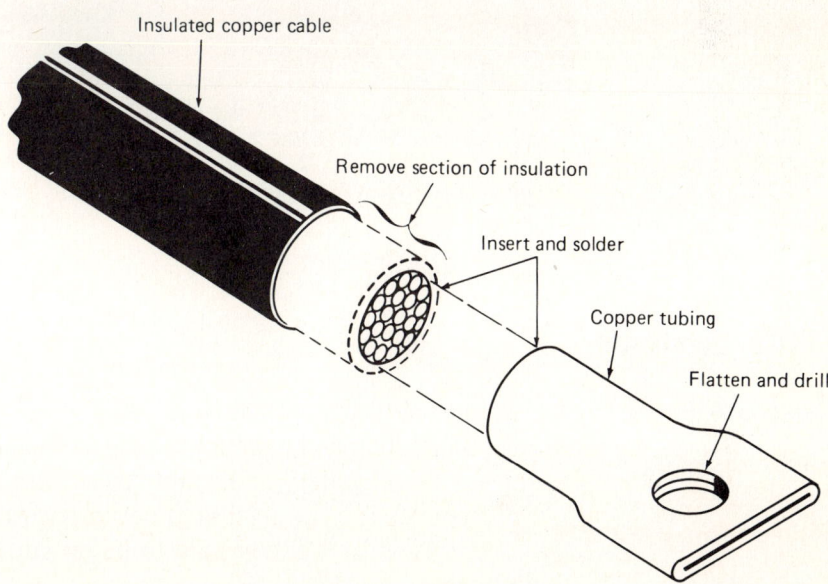

FIGURE 21-10. Battery Cable Ends. *(From a sketch by P. M. Uhl)*

338 / Diesel Engine Service

FIGURE 21-11. Oil Filter for Test. *(Courtesy of Neil Detroit Diesel Co.)*

Oil Filter Systems

The engine has its own full-flow oil filter, which should have a new element. A large bypass filter can be mounted on the dynamometer frame and connected to the engine oil gallery and sump return with flexible tubing. **[Fig. 21-11.]**

Of course, the run-in should be started with new filters and oil.

Note: If several engines are to be tested, the oil should be saved after draining and reused.

Engine Testing / 339

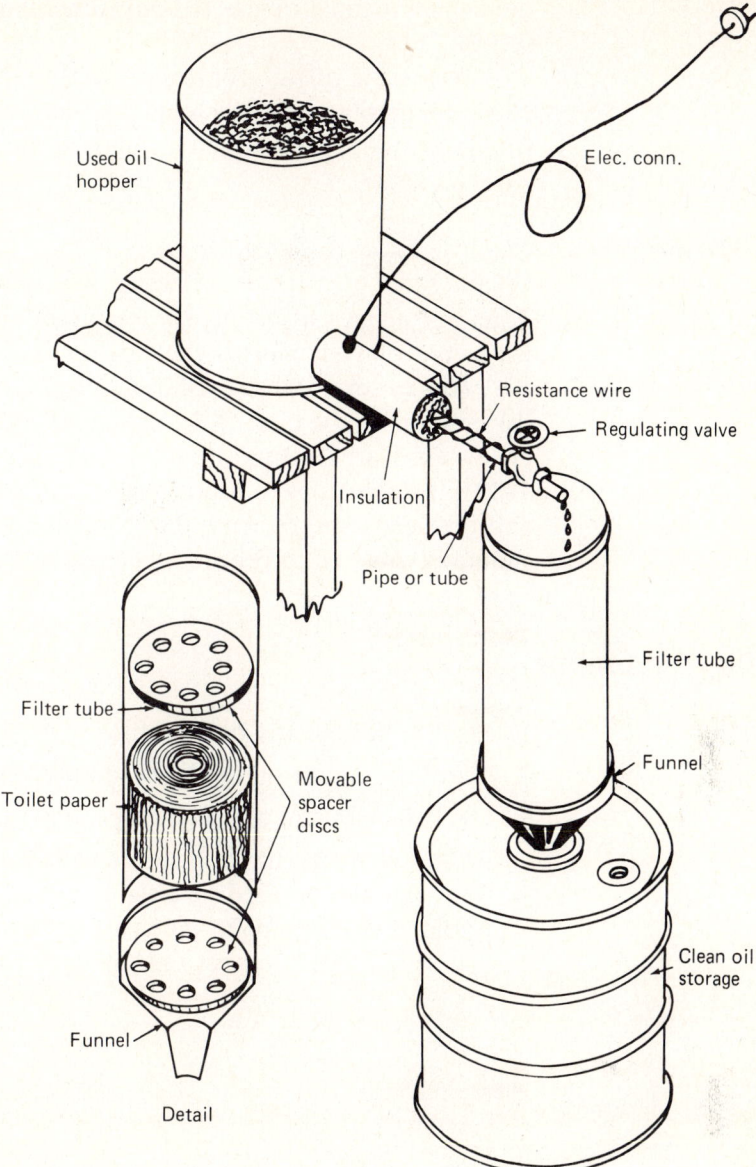

FIGURE 21-12. Batch Oil Filter System. *(From a sketch by P. M. Uhl)*

Batch filtration systems can be used to clean the oil. In fact, much of the oil that is discarded can be salvaged, and such a system can be made locally. [Fig. 21-12]

FINAL ADJUSTMENT, WASH-DOWN, AND PAINTING

Final adjustment of the valve clearance and injectors should be made after run-in. This adjustment can be made using "hot" values. One should refer to the engine manual or use the adjustments included under the text's tuneup section.

Any signs of failure, hose fasteners, hose clamps, etc., should be noticed. If the manufacturer requires that cylinder heads be retightened, this should be done, after which a standard tuneup must be performed.

Wash-Down

After testing and adjusting, the engine will be too oily to hold paint. A spray wash can be given while the engine is supported by a hoist.

Any of several makes of cleaning compound and a steam cleaner or air-solvent gun can be used. All traces of oil and dirt should be washed from the engine exterior. Rinsing should not use any cleaner but a water hose.

It is a good idea to protect all openings, such as air inlets, exhaust outlet, and fuel inlets and returns. Heavy tape should be used. The cranking motor should be covered to prevent water entrance.

Painting the Engine

1. Mask with tape any data plates.
2. Leave the engine on the hoist, and turn it as painting progresses.
3. Use sprayed paint properly to prevent runs. Keep the spray gun moving.
4. Be sure that you have adequate ventilation of the painting area. A good floor fan can be used to blow stray paint droplets away. Wear a mask. Avoid breathing a paint-laden atmosphere. Two light coats are better than one heavy coat. Go over the engine twice.

22
AUXILIARY VEHICLE BRAKING SYSTEM

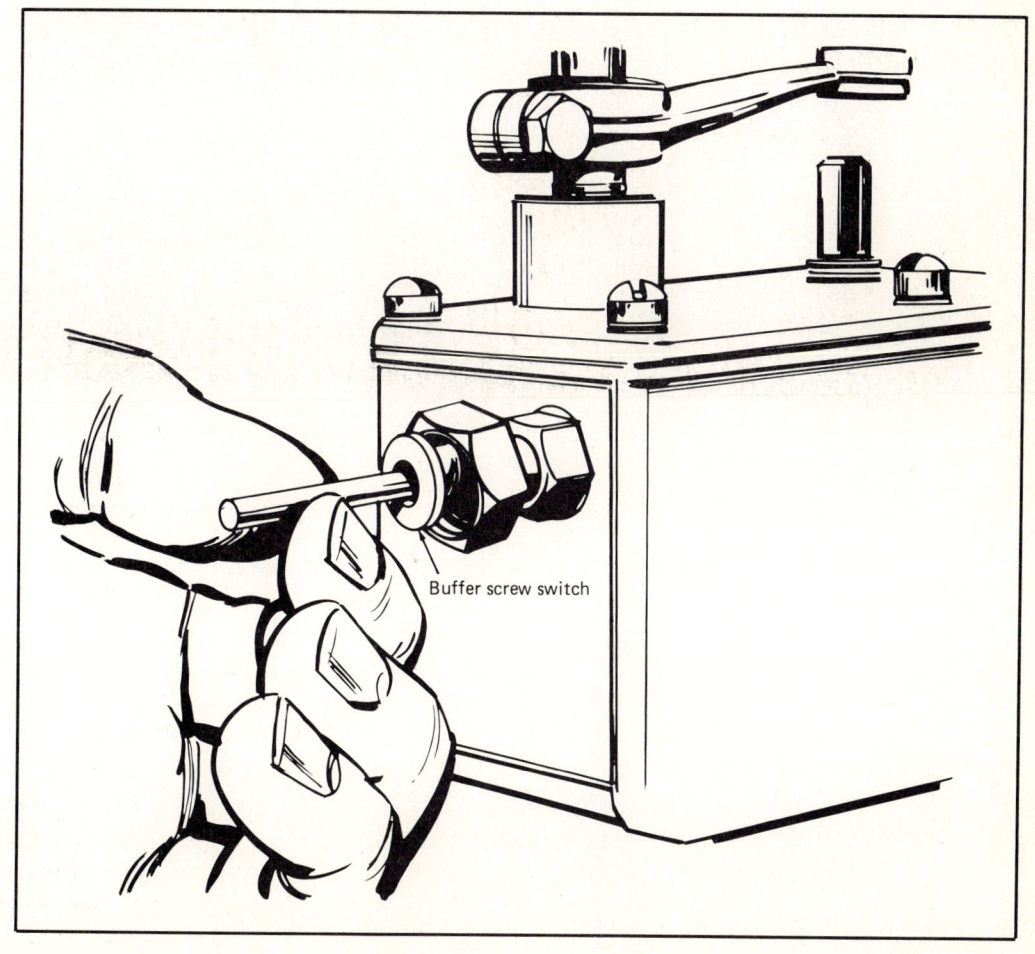

Buffer screw switch

ENGINE RETARDERS

Various devices have been used to retard vehicles in downhill operation, besides the standard service brakes. These can be divided into two categories:

1. systems that make the engine harder to turn, thus retarding the vehicle
2. systems that increase vehicle drag, reducing the energy of motion (kinetic) and acting like brakes to control speed.

The first systems considered are those that act on the engine. They are

1. the Jacobs engine brake and
2. the Williams exhaust brake.

JACOBS BRAKE

This system was the invention of Clessie Cummins, who sold manufacturing rights to the Jacobs Mfg. Co., the makers of drill chucks and other equipment.

The system was first designed for Cummins engines and later was revised to fit other engines. Its essential action is to convert the engine into a power-absorbing air compressor, thus retarding the vehicle.

In the sequence in which the push rods move, during the compression stroke, both intake and exhaust valves are closed, and the injector push rod is moved only at the end of the compression stroke. [Fig. 22-1] Thus the first rocker lever to move as compression ends is the injector rocker lever.

The Jacobs engine brake uses the movement of the injector rocker lever to move a master piston, which is connected to a slave piston, which is positioned to open the exhaust valve.

An electric solenoid valve controls oil flow from the engine oil system into the master piston and slave piston passage. Current to the solenoid is routed through three switches, all of which must be closed to operate the brake:

1. A switch on the fuel pump throttle lever.
2. A switch on the clutch pedal.
3. A toggle switch on the dash panel. The solenoid is spring loaded to open the brake passages to dump oil into the crankcase.

Description of Operation

When the dash switch is turned on, current can flow to the solenoid, provided that the throttle is *not* opened and the clutch is *not* disengaged. [Fig. 22-2]

Oil flows from engine passages into the master-slave circuit and is re-

344 / Diesel Engine Service

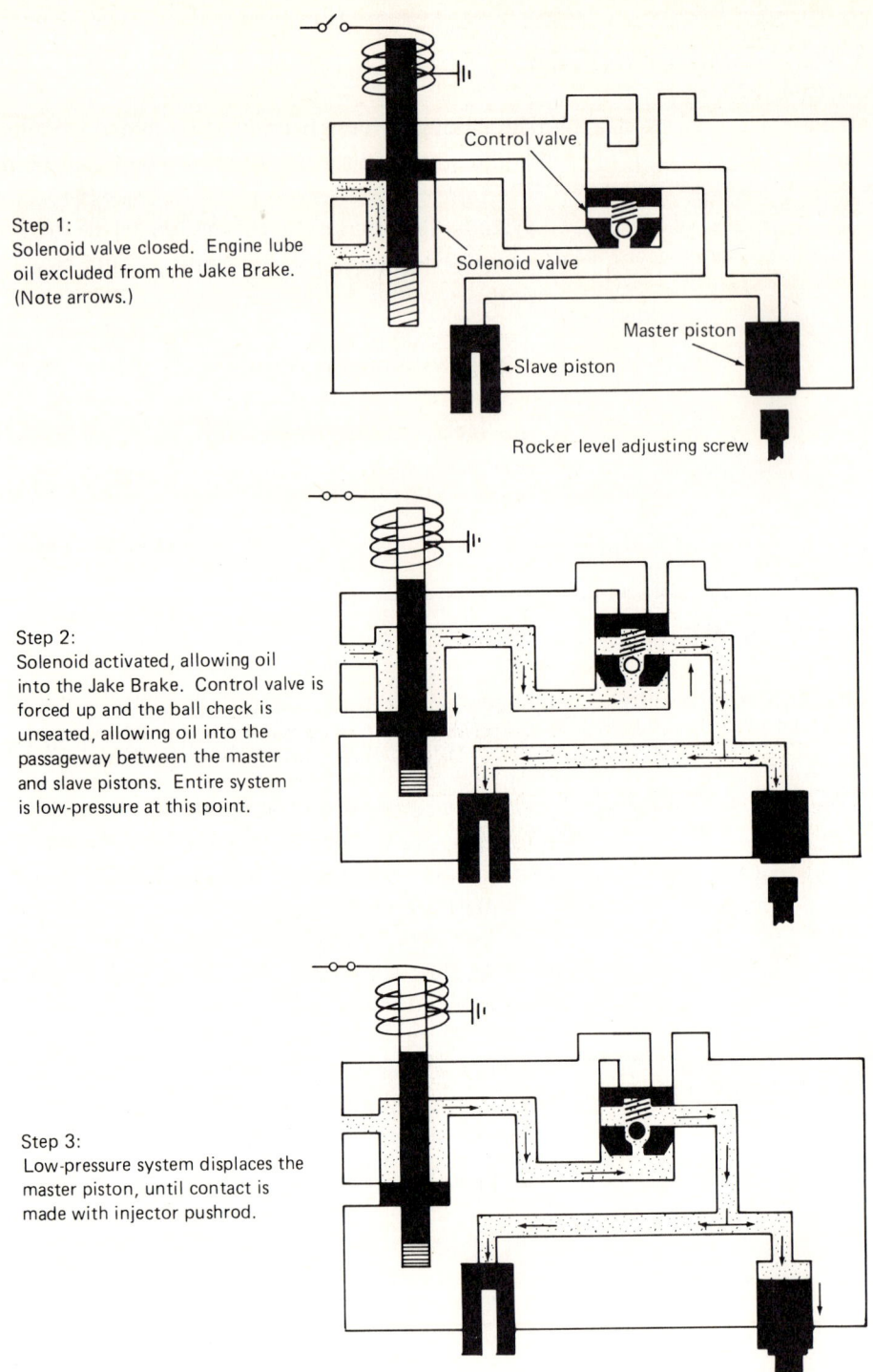

Step 1:
Solenoid valve closed. Engine lube oil excluded from the Jake Brake. (Note arrows.)

Step 2:
Solenoid activated, allowing oil into the Jake Brake. Control valve is forced up and the ball check is unseated, allowing oil into the passageway between the master and slave pistons. Entire system is low-pressure at this point.

Step 3:
Low-pressure system displaces the master piston, until contact is made with injector pushrod.

FIGURE 22-1. Jacobs Brake System. *(Courtesy of Jacobs Manufacturing Co.)*

tained by a check valve until the solenoid valve opens the passage, releasing the oil. A separate passage between the brake circuit and the solenoid allows instant release. [Fig. 22-2]

With the system charged, and the dash switch on, the master piston is pushed up by the injector rocker lever, forcing oil to the slave piston, pushing

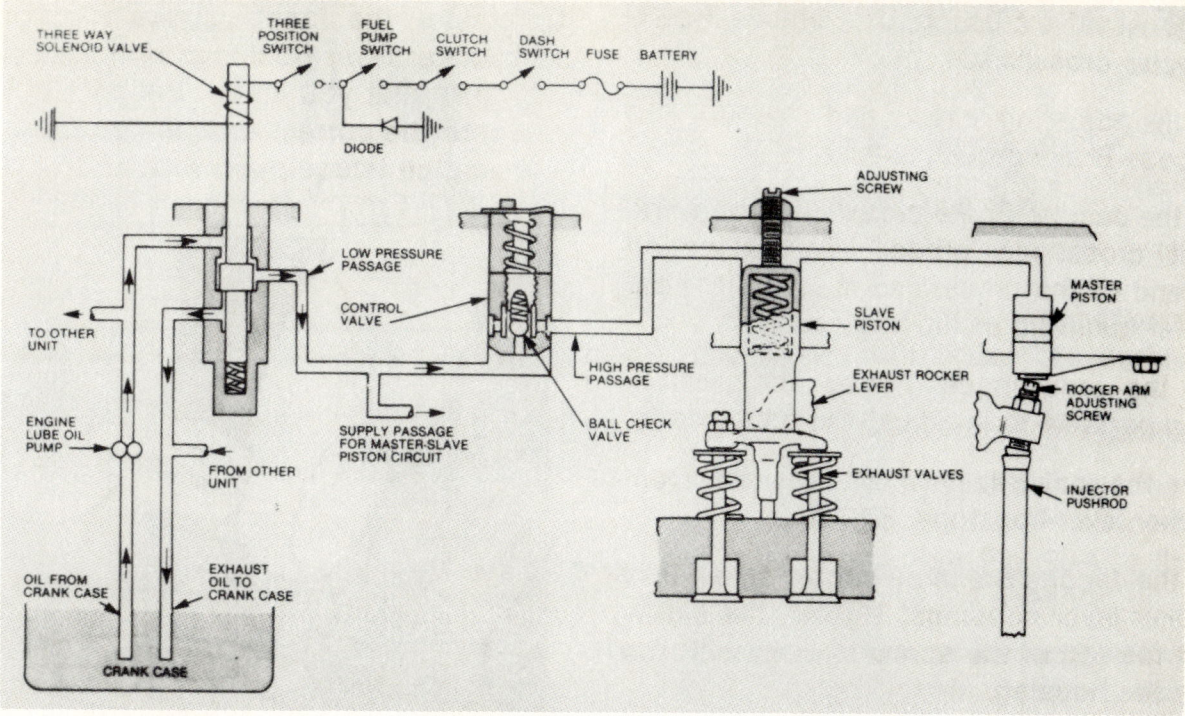

FIGURE 22-2(a). Jacobs Brake Operation—Cummins. *(Courtesy of Jacobs Manufacturing Co.)*

it down to open the exhaust valves. Thus compression in the cylinder is dumped to exhaust, and the energy used to compress the air is not returned to the crankshaft during the following downstroke, which would be the power stroke if fuel were being burned.

The engine thus becomes an air compressor, and power is absorbed.
Note: Special parts are required to install the Jacobs brake. A mechanic must consult the engine builder or the distributor.

Jacobs Brake Service

As long as the brake operates normally, no service is required. During engine overhaul the brake units can be easily disassembled and worn parts replaced.

1. With the brake assemblies off the engine, clean them with a brush and mineral spirits or clean fuel. [**Fig. 22-3**]
2. Remove the control valve cover plate carefully. The spring is normally compressed and may fly up and hurt you.
3. With the cover plate off, use needle nose pliers to take out the spring and control valve.
   Note: As with other parts, store these components in a clean pan.
4. Check the movement of the valve in the bore. These parts are available for replacement if scored or worn. Check the spring for distortion.
5. Remove the wires and solenoid valve. This valve screws into place, and a special wrench is used to turn it. [**Fig. 22-4**]

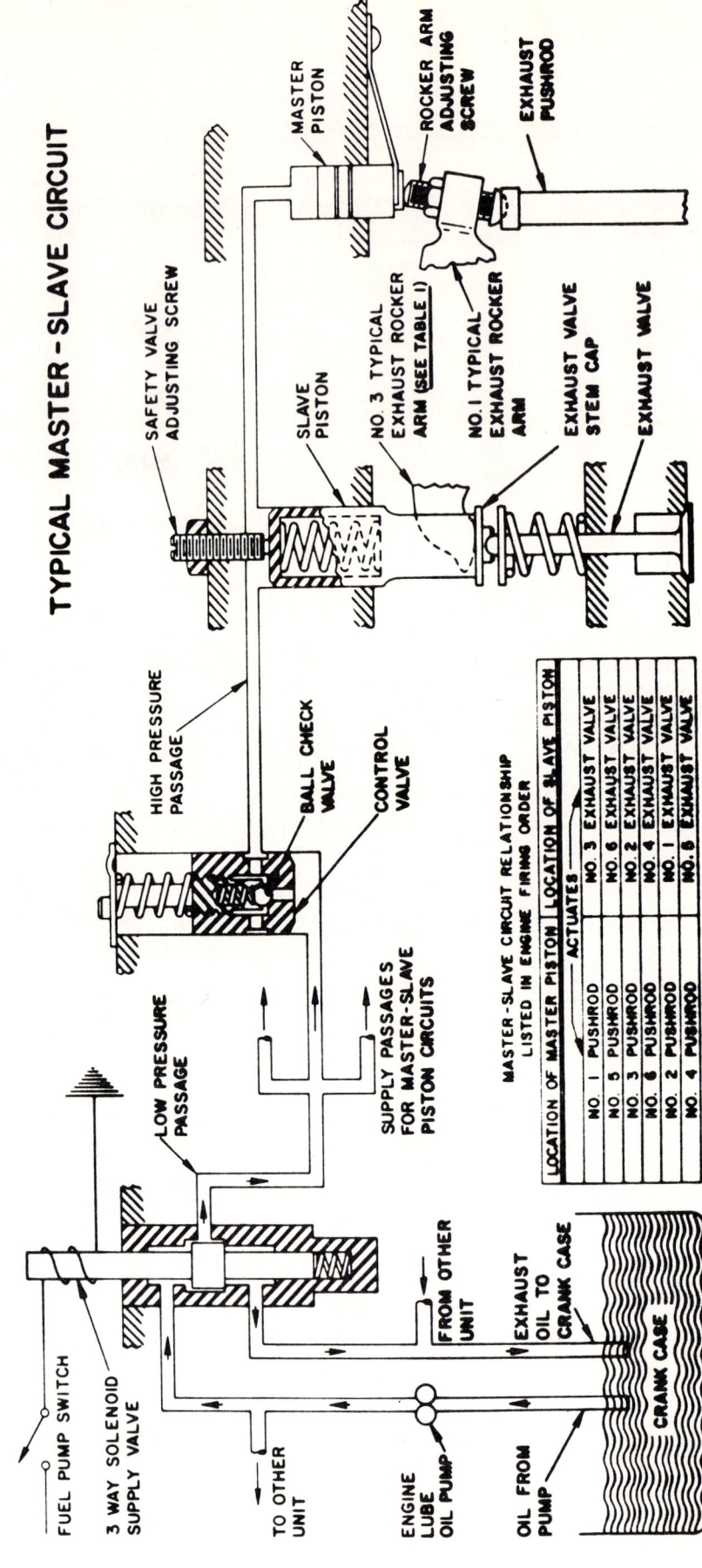

FIGURE 22-2(b). Typical Operating System for Mack and Caterpillar. (Courtesy of Jacobs Manufacturing Co.)

Auxiliary Vehicle Braking System / 347

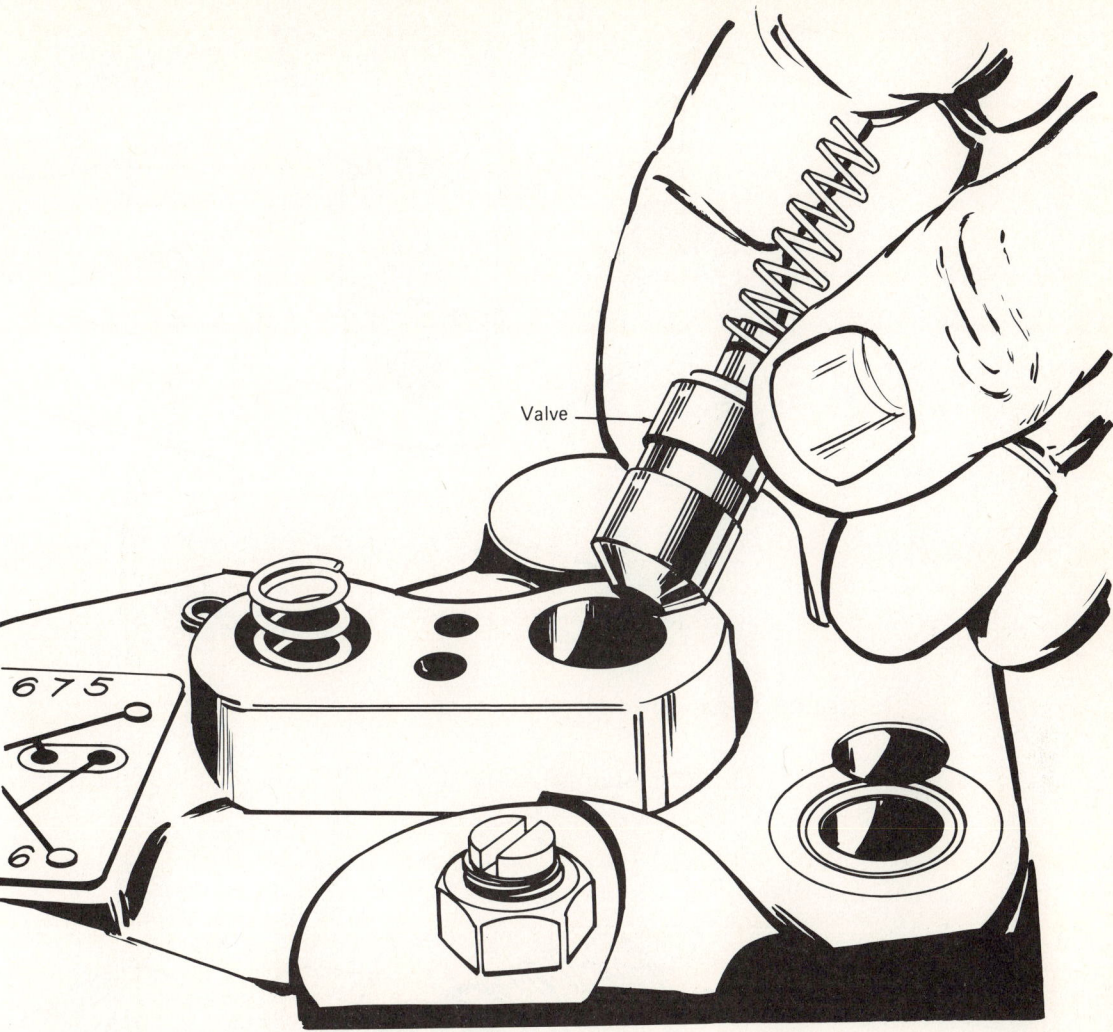

FIGURE 22-3. Control Valve. *(Courtesy of Jacobs Manufacturing Co.)*

6. Operation of the solenoid can be checked by connecting its terminal to a 12-volt source and grounding the valve on the engine.
7. Remove the capscrew and flat spring from the master piston. Some models use a spring retainer. [**Fig. 22-5**] Notice the position of the spring on the top of the master piston, or mark the position with a scribe. Both the flat spring and a spring retainer must be installed in the same position as before removal.
8. Check the master piston for freedom of movement in the bore. This assembly cannot be adjusted but can be replaced as a unit.
9. The slave piston is under heavy spring tension. Loosen the adjusting screw locknut, then turn the brake assembly over, bottom side up.
10. Place the assembly under an arbor press, and carefully press down on the spring retainer. Remove the snap ring. [**Fig. 22-6**]
11. Release the press carefully to take all tension off of the spring.
12. Check all parts for wear or scoring. Store them in a clean container.

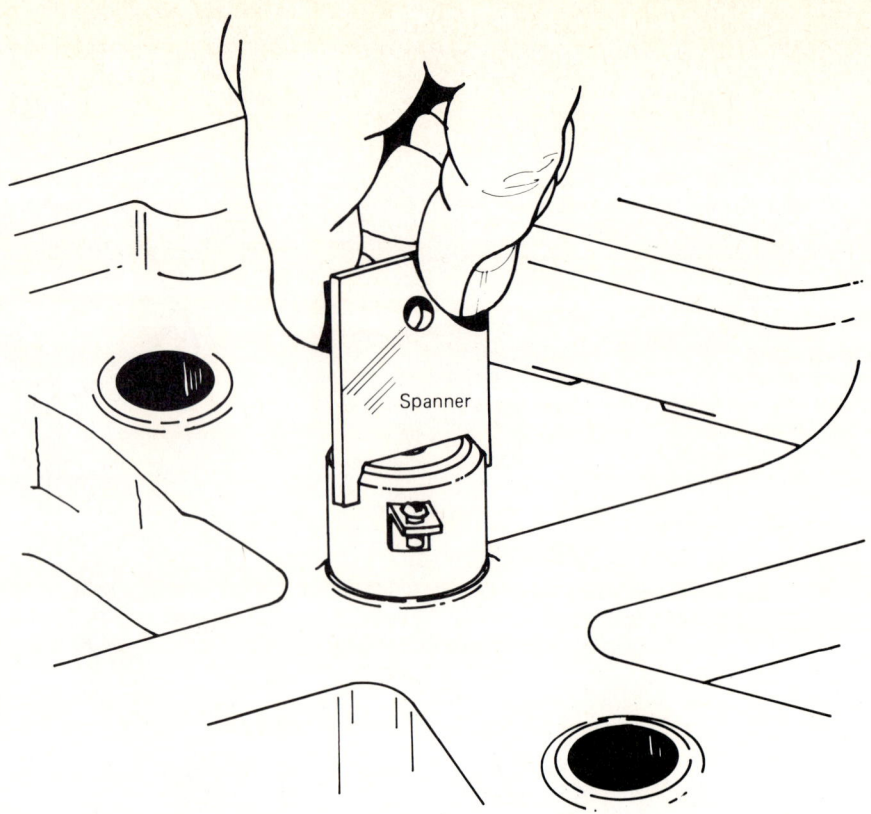

FIGURE 22-4(a). Solenoid Valve. *(Courtesy of Jacobs Manufacturing Co.)*

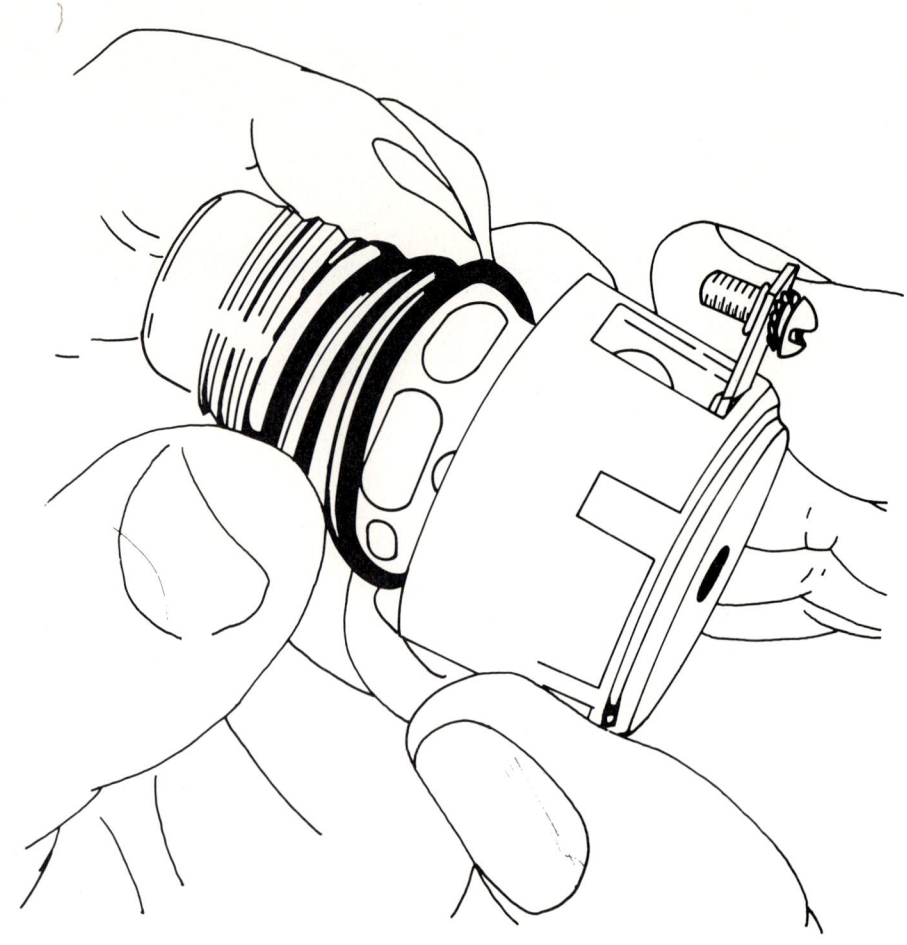

FIGURE 22-4(b). Renew Seals on Solenoid Valve. *(Courtesy of Jacobs Manufacturing Co.)*

Auxiliary Vehicle Braking System / 349

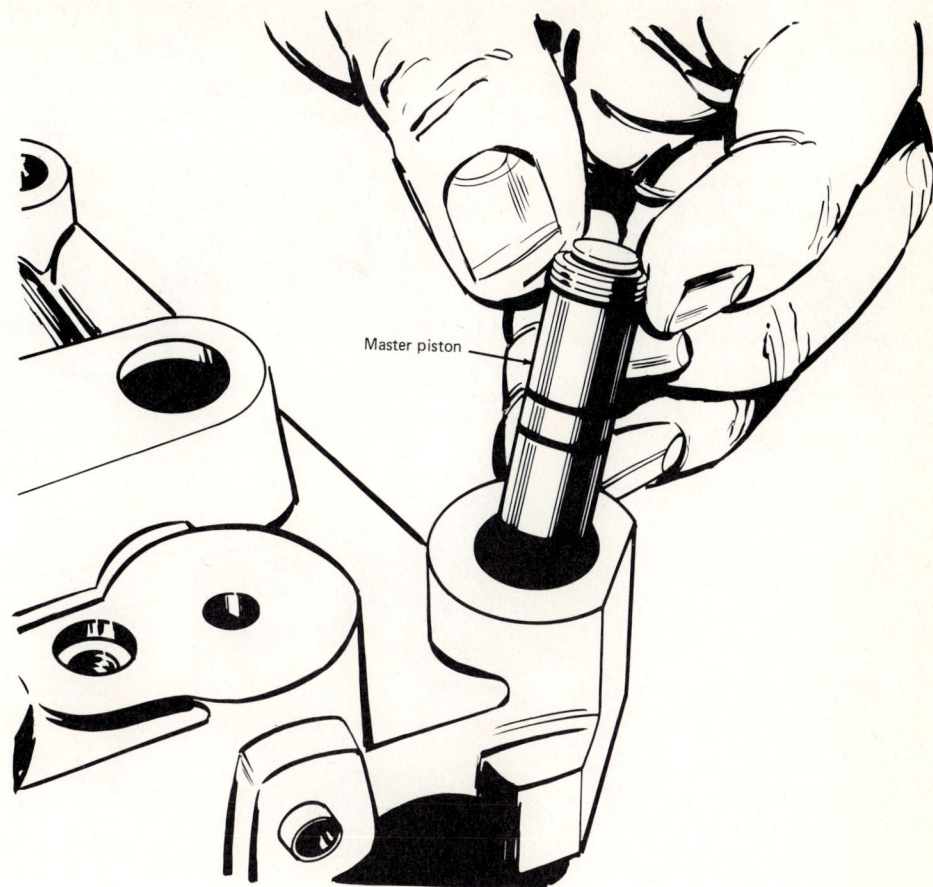

FIGURE 22-5. Master Piston. *(Courtesy of Jacobs Manufacturing Co.)*

Note: All of the components except the master piston and special valve operators can be removed, inspected, and reinstalled with the unit on the engine. These would be service operations following a complaint of functional difficulties or failures.

The following rules must be observed:

1. Replace all seal rings before reassembly. Wet them with oil.
2. Jacobs models 20, 25, 25A and 25B use one spring on the control valve. Models 44A and 44B use two springs.
3. Models 44A and 44B contain an automatic lash adjuster under the slave piston adjusting screw.
4. Be sure that the throttle switch and the clutch switch are properly adjusted.
5. Do not confuse engine valve train faults with brake problems.

In operation, oil pressure in the lash unit forces the slave piston down to open the exhaust valves for brake operation. [**Fig. 22-7**]

Color coding of the top of the lash unit distinguishes between the differ-

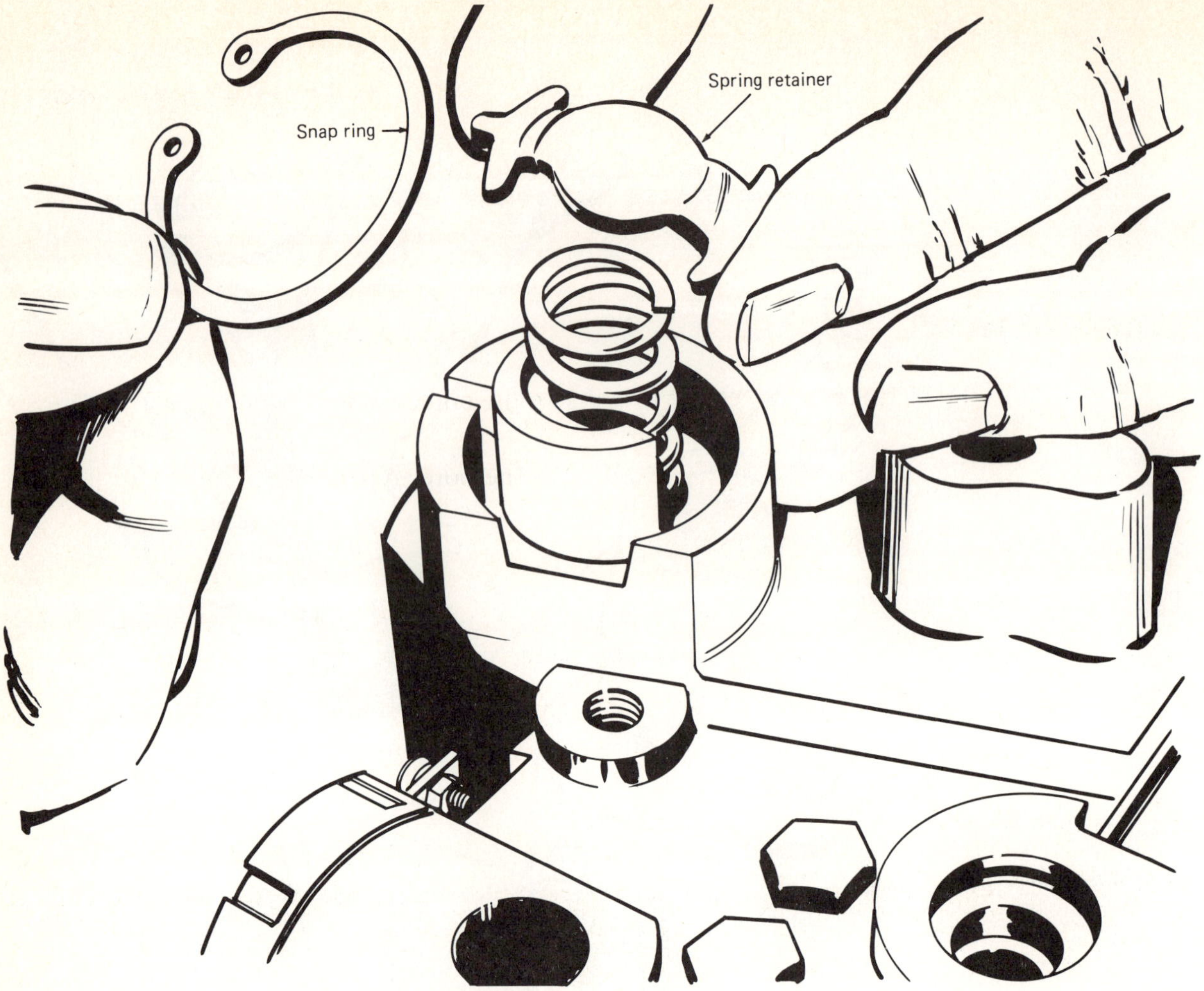

FIGURE 22-6. Slave Piston. *(Courtesy of Jacobs Manufacturing Co.)*

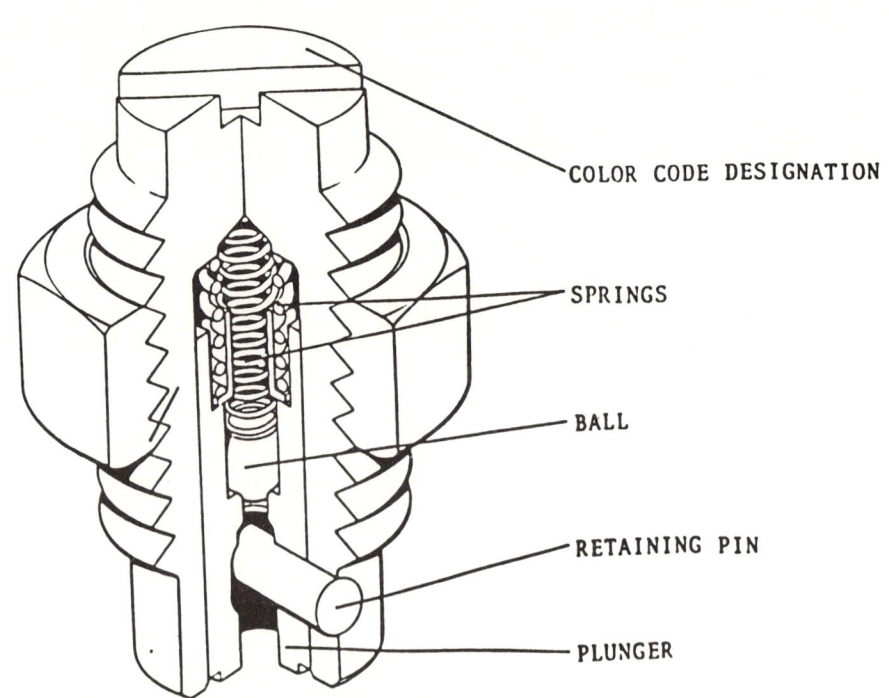

FIGURE 22-7. Lash Adjuster. *(Courtesy of Jacobs Manufacturing Co.)*

ence in plunger protrusion between the brake models it is used in. These units are furnished as complete assemblies. They are not repairable in the field.

Assembly of Jacobs Brake Unit

1. Install the slave piston into the bore, using an arbor press or a clamp to compress the spring. Install the snap ring.
2. Put the master piston into the bore. Assemble the flat spring and capscrew, or the spring retainer. Secure the capscrew, if used.
   Note: Be sure that the tangs of the flat spring do not touch the raised center over the piston.
3. Put new seal rings on the solenoid valve, and start it into the threads by hand. Use the special spanner wrench to tighten the valve in place.
4. Install the control valve and springs. Put the cover plate on.

Adjustment of Slave Piston

This adjustment can be made after the brake units are installed and while injectors and valves are being adjusted. The same engine positions are used for both.

Note: Models 25C and 44 have a hollow hex adjusting screw. Models 44A and 44B have a screwdriver slot. The latter models have a hydraulic lash adjuster.

It is important to select the cylinders by the same method that is used when injectors are adjusted by the indicator method. To adjust the slave piston:

1. Start with the same cylinder on which valve clearance is adjusted. If the valves were just adjusted on No. 3, adjust the slave cylinder on it after valve clearance is set.
2. Loosen the adjusting screw locknut, and turn the screw left to allow the piston to move up to its stop in the bore.
3. Place an 0.018-inch (0.46-millimeter) feeler gauge between the piston and crosshead.
4. Carefully, with finger tips, turn the screw down until the piston contacts the feeler gauge. Do not tighten it further.
5. Tighten the locknut to 15 to 18 pound-foot (20.3 to 24.4 N-m).
6. As you adjust the valves on other cylinders, adjust the slave pistons in the same way.
   Note: Refer to the method of adjusting injectors and valves by the indicator method.
7. When all adjustments are completed, run the engine at idle for about 5 minutes. Then increase the speed to about 1,800 rpm for about 1 minute. Decrease to idle.

352 / Diesel Engine Service

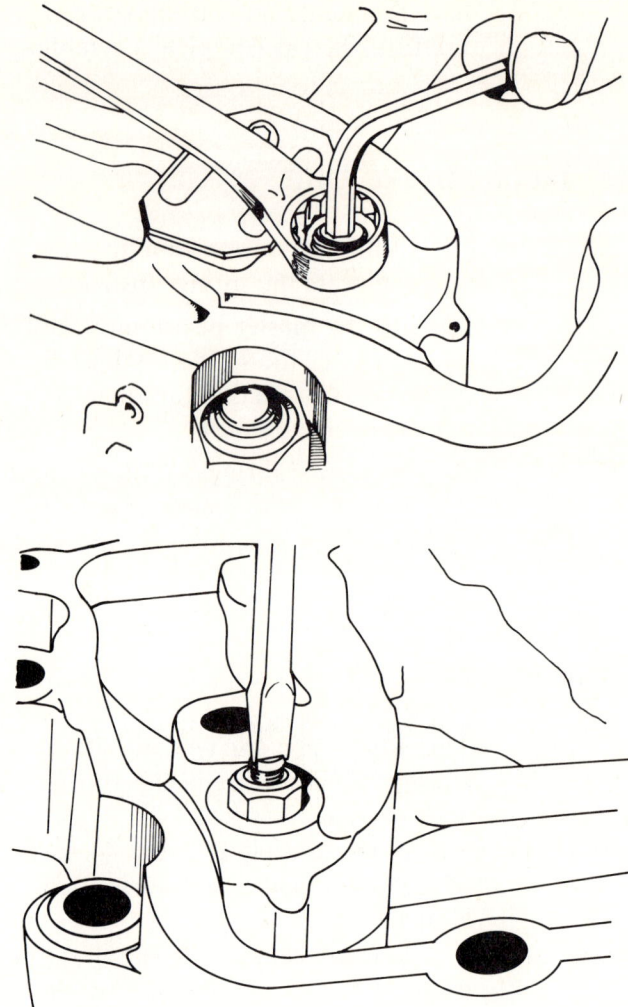

FIGURE 22-8. Adjust Slave Piston. *(Courtesy of Jacobs Manufacturing Co.)*

8. Push down on the center of each solenoid five or six times to fill the brake systems.
9. Stop the engine and install the covers with new gaskets.

Note: These instructions apply to Jacobs brakes installed on Cummins engines. The hydraulic valve lash adjuster referred to is contained in the slave piston adjusting screw. It must be adjusted to the same clearance (0.018 inch), with the slave piston retracted, which is the position for normal operation. [Fig. 22–8]

JACOBS BRAKES ON DETROIT DIESEL ENGINES

These devices have been successfully applied to other engines, including the two-stroke-cycle Detroit Diesels. The principle of operation is identical to that on Cummins engines.

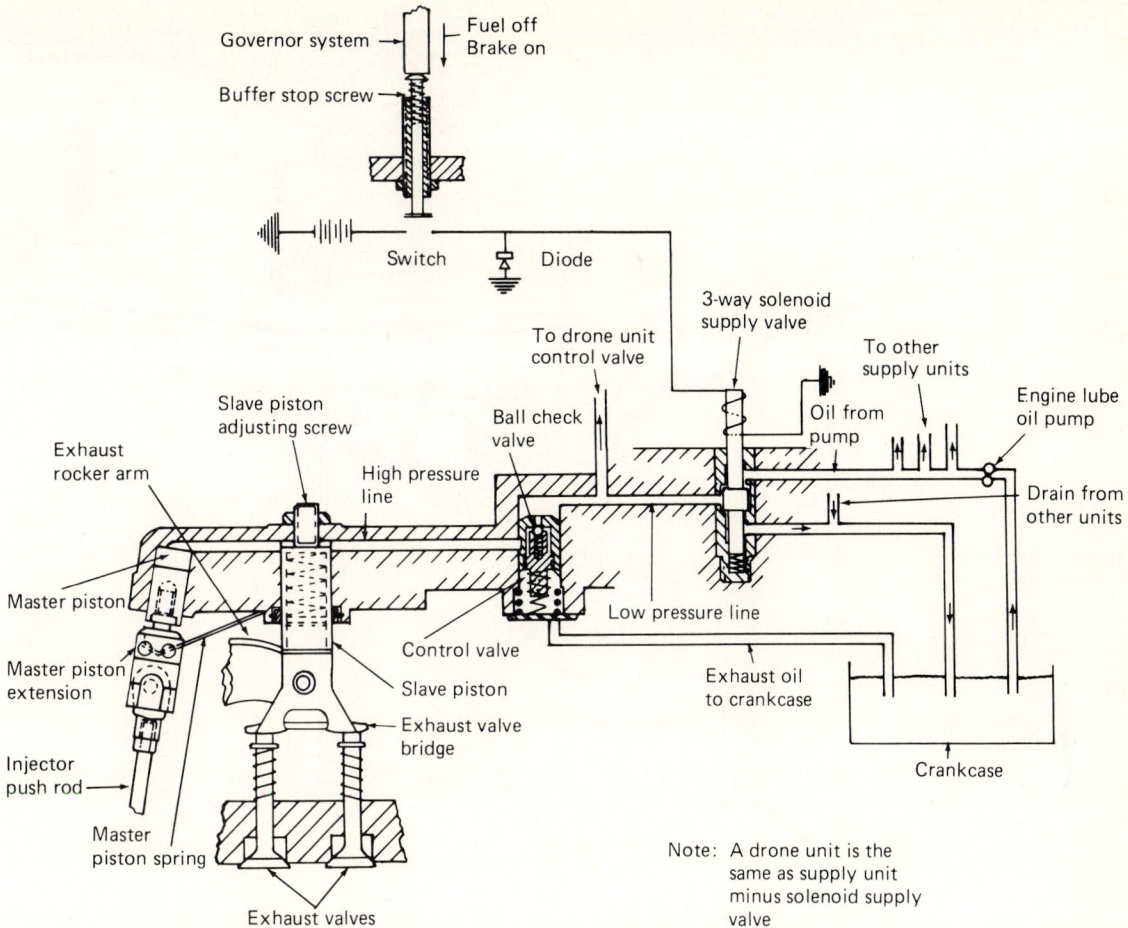

FIGURE 22-9. **Jacobs System on General Motors Engine.** *(Courtesy of Jacobs Manufacturing Co.)*

In the DDA engine, a special master piston and extension is installed on injector push rods. The extension bears on the master piston in the brake housing. [**Fig. 22-9**]

The slave piston bears on a special valve bridge, which of course can open the exhaust valves.

The layout of the system is the same as on a Cummins, with the same type of solenoid valve, control check valve, and other parts.

The drain line from the control valve assures prompt release of the slave piston when the solenoid valve circuit is opened.

Special valve springs are required and are color-coded for each engine model. Be sure that the valve springs in use are correctly coded.

A special buffer screw is used on the governor, which contains a switch in the circuit to the brake. This switch performs the same as the throttle switch on a Cummins. [**Fig. 22-10**]

The slave piston is retained by a heavy spring, retainer, and snap ring. The adjusting screw is turned down until a 0.002-inch feeler gauge placed between the valve bridge pad and the valve stem is contacted. This is a light, finger-tip adjustment.

After contact is established, the screw is turned out 1½ turns. The locknut is tightened to 20 to 25 pound-foot (221 to 339 N-m). This results in about

354 / Diesel Engine Service

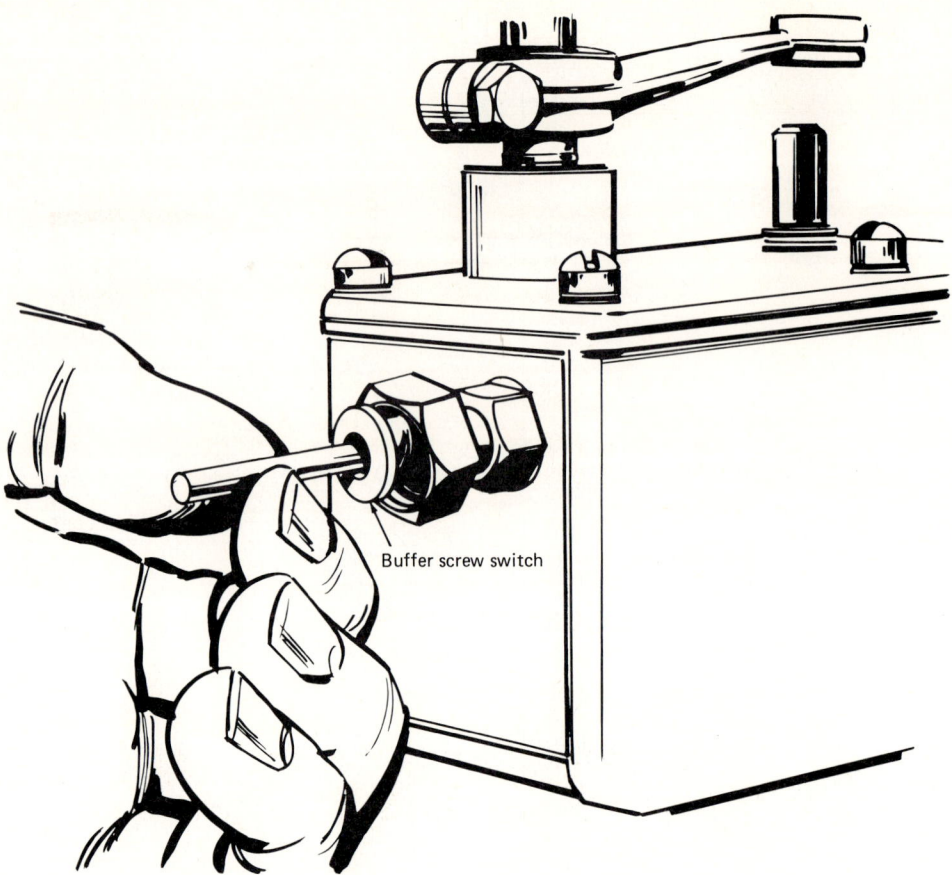

FIGURE 22-10. Jacobs Buffer Screw Switch. *(Courtesy of Jacobs Manufacturing Co.)*

0.064-inch operating clearance. This adjustment must be made when the exhaust valves are closed, off cam.

Servicing the Jacobs Brake on DDA Engines

Because the brake assemblies must be removed to perform tuneup adjustments on these engines, some installation methods must be described.

1. Arrange the brake units on the bench in the order that they will be placed on the engine.
2. The DDA valve bridges on the left side of each assembly have been replaced with a special Jacobs bridge. Check for wear on the pivot pin. Replace as needed. [Fig. 22-11]
3. Supply brake units are installed on the center cylinder of three cylinder heads, and on the second cylinder from the front on four-cylinder heads.
4. An oil connector screw is installed between units on adjacent cylinders. It is a good idea to replace the seal rings before installing the brake units. The outer side oil drilling on the end units is plugged. After renewing the seal, turn the screw in about ½ inch. Leave the locknut loose.

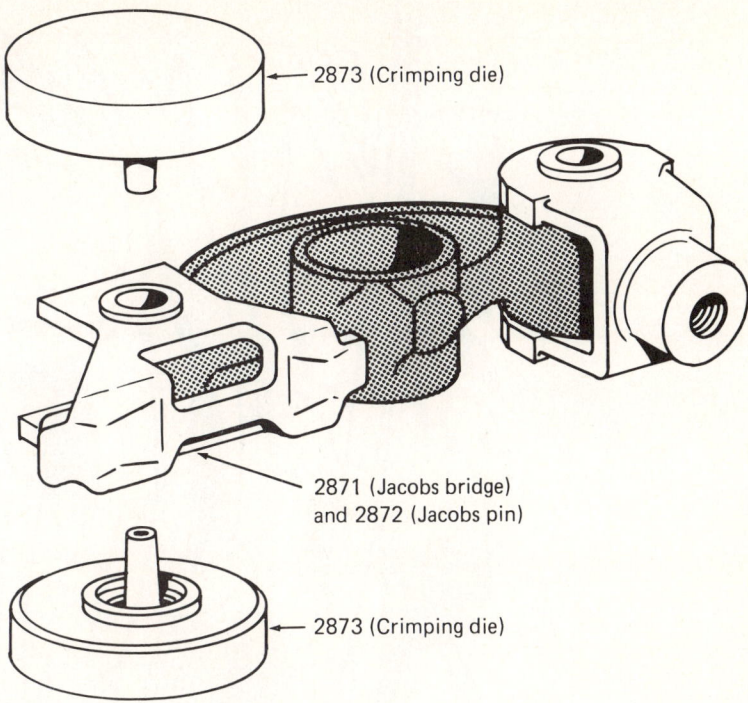

FIGURE 22-11. Jacobs Valve Bridge. *(Courtesy of Jacobs Manufacturing Co.)*

5. Place the brake units over the rocker levers on each cylinder. Install the special long bolts through the brake units and brackets, and tighten them evenly to 55 pound-foot (74.6 N-m). Then check for clearance between the housings and fuel tubes. Fuel tubes can be bent using a Jacobs bending gauge. [**Fig. 22-12**]

6. Position the oil seal ring in the head of the jumper screw, and turn the screw out until it contacts the pad on the next brake housing. After metal-to-metal contact, turn the screw back about one hex to provide clearance. Tighten the locknut.

7. After the engine has been adjusted and all brake units installed, start the engine. Run it at idle for a few minutes, then turn the brake switch on, and raise engine speed to 1,500 to 1,800 rpm. Drop the speed to idle quickly, and repeat this operation until the engine responds to the brake by decelerating rapidly when the throttle is closed.

8. If it does not respond after repeated trials, check the oil supply connection to the brakes. On DDA engines, the oil supply must be taken from the main oil gallery, and cold oil pressure must be below 80 psi. Check the brake manual for proper location of this connection. [**Fig. 22-13**]

9. The buffer switch and screw should have been adjusted during engine tuneup. This switch is normally open but is closed when the dash brake switch is turned on. A small diode is connected to reduce arcing at the switch contacts. Checks of diodes have been described.

10. As with Jacobs brake units on other engines, the components are replaceable if they fail.

FIGURE 22-12. Jacobs Bending Gauge. *(Courtesy of Jacobs Manufacturing Co.)*

JACOBS BRAKES ON MACK AND CATERPILLAR ENGINES

The Jacobs engine brake is applied to engines that do not have cam-operated injectors. The same brake system layout is used, except that an exhaust valve push rod operates the master cylinder.

Any exhaust rocker lever that is moving to open the valve can operate the master piston to operate the slave piston over the exhaust rocker lever on the cylinder on compression. [Fig. 22-14]

A few differences exist between these brake systems and those on DDA and Cummins engines.

Service and Adjustment

A three-position dash switch is used so that all or half of the cylinders can be used as brakes. It is marked Hi, Lo, and Off. [Fig. 22-15]

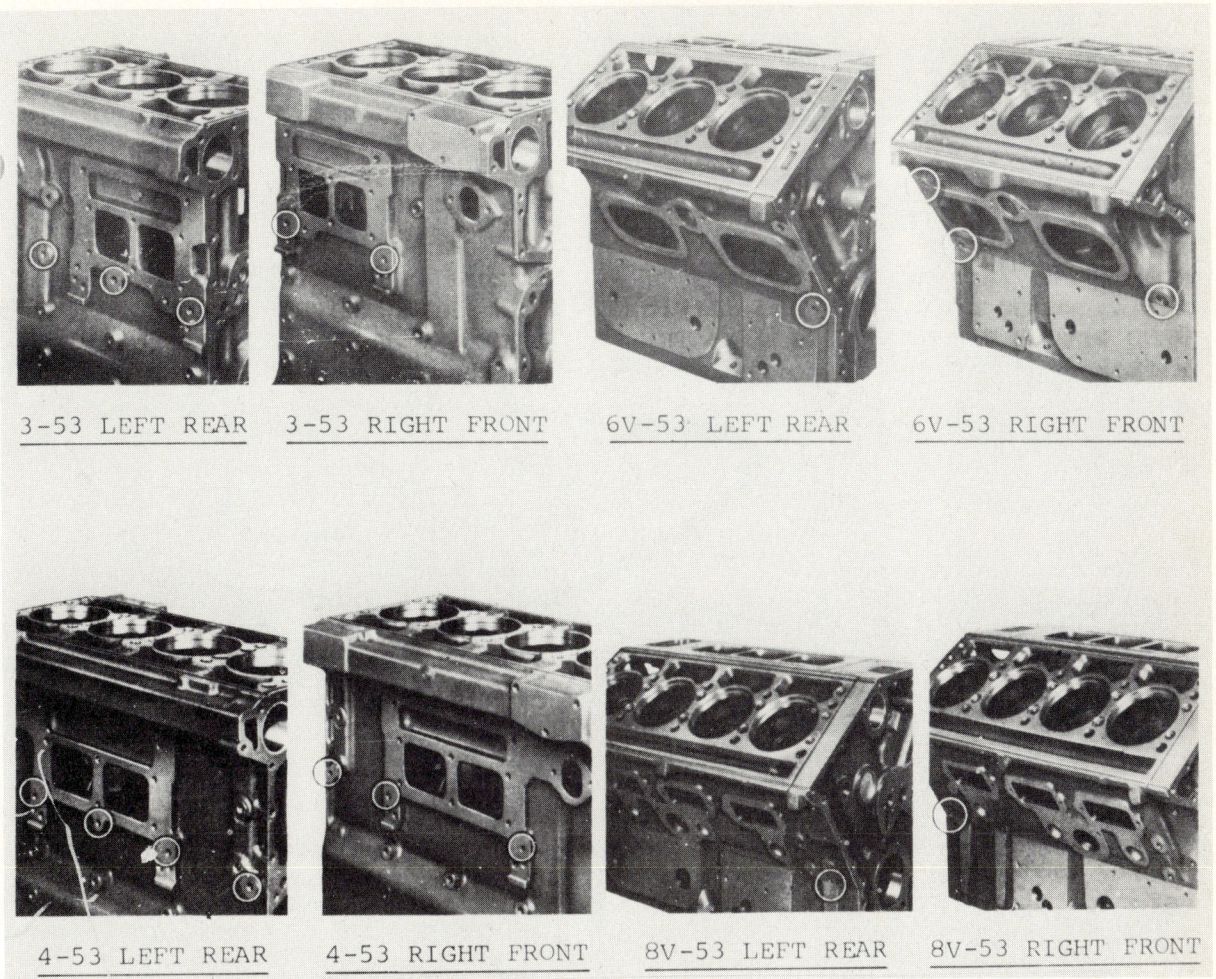

FIGURE 22-13(a). Oil Supply Sources. *(Courtesy of Jacobs Manufacturing Co.)*

The slave piston adjusting screw is turned to contact the *closed* exhaust valve spring cap. When adjusting the Jacobs Model 675 brake slave piston on Mack engines, one should place a 0.030-inch (0.762 millimeter) feeler leaf under each foot of the brake slave piston, then turn the adjusting screw down to contact the gauge. It should not be tightened further. A special Jacobs 0.060-inch feeler gauge is supplied to adjust the slave piston to valve bridge clearance on Caterpillar engines; it should be placed under the slave piston and adjusted as described.

The Jacobs brake on these engines allows valve clearance adjustments to be made without removing the brake units. Again adjustments using cold values on cold engines are recommended.

Brake system priming is done by several accelerations of the warmed-up engine with the brake turned on. The solenoid on Mack engines should not be manually depressed. Acceleration methods should be primed as described.

Note: As the engine decelerates due to brake action, the throttle should be touched lightly to prevent stalling from idle speed.

A C-clamp can be used to control the spring of the slave piston. A special jaw fork for this use can be made from a flat washer. This method can be used on all brake models when the slave piston must be removed in the field.

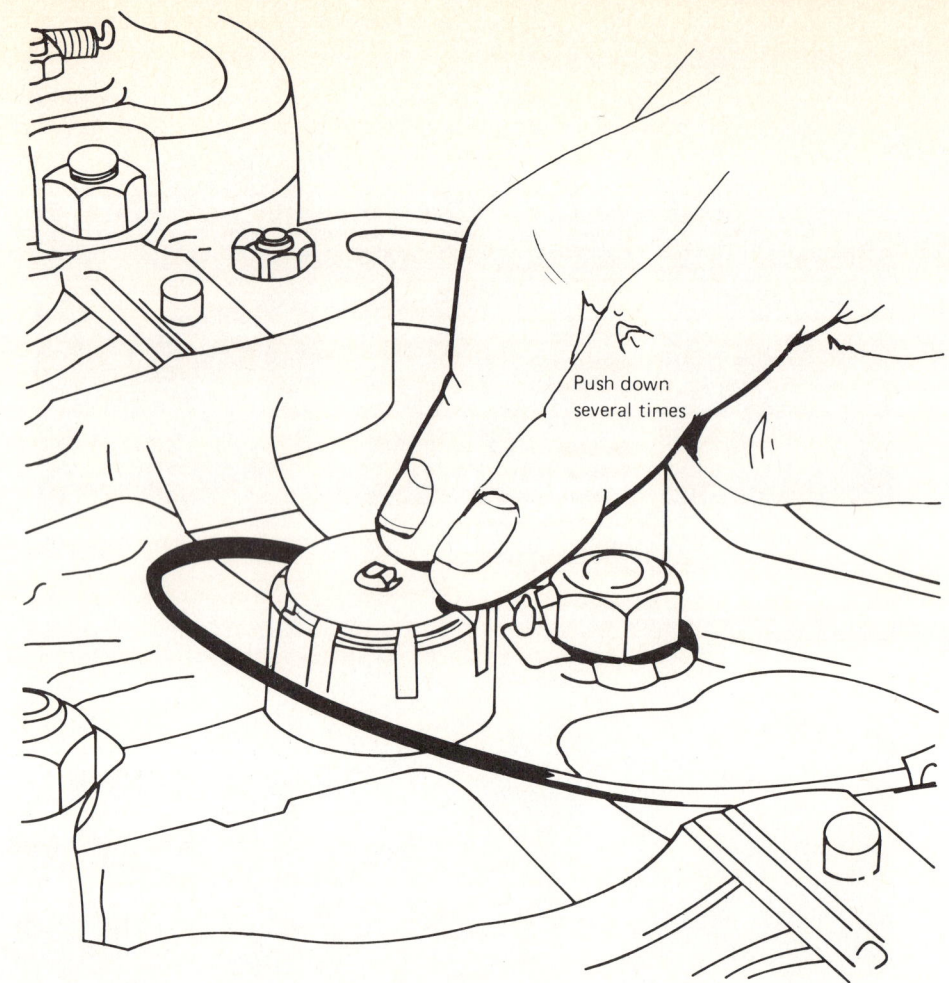

FIGURE 22-13(b). Prime Solenoid Valve. *(Courtesy of Jacobs Manufacturing Co.)*

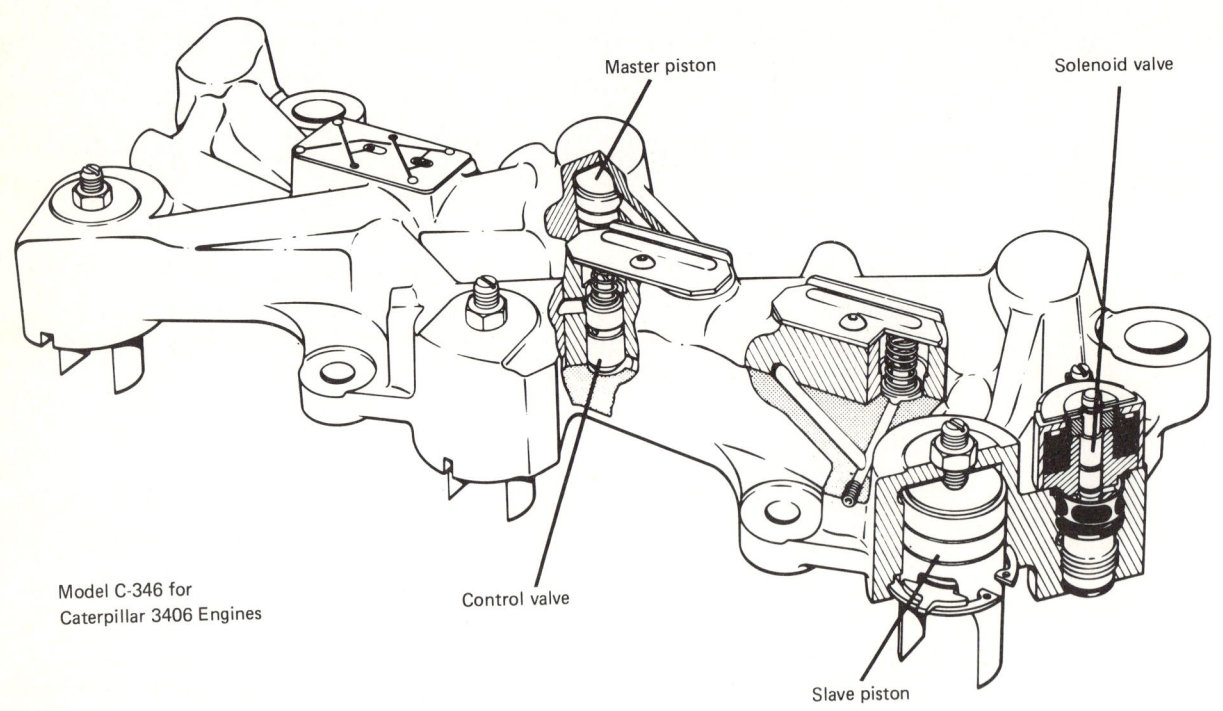

FIGURE 22-14. Jacobs System on Caterpillar Engine. *(Courtesy of Jacobs Manufacturing Co.)*

	SET EXH. VALVE CYL.	SET INTAKE VALVE CYL.
CYL. 1 TDC	1, 3, 5	1, 2, 4
CYL. 6 TDC	2, 4, 6	3, 5, 6

Adjust the intake and exhaust valves on each cylinder. (See Table 1). After setting the exhaust valve clearance and while the exhaust valve is closed, and the bridge loose the slave piston clearance can be set.

Adjust the intake and exhaust rocker lever clearance to Caterpillar specifications: Set lash at .015 in. (0.38mm) for intake and .030 in. (0.76 mm) for exhaust.
Torque rocker adjusting screw lock nuts to 22 lb. ft. (30 N•m).

FIGURE 22-14(b). Adjustment Specifications for Caterpillar. *(Courtesy of Jacobs Manufacturing Co.)*

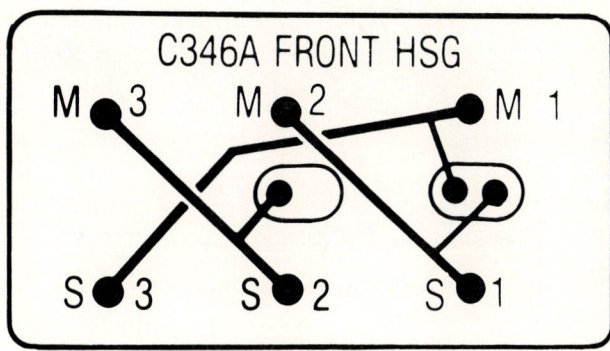

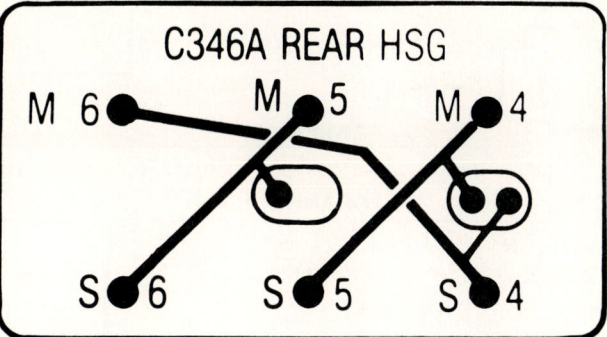

FLOW PLATES

Note the flow plate on the engine brake housing. Housing marked front must be used on front three cylinders. Housing marked rear must be used on rear three cylinders.

Carefully install the brake housing; pay particular attention to the oil supply adapter to insure proper alignment.

FIGURE 22-15. Flow Plates. *(Courtesy of Jacobs Manufacturing Co.)*

Trouble-shooting Jacobs Brakes on Mack and Caterpillar Engines

These units are relatively trouble-free. However, wear and dirt in the oil can cause trouble. The brake units should be rebuilt in a properly equipped shop at engine overhaul periods. [Fig. 22–16]

The following problems are described in the order in which they are most likely.

1. Failure to function
   a. Any open circuit in the solenoid and switch wiring will keep the brake from operating. Check the electrical circuit first. Check the fuse. Check all switches to see that they operate.
   b. Low oil pressure will result in partial braking but may cause complete failure to function.
2. Exhaust valve damage
   a. This can occur if oil supply pressure is too high when the oil is cold. A relief valve can be installed in the oil supply line to prevent this. It should be set to open at 50 to 60 psi. (344.55 to 413.7 kPa).

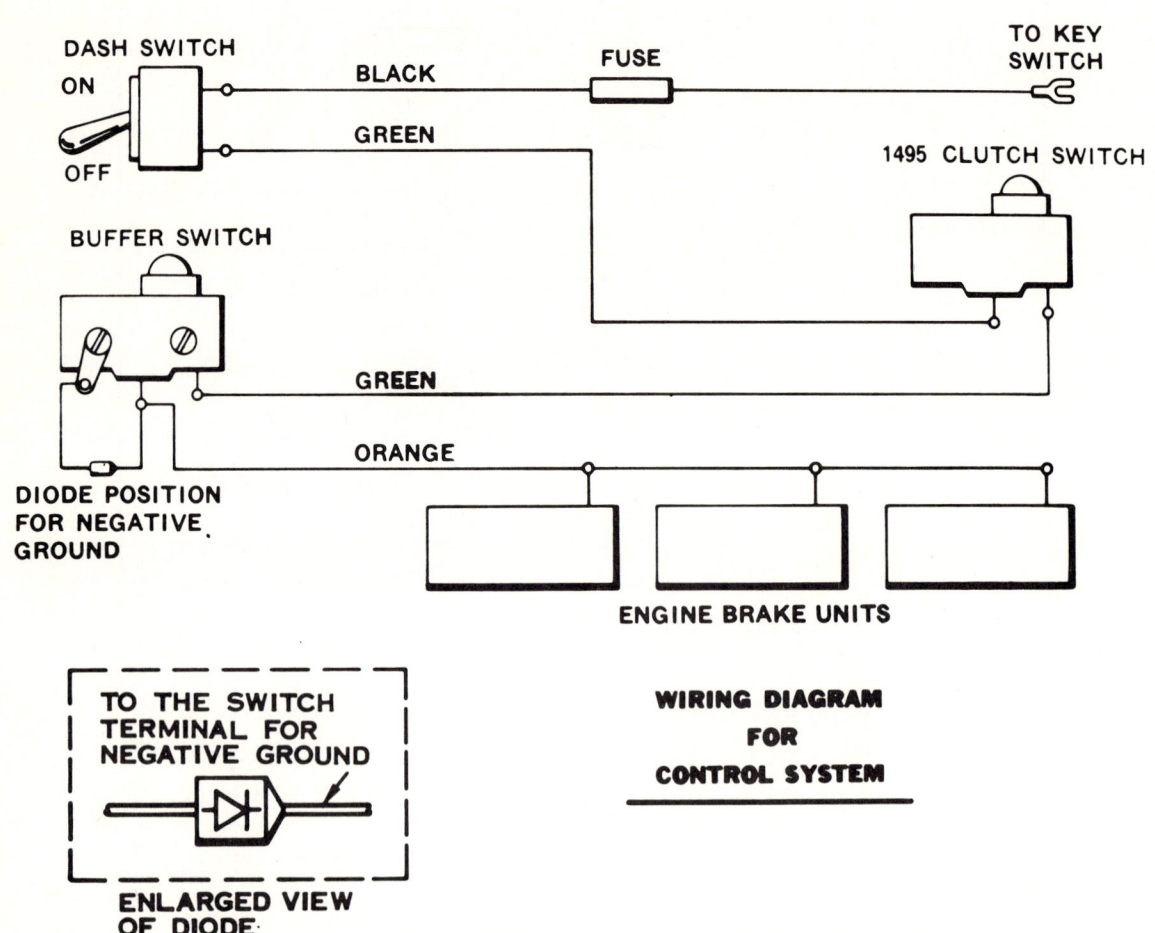

FIGURE 22-16(a). Jacobs Electrical System on Cummins Engine. *(Courtesy of Jacobs Manufacturing Co.)*

Auxiliary Vehicle Braking System / 361

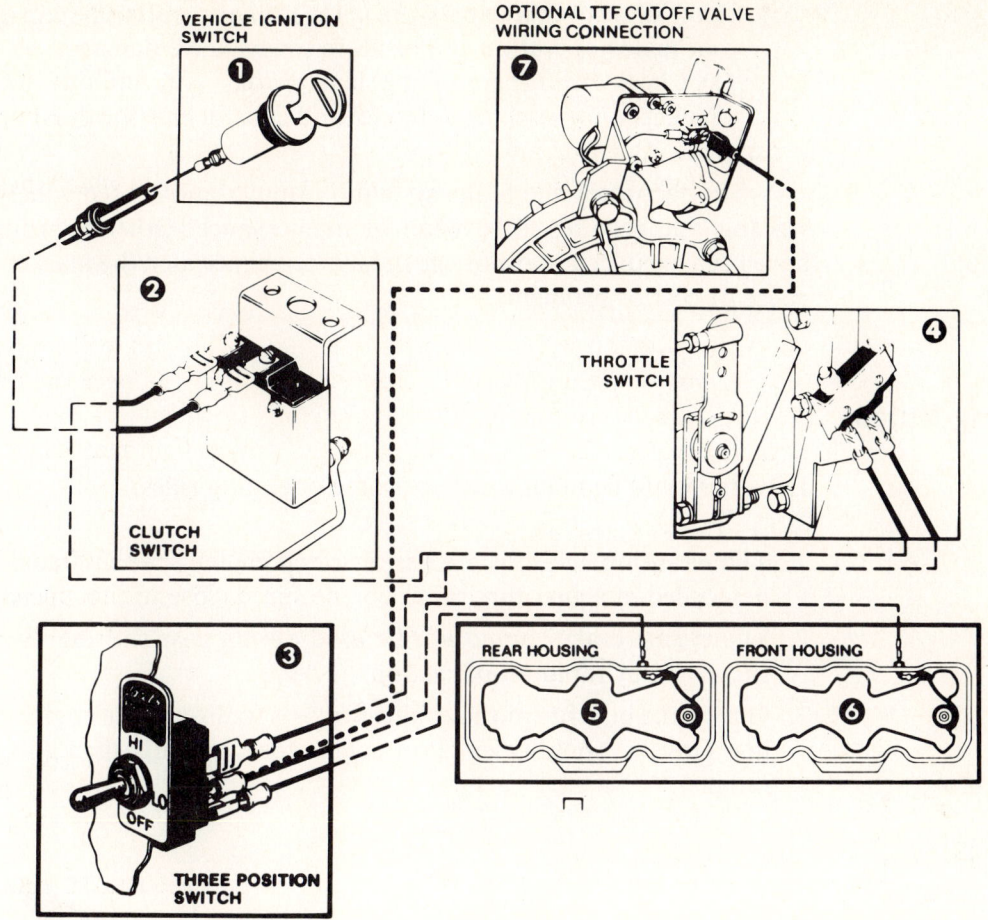

FIGURE 22-16(b). Jacobs Electrical System on Mack Engine. *(Courtesy of Jacobs Manufacturing Co.)*

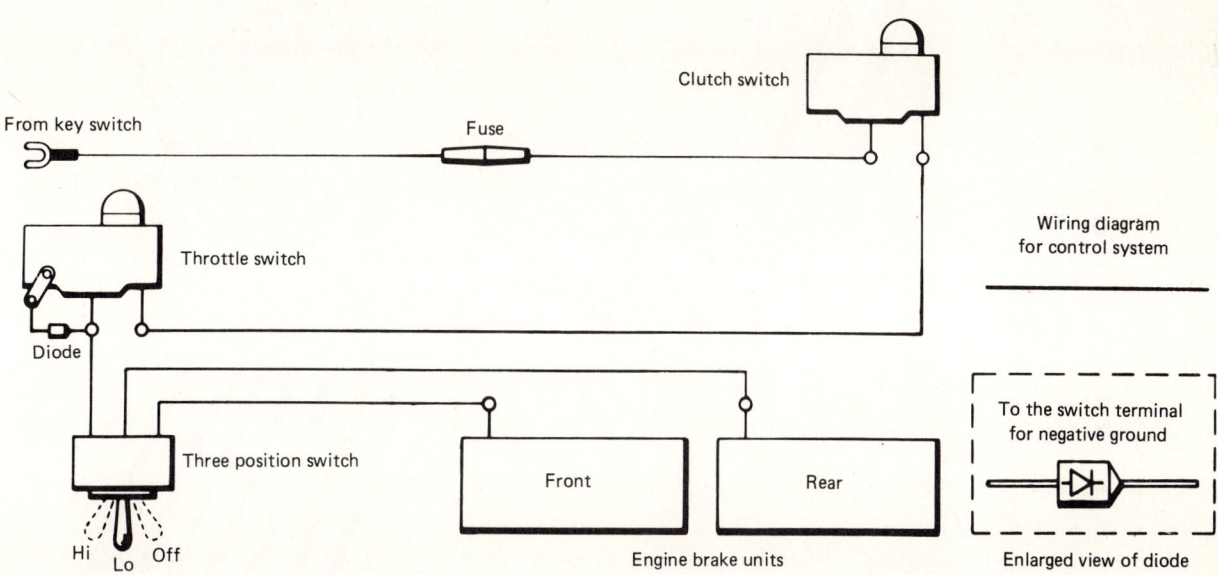

FIGURE 22-16(c). Jacobs Electrical System on Caterpillar Engine. *(Courtesy of Jacobs Manufacturing Co.)*

362 / Diesel Engine Service

b. Any failure to relieve pressure in the brake operating system during normal fueled operation can result in severe valve damage.
 (1) Check adjustment of the slave piston.
 (2) Check for sticking solenoid and control check valves.

The Jacobs engine brake system is popular and is used widely. It is serviceable and requires little repair or maintenance. Other retarding systems have been produced, most of which have not penetrated the market widely because of cost or problems.

Operating Rules

Drivers should be made aware of some operating rules:

1. The retarding action is greatest at rated engine speed. Grades should be descended in a gear that raises engine speed close to that speed.
2. The engine brake should *not* be used during gear shifting. Switch life is shortened by rapid on-off action.
3. One of the benefits of the engine brake is to discourage engine overspeeding. Drivers should never allow engine speeds greater than governed maximums.

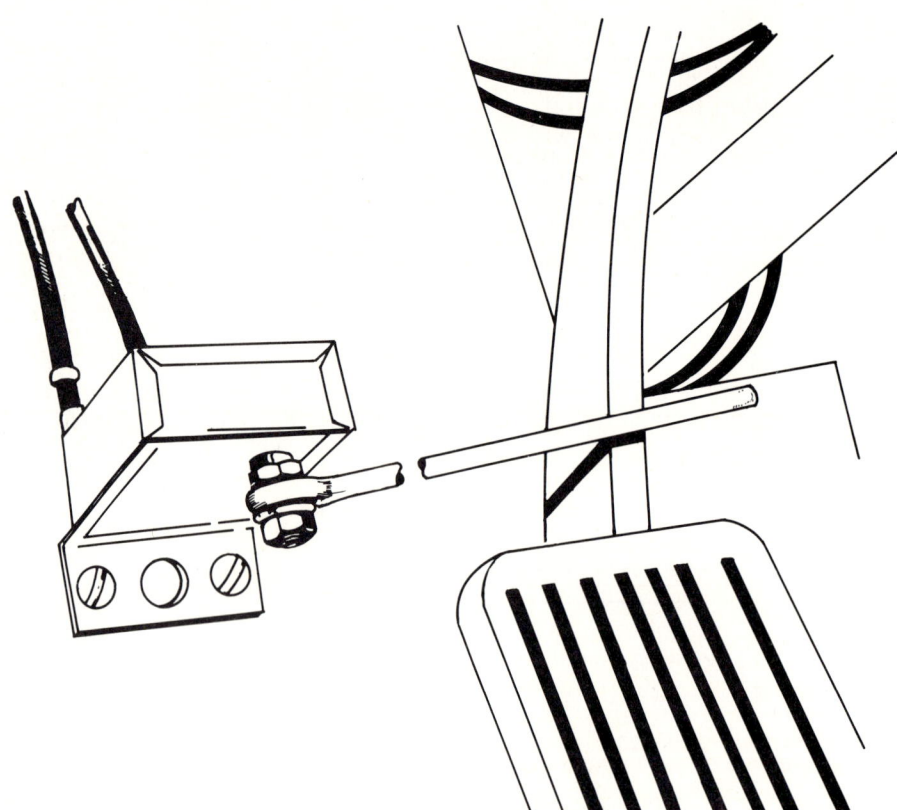

FIGURE 22-17. Clutch Switch—All Models. *(Courtesy of Jacobs Manufacturing Co.)*

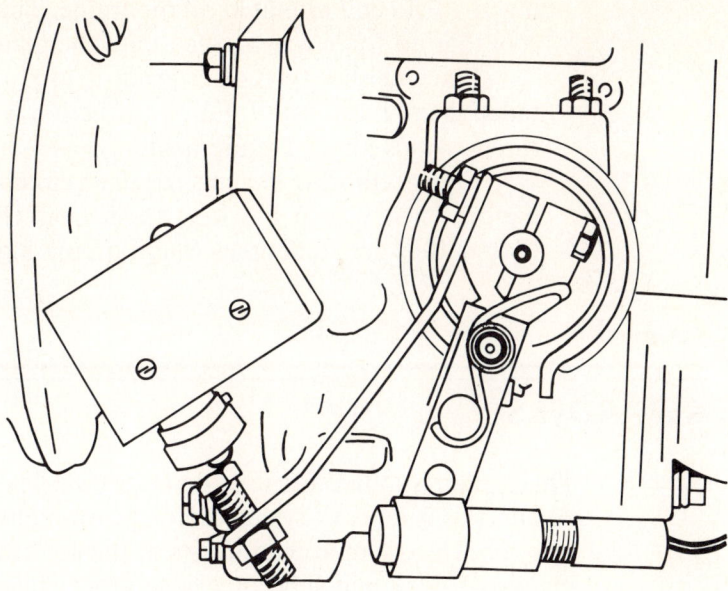

FIGURE 22-18. The Fuel Pump Switch. *(Courtesy of Jacobs Manufacturing Co.)*

OTHER ENGINE RETARDING SYSTEMS

1. The Exhaust Brake
   These units were used to close the exhaust pipe, causing back pressure in the engine. They were simply a butterfly valve in an iron section in the exhaust pipe, which was operated by an air cylinder when the driver turns on the control.

   A limiting screw prevented back pressure exceeding 45 psi. Special valve springs were used. Because the back pressure created by the exhaust brake was pulsulated by cylinder action, an intake suppressor must be used between the air cleaner and the engine to prevent damage to the air cleaner element. Total cost of this retarder resulted in lost popularity.

RETARDERS ON THE DRIVE LINE

Several such devices have been used, and some may still be encountered. They are of two types:

1. Hydraulic Retarders
   Hydraulic systems applied on the drive shaft have been used for some years. They have about the same form as a dynamometer using water to absorb energy.

2. Electric Retarders
   This type of retarder is used on some heavy equipment. It is considered too heavy for on-highway use. It is usually an armature on the driveshaft, turn-

364 / Diesel Engine Service

ing in a field coil mounted on the frame. This is like an electric motor, except that no force is generated until the manual switch is operated.

Since neither type of retarder is used on-highway, we have not detailed them in this book. Full information is available from the manufacturers of this kind of equipment.

Other retarding systems require a circulating system and a cooling radiator. Their weight has kept them from the on-highway truck market, though they are sometimes found on off-highway vehicles.

ELECTRIC RETARDERS

This system is sometimes used on large electrically driven off-highway trucks. Again, this is a heavy system not used on-highway.

The electric retarder consists of the use of the drive motors on large off-highway trucks, which become generators while the vehicle is going downhill. Excessive current thus generated may be dissipated by resistors or may be routed through the engine-driven generator to convert it to a motor, which is loaded by the engine when the throttle is closed in descending a grade. The action is usually automatic.

A description of this system and its service requirements is outside the scope of this manual.

23

ELECTRICAL TROUBLE-SHOOTING

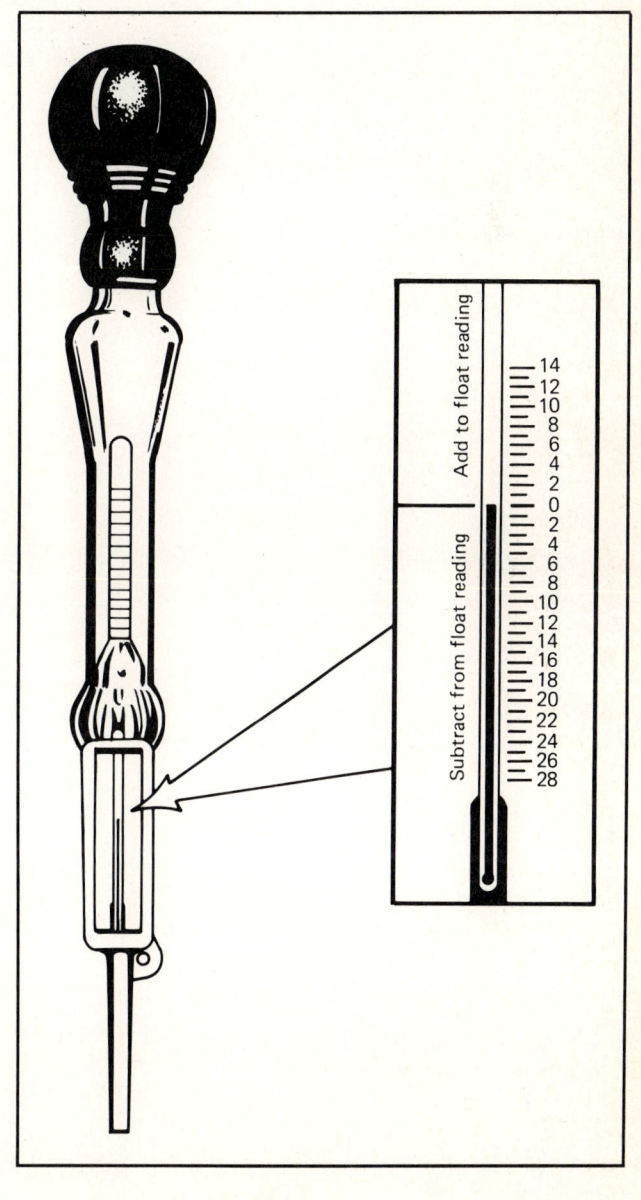

The electrical system for diesel engine operation is confined to the cranking, battery charging, and pre-heater system, if used. Because a diesel engine has no ignition system, no distributor nor spark plugs, etc., are needed.

EXPLANATION OF OHM'S LAW

Some understanding of Ohm's law, a basic formula, is needed by every mechanic. Electrical trouble-shooting involves measurements that can be made and interpreted only by one who has knowledge of the relationships between current, voltage, and resistance. Let us start by describing the terms used.

Intensity (I) is measured in amperes. Voltage (E) is pressure measured in volts. Resistance (R) is measured in ohms.

These factors are set into a formula called Ohm's law:

$$I = \frac{E}{R}$$

I (Current) equals E (voltage) divided by R (resistance)

This is a useful formula, which can be used to find any of the factors when two are known.

Look at the circuit in Figure 23-1. Battery (A), the source of current, is producing 12 volts. We can measure the resistor R-1 and find that it has a resistance of 12 ohms. If we divide 12 volts by 12 ohms, we get 1 ampere, which is the current that can flow in this circuit.

If we want to find the resistance, we can divide the voltage by the current, and get the resistance in ohms. We can find the voltage by multiplying the current by the resistance.

As an example, look at Figure 23-2. If we find the current to be 2 amperes, and the resistance to be 6 ohms, the voltage will be 2 times 6 = 12 volts.

Further, the voltage of any battery is the sum of the voltage of the cells. Resistances connected in series add to make the total resistance.

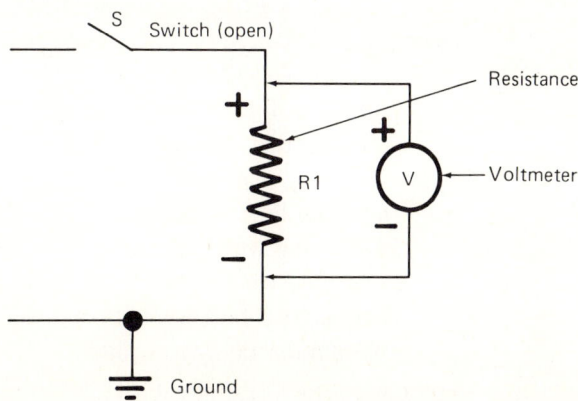

FIGURE 23-1. Simple Circuit. *(From a sketch by P. M. Uhl)*

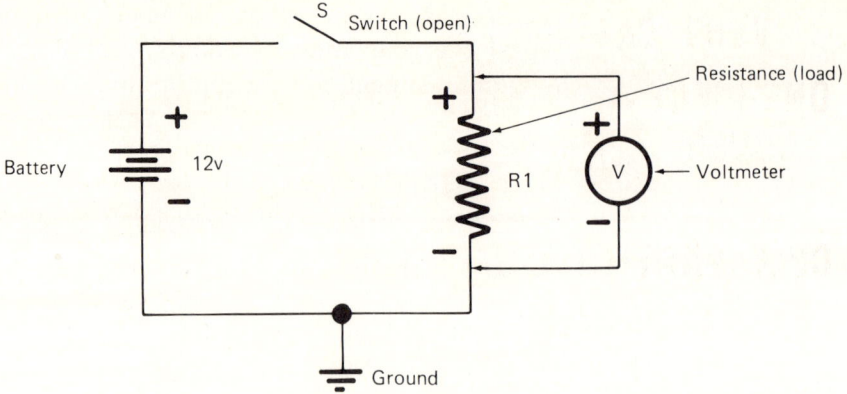

FIGURE 23-2. Simple Circuit—Battery in Series. *(From a sketch by P. M. Uhl)*

BATTERY DESCRIPTION AND SERVICE

The battery is the source of all electrical current for cranking, lights, and all other electrically operated apparatus. It is an energy storage device, using chemical action to produce current flow on demand.

The active elements in any battery are

1. The positive plates.
2. The negative plates
3. The electrolyte, a mixture of 36 percent sulfuric acid and distilled (pure) water. This proportion is by weight, rather than volume.

The plates are open grids, and the positive plate grids contain a compound known as lead peroxide. The negative plates contain a form of lead sponge.

In any one cell, all positive plates are joined by welding, as are all negative plates. The two sets are meshed together, separated by nonconducting thin plates of various materials.

When a set of positive and negative plates is contained in a case, we call it a *cell*. All lead-acid batteries can produce from 1.6 to 2.3 volts per cell, regardless of size. [Fig. 23-3]

Thus, a battery composed of 6 cells will produce about 12 volts.

The battery service is discussed first, because it is the current source.

1. The trouble-shooting section discussed some of the causes of a failure to crank. One was the tendency of a battery to self-discharge due to dirt and electrolyte on its top. This is easily removed and the acid neutralized by cleaning the battery with a brush and soda water. [Fig. 23-4] The cell covers must be left in place.

2. Clean the terminals with a brush made for this purpose. Soda water can be used on the cable ends.

Electrical Trouble-Shooting / 369

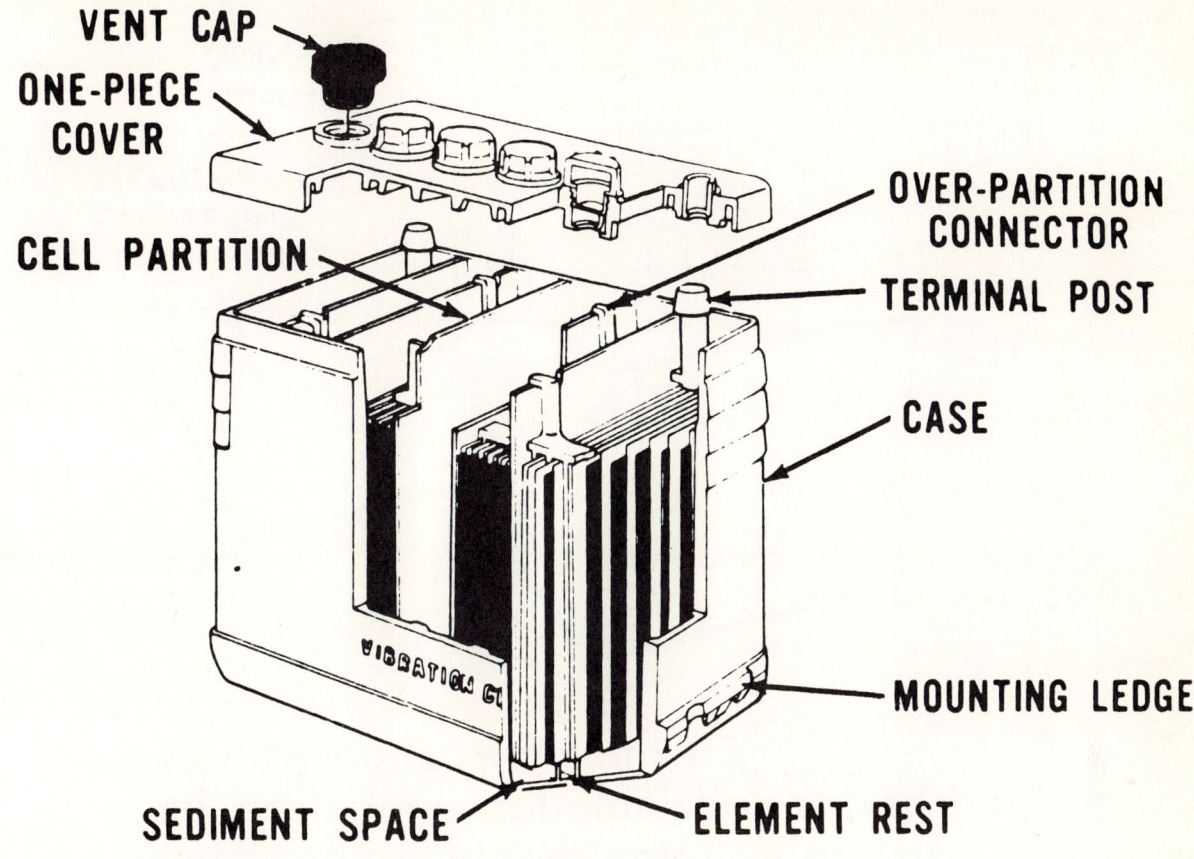

FIGURE 23-3. Battery Cutaway. *(Courtesy of General Motors Corp.)*

3. Use a good hydrometer to check the gravity (weight) of solution. Acid is heavier than water, and the gravity reading must be above 1.200 to be capable of cranking current. All readings must be even. [**Fig. 23-5**]

4. Use a battery charger to charge the battery. A slow charge of about 5 amperes for 12 hours will recharge a battery better than a quick charge of 35 amperes for 1 hour.

5. Remember that the addition of water to the cells will require a period of charging to mix it thoroughly with the rest of the electrolyte.

6. Never smoke around a battery. The gas that it gives off is hydrogen, a flammable gas. See that no flame or welding is near the battery. Avoid striking an arc at the terminals. Work safely.

7. Jumper cables must be connected carefully. Connect the negative terminals last, the positive cable first. Keep the cable ends from contacting during the hook-up.

Battery Charging

8. A 12-volt charger can charge several batteries if they are connected in parallel, with all positive terminals connected together and all negative sides connected together. The voltage of the circuit will be 12 volts. [**Fig. 23-6**]

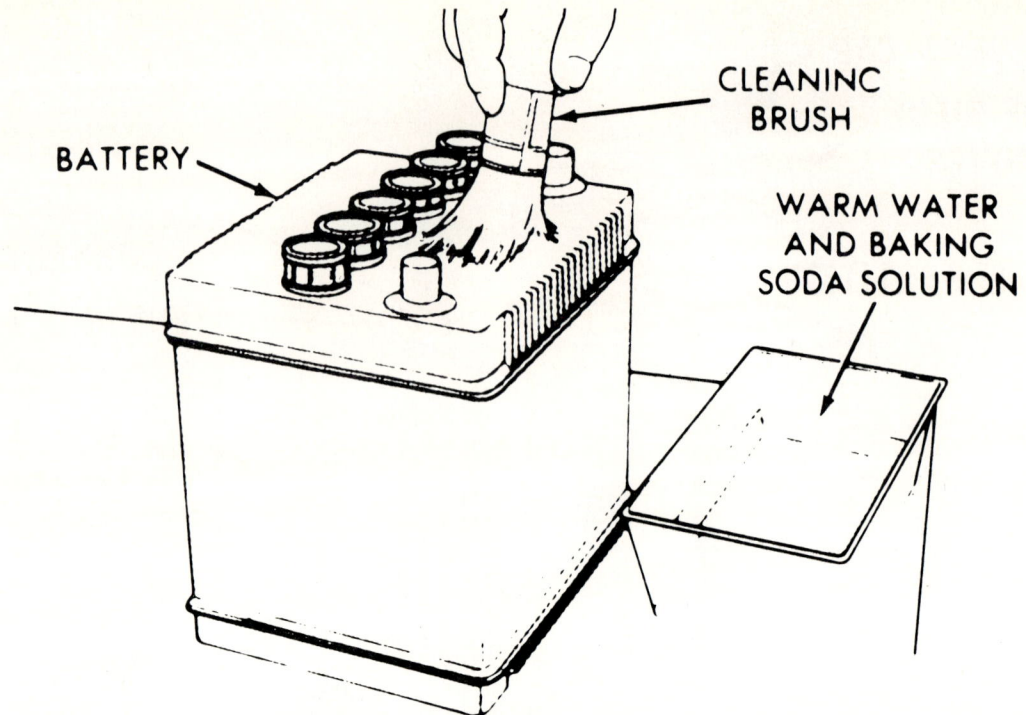

Cleaning the battery with a water and baking soda solution neutralizes battery electrolyte. Solution should not be allowed to enter the battery.

FIGURE 23-4. Clean Battery. *(Courtesy of Chrysler Corp.)*

9. If batteries must be charged in series, with positive to negative connections, the charger voltage must be the sum of the voltage of all batteries. Thus, two 12-volt batteries require a 24-volt charger. [Fig. 23-7]

ELECTRICAL SYSTEM SERVICE

Two basic faults occur in electrial circuits. All service problems fit one of these two faults:

1. the open circuit—broken wires, loose connections, etc.
2. short circuits—worn insulation, wires that dangle into contact with other parts, and abraded clips, etc.

With open circuits, the driven unit does not run. The best way to find the open, other than visual checking, is to connect the part directly to the battery

Electrical Trouble-Shooting / 371

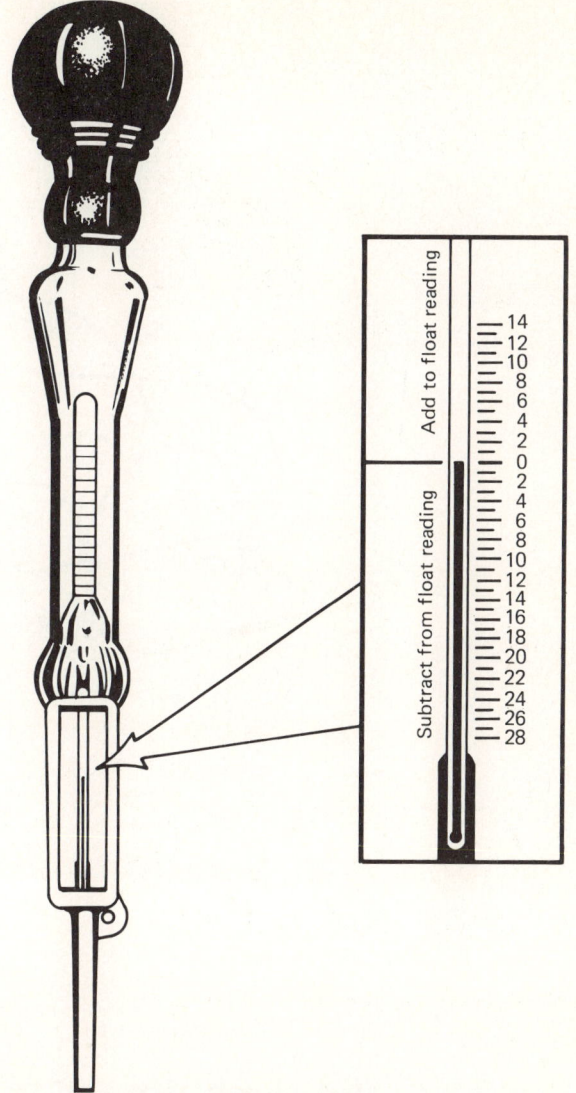

FIGURE 23-5. Hydrometer. *(Courtesy of Ford Motor Co.)*

positive. If the unit now runs, one can be sure that it is OK. The trouble will be found in the wires, switches, controls, or other units in the circuit. The ground must be in good shape. It is just as important as the wired side of the circuit.

Short circuits are always hot. Any short will cause arcing and sparks. Fires may be started by shorts close to flammable material. Most shorts can be found by looking for signs of heat. Discoloration, burned insulation, etc., enable one to discover the location.

It is possible for batteries to "leak" current over the surface when it is damp with the battery liquid. The best way to eliminate such leakage is to clean the top of the battery with soda water and use terminal protecters under the cable ends.

Faults in the battery can be found by doing a hydrometer test of each cell, or using a voltmeter to read battery voltage. A good 12-volt battery will read from 12 to 13 volts. A lower reading indicates deterioration or low charge. This reading should be taken with one terminal cable disconnected.

372 / Diesel Engine Service

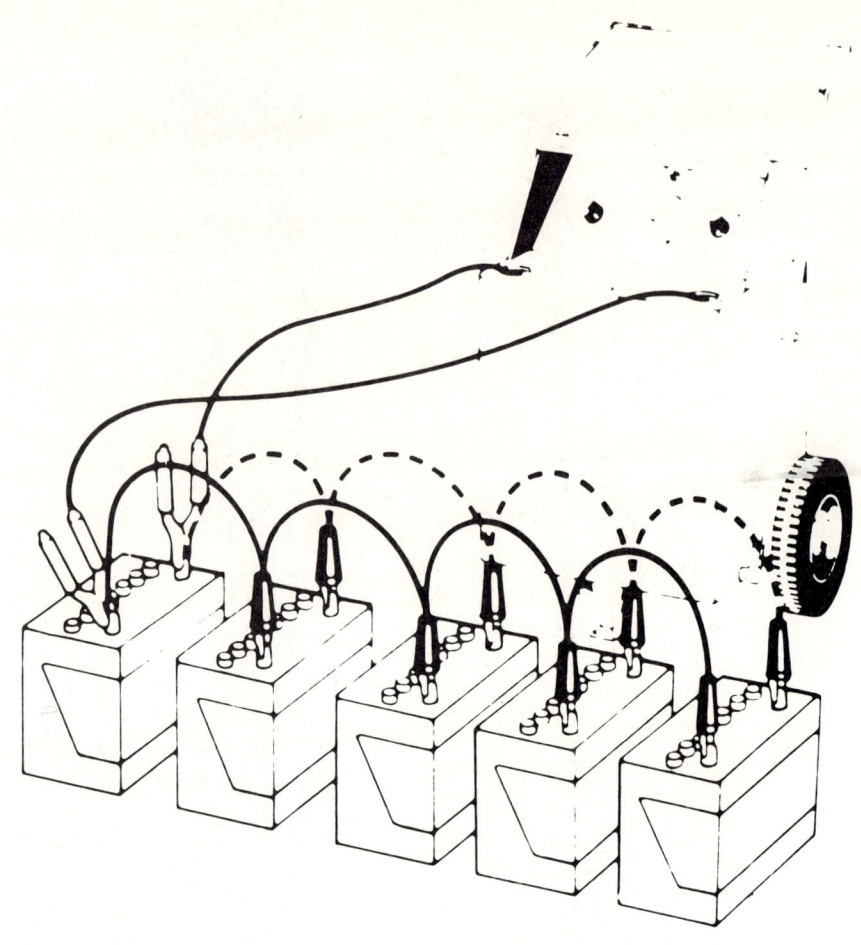

MULTIPLE HOOK-UP

Charging a number of batteries at the same time. Batteries are connected in parallel so combined battery voltage remains at 12 volts.

FIGURE 23-6. Parallel Charging Circuit. *(Courtesy of Ford Motor Co.)*

The voltage will be influenced by the resistance of the meter versus the resistance of the connected circuit. Some difference will show. A voltmeter is also useful for checking circuits, to show when some fault occurs in the wiring to the unit.

The ohmeter is used to check resistance, whether it is in a switch, due to burned contacts or a loose connection. The appearance of any loose connection or contact is granular when perfect contact is not made. Such an appearance on battery terminals will cause poor cranking performance and should be checked first, before one thinks that the cranking motor is bad.

Short-circuited elements, such as solenoids, are checked by connecting the input terminal with the case, using a voltmeter to make the connection.

Electrical Trouble-Shooting / 373

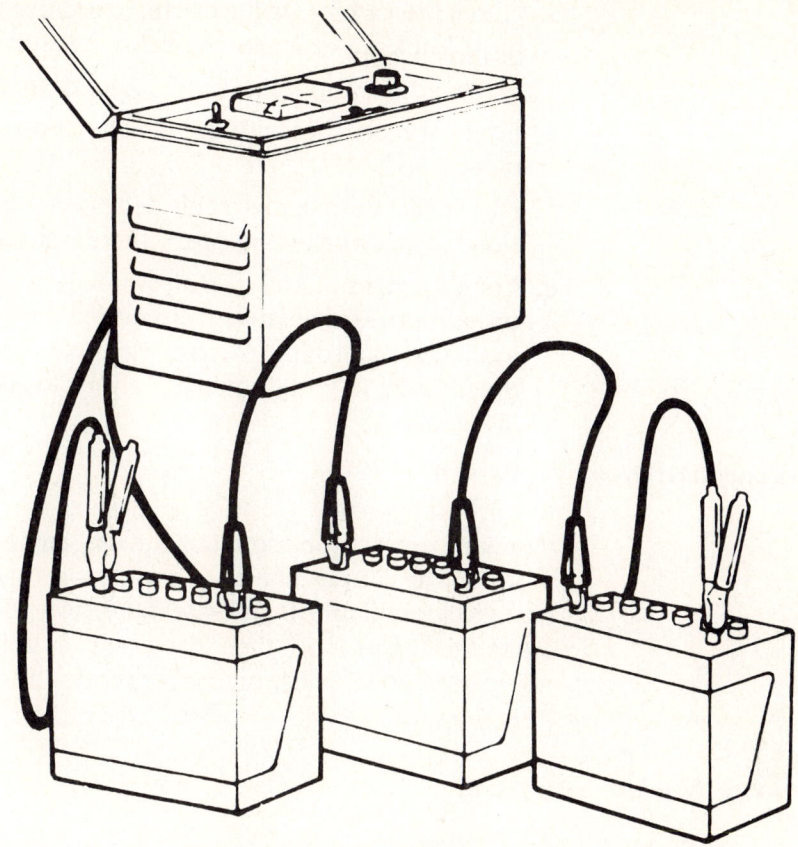

CONSTANT-CURRENT SLOW CHARGING

Many types of battery slow chargers are available. Always follow equipment manufacturer's instructions for hookup and charging rates. These batteries are connected in series.

FIGURE 23-7. Series Charging Circuit. *(Courtesy of Ford Motor Co.)*

Any current flow indicates a shorted coil. Normally open switches can be tested in the same manner by connecting a voltmeter across the open switch. Less than full source voltage means a shorted switch. Usually it will show heat signs.

Poor Cranking Performance

Such a complaint requires thoughtful checking to find the true cause.

1. After a trial crank, feel the battery cables. If they are warm, they have high resistance. Look at the ends and consider the cable size. Is it adequate?

374 / Diesel Engine Service

2. Check the battery. Is the electrolyte above the plates? How old is the battery? Look for cells reading below the others.
3. Is the cranking motor warm? Failures are rare but do happen.
4. Check the switches with an ohmeter. Poor contacts may cause high resistance.
5. Disconnect the negative cable and check for sparks, which indicate current flow. If all switches are open, there is a short somewhere.
6. Look for signs of arcing in all switches in the cranking circuit. Faults show up as burned insulation, discolored switch housings, etc. By now the mechanic has found the trouble. Batteries may have internal shorts or other faults. A dirty battery is always suspect.

Cranking Motors

Cranking motors for commercial diesel engines are heavy-duty series wound motors. The current that they draw is about all that the battery can furnish; extended cranking will generate much heat. In general, one should try not to crank the engine more than 30 seconds at a time. When a diesel engine does not start in 30 seconds, one needs to look for other reasons.

Cranking Motor Solenoid

A *solenoid* is an electromagnet with a movable core or *armature*. When a current is passed through the coil windings, the core is drawn endwise to move a load, such as the cranking motor pinion gear. The energy required to move the core is greater than the energy needed to hold it in its engaged position.

The cranking motor solenoid is used to engage the motor drive gear or pinion with the teeth of the flywheel ring gear. As this is done, a heavy contact washer closes the main switch, energizing the motor. [**Fig. 23-8**]

It is necessary that this system is used to control cranking motor current, because the amperage is too high to handle with a manual starting switch. The manual switch controls the circuit through the coils of the solenoid.

The main terminals of the solenoid are connected between the battery and cranking motor. [**Fig. 23-8**]

Cranking Circuit Connections

It is important to have all fasteners and connections tight before starting. As pointed out, one should never try to use a cable end as a switch. Such action will always burn the cable end and contact.

Cranking Motor Drives

There are at least two ways of engaging the cranking motor pinion with the flywheel ring gear. The common Bendix drive uses a weighted pinion that travels along the threaded shaft as the shaft turns when the motor is energized.

Electrical Trouble-Shooting / 375

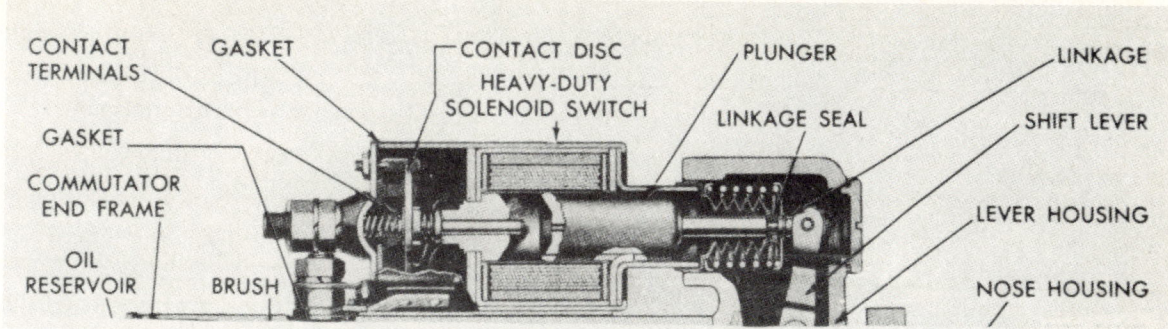

FIGURE 23-8(a). Cranking Motor Solenoid. *(Courtesy of General Motors Corp.)*

As long as the motor is cranking the engine, the pinion stays engaged. Where the engine fires, the ring gear turns faster than the pinion, and the pinion is thrown back along the screw threads, out of engagement.

This type of drive does not use a solenoid engagement. It is used on lighter engines and is common.

The need for greater cranking capacity and more powerful motors led to

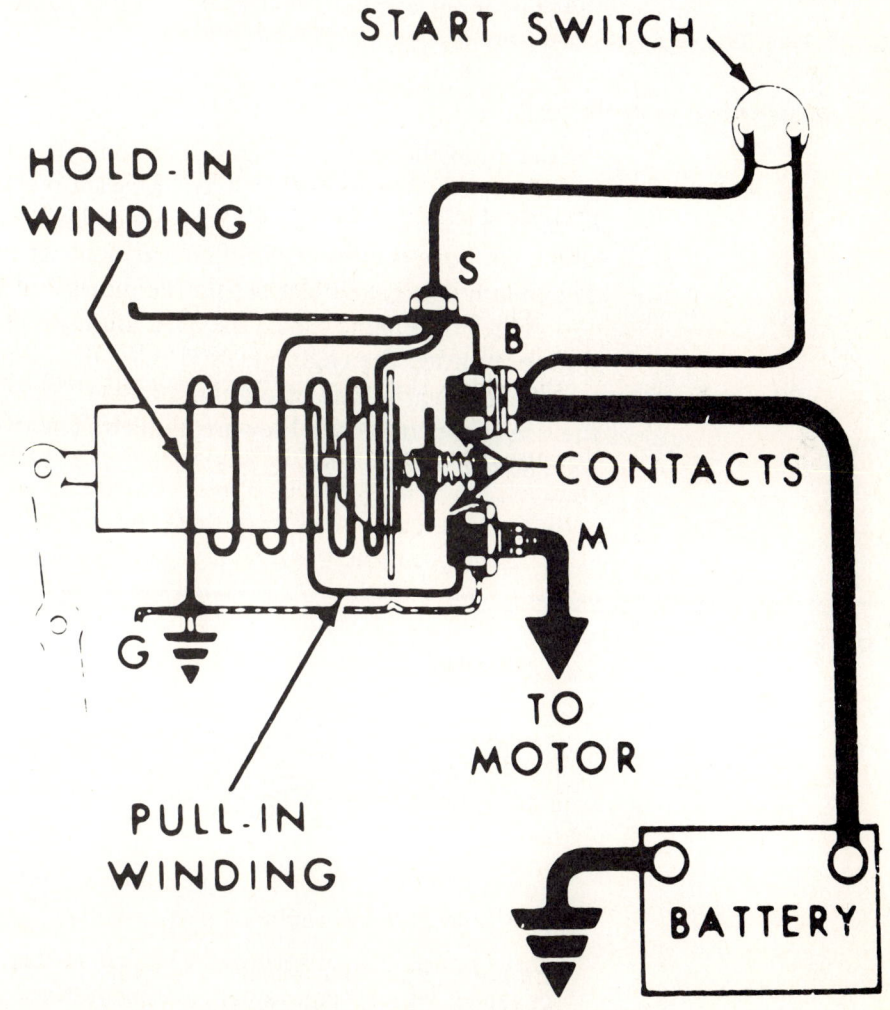

FIGURE 23-8(b). Solenoid Circuit. *(Courtesy of General Motors Corp.)*

376 / Diesel Engine Service

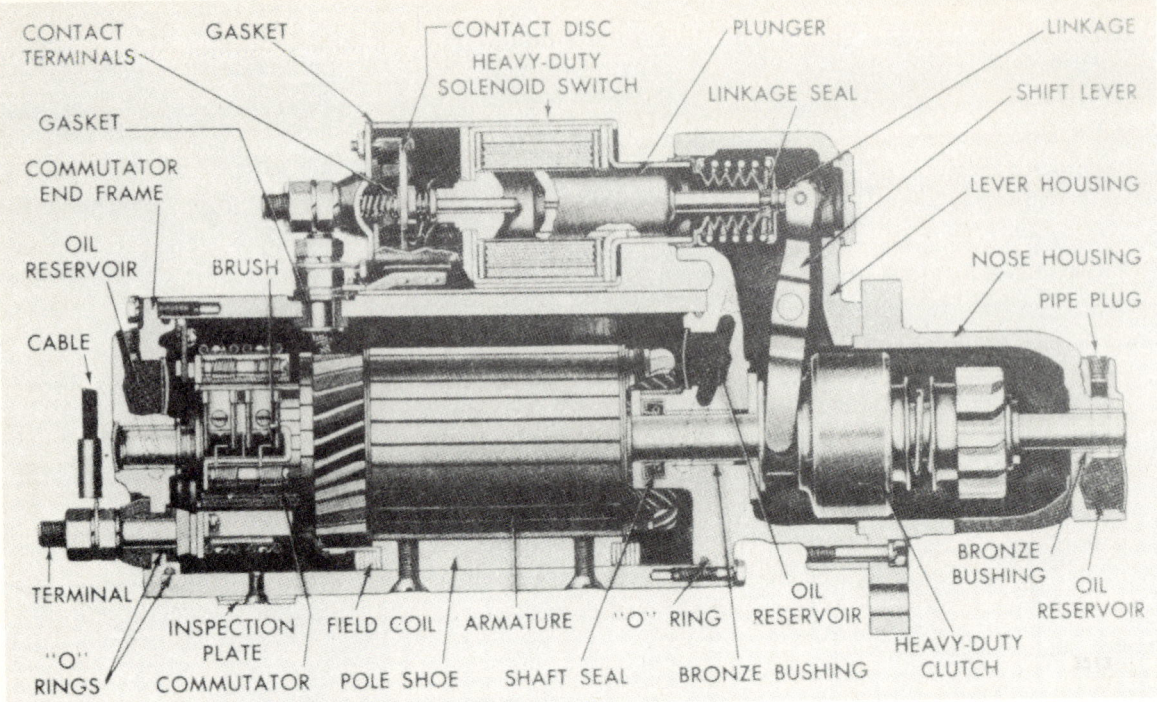

FIGURE 23-9. Cranking Motor and Drive. *(Courtesy of General Motors Corp.)*

the design of the solenoid-engaged cranking pinion drive. Figure 23-9 shows such a system. The pinion is moved along the motor shaft on splines. The shift lever on the solenoid plunger engages a collar, which acts on the overrunning clutch to move the pinion into engagement when the solenoid is energized. The pinion is engaged just before the motor contacts close.

When the engine starts, the overrunning clutch permits free pinion rotation, preventing the engine from driving the motor.

Several variations of cranking motor drives are current, depending on the size engine on which they are applied. The two types described dominate the market.

CRANKING MOTOR SERVICE

Service to cranking motors is often required of the field mechanic. Operators can cause damage by improper use of starting switches, and the heavy current draw can quickly discharge batteries.

The points to check follow:

1. Solenoid contact points. These points carry full cranking current and need to be checked and replaced occasionally.
2. Drive pinion return springs. A broken spring will allow the pinion to stay engaged, with severe damage to the motor.

Electrical Trouble-Shooting / 377

3. Motor brushes and commutator. Brushes must never be allowed to wear short enough to cause arcing. Service at engine overhaul periods is a good preventive of expensive wear.

4. General inspection during tuneup can reveal loose fasteners, frayed cable insulation, and poor ground connections. All of these will lead to trouble if neglected.

5. Finally, check the battery condition. Cranking motors require adequate voltage; a "chattering" starter means low battery current.

USE OF METERS

Three common test meters are used to check electrical circuits. Besides the continuity checks that can be made with a simple test light, one needs to determine such factors as battery voltage, the current in amperes that a load draws, and the resistance of various units.

All of these values help to find the cause of a problem. When a value is above or below the designed value, the mechanic can point to the faulty unit for repair or replacement.

1. Voltmeters. These instruments read the system pressure, or voltage. They are always connected across the circuit, in parallel, to read the voltage applied. [**Fig. 23-10**] Resistors in the meter allow readings of circuits of various voltages.

2. Ammeters. These instruments are used to find the volume of current flow, or amperage. One connects the meter in series, using a resistor in the line to direct the current through the meter. Resistors in the meter allow readings of various circuits.

3. Ohmmeters. Resistance of a circuit is measured in ohms. While both voltmeters and ammeters are used on live circuits, the ohmmeter must be used to measure the resistance of a part of the circuit, and the meter is equipped with a battery. [**Fig. 23-11**] The unit to be measured is disconnected from its current supply.

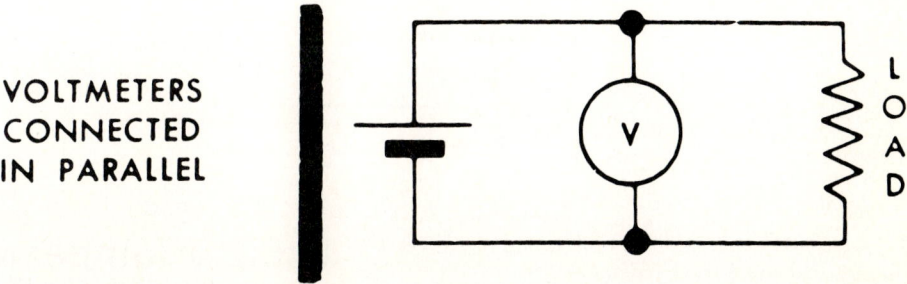

VOLTMETERS CONNECTED IN PARALLEL

FIGURE 23-10. Voltmeter in Circuit. *(Courtesy of General Motors Corp.)*

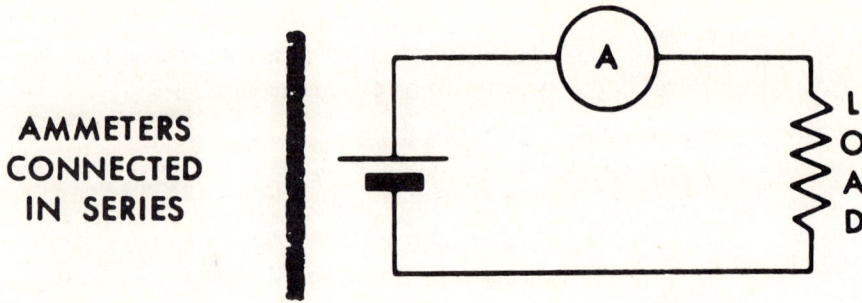

FIGURE 23-11. Ammeter in Circuit. *(Courtesy of General Motors Corp.)*

4. Continuity tests for various circuits are simple, using a test light and wires or probes. The test of a switch is shown in Figure 23-12. A wire is used to connect the switch terminals, bypassing the switch. If the circuit is closed by the jumper, the test light will glow. This would indicate a faulty switch. Any unit in the circuit can be tested in the same way, by disconnecting the unit from its ground connection and connecting the test light in place of the ground. If the light glows, the circuit through the unit is complete.
Note: A battery-powered test light can be used to test for broken wires or loose connections by isolating the section, and connecting the test light to each end.

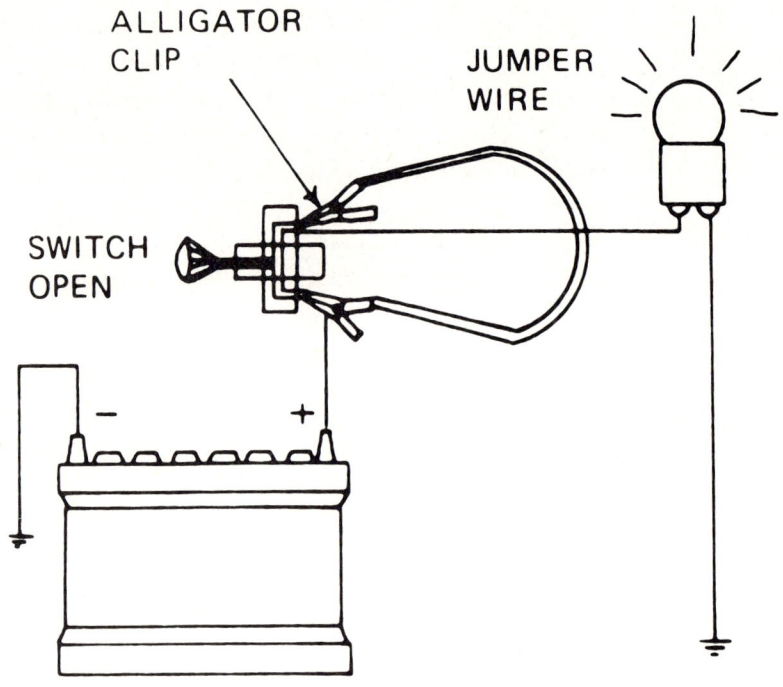

Using a jumper wire to bypass a switch that is suspected of being defective.

FIGURE 23-12(a). Switch Test. *(Courtesy of Chrysler Corp.)*

Electrical Trouble-Shooting / 379

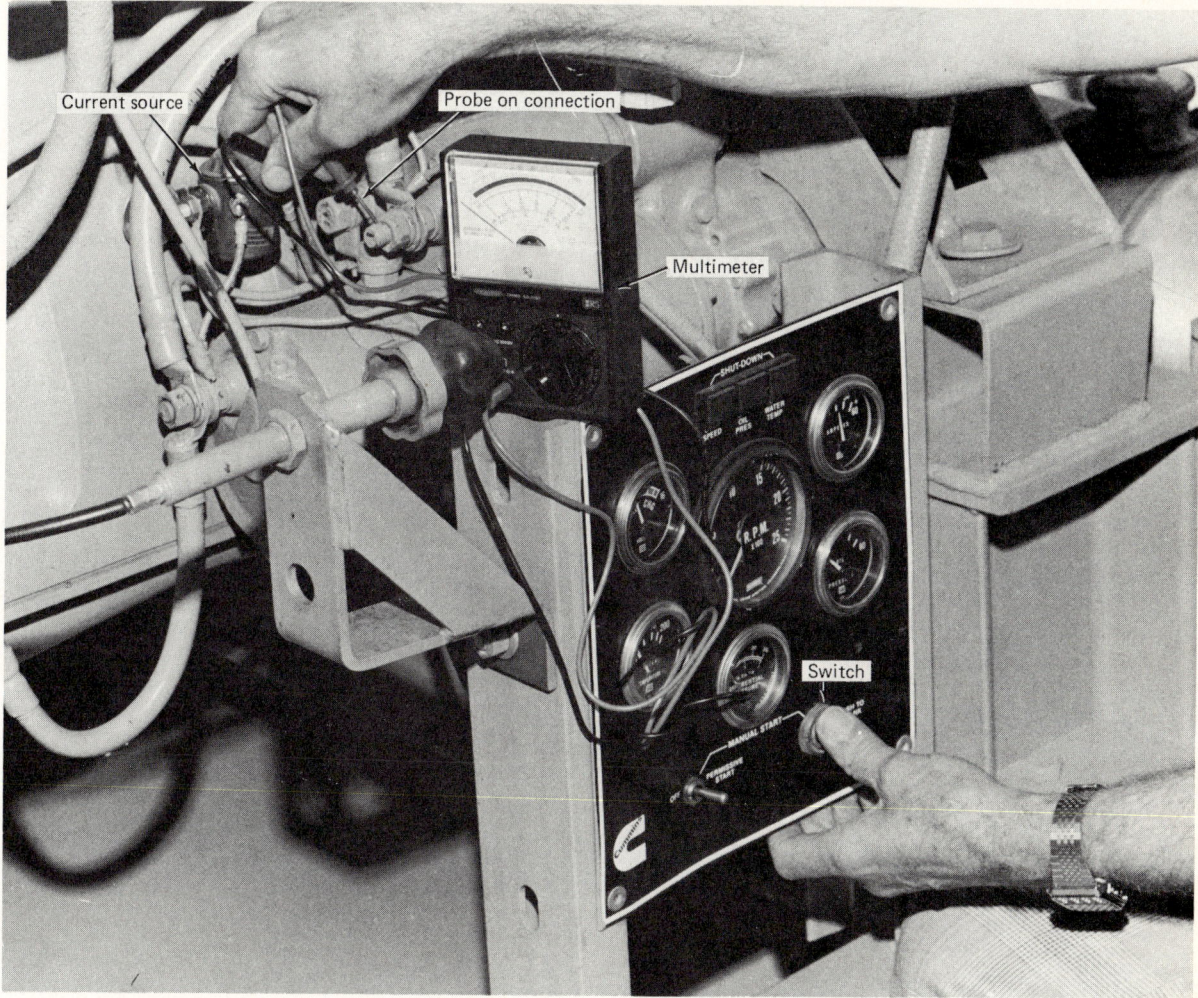

FIGURE 23-12(b). Switch Test with Multimeter. *(Courtesy of Cummins Arizona Diesel, Inc.)*

GENERAL ANALYSIS OF ELECTRICAL PROBLEMS

The following rules apply:

1. Lower than normal resistance can indicate a shorted circuit. Heat is nearly always generated.
2. Higher than normal resistance indicates open circuits, broken wires, loose connections, bad ground connections, or burned-out bulbs. The unit does not function.
3. One must decide whether the circuit is normally open or closed. Failure to function indicates some deviation from the normal condition.

AIR CRANKING SYSTEMS

Several engine builders use cranking systems powered by compressed air. These systems give high cranking speeds, making starting easier. [**Fig. 23-13**]

Their disadvantage is the limited cranking time that they provide. Thus, it is necessary to see that the engine is in good condition to start, with the fuel system primed and ready to inject fuel. Engine maintenance becomes imperative when using air to power the cranking motor. Any starting aids must be in operating condition.

Several improvements have been made to reduce noise and improve service life.

1. An exhaust air muffler is used to reduce the severity of noise from the starter.
2. An injection of lubricant into the air stream serves to lubricate the motor parts.
3. The problem of violent engagement of the motor pinion has been reduced by using an air-actuated engagement, in which an air cylinder replaces the

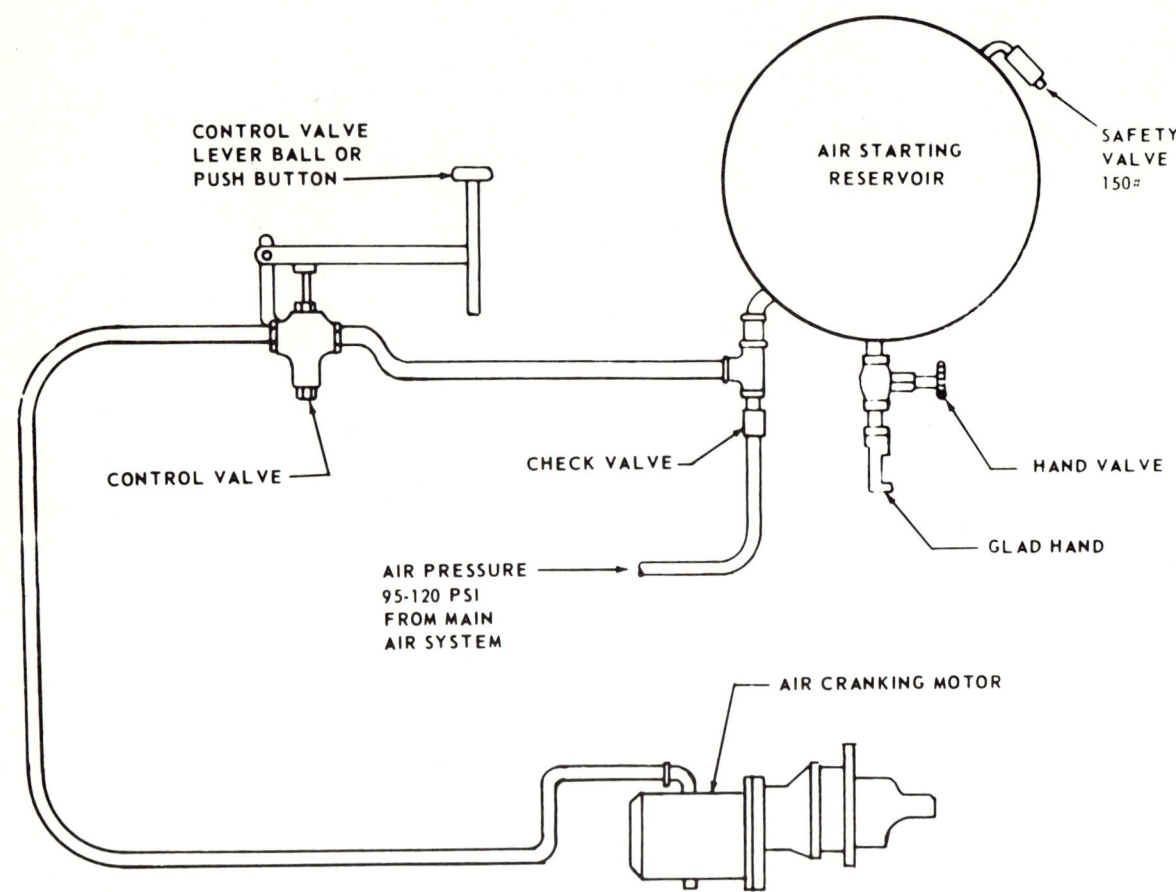

FIGURE 23-13(a). Air Cranking System. *(Courtesy of Ingersoll Rand Corp.)*

Electrical Trouble-Shooting / 381

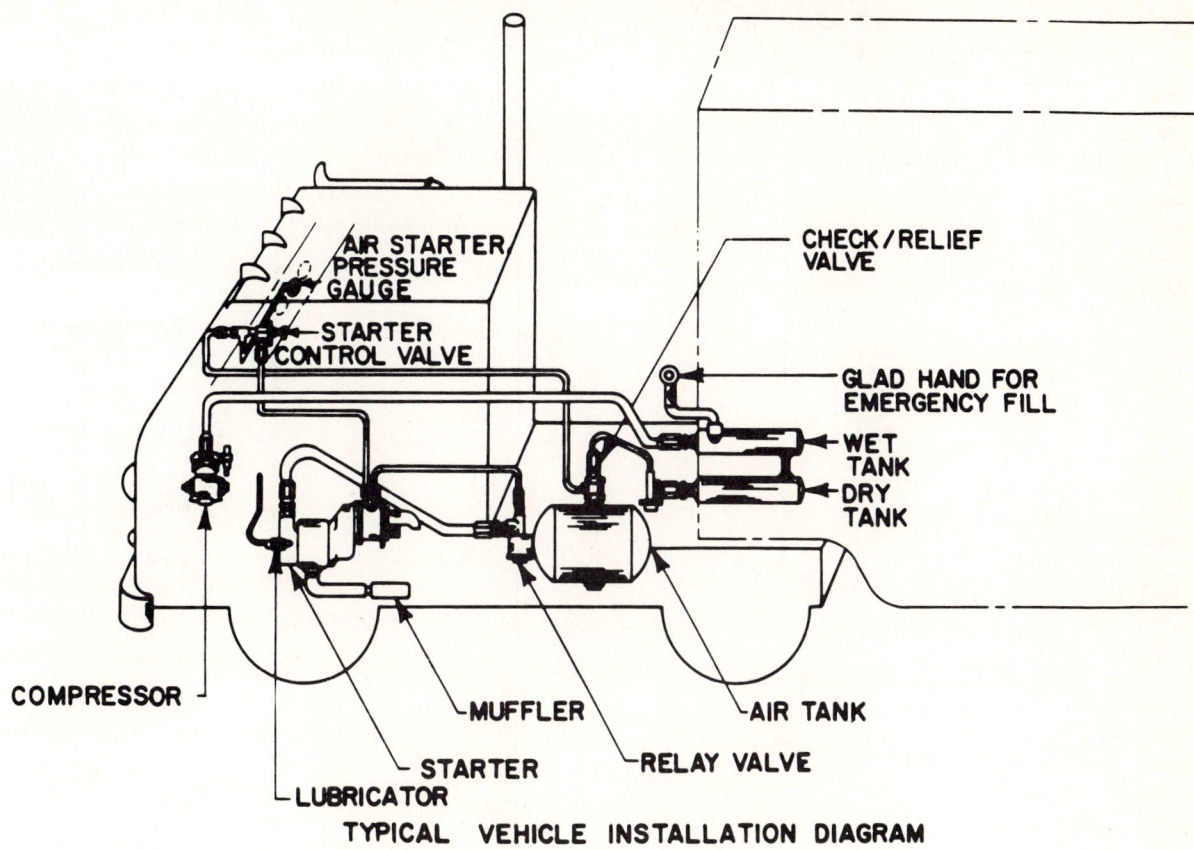

FIGURE 23-13(b). Air Cranking System on Vehicle. *(Courtesy of Ingersoll Rand Corp.)*

solenoid engaging system. This is operated by a manual valve to admit compressed air to the air cylinder to engage the pinion with the flywheel ring gear. The end of the air cylinder plunger opens a relay valve to admit air in large volume to the motor. Thus, a violent engagement is prevented, and pinion and ring gear life is extended.

Servicing the Air Cranking System

Air pressure is supplied to the reservoir through a check valve, from the service air system. [**Fig. 23-13**]

A standard trailer connection allows starting air to be furnished from a separate source. This permits air to be supplied from shop air or another vehicle in a field situation.

A safety valve on the reservoir is set to open at a safe limit.

The only normal service required is to prevent air leaks and see that all units function properly. Failures are rare. These air cranking systems are often used for engine test cells, because no battery maintenance is needed.

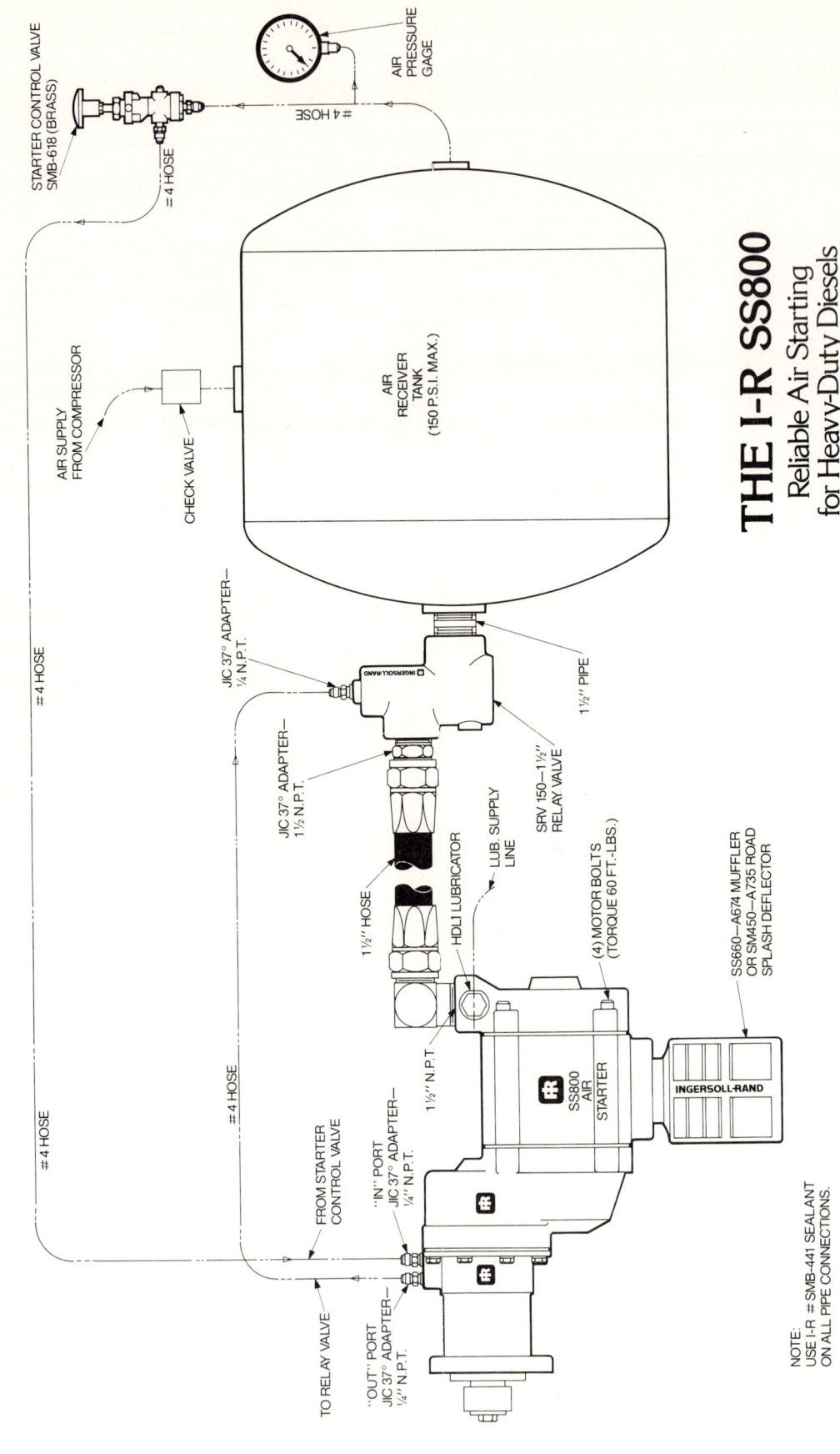

FIGURE 23-13(c). Air Cranking System Schematic. *(Courtesy of Ingersoll Rand Corp.)*

Electrical Trouble-Shooting / 383

BATTERY CHARGING ON VEHICLES

Description of Function

All vehicles that depend on batteries for cranking and other electrically operated units have an engine-driven system to provide current for battery charging and other loads while the engine is running. Modern vehicles use alternators for this purpose. We describe the function of the components of the charging system and give some of the service and trouble-shooting information that applies.

To "charge" the battery, one must reverse the current flow through it, which reverses the chemical process. This requires a charging system that can produce a voltage slightly higher than battery voltage. In practice a 12-volt system gives a charging voltage of 13.5 to 14.5 volts.

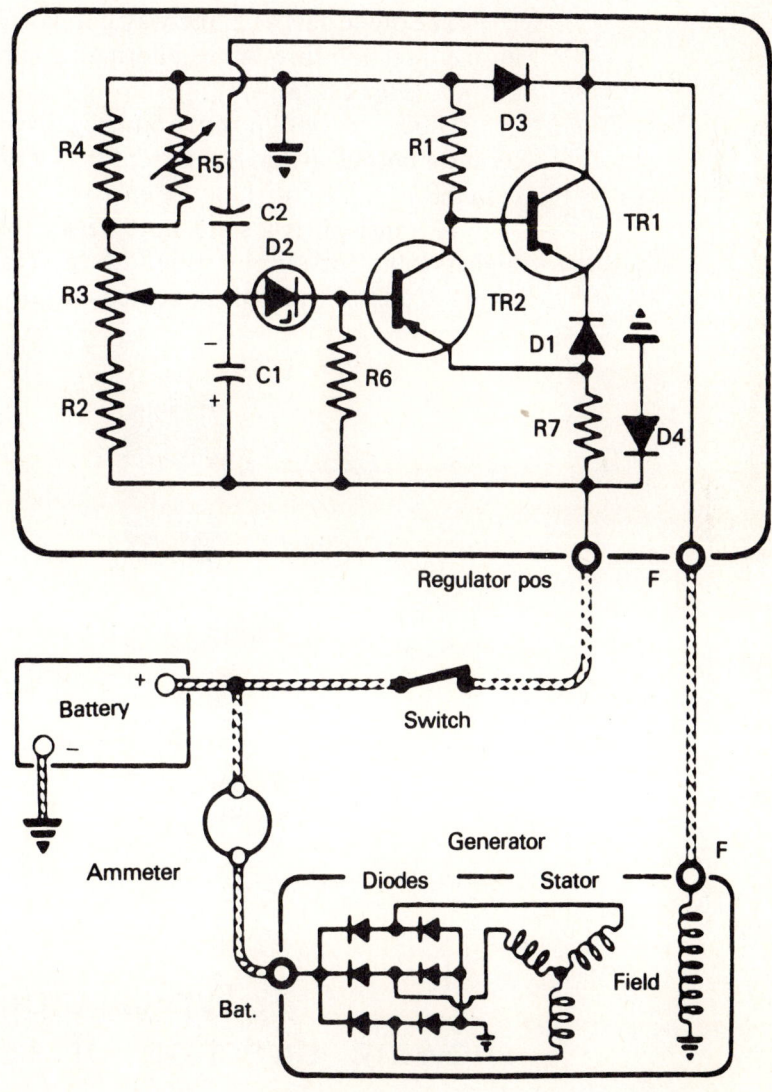

FIGURE 23-14. Charging Circuit on Vehicle. *(Courtesy of General Motors Corp.)*

384 / Diesel Engine Service

The battery and electrical systems of all vehicles are operated by direct current, which flows in one direction only. [Fig. 23-14]

Alternator

The charging generator produces direct current; the modern alternator produces current, which changes direction many times a second. This alternating current must be changed to direct current, and much of the apparatus in the charging alternator is used for that purpose. [Fig. 23-14]

Diodes and Transistors

These units act to change the alternator current to direct current (diode) and to increase current flow (transistor).

The diode acts as a one-way check valve. Because current can flow only in one direction through it, alternating current input becomes direct current output. [Fig. 23-15]

Diodes are both positive and negative in polarity. They are used in the alternator output so that the current flow is always of the proper polarity for the system.

A transistor is a solid-state device that can increase input current to a higher value for output. [Fig. 23-16] Transistors are often used to act as switches and to transmit increased current to the alternator fields.

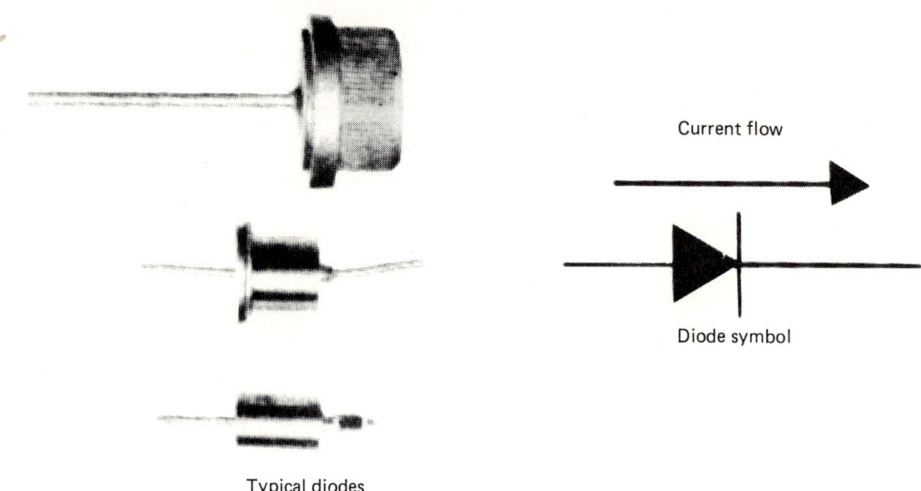

Typical diodes and diode symbol showing direction of current.

FIGURE 23-15. Diode and Symbol. *(Courtesy of General Motors Corp.)*

Electrical Trouble-Shooting / 385

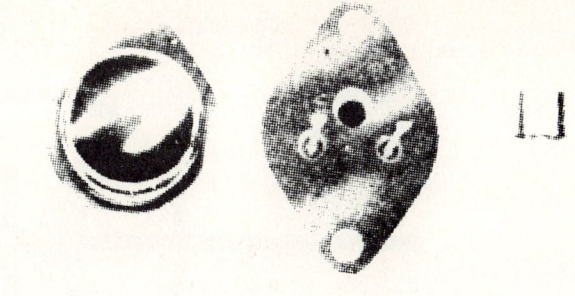

Typical transistors

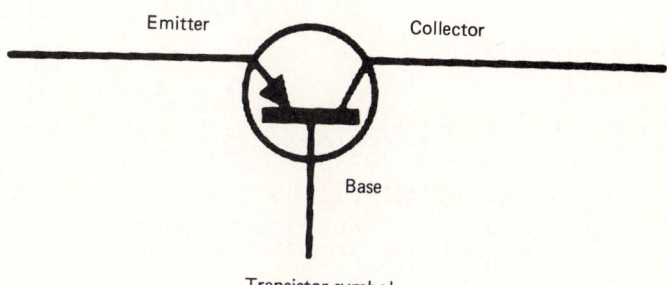

Transistor symbol

Typical transistor and transistor symbols. A small base current from the emitter to the base turns the transistor on, Allowing a larger current from emitter to collector.

FIGURE 23-16. Transistor and Symbol. *(Courtesy of General Motors Corp.)*

They are turned on by a control circuit that senses output voltage and turned off when that voltage reaches the limit for which the unit is designed. This action occurs many times a second.

Both of these solid-state devices are used to change alternating current into direct current and to limit the alternator output voltage. They are represented by symbols in wiring diagrams. [**Fig. 23-16 and 23-17**]

A Zener diode will transmit current in the opposite direction when a specified voltage is applied. This is useful in the control system.

VOLTAGE REGULATOR

An electronic voltage regulator circuit is shown and the action of these units is described in **Figure 23-18.**

There are no moving parts in electronic regulators. The solid-state devices are not repairable but can be replaced when tests reveal failure.

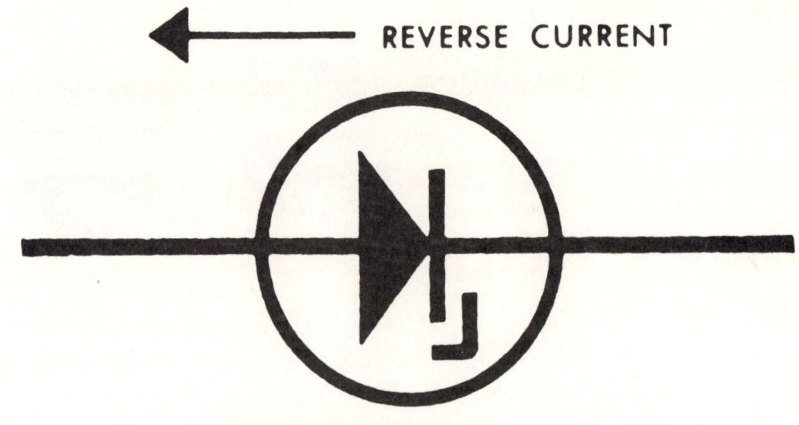

The Zener diode will allow current in the reverse direction when specified voltage is imposed.

FIGURE 23-17. Zener Diode and Symbol. *(Courtesy of General Motors Corp.)*

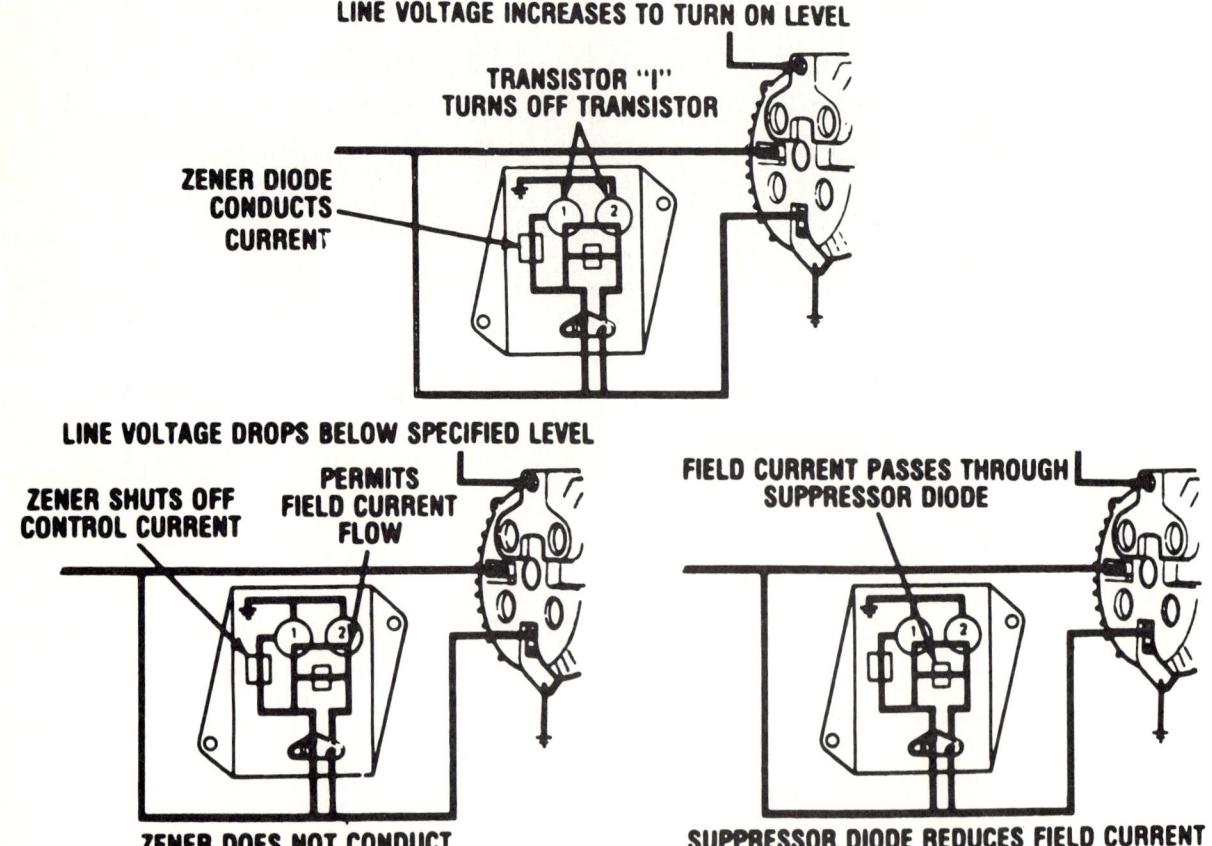

FIGURE 23-18. Electronic Voltage Regulator Circuit. *(Courtesy of General Motors Corp.)*

Heat Sink

Diodes and transistors are mounted on metal members called *heat sinks*. They heat in operation, and the additional surface of the mount increases the area from which heat can be radiated.

Capacitors

The term *capacitor* refers to a unit that can take a current charge while energized, then discharge in the opposite direction when the circuit is opened. This action is useful to prevent arcing at contact points. An example is the *condenser* in distributors of spark-ignited engines. [**Fig. 23-19**]

SERVICE ON THE CHARGING SYSTEM

It is likely that a mechanic will be called on to correct problems of failure to keep batteries charged more than most other problems. A good multimeter and a continuity tester or test light are essential tools. The multimeter includes a voltmeter, ammeter, and ohmmeter.

Electrical service always starts with a battery check. There is no use in testing the charging system if the battery is ready for replacement. [**Fig. 23-20**]

The following items must be observed:

1. Be sure the battery is connected correctly. Most systems are negative ground, but some are positive ground. Check before you connect.
2. Disconnect the battery before installing an exchange alternator.
3. Never ground any terminals on the alternator.

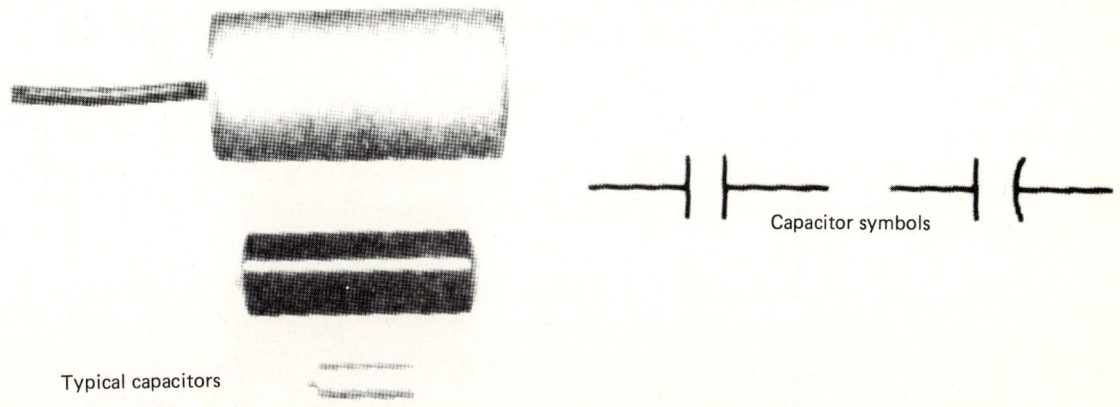

FIGURE 23-19. Capacitor Symbol. *(Courtesy of General Motors Corp.)*

388 / Diesel Engine Service

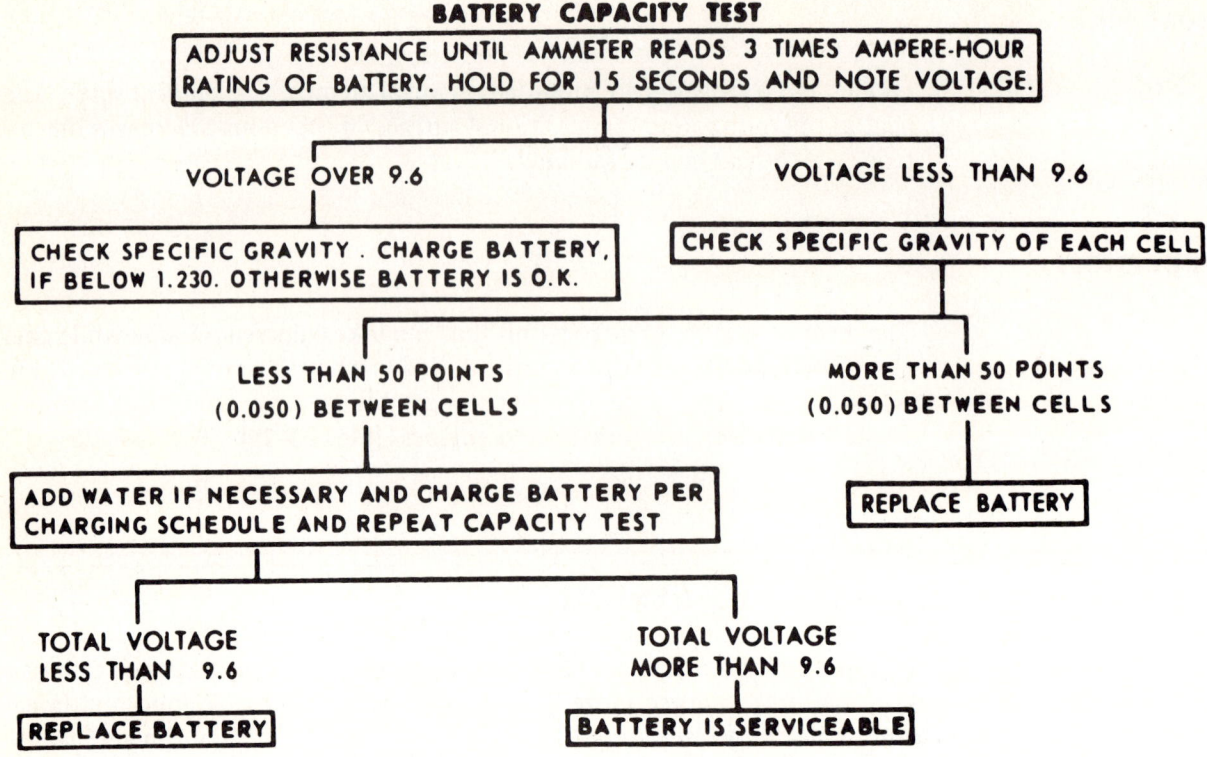

FIGURE 23-20. Battery Test Standards. *(Courtesy of Ford Motor Co.)*

4. Never drive an alternator while it is disconnected. There are three possible complaints:
   a. insufficient charging rate
   b. charging rate too high
   c. noisy operation

Insufficient Charging Rate

1. The most common cause of this complaint is a loose drive belt. Tighten or replace the belt.
2. Most alternators can be "full fielded" by using a jumper wire to connect the BAT (battery) terminal with the F (field terminal). This will result in maximum output as the engine is run at idle speed. One can then tell whether the alternator is at fault or whether other units have failed.
3. Connect an ohmmeter between the rotor shaft and slip ring. Any reading other than infinity indicates a shorted winding.
4. You can use a test light to check the same problem.
5. Use the test light between the two slip rings. It should glow. If not, the rotor winding is broken.
6. Connect the ohmmeter from a stator terminal to the frame. An infinity reading is normal. Any other reading indicates a short.
7. Use the test light in the same way. It should not glow.

8. Connect the ohmmeter across a diode, then reverse the leads. One reading should be higher than the other. If both readings are equal or low, the diode is shorted. If both readings are high, the diode is open.
9. Use the test light in this test. Normally it will glow on one connection but not on both. If it glows on both connections, the diode is shorted.

These tests can be applied to all diodes and transistors, whether they are in the alternator or regulator.

Electromechanical Regulators

These units are common on most older systems using alternators for charging. They are similar to regulators used with generators but have no cut-out relay or current limiter unit. Alternators are self-limiting in current (amperes) output.

Although there are several designs, the chapter describes a typical system of electromechanical regulation. [Fig. 23-21]

Battery current is directed through the ignition switch, charge indicator lamp, and a resistor to the solenoid-operated contacts of the regulator. When the switch is turned on, current flows to the field terminal of the alternator, of value enough to provide start-up output.

Alternator output voltage is controlled by limiting the voltage through the fields. As the engine speed increases to idle speed from cranking speed, the alternator increases its output voltage to the solenoid-operated field relay, turning off the indicator light, and connects battery voltage to the voltage regulator and alternator field.

At low speeds or during the makeup charge from cranking, the regulator armature connects resistance into the field circuit as the lower contacts close

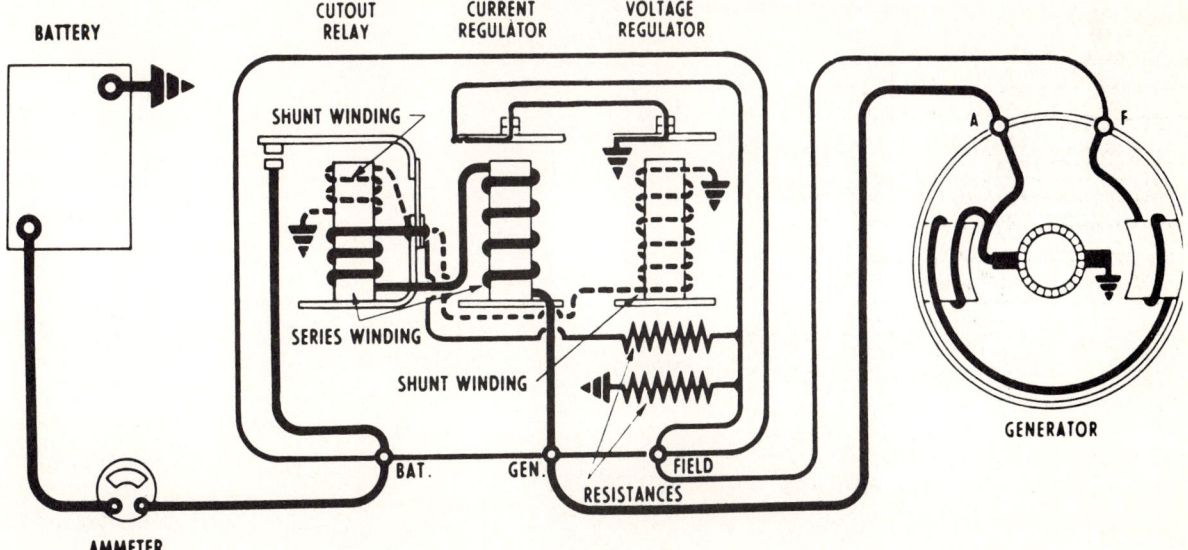

FIGURE 23-21. **Electromechanical System.** *(Courtesy of General Motors Corp.)*

and open as the solenoid current is interrupted. This allows a high charge rate at low speeds and heavy charge loads.

When the engine speed increases in operation, the alternator voltage builds up to a value that makes the solenoid pull the armature down, opening the upper contacts. This action opens the alternator field circuit, stopping current production.

The voltage falls, and the contacts close again, imposing full voltage on the fields. This action occurs many times a second, and alternator output voltage is then controlled.

Thus the alternator fields may receive current through the lower contacts and resistor or through the upper contacts without a resistor, but not both. Changes in speed or load will cause the system to change, but the alternator fields will always receive current, regardless of the regulator mode, as long as the ignition switch is on.

GENERAL NOTES

Charging regulators fail more often than alternators or generators.

A loose drive belt can cause a low charge rate.

Batteries on a bench charger should receive at least 24 hours at a low rate. The fast charge common in service stations will not bring the battery to a full charge.

Poor switch contacts often cause intermittent performance.

24
DO-IT-YOURSELF TOOLS

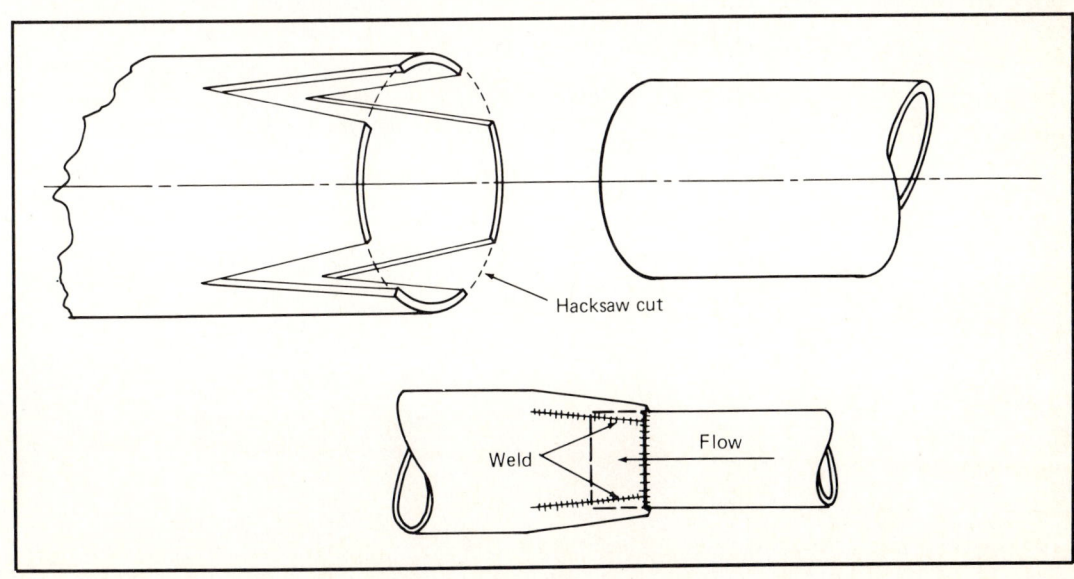

INTRODUCTION

This chapter presents simplified ways and devices to accomplish the required results. In no case has accuracy been sacrificed nor precision reduced. The aim is to make the work easier by offering the accumulated experience of many years in the trade.

These tips have been drawn from many sources, as well as the experience of the author. That they do work has been demonstrated by actual use, and no research will be required if the directions are followed.

At the risk of duplicating some of the instructions in previous chapters, this chapter offers greater detail, both of tools and principles, that every mechanic should know.

Maintenance is little understood and often poorly followed. It is *not* breakdown repair. It is the series of adjustments, filter service, and care that allow a machine to give the longest service life. Without it, unnecessary downtime and repair are sure to follow, with increased costs. Proper adjustment and service at regular intervals are required, and nothing less than this will ensure low-cost operation.

Money spent on maintenance compares with breakdown repair cost as 1 to 3. Manufacturers' recommendations are usually conservative, and it is possible to extend some service intervals after experience and testing. Such extensions should be approached carefully and supported by laboratory testing of oil, regular inspections, and an attitude of respect for the engineers who set up the schedules. They had reasons why they set service intervals.

When one considers that parts replaced during repair may replace as little as 4 ounces of worn metal, such repair comes into perspective.

USE OF TORQUE WRENCHES

Nearly all fasteners have a specified torque or maximum tension to which they can be tightened without excessive strain. Special wrenches for tightening fasteners are made, some of high quality and rather costly, some of lower quality and cost. All torque wrenches are precision tools and should be cared for as such. They must be used in the correct manner to ensure accuracy. [**Fig. 24-1**]

1. Keep torque wrenches in a safe place, protected from contact with other tools.
2. Use the right size wrench. Do not try to read pound-inches on a pound-foot wrench.
3. Tighten multiple fasteners in the proper sequence. For example, cylinder heads must be tightened in four passes, starting at the center and alternating toward the ends.
4. Pull the wrench handle straight and smoothly. Many torque wrenches cannot be read accurately when the handle is jerked to tighten.

394 / Diesel Engine Service

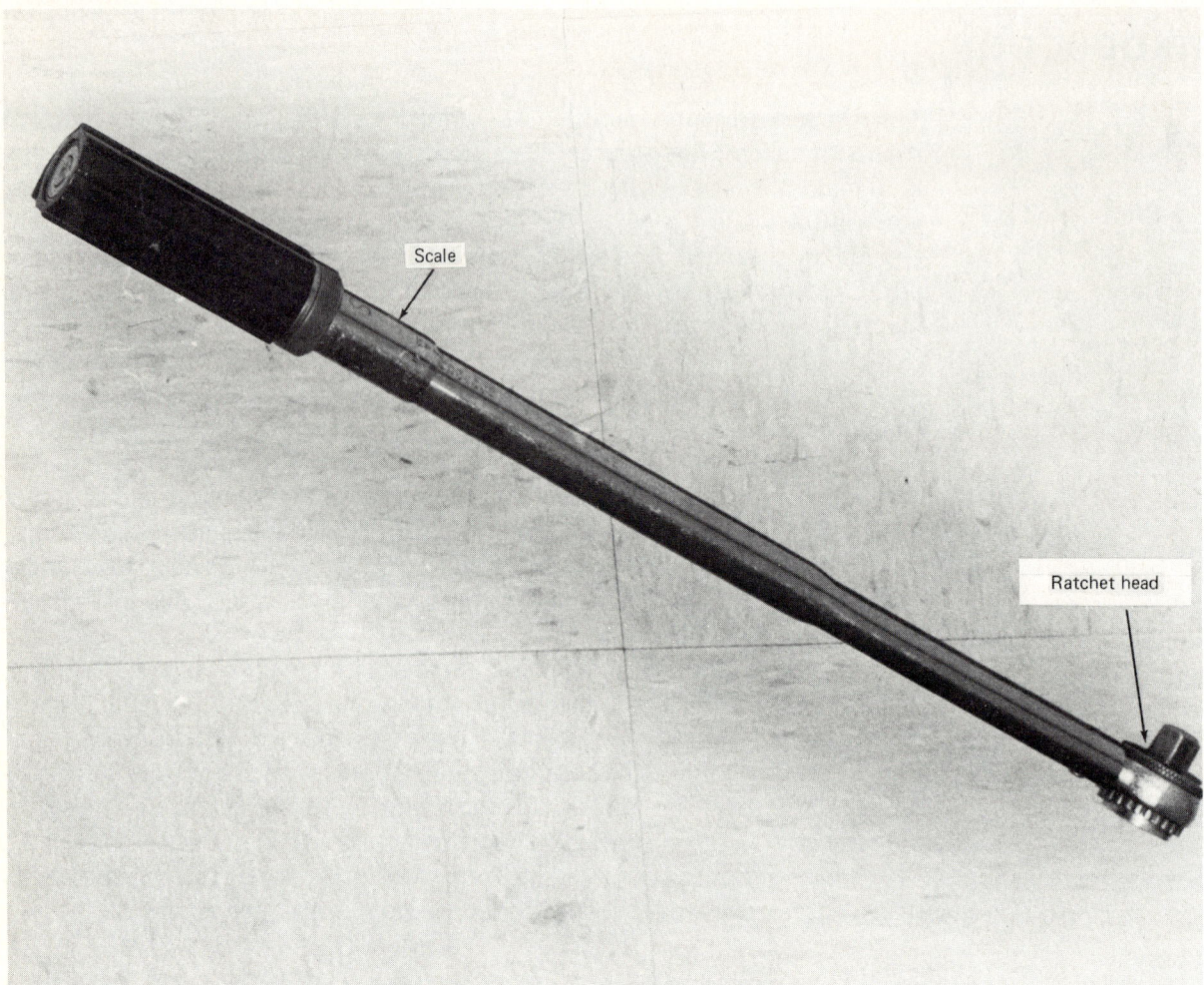

FIGURE 24-1. Torque Wrench.

5. Never exceed specified torque. It does no good to break a part when a glance at the specifications would have saved it.

USE OF POWER WRENCHES

These modern tools are useful to speed the disassembly of machinery. Properly used, they have great value. Carelessly used, they can cause distortion and leaks. [Fig. 24-2]

1. Use of power tools to *remove* fasteners. Do not trust them to exert proper torque in tightening. Use a torque wrench.
2. Be careful. Power tools are quick and positive in action. Never remove the tool from the fastener while it is in motion.
3. Be sure that you are using the correct fasteners. Time saved with power tools is wasted when fasteners have to be sorted from a pile.

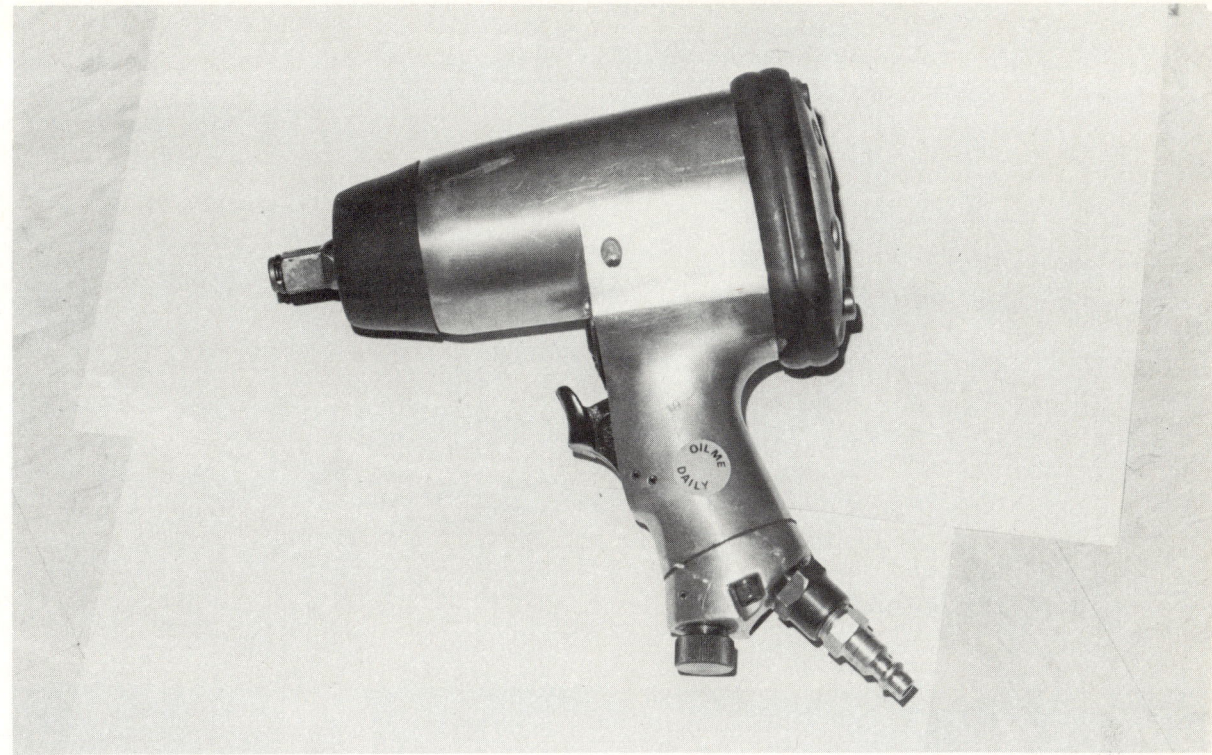

FIGURE 24-2. Power Wrench.

USE OF PRECISION TOOLS

Tools used for fine measurement, such as dial indicators and micrometers, are not intended for any other use. We can't use micrometers as C-clamps. Good mechanics recognize the use and limitations of such tools and treat them with respect. [Fig. 24-3]

1. Never lay 1-inch micrometers aside with the spindle in contact with the anvil. Temperature variations can cause strains.
2. Store precision tools in protective containers. Do not mix them with other tools.
3. A light touch is essential when using any micrometer. Contact with the work is all that is required.
4. Ball micrometers are available for measuring such things as bearing thickness. Balls of 0.250 inch are available to slip over the anvil, so that a plain micrometer can be used. The ball thickness is subtracted from the reading.
5. Dial indicators should be supported as close to the work as possible. Long extensions and friction joints subtract from accuracy. [Fig. 24-4]
6. Inside measurements require care to obtain the true diameter. Set one probe in position, and move the other to the smallest reading. [Fig. 24-5]
7. Use depth micrometers for such measurements as liner counterbore depth. Be careful to base the tool on a clean and even surface.

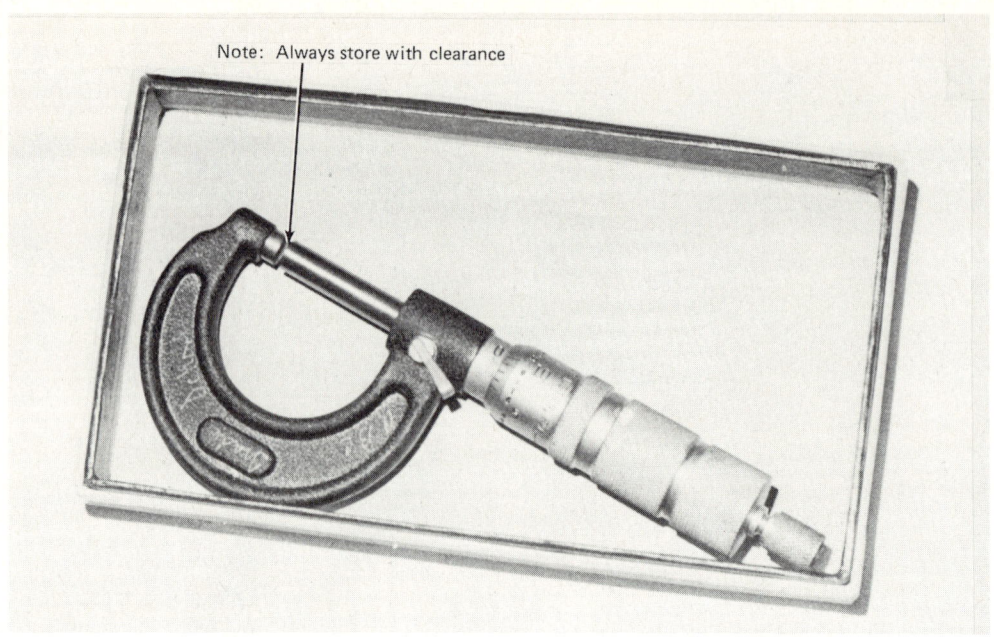

FIGURE 24-3. Micrometer in Case.

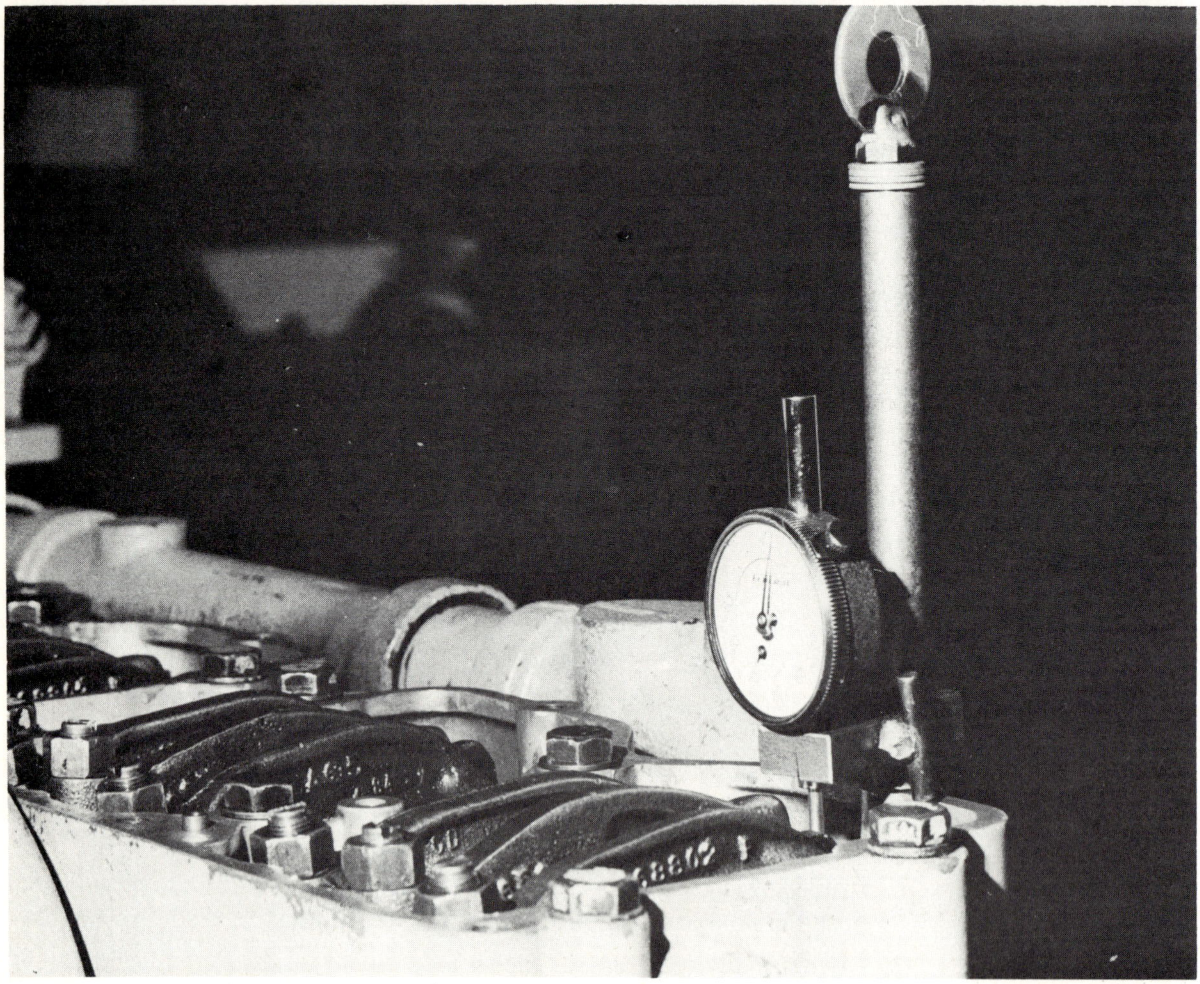

FIGURE 24-4. Indicator on Fixture.

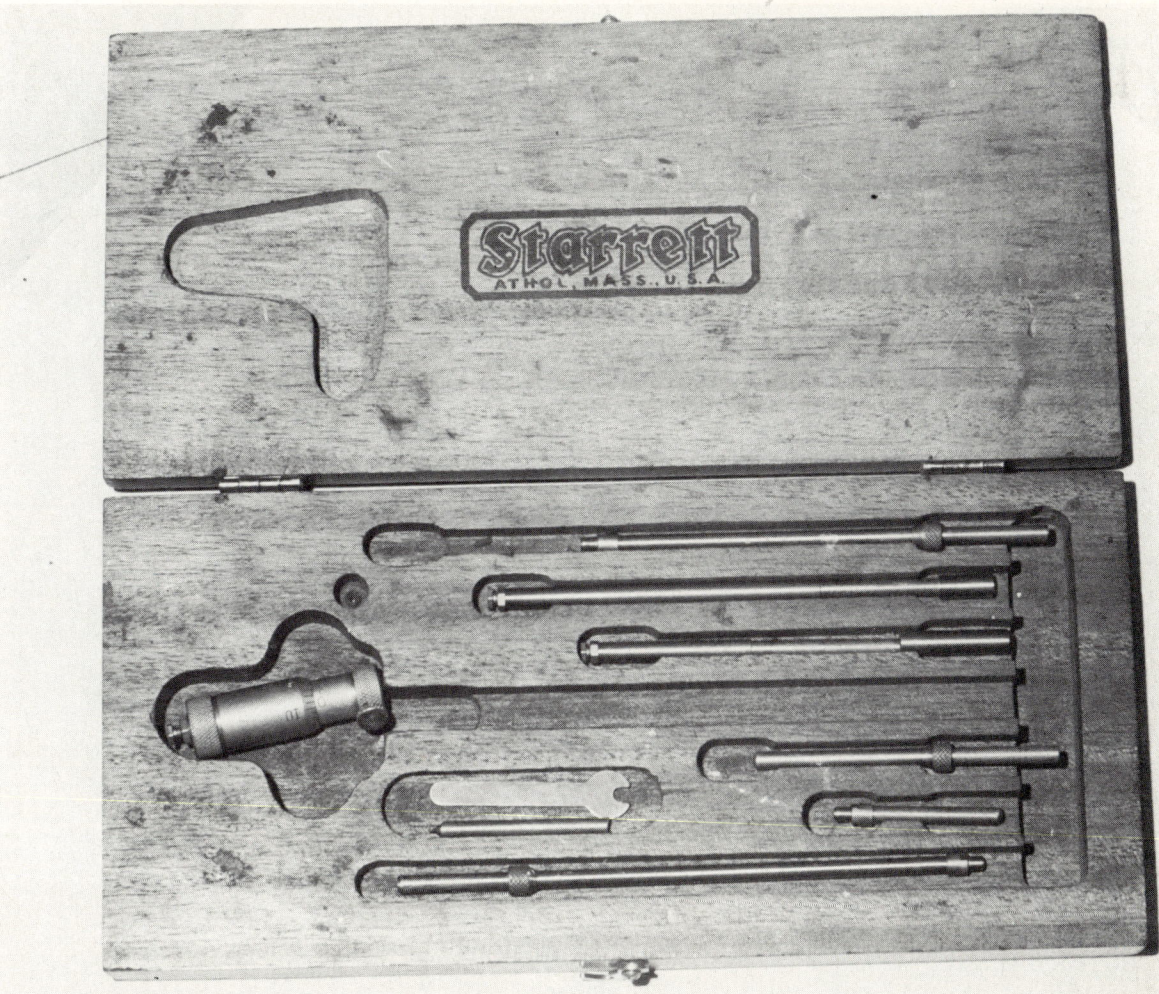

FIGURE 24-5. Inside Micrometers.

COMMON TOOLS

Using Files and Hacksaws

Although power tools are widely used, they are not always available. We find many uses for these common hand tools for deburring, cleaning sharp edges, and general metal-cutting tasks. [Fig. 24-5]

No matter what the sectional shape, files cut in only one direction. The teeth are made to cut as one pushes the file away, not as one draws it back. Contact on the backstroke only dulls the file. [Fig. 24-6]

So too the hacksaw blade has teeth pointing in one direction. Blades are always installed in the frame so that the teeth point away from the handle. Again, the out-stroke does the cutting [Fig. 24-7], and the blade should be lifted slightly on the backstroke. Also, excessive force need not be exerted to force the blade through the work. The blade should do the work. It should not be forced.

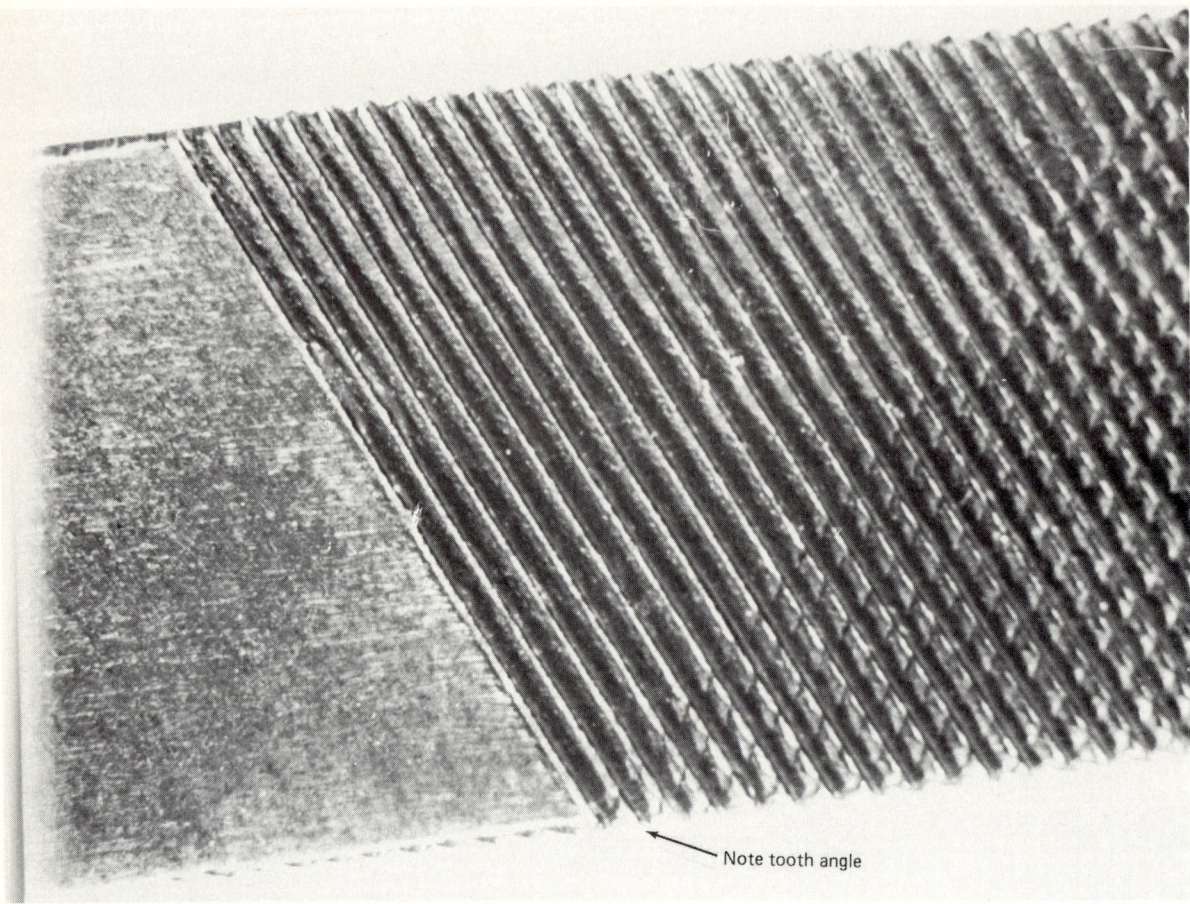

FIGURE 24-6. File Teeth.

Reading a Micrometer

The common micrometer is used in many parts inspection jobs to measure such things as crankshaft journal wear, the thickness of various pieces, and the diameter of wires.

These instruments have been available for many years and are relied on for accurate dimensions of many parts. All good mechanics must know how

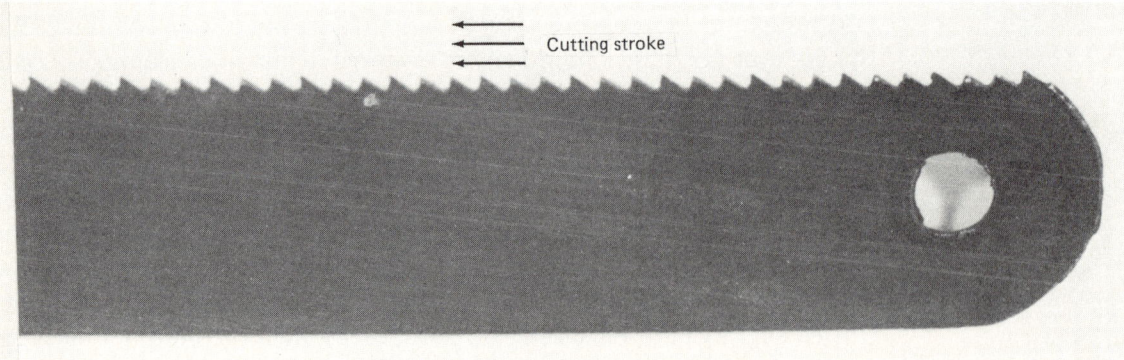

FIGURE 24-7. Hacksaw Teeth.

to use the micrometer movement properly and how to read the instrument accurately. Let us first consider the parts of a typical 1-inch micrometer [**Fig. 24-8**]:

1. frame
2. anvil
3. spindle
4. locknut
5. sleeve
6. thimble
7. ratchet stop, if used

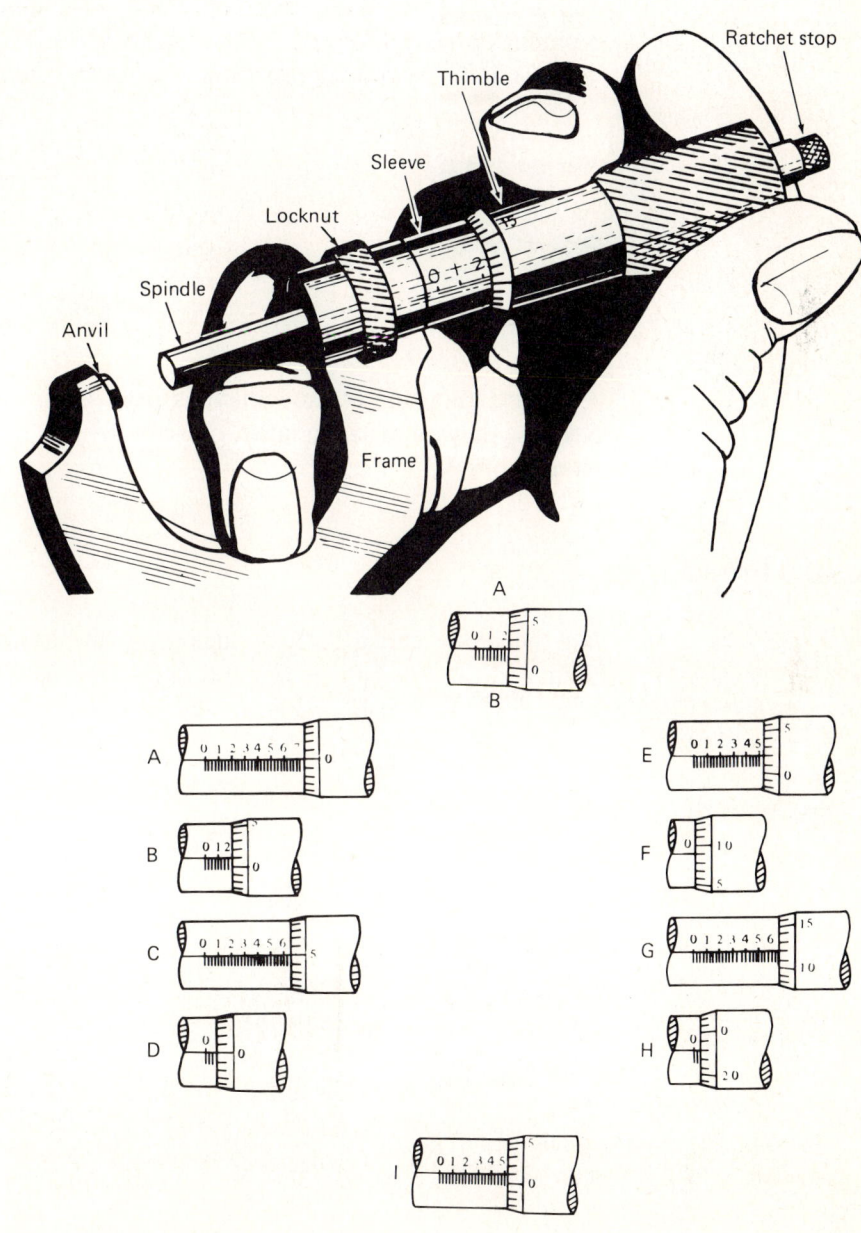

FIGURE 24-8. Micrometer Terms. (*Courtesy of Starrett Corp.*)

The spindle is threaded and attached to the thimble at its inner end. There are 40 threads per inch on both the spindle and the frame through which it screws; $\frac{1}{40}$th of an inch equals 0.025 inch. Thus, one complete turn of the spindle will move it 0.025 inch.

The thimble has 25 marks around it; an index line on the sleeve indicates the 0 point. Thus, when the thimble is moved through one revolution, it is moved 0.25 inch. Each mark on the thimble then indicates 0.001 inch of movement.

Every fourth space on the sleeve is marked with a number for easy reading. The 25 divisions on the edge of the thimble are marked at every 5th division. Thus, when one moves the thimble one mark, it has been turned through 0.001 inch, this being the unit that the micrometer is designed to read.

In the example, one sees the second mark on the sleeve. One can read it 0.2. One sees one small mark past the 2, which indicates 0.025 inch. The second mark on the thimble is aligned with the index line, so we can read it 0.002 inch. When one adds the readings together, one gets:

$$
\begin{array}{r}
0.2 \\
+\ 0.025 \\
+\ 0.002 \\
\hline
0.227 \text{ inch, total}
\end{array}
$$

The micrometer movement is used on both inside and outside tools. Because they all read the same, one can use any micrometer if one knows how to read one.

Use of Thread Files

No mechanic can afford to lack this simple and inexpensive tool. Many damaged threads can be cleaned and made usable by treating the damaged area with a thread file. [**Fig. 24-9**]

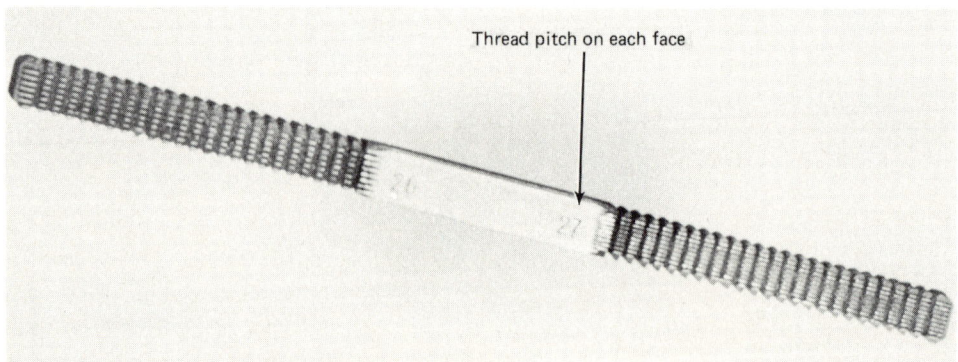

FIGURE 24-9. Thread File.

Each file has eight thread pitches and can be used as a thread gauge. In use the file can be "rubbed" over the damaged threads, because the teeth are relieved to permit backward contact without damage.

Use of Wrench Handle Extensions

Wrench handle extensions, or cheaters, are often used to give increased leverage for loosening a tight fastener. There is nothing wrong with this, as long as the wrench is adequate to stand the extra strain. One must use the size required by the fastener and not try to turn a lug nut with a ⅜-inch drive handle. In no case can one substitute an extension for a torque wrench.

PARTS INSPECTION FOR WEAR

Although many gauges are available for inspection purposes, worn parts can usually be detected by visual examination. Some parts, such as crankshafts and liners, require the use of gauges or other precision tools to detect the few thousandths of wear that can render them unfit for further service.

Gears

Broken or chipped teeth are noticeable. One should look for wear patterns on the tooth faces and compare their appearance with that of a new gear.

Camshafts

Camshafts fail by lobe breakdown. No gauge measurements are needed.

Crankshafts

Although obvious grooving, scores, or signs of wear are visible, we must rely on micrometers to detect the normal wear in thousandths. Cracks can be detected by magnetic inspection.

Cylinder Liners

Damage, such as scores, roughness, or external pits, may show the need for replacement. Internal signs of distress can be seen without removing the liner. Pits on the outer surface can be repositioned in the block to present an unpitted surface to the exhaust side, where most pits occur.

Liners to be reused must be honed. This is not intended to resize the bore but to present a new surface for ring seating.

Pistons

Pistons are rejected for ring groove wear, cracks, scores, etc. They are seldom replaced for diametrical wear. All of these faults can be detected by visual inspection. Ring groove wear is detected by installing a new ring in the cleaned groove and inserting a thickness gauge beside it when the ring is held flush with the piston surface. Over 0.002-inch side clearance is cause for rejection of the piston.

Bearings

Although some feel that new bearings must be installed at some arbitrary service period, bearings are fit for further service unless scored, grooved, and are free of foreign particles, and generally smooth. Many bearings have a gray coating, which reveals wear when the underlying copper is visible.

If bearings are to be reused, they should be spread slightly to fit in the bore. Copper lead bearings tend to cling to the shaft and must be altered to seat in the bore.

To spread a bearing, one should place it on a solid flat surface, ends down, then press with the palm until a slight yield is felt. Little spring is needed; if one presses too hard, the bearing will break. [**Fig. 24-10**]

One should try the bearing in the bore. It should fit snugly but not need much force to seat.

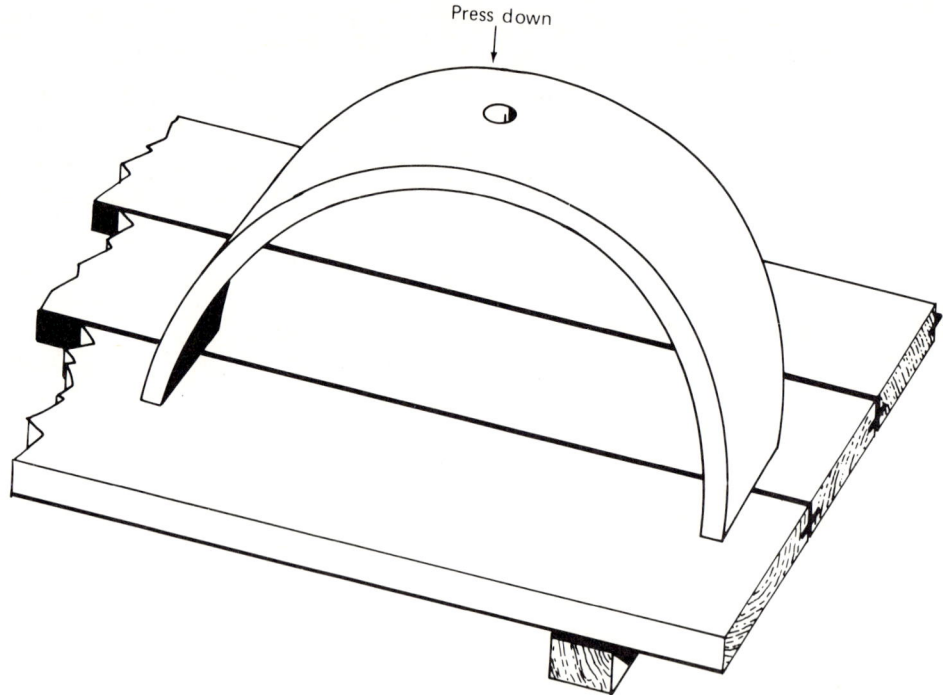

FIGURE 24-10. Spreading a Bearing. *(From a sketch by P. M. Uhl)*

TOOLS TO MAKE

A Ring Compressor

The common band-type ring compressor is well known. These tools are often damaged and can make a long job of piston installation. Assuming that the engine has removable cylinder liners, a handy ring compressor can be made from a used liner. [Fig. 24-11]

Given: At least one used liner, worn but unbroken. Access to a good machine shop.

1. Lathe-cut the lower 4 inches from the liner. This end is not worn.
2. Cut the internal bevel from the lower end. Make the end straight and smooth.
3. Taper-bore the top of the section to within 1 inch of the bottom. Make the taper as smooth as possible.

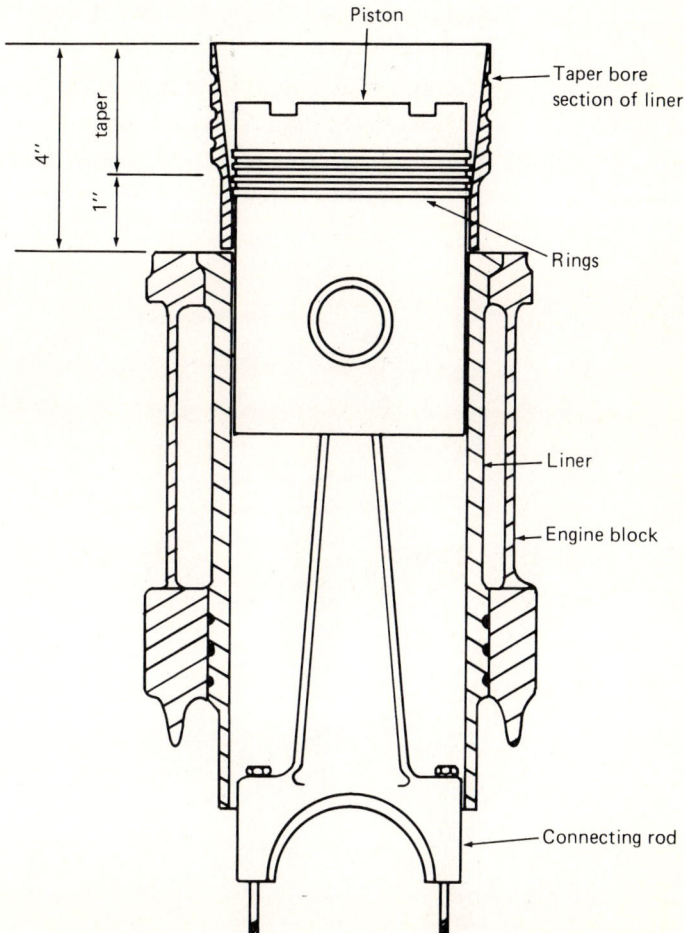

FIGURE 24-11. Making a Ring Compressor. *(From a sketch by P. M. Uhl)*

4. To use, slip the tapered section over the rod end and piston, centering the rings.
5. Insert the assembly into the cylinder, allowing the rings to hold it.
6. Working from the bottom, grasp the rod end and pull the piston and rings through the compressor. Seat the bearing on the crankshaft journal. Secure the rod bolts to the specified tension.

Ring Expanders

Special ring expanders are available; they make ring installation easy and without distortion. These are fairly costly tools and may fit only one size ring. The shop tool may be broken, lost, or used by others. Because the object is to install the rings without scratching the piston or distorting the ring, the simple way described will be useful.

Given: Pistons cleaned and assembled to the rod. New rings, and a clean shop rag. [**Fig. 24-12**]

1. Set the rod in the vise allowing the piston bottom to rest on the vise.
2. Place the oil ring expander in its groove. Be sure that the ends abut, not overlap.
3. Lay the oil ring on the piston top, and fold the rag to make a loop in each side of the hem. [**Fig. 24-12**]
4. Engage the ring ends in the loops, holding the rag with your thumb and second finger. Extend your first finger along the ring to support it.

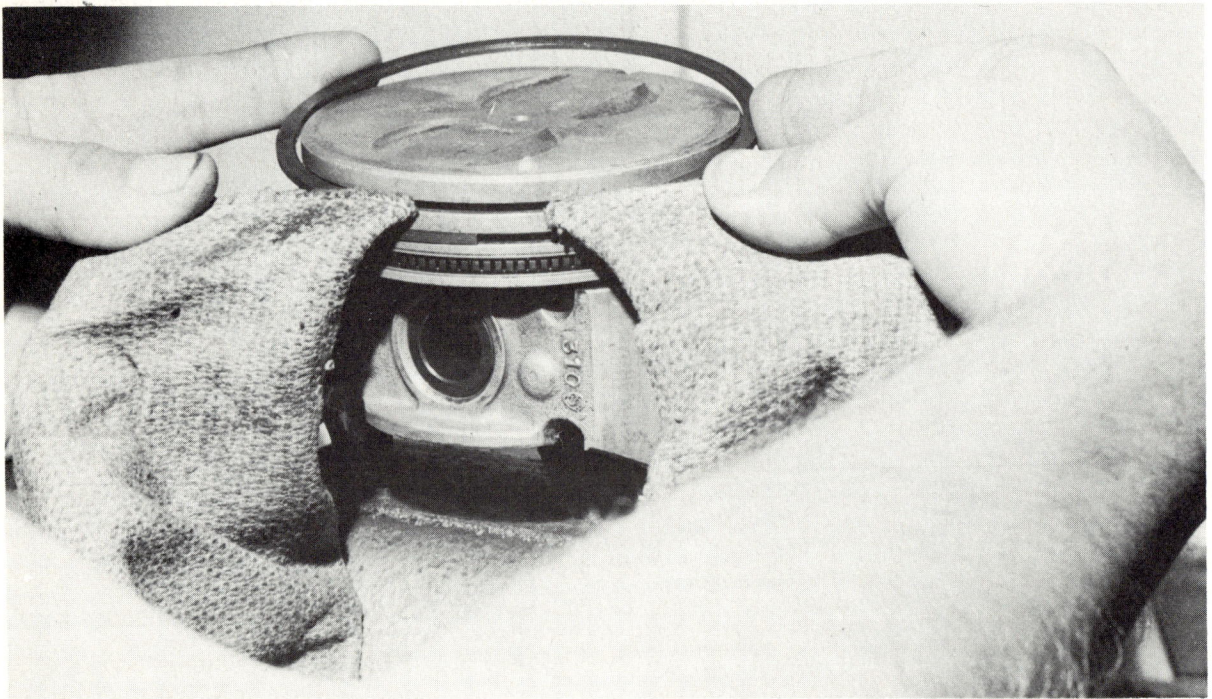

FIGURE 24-12. Installing Rings with a Rag.

5. Pull the loops and ring ends just enough to clear the piston. Place the ring in its groove, and disengage the rag.
6. Install the other rings in the same way. Do not stretch them more than necessary.

Liner Puller

There are several good makes of such pullers. If on a job without such a tool and needing to pull the liners in an engine so equipped, one can make an adequate puller from material at hand. **[Fig. 24-13]**

Given: 1 steel bar, 1 inch by 1½ inches by 7 inches
       2 pieces of steel, ½ inch by 1½ inches by 4 inches
       1 steel bar, ¾ inch by 1½ inches by size of the liner
       1 forcing bolt, ⅝-inch NC by 18 inches

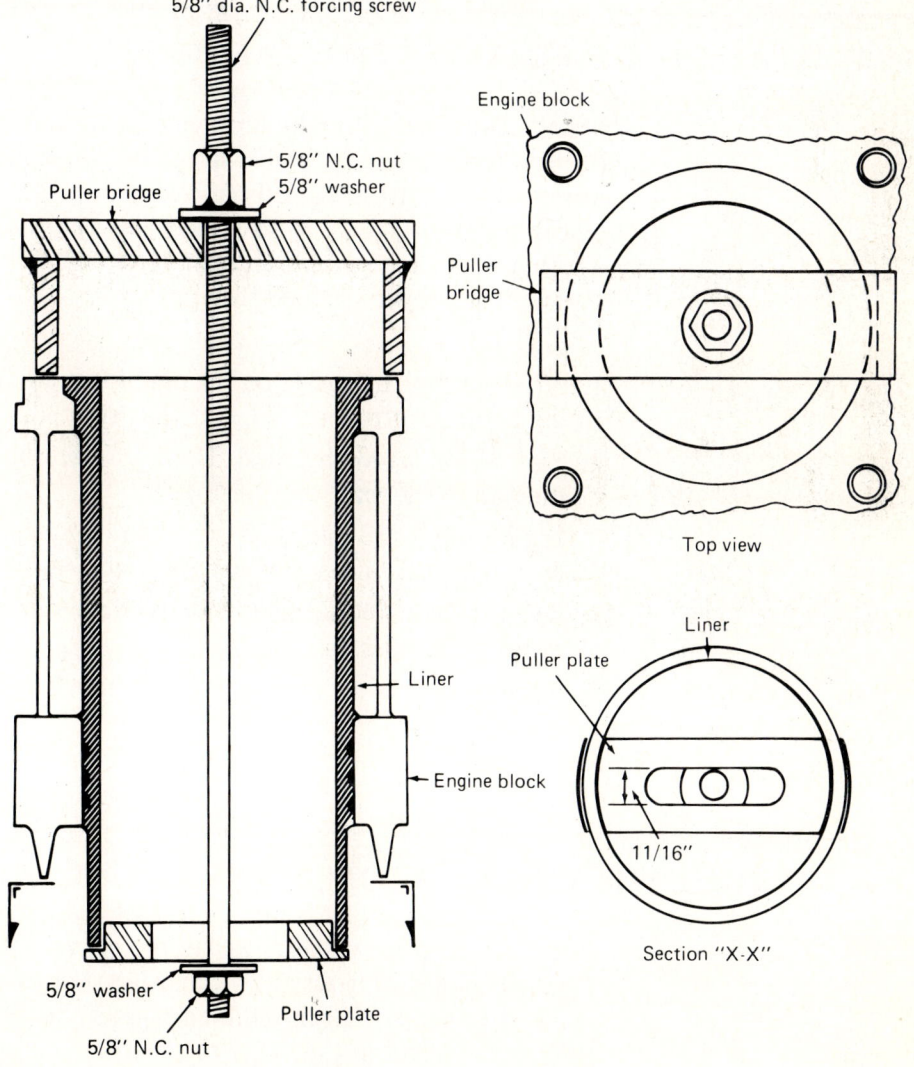

FIGURE 24-13. Making a Liner Puller. *(From a sketch by P. M. Uhl)*

2 nuts, ⅝-inch NC
2 washers for nut bearing

1. Bore a hole ¹¹⁄₁₆ inch in diameter in the center of the 1-inch bridge bar.
2. Join the bridge bar to the 4-inch legs to form a bridge. Weld in place.
3. Flame-cut a slot ¹¹⁄₁₆ inch by 1½ inch in the center of the puller bar.
4. If a machine shop is available, have the puller bar lathe-cut to form a ledge to engage the lower end of the liner.
5. The slot allows the puller bar to be tilted for insertion in the liner, and to fall to the horizontal to engage the liner end.
6. Assemble the tool, and place one of the nuts and washers on each end of the forcing bolt. The lower nut can be welded to the bolt.
7. Insert the tool in the liner, and turn the top nut to exert pulling force.

INSTALLATION OF BUTTRESS-TYPE OIL PANS

Many engines have been built on which the back of the oil pan was bolted to the flywheel housing for added strength. The sequence for securing such pans was never well explained. Although the pan-to-block gasket will compress several thousandths, there is no gasket between the pan face and the flywheel housing. Thus we must allow the pan to move a little while we tighten the capscrews. The proper sequence follows:

1. After placing the gasket on the pan, lift it into position, and hold it by using two vise grip pliers, one on each side.
2. Start all capscrews by hand. Do not forget the four small capscrews from the pan to the rear seal housing.
3. Lightly snug the pan to flywheel housing capscrews. Leave them loose enough to permit pan movement. Lockwasher tension is enough.
4. Secure the pan by alternating from side to side, starting at the center. Make at least three passes to tighten.
5. Finally, tighten the pan-to-flywheel housing capscrews.

PROTECTING THE ENGINE FROM DIRT DURING ASSEMBLY

Small bits of dirt, washers, gasket pieces, and other solid material dropped into an engine during assembly can cause serious damage. Care must be taken while working; partially assembled engines should be covered at the end of the shift. It is much less expensive to prevent such matter from entering than to repair the damage that may result.

Discussion of Filters

All diesel engines are equipped with filters to clean the various fluids. Some are more sophisticated than others, but four systems require filtration:

1. air
2. fuel
3. oil
4. coolant

As described, most modern engines are equipped with paper-type air cleaners. It is important to service these filters often enough to assure a clean and adequate air supply. Some dusty conditions may require daily service. On-highway use may not need service more often than once a week.

Restriction gauges are available to guide the service interval. It is poor economy to try to stretch air cleaners to meet some arbitrary service interval.

Fuel filters are designed to remove small dirt particles from the fuel, thus protecting the finely machined fuel system parts. Some engines are furnished with two or more filters connected in parallel. Such a connection acts as one filter with large capacity. Other engines have a primary suction filter on the fuel pump inlet and a secondary filter after the supply pump.

Some filters are equipped with a water drain, which is convenient for removing moisture from the system. In any case, fuel filters must be serviced often enough to prevent any dirt or water from entering the fuel system.

Fuel filters should always be installed full of fuel. This lessens the chance for air to be drawn in.

Causes of Power Loss

This fairly common complaint has only a few basic causes. These follow, in order of their frequency:

1. fuel filter service
2. wear in the throttle linkage, covered in Chapter 3
3. when accompanied by exhaust smoke, a restricted air supply
4. restriction in the fuel suction lines
5. poor engine adjustment
6. small suction leaks
7. when the tanks are low in fuel, bubbles drawn into the suction line, giving the symptoms of air leaks
8. dragging brakes
9. unrecognized grades

All of these causes can be detected by logical processes. Even the unrecognized grade can be explained to the operator. It is always important to learn of any previous work.

Fuel filters have been discussed.

RESTRICTED AIR SUPPLY. Excessive exhaust smoke always signals this problem. The engine will smoke at all speeds and may even be hard to start. The remedy is obvious.

RESTRICTION IN THE FUEL SUCTION LINES. This problem is not frequent but can be puzzling. The engine may run fine for some distance, then lose power suddenly. One knows that nothing is wrong, such as poor adjustments, or other engine faults.

Suction hoses should be checked for cut linings. These can act as a flapper valve under suction. Such things as rags left in the fuel tank may be found. The engine may run well for a time.

POOR ENGINE ADJUSTMENT. When an engine loses power, some think of the fuel pump first. This is rarely the trouble. Fuel pumps are some of the longest-lived parts. Of course, wear is possible and can be detected by appropriate gauge use.

Much more likely are poorly adjusted injectors on Cummins engines. Many expensive repairs have been made when all that was needed was a tuneup.

SMALL SUCTION LEAKS. Most of such leaks are due to faults in the suction system. Although such things as sight glasses are handy, leaks can be found with an auxiliary hose to bypass the suction system and locate the leak. On older engines, such things as flat throttle shaft seal rings, cracks in the fuel filter head, crossed threads, and loose fittings, can allow suction leaks.

If no leaks are found between the fuel pump and the tank, one must check such things as a worn tachometer shaft seal on Cummins engines. This is done by filling it with fuel with the engine idling. If the seal is leaking, fuel will be drawn into the pump.

If needed, one should remove the housing from the tachometer drive and use a flaring tool and a roll-head bar to pull the drive and seal.

Later pumps have the tachometer drive in the drive cover and will not leak at this point.

AIR DRAWN IN FROM THE RETURN FUEL. When air tanks are low, air bubbles can be picked up by the suction connection in the tank. The return fuel should be directed at least 1 foot from the suction connection. If possible, one should connect the return to the other side of the saddle tanks remote from the suction connection.

DRAGGING BRAKES. Such a fault is always shown by hot drums. Again, the remedy is obvious.

UNRECOGNIZED GRADES. This can be a troublesome problem for the mechanic. Many small grades are not apparent to the operator. It may be necessary to make a run with him using another tractor.

INTERPRETING EXHAUST SMOKE. The results of restricted air supply have been described. But exhaust smoke can reveal other problems that must be considered. In general, blue smoke indicates excessive oil consumption. Black smoke indicates either too much fuel, not enough air, or both.

Turbocharged engines will smoke during acceleration unless a device to modify the fuel supply is used. Cummins used an air-controlled bypass valve for several years. Later engines have this control built into the fuel pump and are called *AFC,* which means air fuel control. Either of these devices was effective in limiting the smoke of turbocharged engines during acceleration. Neither system had any effect on full power. Other engines have similar devices.

Causes of Failure to Start

This problem can be divided between a simple lack of fuel supply and slow ignition due to cold weather. There are several other causes:

1. high resistance in the cranking circuit; failure to achieve proper cranking speed
2. inoperative starting aid
3. failed cells in the battery
4. burned switch contacts
5. shorted or locked cranking motor
6. fuel system dry (must be primed)
7. battery corroded and in bad condition
8. fuel frozen (cold climates)
9. timing (after fuel pump replacement)

INOPERATIVE STARTING AID SYSTEM. An engine that emits a volume of white smoke during a starting attempt is indicating that the starting aid system is not working. Most engines have some system to aid cold starts.

Some engines are equipped with glow plugs; others may use a fuel fed flame in the intake system. Still others have various starting fluid systems. One must find out how the engine is equipped and then check the operation.

When glow plugs fail to heat, one should check the circuit to them and the switch. Glow plugs can be checked after removal by connecting them to the battery positive and grounding the body of the plug. The heating element should not be touched.

Engines with a fuel fed flame system can be checked by removing a plug from the intake manifold and operating the flame system. If a flame is not visible, the fuel supply and the circuit to the igniter should be checked.

Starting fluid systems can be checked by making sure that the fluid supply is adequate and that the sprayer is open. These are fairly obvious.

BATTERY IN BAD CONDITION. The battery should always be checked when the engine cranks slowly or does not turn. Corroded terminals, dirt on the top of the cells, frayed cables, and other signs of distress should be noted.

The cells should be tested with a hydrometer. No battery will produce cranking voltage if any of its cells read below 1.200 on the hydrometer.

BURNED SWITCH CONTACTS. The main contacts in the cranking motor circuit tend to burn and increase resistance as they age. They can be found in the solenoid on the motor. They can usually be repaired by replacing the contact disc.

Cranking motors seldom fail but can be locked by faulty ring gears or dry bearings in the motor. It is possible for the pinion to lock in the flywheel gear teeth.

FUEL SYSTEM DRY. When a fuel pump is replaced, it may be necessary to perform a priming procedure. On a Cummins the pump should be filled with fuel and the filter filled. It helps to squirt lubricating oil into the pump inlet fitting. With other fuel systems, a priming pump may be provided, and air bleed plugs located on the fuel pump. The manufacturer's priming instructions should be followed. In any case, solid fuel must be delivered to the injectors before the engine can run.

FUEL FROZEN IN COLD CLIMATES. Some fuels will become cloudy with wax in low temperatures. Under these conditions the fuel will not flow readily and may fail to go through the filter. The tanks can be heated or the fuel diluted with No. 1 furnace oil when the climate is adverse.

TIMING AFTER PUMP REPLACEMENT. Engines equipped with Bosch-type fuel systems, such as Roosa Master, Lucas, and Caterpillar, must be timed to the drive system when installed. Unless this is done according to the manufacturer's manual, the engine may not run. A common mistake is to set the engine to the wrong stroke. The engine must be in the right position when the fuel pump is installed; the steps described in the manual must be followed.

PARTS FAILURES. Although such happenings are rare, they are possible. One should always check to be sure the pump is turning when the engine is cranked. Broken drive parts will fail to drive the pump, and the engine will not run.

Failures to start should never be the cause of major repairs unless an obvious disaster has happened.

TROUBLE-SHOOTING HINTS

Before using a flaring tool, it is often necessary to soften the end of the copper tube to prevent splitting. This is done by heating the tube to red heat, then quenching it in water.

This is the opposite of steel. Copper will soften readily when treated in this manner.

Installing Lip-Type Seals

First, be sure that all residue is cleaned from the seal bore. Check for burrs and other damage, and correct before trying to install the new seal.

1. Obtain a proper mandrel that fits the seal, if one is available.
2. When installing over a shaft, use a seal guide, or at least a piece of thin stock to help the seal over the shaft.
3. Lubricate the seal lips, and the outside of the seal shell.
   Note: Some seals are installed dry. Observe the package.
4. Start the seal carefully in its bore. If no mandrel is available, use any solid object that fits the seal, such as a heavy socket or clean tubing.
5. If no mandrel is available, use a soft hammer on the outer edge *only* of the seal to drive it to its seat. Work around the seal evenly.

Grinding Drills

The skill of grinding drills is almost lost. But there will always be times when a drill must be ground by hand or discarded for a new one.

The secret of renewing the cutting edge and providing a properly shaped point is to hold the drill at the correct angle as it contacts the grinding wheel. The wheel should be of fine grit and dressed to present a square, ungrooved surface.

The drill is held so that the cutting edge contacts the wheel squarely. The drill is then rotated slightly and moved down a little. The drill stem must point across the wheel. Each cutting lip should be ground the same number of passes to keep the cutting lips even. The point should be cooled often enough to prevent burning. [**Fig. 24-14**]

Modifying a Drill to Enlarge a Pilot Hole

Most mechanics have experienced grabbing drills, broken drill tips, and difficulty in holding the drill motor when enlarging pilot holes. The following modification will solve the problem:
Given: A square and well-dressed wheel.

One should bring the drill against the side of the grinding wheel with the cutting edge in square contact. The drill should be held so that its axis points across the wheel. [**Fig. 24-15**]

A light grind should be made on each cutting edge. The drill will not grab and will cut smoothly.

FIGURE 24-14. Grinding a Drill.

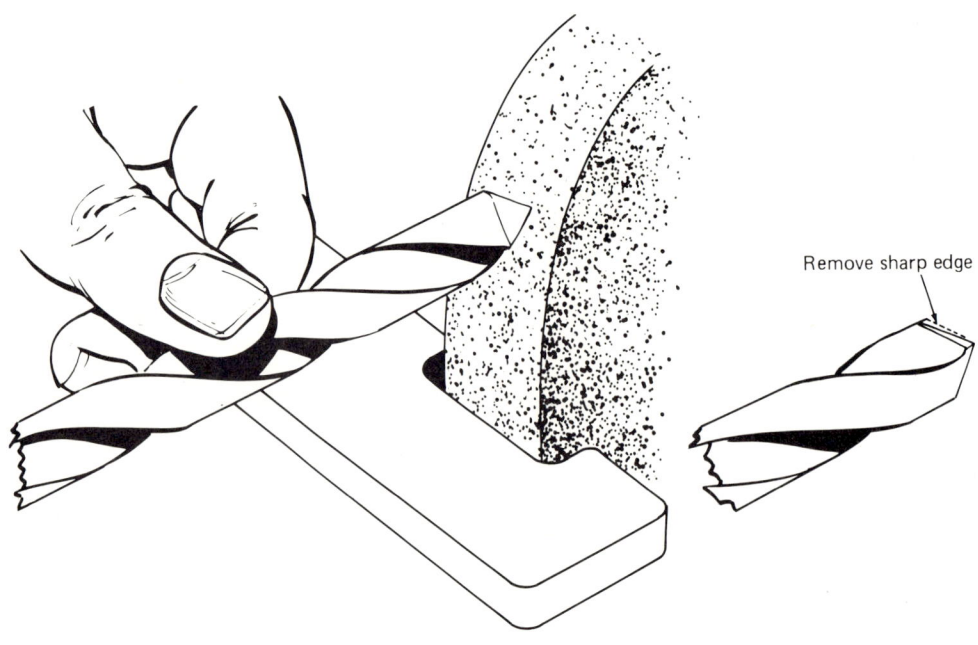

FIGURE 24-15. Modifying Drills.

Protecting Fasteners from Loss

It has been estimated that at least three hours per engine are lost in trying to select the right fasteners for each assembly. At the same time, mechanics throw away many small cans that could be used to hold the capscrews, nuts, washers, etc., from each subassembly, so that they could be ready for use during assembly. [**Fig. 24-16**] Supervision should encourage the use of such containers to receive fasteners during teardown from each subassembly. Much time can be saved by this method, and the expense of lost fasteners reduced.

OTHER TOOLS

Clutch Disc Pilot

It is often necessary to install a new clutch when no pilot tool is available. The clutch discs must be aligned with the pilot bearing and clutch shaft.

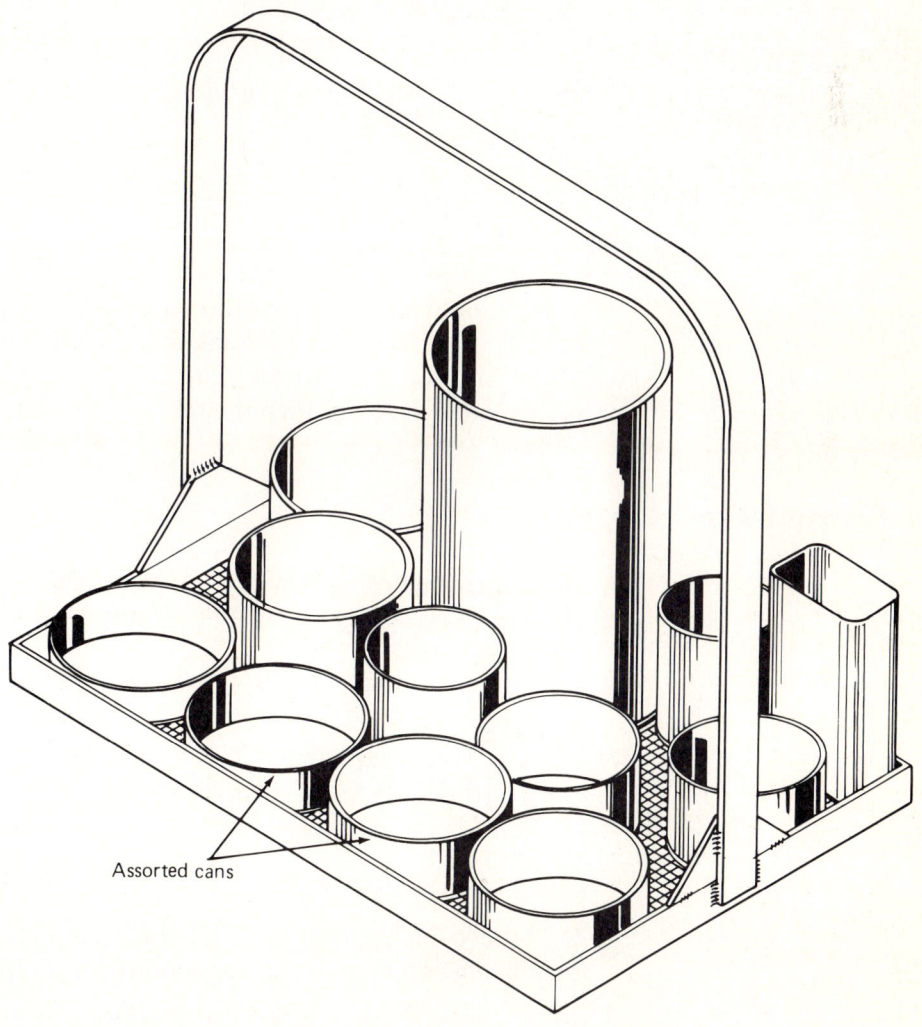

FIGURE 24-16. Tray for Fasteners. *(From a sketch by P. M. Uhl)*

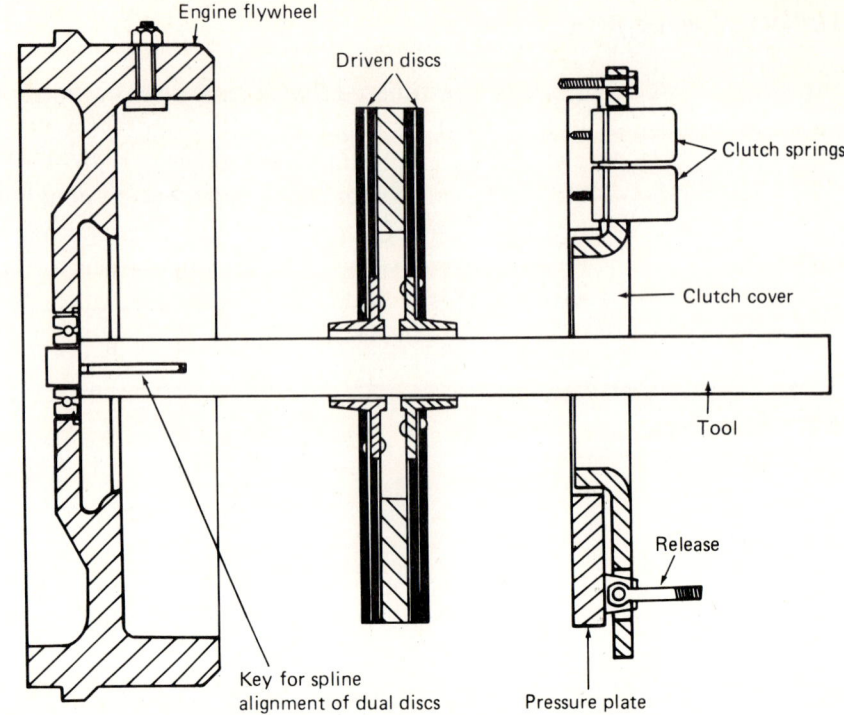

FIGURE 24-17. Clutch Alignment Tool. *(From a sketch by P. M. Uhl)*

Given: A short piece of steel tubing, the same size as the disc spline opening.

A bolt or bar the same size as the pilot bearing ID. **[Fig. 24-17]**

One should insert the tube through the clutch discs, aligning them with a small piece of square stock, or a small screwdriver. The end of the bar or bolt is slipped into the pilot bearing; the discs are applied to the flywheel. The pressure plate is secured; then the tool is removed.

A Transmission Hanger

A convenient way of suspending the transmission from the frame will save much time when an engine must be removed for rebuilding.

Given:
    A length of 2-inch channel
    A piece of steel plate, ½ inch by 8 inches by 12 inches
    2 ⅝-inch NC by 6-inch capscrews

1. Cut and weld the 2-inch channel to make a U-shaped hanger. Weld short pieces to each side to engage the frame rails. **[Fig. 24-18]**

2. Drill the ½-inch plate in two diagonal corners to take the capscrews. Weld the plate lengthwise to the horizontal hanger channel.

3. In use, slide the hanger under the transmission, keeping the ends on the frame rails. Use the screws to raise the transmission just enough to clear the engine. When the engine is removed, the chassis can be moved away.

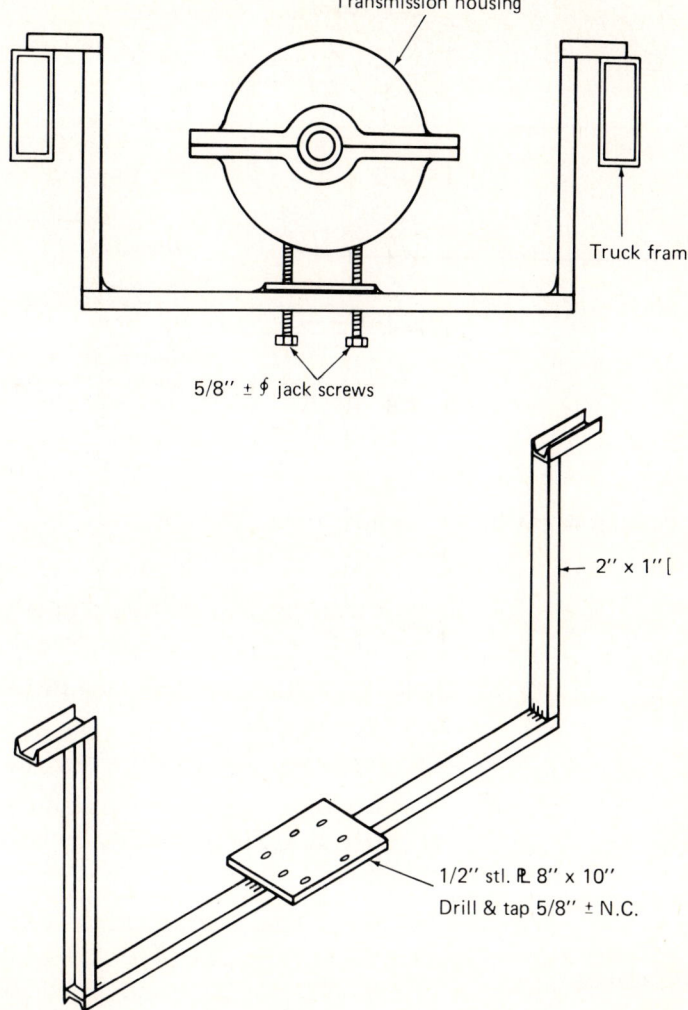

FIGURE 24-18. Transmission Support. *(From a sketch by P. M. Uhl)*

Match of Two Different Sizes of Exhaust Tubing

When two exhaust pipes of more than 1-inch difference in size must be joined, an adapter of close to the size required can be used to complete the joint.

Given: An adapter that matches the larger pipe.
A gas welding rig.

1. Set the adapter in a vise lightly, and use a hacksaw to make several V-shaped cuts in the small end. The cuts should be kept even, and only enough made to allow them to be bent to the size of the smaller pipe. Insert the smaller pipe, and bend the Vs to hold it snugly. The V openings will close in as you bend the points in. **[Fig. 24-19]**
2. When the fit is good, weld the openings, and weld the smaller pipe in place.

Balancing a Driveshaft

Any shaft out of balance will vibrate when running and may cause damage. A "quick fix" may be useful in some cases.

416 / Diesel Engine Service

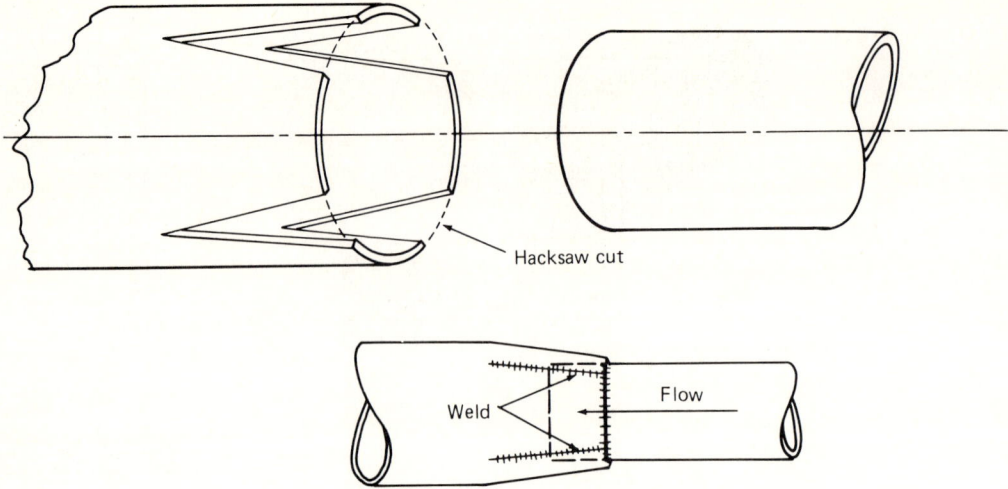

FIGURE 24-19. Matching Different Pipe Sizes. *(From a sketch by P. M. Uhl)*

1. Be sure that the driveshaft is properly installed with the U-joints in line. Check the operating angles and see that the U-joints have not failed.
2. Apply two screw-type hose clamps to the shaft, placing them so that the screws are together.
3. Jack up the rear axle to allow test runs.
4. After making a run, move the clamp screws apart and test to find the setting where vibration is lowest. By moving the clamps and testing, a point will be found where vibration is cancelled.

Battery Cables

Scrap 2-0 cable can be used to make good battery cables. [**Fig. 24-20**]

Given: Some copper tubing sized to fit the stripped cable end.
Solder and a torch.

1. Cut short lengths of the tubing, about 2 inches.
2. Flatten one end of the cut lengths, and drill a ⅜-inch hole in the flat. Set these ends in a vise, and melt solder with the torch to fill the open tube end.
3. Tin the cable, and apply some flux. Insert it into the melted solder, and allow it to freeze.

One can make up cables for many uses in this manner.

Removal of a Tight Key

Most keys fit tightly in their keyways. The easy way to remove them is to grasp one end with a diagonal cutting plier. This can be used to pry the key from its seat. [**Fig. 24-21**]

Do-It-Yourself Tools / 417

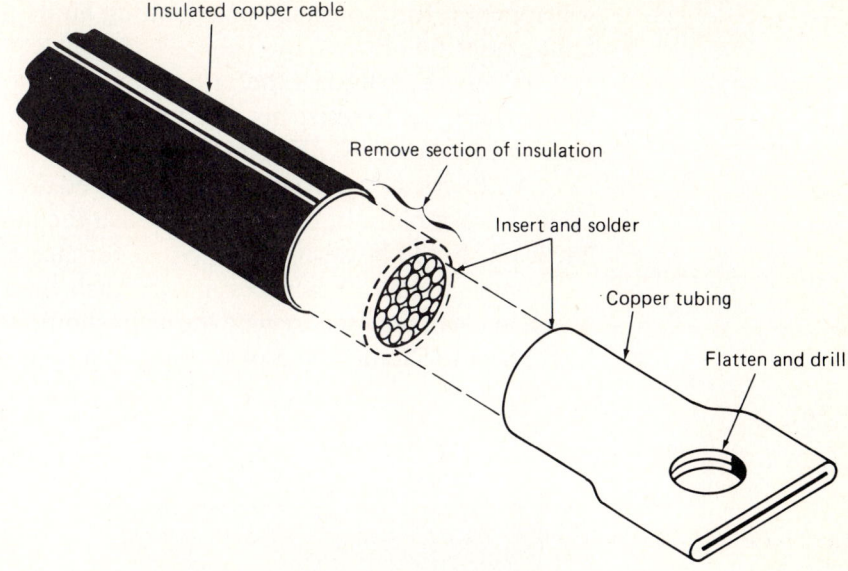

FIGURE 24-20. Cable Ends. *(From a sketch by P. M. Uhl)*

Honing Cylinders

All engine parts, when new, are somewhat imperfect. Cylinders are not perfectly round, rings do not exactly seal when new, and all parts need some wear to fit together properly. This initial wear is called *break-in*.

Cylinder walls are honed to a specific surface pattern; rings are made with an interrupted face. These two surfaces will wear to a perfect contact and

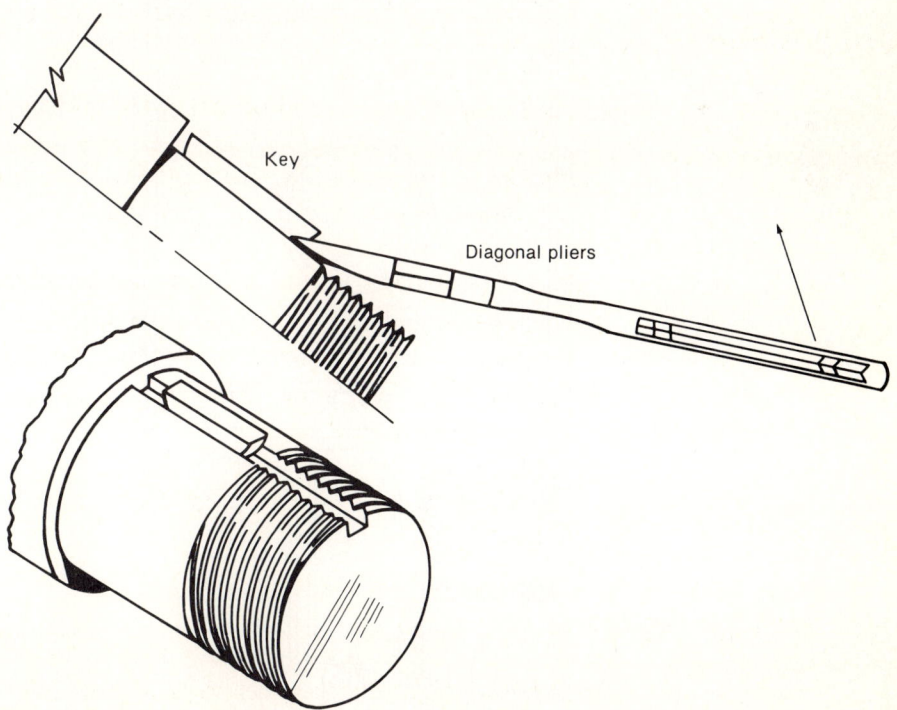

FIGURE 24-21. Removing a Tight Key. *(From a sketch by P. M. Uhl)*

418 / Diesel Engine Service

will function for thousands of operating hours unless some outside force or fault causes failure.

In service, cylinders that are not worn beyond limits can be honed to a new surface, not to resize but to establish the honing pattern that permits new rings to seat properly. [**Fig. 24-22**]

A 220-grit stone, driven by a slow-speed drill motor, and kept in constant up-and-down motion, can soon establish the desired angular pattern. Stone lubricant should be kerosene or No. 1 furnace oil.

After honing, it is necessary to wash the cylinder with detergent and water. A clean rag used to wipe the walls should show no dirt. The best way to hold liners for honing is to place them in an old block with one packing ring.

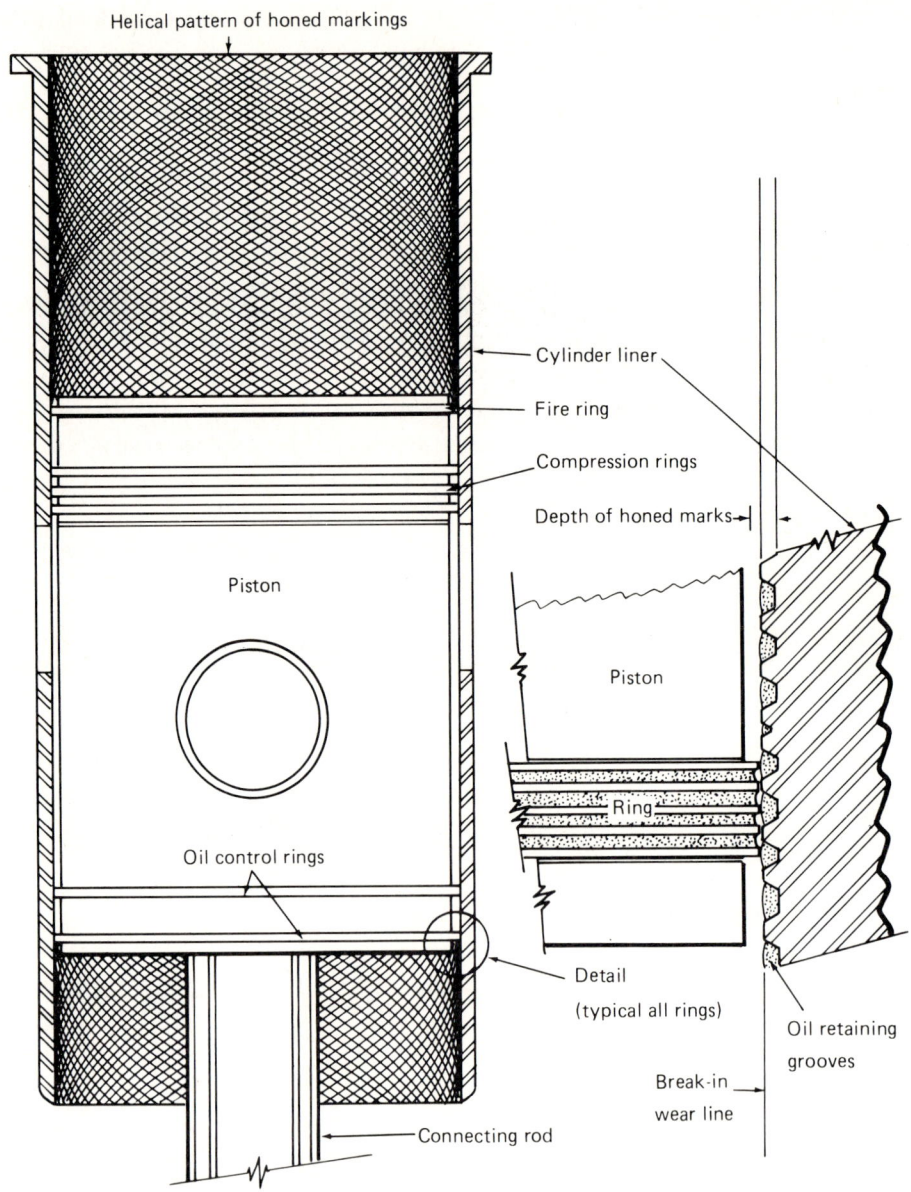

FIGURE 24-22. Honed Cylinder. *(From a sketch by P. M. Uhl)*

SALVAGING USED ENGINE OIL

It has been proven that lubricating oil does not wear out. It does, however, pick up carbon, dirt, water vapor, and other material that can cause damage to an engine.

Some efficient filtering systems are made. Railroads, ships at sea, and other large users of oil have long been using such systems and add only the amount consumed. There are ways to salvage oil without a large investment in apparatus, and we present a low-cost way of reclaiming oil that can be fabricated on site.

Oil must be filtered *hot*. Thus, a simple heating device is shown, and the filter medium is readily available. Nothing in this idea modifies the need to change oil regularly. The purpose is to make controlled salvage of drained oil practical. Tests have revealed that this system works and that it is safe.

Given: A tank of suitable size for the amount of oil ordinarily drained. **[Fig. 24-23]**
A piece of PVC tubing about 3 feet long, of 4½-inch diameter to accept a roll of toilet tissue. This is an excellent filter material.
A piece of PVC tube of the diameter of the inside of the toilet paper roll, and about 3½ feet long.
A piece of steel screen to be attached to the small PCV tube, to permit the roll of paper to be withdrawn.
Suitable piping and a valve.

One should make up the apparatus as shown in the sketch. Capacities and dimensions should suit one's need. The piping should be wrapped with heating tape after assembling.

It is not necessary to heat the oil in storage. Heat should be applied to the pipes, which will allow flow and be conducted into the reservoir.

Note that oil is allowed to "soak" through the filter roll. Flow should be adjusted to give a fast drip, but oil should not be made to pool on the end of the roll. The valve in the pipe allows the filter roll to be changed by removing the small PVC tube.

This apparatus can be "ganged" to filter a larger amount of oil. A sketch is included to show this arrangement. These are all suggested designs. The local requirement will modify these devices to suit. **[Fig. 24-23]**

Use of Additives in Oil

Some years ago a large industrial laboratory tested some 150 brands of popular additives that are sold as magic remedies for all kinds of engine ills. The results of these tests bear out the saying, "let the buyer beware."

The contents of solvent-type additives showed up as 95 to 98 percent light mineral oil, with aromatics and coloring matter. The same results come with clean fuel oil. Graphited additives were shown to contain about one half of 1 percent of graphite, which is enough to color the product but not enough to add to the lubricating value of the oil.

420 / Diesel Engine Service

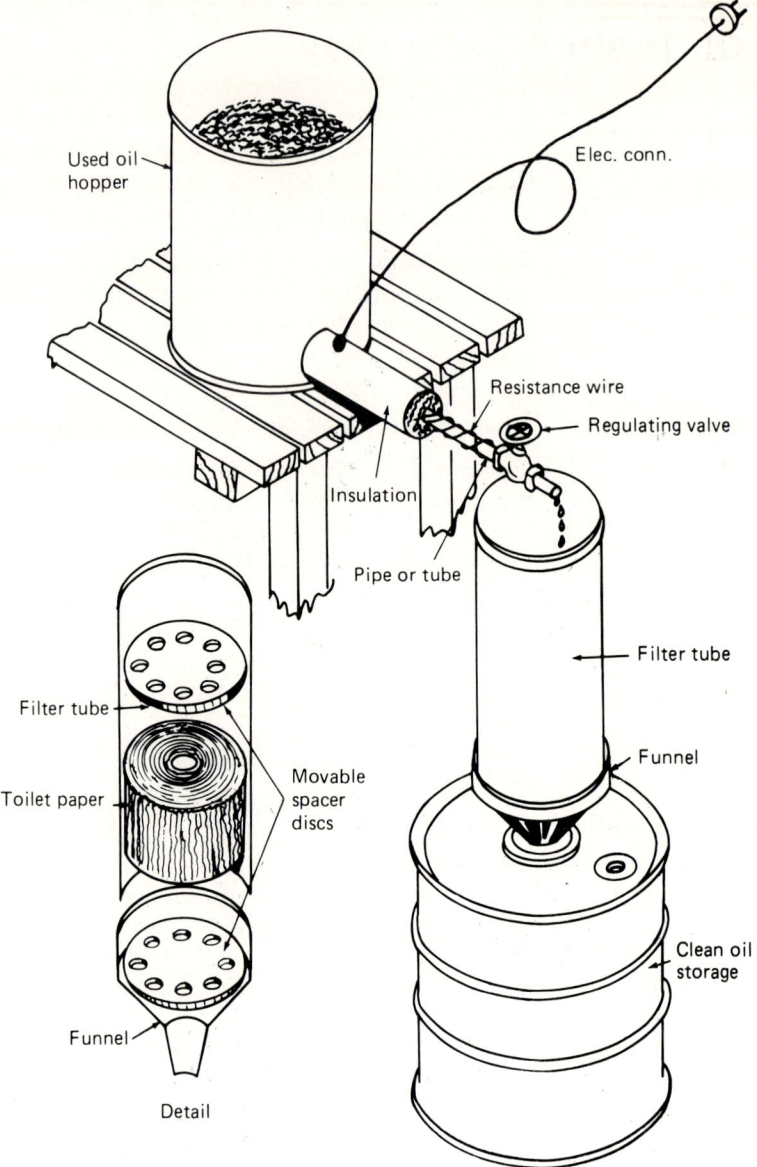

FIGURE 24-23. Salvage System for Used Engine Oil. *(From a sketch by P. M. Uhl)*

Thus, so-called magic fluids are intended to attract the dollars of the gullible, while adding nothing to the service life of the engine. If one wants to flush the crankcase, here is a good method:

1. Drain the oil while hot, and use a new filter. Keep this filter for flushing.
2. Fill the crankcase with about half-and-half fuel oil and engine oil.
3. Start the engine, and allow it to run long enough to warm up. Do not snap accelerate or load the engine.
4. After about 15 minutes, shut it down and drain the crankcase, letting it drip for at least a half hour.
5. Save the filter for the next flush. Put on a new filter and fill the crankcase with new oil.

USES OF TUBING WRENCHES

Tubing wrenches for use on tube nuts are made with an open box end. In many cases it is necessary to bend the tube slightly to align it with the fitting. Inverted nuts are quite likely to need some help, because they must be started by hand to prevent crossthreading.

By slipping the wrench over the nut, leverage may be applied to make the small adjustments that are necessary. No other tool works as well.

Supports for Tubing

Clips used to support tubing are often lost, distorted, or for some reason unusable. This does not mean that the tube can be left without support. Vibration damage is likely in such cases.

Using the perforated strap called *plumbers' tape,* serviceable supports can be made. It is well to protect the tube with a few wraps of tape or light rubber hose.

A KEYWAY

We are sometimes faced with the need for a keyway in the end of a shaft, which would be hard to remove for machining. Such a keyway can be made in the following manner.

Given: A combination square
A ¼-inch drill motor
A medium weight hammer
Suitable sized chisels
A sharp scribe
A center punch
A good small file
A short piece of ¼-inch tubing.

1. Make sure the end of the shaft is square.
2. Using the combination square, scribe a line parallel to the axis of the shaft. Use this line for the center of your keyway.
3. Lay out the shape of the keyway from the centerline. Scribe lines on each side of the keyway.
4. Make center punch marks on each scribed line, staying just inside of the lines. Keep the marks about ⅛ inch apart.

5. Slip a piece of ¼-inch tubing over the drill to limit its depth. Drill to half the thickness of the key.
6. Make punch marks in the space inside the lines, so that you can remove as much metal as possible with the drill.
7. You now have a rough key slot. Use the chisel to clean it up and file it to the key size. The key should fit snugly in the slot without any wobble.

GLOSSARY

ALLOY. Some materials can be blended by melting so that the properties of all are combined to increase strength, resist corrosion, or have other desirable properties.

AMALGAM. This is a mixture of materials that do not blend as completely as in an alloy.

BMEP. Break mean effective pressure. The difference between compression pressure and average firing pressure. The higher the BMEP, the better the engine design.

CETANE. A number used to describe the ignitability of diesel fuel. The higher numbers will start more easily but may not have high enough heat value.

COMPOUNDED OILS. Petroleum oils with various chemical additives. The object is to gain film strength, suspend carbons, and improve viscosity range. Some oils are as much as 30 percent chemicals.

COMPRESSION RATIO. The ratio of cylinder volume at bottom center (BDC) with that at top center (TDC). Thus a cylinder that has 100 cubic inches volume at BDC and 10 cubic inches at TDC, has a compression ratio of 10 to 1. Diesel engines commonly have compression ratios of 13 to 22 to 1. Ability to start depends on compression ratio. Many engines have starting aids to provide better cold starting ability.

CORROSION. Deterioration of surfaces exposed to chemical or electro-chemical attack.

DENSITY. Ratio of weight to volume.

EROSION. Wearing away of a surface due to action of air, water, or any abrasive.

FLASH POINT. Temperature at which a surface flash occurs when flame is applied.

FORCE. Applied energy doing mechanical work. Expressed as pound feet.

FUEL OILS. Fuels for diesel engines are made from distillation of petroleum. They must have good flow characteristics and be free of water and excessive sulphur. The lighter fuels, such as furnace oil, burn readily but have less heat value. Most engines operate on No. 2. Some few engines do well on No. 1. No. 2 has more heat value per pound.

HEAT ENERGY. Expressed as Btu (British thermal units). 1 BTU is the amount of heat required to raise 1 pound of water 1° F.

HORSEPOWER. The English unit of power. 1 hp (horsepower) = 0.746 Kw.

KILOWATT. The metric unit of power. 1000 watts = 1.34 HP.

LEVER. The oldest and most widely used type of mechanical advantage. A lever multiplies force by allowing the driving member to move farther than the driven member.

MAGNETIC HARDENING. A method of hardening the surface of a part by exposing it to a strong magnetic field. The depth of hardness is controlled by the intensity of the field. Widely used on crankshafts, camshafts, and other wearing parts.

MULTIGRADE OILS. Compounded oils having a wide range of flow ability. They are essentially light oils, with viscosity and film strength improvers.

OCTANE. A number used to describe the antiknock quality of gasoline. The higher the number is the less knock. Such fuels burn slower and more smoothly.

PASCAL'S LAW. Confined liquids transmit applied force equally in all directions. This is the basic law of hydraulics and is applied to all hydraulically operated devices. The Pascal is the metric unit of pressure corresponding to lb./sq.in.

PISTON SPEED. The distance the piston moves in feet at rated engine speed. The inertia of the piston and rod limit the maximum engine speed.

POUR POINT. Lowest temperature at which liquid will flow.

SINTERING. A method of molding complex shapes by compressing powdered metals, then heating them to produce a stable shape.

SOLUBLE OILS. Mineral oils that mix with water, used to cool cutting tools. They are *not* lubricating oils.

SPECIFIC GRAVITY. Ratio of substance density to that of pure water.

STEEL. This material is iron from which the carbon has been removed by heating. There are several ways to make this conversion.

STROKE. The distance the piston travels in the cylinder. An engine with a longer stroke than bore diameter will produce higher force, or torque, at lower speed than a short-stroke engine.

SYNTHETIC LUBRICATING OILS. Made by chemical means, not based on petroleum. If well-made, they have a long life and good quality.

VISCOSITY. Resistance to flow. Used to grade lubricating oils.

APPENDIX

Decimal Equivalents

1/64 – .0156	17/64 – .2656	33/64 – .5156	49/64 – .7656
1/32 – .03125	9/32 – .281	17/32 – .531	24/32 – .781
3/64 – .0469	19/64 – .2968	35/64 – .5469	51/64 – .7969
1/16 – .0625	5/16 – .3125	9/16 – .5625	13/16 – .8125
5/64 – .0781	21/64 – .3281	37/64 – .5781	53/64 – .8281
3/32 – .0937	11/32 – .344	19/32 – .594	27/32 – .844
7/64 – .1094	23/64 – .3593	39/64 – .6094	55/64 – .8594
1/8 – .125	3/8 – .375	5/8 – .625	7/8 – .8750
9/64 – .1406	25/64 – .3903	41/64 – .6406	57/64 – .8906
5/32 – .156	13/32 – .4062	21/32 – .656	29/32 – .9065
11/64 – .172	27/64 – .4219	43/64 – .6719	59/64 – .9219
3/16 – .1875	7/16 – .4375	11/16 – .6875	15/16 – .9375
13/64 – .2031	29/64 – .4531	45/64 – .703	61/64 – .9531
7/32 – .219	15/32 – .469	23/32 – .719	31/32 – .969
15/64 – .2344	31/64 – .4844	47/64 – .734	63/64 – .9844
1/4 – .250	1/2 – .5000	3/4 – .750	1.00 – 1.000

For practical use, these values have been rounded to the nearest ten-thousandth.

Pressure Conversion Chart

1″ water	=	.0735″ mercury
1″ water	=	.0361 psi
1″ mercury	=	13.6000″ water
1″ mercury	=	.4910 psi
1 psi	=	27.7000″ water
1 psi	=	2.0360″ mercury
1 psi	=	6.895 kPa
1 kPa	=	.145 psi

Metric Torque Conversion

Newton Meters (N-m)	to	Pound Feet (lb.ft.)	Pound Feet (lb.ft.)	to	Newton Meters (N-m)
Divide by:			Multiply by:		
1		0.7376	1		1.356
2		1.5	2		2.7
3		2.2	3		4.0
4		3.0	4		5.4
5		3.7	5		6.8
6		4.4	6		8.1
7		5.2	7		9.5
8		5.9	8		10.8
9		6.6	9		12.2
10		7.4	10		13.6
15		11.1	15		20.3
20		14.8	20		27.1
25		18.4	25		33.9
30		22.1	30		40.7
35		25.8	35		47.5
40		29.5	40		54.2
50		36.9	45		61.0
60		44.3	50		67.8
70		51.6	55		74.6
80		59.0	60		81.4
90		66.4	65		88.1
100		73.8	70		94.9
110		81.1	75		101.7
120		88.5	80		108.5
130		95.9	90		122.0
140		103.3	100		135.6
150		110.6	110		149.1
160		118.0	120		162.7
170		125.4	130		176.3
180		132.8	140		189.8
190		140.1	150		203.4
200		147.5	160		216.9
225		166.0	170		230.5
250		184.4	180		244.0

Conversion Table

Given	Multiplication Factor	Converts To:
TORQUE		
lb-in. (pound-inch)	× 0.113	= N-m (Newton meters)
lb-ft. (pound-foot)	× 1.356	= N-m (Newton meters)
POWER		
hp (horsepower)	× 0.746	= kw (kilowatts)
PRESSURE		
" (in.) H_2O (inches of water)	× 0.2488	= kPa (kilopascals)
" Hg (inches of mercury)	× 3.28	= kPa (kilopascals)
lb./sq.in. (pounds per square inch)	× 6.895	+ kPa (kilopascals)
FUEL PERFORMANCE		
mpg (miles per gallon)	× 0.425	= km/1 (kilometers per liter)
VELOCITY (SPEED)		
mph (miles per hour)	× 1.609	= km/h (kilometers per hour)
LENGTH		
" (inch,) in.	× 25.4	= mm (millimeters)
' (foot,) ft.	× 0.305	= m (meters)
yd. (yard)	× 0.914	= m (meters)
mi. (mile)	× 1.609	= km (kilometers)
AREA		
sq.in. (square inches)	× 645.2	= mm^2 (square millimeters)
sq.ft. (square feet)	× 0.093	= m^2 (square meters)
sq.yd. (square yards)	× 0.836	= m^2 (square meters)
VOLUME		
$in.^3$ (cubic inches)	× 164	= mm^3 (cubic millimeters)
qt. (quart)	× 0.946	= 1 (liters)
gal. (gallon)	÷ 3.785	= 1 (liters)
MASS (WEIGHT)		
lb. (pound)	× 0.454	= kg (kilograms)
FORCE		
lb. (pound)	× 4.448	= N (Newtons)
TEMPERATURE		
°C (degrees Centigrade)	× 1.8 + 32	= °F (degrees Fahrenheit)
°F (degrees Fahrenheit)	− 32 ÷ 1.8	= °C (degrees Centigrade)

432 / Diesel Engine Service

Conversion Table

Given	Multiplication Factor	Converts To:
PROPERTIES OF THE CIRCLE		
D (diameter)	D × pi (diameter × 3.1416)	= circumference
R (radius)	R² × pi (radius² × 3.1416)	= A (area)
METRIC PREFIXES		
K = kilo-thousand	1 kilometer	= 1000 meters
C = centi-hundredth	1 centimeter	= 00.01 meter
M = milli-thousandth	1 millimeter	= 0.001 meter

Circumference = distance around the circle
Diameter = distance across the circle
Radius = one-half the diameter

These are the most commonly used prefixes. The metric system is based on 10 as the difference between a unit and the next larger or smaller unit. Thus, such differences can be expressed by moving the decimal point to the right or left, as the case requires.

TORQUE FOR FLARED AND O-RING FITTINGS

The torques shown in the chart that follows are to be used on the part of 37° Flared, 45° Flared and Inverted Flared fittings (when used with steel tubing), O-ring plugs and O-ring fittings.

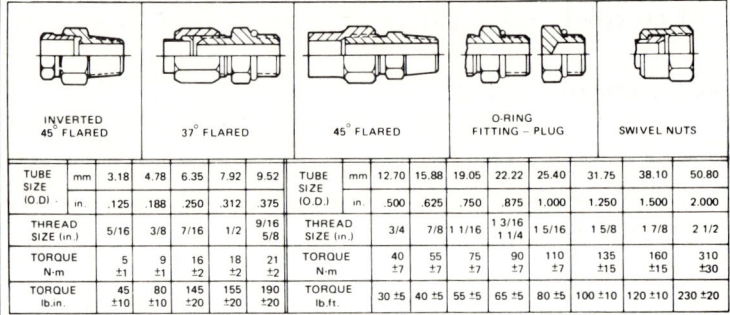

		INVERTED 45° FLARED					37° FLARED		45° FLARED		O-RING FITTING – PLUG		SWIVEL NUTS	
TUBE SIZE (O.D.)	mm	3.18	4.78	6.35	7.92	9.52	TUBE SIZE (O.D.) mm	12.70	15.88	19.05	22.22	25.40	31.75	38.10	50.80
	in.	.125	.188	.250	.312	.375	in.	.500	.625	.750	.875	1.000	1.250	1.500	2.000
THREAD SIZE (in.)		5/16	3/8	7/16	1/2	9/16 5/8	THREAD SIZE (in.)	3/4	7/8	1 1/16	1 3/16 1 1/4	1 5/16	1 5/8	1 7/8	2 1/2
TORQUE N·m		5 ±1	9 ±1	16 ±2	18 ±2	21 ±2	TORQUE N·m	40 ±7	55 ±7	75 ±7	90 ±7	110 ±7	135 ±15	160 ±15	310 ±30
TORQUE lb.in.		45 ±10	80 ±10	145 ±20	155 ±20	190 ±20	TORQUE lb.ft.	30 ±5	40 ±5	55 ±5	65 ±5	80 ±5	100 ±10	120 ±10	230 ±20

ASSEMBLY OF FITTINGS WITH STRAIGHT THREADS AND O-RING SEALS

1. Put locknut (3), backup washer (4) and O-ring seal (5) as far back on fitting body (2) as possible. Hold these components in this position. Turn the fitting into the part it is used on, until backup washer (4) just makes contact with the face of the part it is used on.

NOTE: If the fitting is a connector (straight fitting) or plug, the hex on the body takes the place of the locknut. To install this type fitting tighten the hex against the face of the part it goes into.

2. To put the fitting assembly in its correct position turn the fitting body (2) out (counterclockwise) a maximum of 359°. Tighten locknut (3) to the torque shown in the chart.

A71009

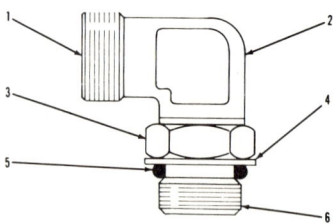

ELBOW BODY ASSEMBLY
1. End of fitting body (connects to tube). 2. Fitting body. 3. Locknut. 4. Backup washer. 5. O-ring seal. 6. End of fitting that goes into other part.

INDEX

A

Adjust rack control lever screws, 75
Adjust valve crossheads, 213
Adjusting the governor, 78
Aftercooler, 113, 120
Aftercooler service, 267
Air cranking systems, 380
Alignment marks, 241
Alignment of flywheel housing, 303
American Bosch, 27
A ring compressor, 403
Assembly suggestions, 214
Assembly of aftercooler, 270
Assembly of crossbolt aftercooler, 273
Assembling pistons on rods, 159

B

Balancing a driveshaft, 415
Battery description, 368
Battery charging, 369
Battery charging on vehicles, 383
Bearing description and lubrication, 151
Bearing numbering system, 153
Bleed valve, 51
Block description, 139
Block service, 141, 142
Bosch fuel system, 11
Bushings, bearings, and seals, 281

C

Cab heater, 120
Camshaft bushings, 230
Camshaft inspection, 232
Cam followers, 210
Capacitors, 387
Caterpillar fuel system, 50
Caterpillar tune-up, 309
Causes for head removal, 146
Causes of power loss, 407
Causes of failure to start, 409
Checks on engine, 48
Checking rack freedom, 76
Checking valve clearance, 73
Connecting rod inspection, 159
Connecting rod service, 153
Coolant flow, 112
Coolant filter installation, 255
Cooling system service, 101
Cleaning exhaust manifolds, 260, 262
Cleaning liners after honing, 157
Cleaning used pistons, 158
Cleaning air filter elements, 318
Clutch service, 294, 296, 297, 299, 300
Cranking motor and drive service, 374, 376
Crankshaft inspection and service, 150, 152
Crossflow radiator, 104
Crossbolt aftercooler, 272
Cummins fuel system, 84
Cummins adjustments, 311
Cylinder misfiring, 41
Cylinder liner service, 156, 157

D

Diesel, 2
Dense exhaust smoke, 45

Descriptions, 50, 17, 68, 84, 227, 278, 343, 383
Detroit diesel engine aftercoolers, 274, 276
Diodes, 284
Double dampers, 236
Dynamometer testing and use, 323, 326, 332

E

Electric dynamometer, 328
Electric retarders, 364
Electromechanical regulators, 389
Engine block service, 142
Engine cooling during test, 329, 330
Engine tune-up, 304
Engine types, 3
Engine retarders, 343
Excessive detonation-knock, 42
Excessive fuel consumption, 45

F

Failure to reach governed speed, 42
Failure to run at idle speed, 44
Field repairs and adjustments, 35
Filters, a discussion of, 407
Final adjustment and painting, 339
Flywheel inspection, 294
Front seal installation, 242
Front cover inspection, 283
Fuel control, 14
Fuel lines, 19
Fuel pump installation, 59
Fuel ratio control, 53
Fuel supply pump, 38
Fuel systems, 5, 11, 27, 46, 50, 68, 84
Fuel system complaints, 40
Functional descriptions, 95, 103, 127, 141, 146, 236

G

Gear inspection, 234
Gear fasteners, 235
Gear train service, 229
General analysis of electrical problems, 379
General descriptions, 227, 278

Governor action, 18, 70
Governor modifications, 83
Grinding drills, 411

H

Head service, 145, 148
Honing cylinders, 417
Hoses, 121
Hydraulic governor, 82

I

Injector, 70
Injection advance device, 32
Injection advance system, 18
Injection timing on Cummins engines, 216
Injection nozzles, 49
Installation of fuel system parts, 64, 65, 66
Installing rings on pistons, 160
Installing piston and rod assemblies, 161
Installing oil pan, 154
Installing water pump, 249
Installing exhaust manifold, 260
Installing aftercooler elements, 276, 277

J

Jacobs brake service, 343, 345
Jacobs brakes on Mack and Caterpillar engines, 356, 360, 362

L

Liner cracks, 157
Liner inspection, 163
Liner puller, 405
Load stop, 53
Low power complaints, 42

M

Machining repairs to blocks, 143
Main bearing assembly, 152
Making battery cables, 416
Making a clutch disc pilot, 413
Making a keyway, 421

Making a transmission hanger, 413
Manual priming pump, 51
Mechanical variable timing, 221, 224
Motor driven fan, 118
Multigroove belts, 119

O

Ohm's law, 367
Oil cooler, 131
Oil filter, 131
Oil filter system for test, 338
Oil pump, 129
Oil specifications, 135
Oil suction tubes, 131
Oil sump or pan, 129
Operation of the cooling system, 103
Operating rules, 362
Other governors, 82
Other oiled parts, 138
Other retarding systems, 363
Overflow valve, 37

R

Rocker lever positions, 73

S

Seal removal, 242
Service operations, Bosch fuel system, 24
Service main bearing caps, 164
Shut-down lever, 37
Shutters, 115
Single weight governor, 79
Solenoid shut-off, 52
Speed adjusting screw, 56
Speed adjustment, 35
Spray valves, 19
Starting aids, 22
Starting fuel system, 37
Starting system for testing, 337

T

Taylor dynamometer, 327
Tension specifications, 261
Test principles, 323

Test procedure, 334
Testing pressure caps, 120
Thermostat, 111
Thermostat test, 252
Throttle linkage, 35
Thrust bearings, 151
Thrust washers, 234
Timing injectors, 74
Timing and adjustments, 34
Tools to make, 403
Troubleshooting fuel systems, 38
Troubleshooting hints, 410
Tune-up, 72
Turning device on pulley, 73
Turbochargers, 97
Turbocharger oil supply and drain, 135
Turbocharger installation, 262

U

Use of additives in oil, 419
Use of meters, 377
Uses of oil pressure, 128
Use of tubing wrenches, 420
Using the chassis dynamometer, 334
Using files and hacksaws, 397
Using power wrenches, 394
Using precision tools, 395
Using thread files, 400
Using wrench handle extensions, 401

V

Vibration dampers, 235
Vibration damper service, 240
Viscous dampers, 240

W

Water cooled engines, 103
Water cooled exhaust manifold, 261
Water manifold installation, 252
Water manifold repair, 250
Water pumps, 108
Water pump assembly, 249
Water separator, 68
Wet clutches, 297